Millets

Millets are diverse small-seeded crops which are resilient to climatic stress, pests, and diseases. They can be grown in rainfed conditions with minimal agricultural inputs. They are considered nutritionally superior to other major cereals like rice, wheat, and maize as they contain a significant amount of protein, dietary fibre, and minerals. Today, millets are recognized and considered as integral components of a sustainable food system. Millets have a low glycaemic index and are beneficial for diabetics. Millet protein is devoid of gluten, making it a better option than wheat for patients with gluten sensitivity.

Millets: Cultivation, Processing, and Utilization covers information on taxonomy, morphology, germplasm accessions, cultivation practices, harvesting methods, threshing, cleaning, storage, milling, structural and engineering properties, nutritional and anti-nutritional values, health benefits, food applications, by-products, non-food applications, quality standards, and prospects in millet processing. This book's 16 in-depth chapters give readers complete information on all facets of millet cultivation, processing, and use. It is a unique compilation of information on millets from farm to fork and beyond. This book will be useful for students, researchers, farmers, and entrepreneurs in understanding millets and their applications. It has been compiled by experts in the field and can serve as a guidance document for stakeholders.

Millets
Cultivation, Processing, and Utilization

Ashwani Kumar, Vidisha Tomer, Mukul Kumar
and Prince Chawla

CRC Press
Taylor & Francis Group
Boca Raton London New York

CRC Press is an imprint of the
Taylor & Francis Group, an **informa** business

First edition published 2024
by CRC Press
2385 Executive Center Drive, Suite 320, Boca Raton, FL 33431

and by CRC Press
4 Park Square, Milton Park, Abingdon, Oxon, OX14 4RN

CRC Press is an imprint of Taylor & Francis Group, LLC

© 2024 Ashwani Kumar, Vidisha Tomer, Mukul Kumar and Prince Chawla

Reasonable efforts have been made to publish reliable data and information, but the author and publisher cannot assume responsibility for the validity of all materials or the consequences of their use. The authors and publishers have attempted to trace the copyright holders of all material reproduced in this publication and apologize to copyright holders if permission to publish in this form has not been obtained. If any copyright material has not been acknowledged please write and let us know so we may rectify in any future reprint.

Except as permitted under U.S. Copyright Law, no part of this book may be reprinted, reproduced, transmitted, or utilized in any form by any electronic, mechanical, or other means, now known or hereafter invented, including photocopying, microfilming, and recording, or in any information storage or retrieval system, without written permission from the publishers.

For permission to photocopy or use material electronically from this work, access www.copyright.com or contact the Copyright Clearance Center, Inc. (CCC), 222 Rosewood Drive, Danvers, MA 01923, 978-750-8400. For works that are not available on CCC please contact mpkbookspermissions@tandf.co.uk

Trademark notice: Product or corporate names may be trademarks or registered trademarks and are used only for identification and explanation without intent to infringe.

Library of Congress Cataloging-in-Publication Data
Names: Kumar, Ashwani (Food scientist), author. | Tomer, Vidisha, author. |
Kumar, Mukul (Food scientist), author. | Chawla, Prince, author.
Title: Millets : cultivation, processing, and utilization / authored by
Ashwani Kumar, Vidisha Tomer, Mukul Kumar, and Prince Chawla.
Description: First edition | Boca Raton, FL : CRC Press, 2024 | Includes
bibliographical references and index.
Identifiers: LCCN 2023039706 | ISBN 9780367748623 (hardback) |
ISBN 9780367709006 (paperback) | ISBN 9781003159902 (ebook)
Subjects: LCSH: Millets. | Millets--Utilization. | Millets--Nutrition.
Classification: LCC SB191.M5 K86 2024 | DDC 633.1/71--dc23/eng/20231213
LC record available at https://lccn.loc.gov/2023039706

ISBN: 978-0-367-74862-3 (hbk)
ISBN: 978-0-367-70900-6 (pbk)
ISBN: 978-1-003-15990-2 (ebk)

DOI: 10.1201/9781003159902

Typeset in Kepler Std
by Deanta Global Publishing Services, Chennai, India

This book is dedicated to all tribes of the world who have preserved their culture and who are the ambassadors of sustainable development. We also owe gratitude to the food and agricultural scientists for their dedicated and tireless efforts, whose innovations have made it possible to feed the world.

Contents

Foreword	xiii
Preface	xv
About the Authors	xvii

1 Origin, Taxonomy, Botanical Description, Biodiversity, Cultivation, Agrarian Requirements, and Production of Millets — 1
- 1.1 Introduction — 1
- 1.2 Origin — 2
- 1.3 Taxonomy — 3
- 1.4 Growth Cycle — 3
- 1.5 Botanical Description — 3
- 1.6 Germplasm Accessions of Millets — 9
- 1.7 Cultivation and Agrarian Requirements — 11
- 1.8 Production — 14
- 1.9 Future Prospects — 17
- References — 18

2 Harvesting, Threshing, and Storage of Millets — 23
- 2.1 Introduction — 23
- 2.2 Harvesting of the Millets — 24
  - 2.2.1 Cutting/Reaping of Millet — 24
  - 2.2.2 Threshing — 25
- 2.3 Curing and Drying — 26
- 2.4 Yield of Millets — 26
- 2.5 Primary Processing of Millets — 26
  - 2.5.1 De-Husking/Dehulling — 26
  - 2.5.2 Cleaning — 27
  - 2.5.3 Sorting and Grading — 27
- 2.6 Storage of Millet Grains — 28
  - 2.6.1 Mud-Based Bins — 28
  - 2.6.2 Wood-Based Storage Structures — 30
  - 2.6.3 Bamboo Bins — 30
  - 2.6.4 Composite Structures — 30
  - 2.6.5 Other Structures — 31
  - 2.6.6 Improved Structures for Household- and Farmhouse-Level Storage — 31
  - 2.6.7 Commercial Storage Structures — 32
- 2.7 Conclusions — 32
- References — 33

3 Milling and Secondary Processing of Millets — 36
- 3.1 Introduction — 36
- 3.2 Milling — 36
  - 3.2.1 Conditioning/Tempering — 37
  - 3.2.2 Debranning — 37
  - 3.2.3 Polishing/Pearling — 38
  - 3.2.4 Grinding/Pulverization — 38
  - 3.2.5 Packaging of Flour — 38
- 3.3 Secondary Processing of Millets — 39
  - 3.3.1 Roasting — 39
  - 3.3.2 Popping/Puffing — 40
  - 3.3.3 Flaking — 42
  - 3.3.4 Frying — 42
  - 3.3.5 Malting — 43
  - 3.3.6 Baking — 44
  - 3.3.7 Extrusion — 45
  - 3.3.8 Fermentation — 46
- 3.4 Conclusions — 47
- References — 47

4 Machines for Processing Millets — 53
- 4.1 Introduction — 53
- 4.2 Traditional Equipment — 53
- 4.3 Machines Used in Primary Processing of Millets — 55
  - 4.3.1 Harvesters — 55
  - 4.3.2 Threshers — 58
  - 4.3.3 Cleaners — 61
  - 4.3.4 Sorting/Grading Machines — 62
  - 4.3.5 Milling — 66
- 4.4 Secondary Processing of Millets — 71
- 4.5 Conclusions — 71
- References — 72

5 Structural Composition and Engineering Properties of Millets — 75
- 5.1 Introduction — 75
- 5.2 Type of Millet Seeds — 75
- 5.3 Shape of Millet Grains — 76
- 5.4 Structure and Composition of Millet Grains — 76
  - 5.4.1 Bran — 77
  - 5.4.2 Endosperm — 81
  - 5.4.3 Germ — 83
- 5.5 Physical Characteristics of Millets — 83
  - 5.5.1 Colour — 83
  - 5.5.2 Hardness — 84
- 5.6 Engineering Properties — 84
  - 5.6.1 Principle Axial Dimensions: Length, Width, and Thickness — 84
  - 5.6.2 Diameter — 86
  - 5.6.3 Sphericity — 87

	5.6.4	Thousand Kernel Weight	88
	5.6.5	Bulk Density	88
	5.6.6	True Density	89
	5.6.7	Volume	89
	5.6.8	Interstice	90
	5.6.9	Surface Area	90
	5.6.10	Porosity	90
	5.6.11	Coefficient of Friction	91
	5.6.12	Coefficient of Internal Friction	91
	5.6.13	Coefficient of External Friction	92
	5.6.14	Angle of Repose	93
	5.6.15	Terminal Velocity	93
5.7	Conclusions		94
References			94

6 Nutritional and Functional Composition of Millets — 99

6.1	Introduction		99
6.2	Macronutrients in Millets		99
	6.2.1	Carbohydrates	99
	6.2.2	Fats	103
	6.2.3	Proteins	105
6.3	Micronutrients		109
	6.3.1	Vitamins	110
	6.3.2	Minerals	112
6.4	Conclusion		114
References			114

7 Phytochemicals in Millets and Related Health Benefits — 119

7.1	Introduction		119
7.2	Antioxidant Potential of Millets		119
7.3	Phytochemicals in Millets		120
	7.3.1	Phenolic Compounds	120
	7.3.2	Phytosterols	127
	7.3.3	Bioactive Peptides	128
	7.3.4	Fibre and Its Components	129
7.4	Effect of Processing on Bioavailability and Functionality		130
7.5	Conclusion		132
References			132

8 Anti-Nutritional Factors and Methods to Improve Nutritional Value of Millets — 138

8.1	Introduction	138
8.2	Phytates	138
8.3	Tannins	140
8.4	Oxalates	142
8.5	Non-Starch Polysaccharides	142
8.6	Enzyme Inhibitors	143

	8.7	Methods to Improve Nutritional Value	143
		8.7.1 Soaking	143
		8.7.2 Processing/Milling	146
		8.7.3 Germination	146
		8.7.4 Fermentation	146
		8.7.5 Heat Treatment	147
		8.7.6 Extrusion	148
		8.7.7 Radiation	148
		8.7.8 Exogenous Enzymes	149
	8.8	Conclusion	149
	References		149
9	Role of Millets in Disease Management		154
	9.1	Introduction	154
	9.2	Global Disease Paradigm	154
	9.3	Compounds in Millets with Disease Prevention Potential	155
		9.3.1 Nutrients with Health Potential	155
		9.3.2 Bioactive Compounds with Health Potential	157
		9.3.3 Bioavailability of the Bioactives	159
	9.4	General Mechanism of Health Promotion	159
		9.4.1 Protection against Free Radical Damage	159
		9.4.2 Anti-Obesogenic Effects	162
		9.4.3 Anti-Inflammatory Properties	163
	9.5	Millets and Protection from Diseases	163
		9.5.1 Hypertension	164
		9.5.2 Cardiovascular Diseases	164
		9.5.3 Diabetes	167
		9.5.4 Cancer	168
		9.5.5 Infections	169
	9.6	Conclusions	170
	References		170
10	Extraction and Application of Millet Starch		175
	10.1	Introduction	175
	10.2	Isolation Methods and Composition of Millet Starch	176
	10.3	Morphological Characteristics of Millet Starch	177
	10.4	Crystalline Behaviour of Millet Starch	179
	10.5	Solubility and Swelling Power of Millet Starch	179
	10.6	Rheological and Gelatinization Properties of Millet Starch	180
	10.7	Applications of Millet Starch	181
	10.8	Conclusions	182
	References		182
11	Extraction and Application of Millet Proteins		188
	11.1	Introduction	188
	11.2	Protein Composition of Millets	188

	11.3	Extraction of Millet Proteins	189
	11.4	Characterization of Millet Proteins/Peptides	190
	11.5	Effect of Processing on Millet Protein	192
	11.6	Application of Millet Proteins	194
		11.6.1 Food-Related Applications	194
		11.6.2 Therapeutic Applications	194
		11.6.3 Other Applications	195
	11.7	Conclusions	196
	References		196

12 Mineral Bioavailability and Impact of Processing — 201

- 12.1 Introduction — 201
- 12.2 Factors Affecting Minerals Bioavailability — 203
  - 12.2.1 Chemical Form of Minerals — 203
  - 12.2.2 Presence of Anti-Nutrients — 203
  - 12.2.3 Nutrient Interactions — 203
  - 12.2.4 pH and Digestive Enzymes — 204
  - 12.2.5 Individual Variation in Bioavailability — 204
  - 12.2.6 Physical Properties of Grain — 205
- 12.3 Effect of Processing Techniques on Mineral Bioavailability — 205
  - 12.3.1 Extrusion — 205
  - 12.3.2 Coating — 206
  - 12.3.3 Roasting and Toasting — 206
  - 12.3.4 Milling and Grinding — 206
- 12.4 Techniques to Enhance Mineral Bioavailability — 207
  - 12.4.1 Addition of Chelating Agents — 207
  - 12.4.2 Fermentation and Enzyme Treatment — 207
  - 12.4.3 Fortification and Micronutrient Premixes — 207
  - 12.4.4 Novel Technologies for Mineral Delivery — 208
- 12.5 Evaluating Mineral Bioavailability in Millets — 208
  - 12.5.1 *In-Vitro* Digestibility Methods — 208
  - 12.5.2 *In-Vivo* Studies and Animal Models — 209
  - 12.5.3 Human Intervention Trials — 209
  - 12.5.4 Analytical Techniques for Bioavailability Assessment — 209
- 12.6 Conclusion and Future Perspectives — 210
- References — 211

13 Millet-Based Traditional Foods — 216

- 13.1 Introduction — 216
- 13.2 Fermentation — 216
  - 13.2.1 Lactic Acid Fermented Foods — 217
  - 13.2.2 Alcoholic Beverages — 229
- 13.3 Non-Fermented Traditional Foods — 235
  - 13.3.1 Dambu — 235
  - 13.3.2 Halwa — 235
  - 13.3.3 Kheer/Basundi/Pyasam — 235
  - 13.3.4 Kurrakan Kanda — 237

		13.3.5	Laddu/Ladoo	237
		13.3.6	Masavusvu	237
		13.3.7	Pongal	237
		13.3.8	Roti	238
	13.4	Conclusions		238
	References			238

14 Millet-Based Processed Foods — 243
- 14.1 Introduction — 243
- 14.2 Millet-Based Processed Products — 244
  - 14.2.1 Non-Fermented Products — 244
  - 14.2.2 Fermented Products — 256
- 14.3 Millet-Based Products Commercially Available in the Indian Market — 259
- 14.4 Conclusions and Future Outlook — 261
- References — 262

15 Millet By-Products and Their Food and Non-Food Applications — 267
- 15.1 Introduction — 267
- 15.2 Major Waste Generated from Millet — 267
  - 15.2.1 Stover/Straw — 268
  - 15.2.2 Husk — 271
  - 15.2.3 Bran — 272
  - 15.2.4 Germ — 276
  - 15.2.5 Spent Grains — 276
  - 15.2.6 Non-Food Grade Grains — 277
  - 15.2.7 Prebiotics — 277
- 15.2.3 Conclusions — 278
- References — 279

16 Prospects in the Production, Processing, and Utilization of Millets — 283
- 16.1 Introduction — 283
- 16.2 Future Prospects in the Production of Millets — 285
  - 16.2.1 Advancements in Genetics — 285
  - 16.2.2 Sustainable Agricultural Practices — 285
- 16.3 Future Prospects in the Processing and Utilization of Millets — 287
  - 16.3.1 Diversification of Millet-Based Food Products — 287
  - 16.3.2 Value-Addition and Fortification — 288
  - 16.3.3 Functional Food and Nutraceutical Applications — 288
  - 16.3.4 Beverage Industry — 289
  - 16.3.5 Livestock Feed — 290
  - 16.3.6 Market Development and Consumer Awareness — 291
- 16.4 Quality Standards for Millets — 291
- 16.5 Conclusions — 292
- References — 294

Index — 299

Foreword

It is my pleasure to write the Foreword to this book. I express my sincere appreciation to the team for choosing a topic of significance. Agriculture faces numerous challenges, viz., climate change, erratic and deficient rainfall, and depleting groundwater levels. This is posing serious concern to food and nutritional security. As the world population is expected to reach 9.7 billion by 2050, the issue of ensuring food security will be more challenging. Such scenario necessitates to look beyond the traditional wheat and rice cycles and opt the crops that can be beneficially and sustainably cultivated with external inputs in varied conditions especially under rainfed conditions. Robust crops that require less application of fertilizers and pesticides are also important in protecting our environment. "Millets" have emerged as a sustainable solution to the existing problems.

Millets can grow on marginal soils with limited water supply. Being a C-4 crop, it has better photosynthetic capacity and can help mitigate the problems of climate change. Most millets are also nutritionally superior to major cereals. Being a rich source of dietary fibre, minerals, vitamins, and phytochemicals, millets are also a grain of interest to nutritionists. Its low glycaemic index and gluten-free proteins make the cereals of choice for the development of functional foods. The consumption of millets is reported to control diabetes, obesity, cancer, and cardiovascular diseases, and build immunity. It has also been used for the development of modern-age functional foods like weaning, geriatric, and probiotic foods. Millets have emerged as the most promising cereals in the changing agricultural landscape with the growing demand for healthy foods. Agriculturists, food technologists, and environmentalists from all over the world have acknowledged millets as a crucial crop in addressing current issues.

Worldwide, governments are promoting millets in different ways. India, the largest producer of millets in the world, has dispensed with the use of the nomenclature "coarse cereals" and renamed millets "*Shree Anna*". The current year 2023 is celebrated as the "International Year of Millets" by the United Nations and many global initiatives have helped to discover the lost identity of millets, and more acceptance is observed among the masses for millet cultivation and consumption. The global millet market is expected to grow at a compound annual growth rate of 4.5% from 2021 to 2026, reaching 12 billion USD by 2025.

This book *Millets: Cultivation, Processing, and Utilization* is an updated compendium on millets and covers the important aspects of millets right from their origin to commercial application. I am sure this book will serve as instructional material to readers for many years in understanding the various aspects of millet cultivation, processing, storage, and utilization. This book will be of special help to millet farmers, researchers, processors, entrepreneurs, and consumers. I am confident that it will strengthen the knowledge base to promote millet cultivation, processing, and utilization throughout the world.

Ashok Kumar Singh
Vice Chancellor
Rani Lakshmi Bai Central Agricultural University
Jhansi, Uttar Pradesh, India

Preface

The contribution of agricultural scientists, engineers, and farmers has enabled sufficiency in terms of grain production and the maintenance of adequate food supplies throughout the year. However, major progress is observed in the cultivation, processing, and utilization of three major cereal crops, i.e. rice, wheat, and maize. Approximately 50% of the world's food requirement is dependent on these three staples and a threat to the existence of any of these may hinder the world's food security. So, there is a need to diversify cropping systems to decrease the dependence on these cereals. Additional challenges like fluctuating temperatures, reducing waterbed, intensive farming practices, and the overuse of agro-chemicals have drained the soil capacity and the result is nutrient-deficient grains and hidden hunger. All these challenges can be addressed by bringing millets to mainstream and utilizing their potential in achieving sustainable agricultural and nutritional security. Millets offer a promising addition to the current cropping pattern with the advantage of production with minimum agricultural inputs and resilience towards small temperature fluctuations.

The C-4 photosynthetic pathway of millets makes them a good crop for the environment and their ability to grow on marginal lands under rainfed conditions with limited agricultural inputs makes them the crop of the future. The nutritional composition of millets is comparable to staple cereals. Their mineral content is several folds higher than major cereals and regular consumption of these grains can help to eliminate the diseases caused by deficiencies of iron, calcium, etc. The low glycaemic index and gluten-free proteins of millets make them a good food for diabetics and celiac patients. The high dietary fibre and nutraceutical components like polyphenols, flavonoids, tannins, etc., present in millets also exhibit antioxidant, antidiabetic, antiobesity, antiageing, and anticancer properties.

Many scientific studies have provided a sufficient body of evidence regarding the benefits of millets. Realizing this, the United Nations declared 2023 the "International Year of Millets" to create awareness for millets and their adoption by stakeholders. However, limitations in food applications and low prices discourage farmers from growing millets. To get millet-based products to the mainstream consumer market, robust infrastructure is needed. The major challenge in the food applications of millets is the limited availability of the primary and secondary processing equipment, followed by quality standards, recipes to produce millet-based food products, and a lack of production lines to produce millet-based value-added products.

Millets: Cultivation, Processing, and Utilization comprises 16 chapters which cover the origin, taxonomy, botanical description, biodiversity, cultivation, agrarian requirements, and production of millets. The traditional as well as modern methods and equipment available for the harvesting, cleaning, grading, milling, and storage of millets are also discussed. In addition, structural and engineering properties, nutritional and phytochemical composition, health benefits, major anti-nutritional factors and methods to reduce them, millet-based traditional and modern food products, millet by-products, non-food applications, the extraction and application of millet starch and proteins, and prospects in millet cultivation are also discussed. We believe that the readers will obtain sufficient information on the cultivation, processing, and utilization of millets and this book will serve as instructional material for researchers, entrepreneurs, and food processors wishing to produce and process millets. There may, however, still be a few shortcomings in this book and the editors will be grateful to receive suggestions for incorporation in the next edition of this book.

Ashwani Kumar
Jhansi, Uttar Pradesh, India

Vidisha Tomer
Vellore, Tamil Nadu, India

Mukul Kumar
Jalandhar, Punjab, India

Prince Chawla
Jalandhar, Punjab, India

About the Authors

Dr Ashwani Kumar earned a PhD in Food Technology from Punjab Agricultural University, Ludhiana, Punjab, India. He completed his Master's in Food Technology at Dr YS Parmar University of Horticulture and Forestry, Nauni, Solan, Himachal Pradesh, and his BSc in Food Science and Technology (honours) at DAV College Jalandhar. Presently he works as Teaching-cum-Research Associate, Postharvest Technology, at Rani Lakshmi Bai Central Agricultural University, Jhansi, Uttar Pradesh, India since 3 January 2022. He also has 4.5 years of experience working as an Assistant Professor of Food Technology and Nutrition at Lovely Professional University, Phagwara, India. Dr Kumar received the prestigious Dryland Cereals Scholarship from the Asia Pacific Associations of Agricultural Research Institutions, Bangkok, Thailand; an INSPIRE Fellowship from the Department of Science and Technology, New Delhi, India; and a University Merit Scholarship from Punjab Agricultural University, Ludhiana, India (for PhD) and Dr YS Parmar University of Horticulture and Forestry, Solan, India (for MSc). He has also passed the National Teaching Eligibility Test in Food Technology conducted by the Indian Council of Agricultural Research. Dr Kumar's research is focused on the development of millet-based value-added products. He has also presented his work at national and international conferences. He has authored 43 research publications in indexed journals and 6 book chapters. He is also a reviewer of many esteemed journals.

Dr Vidisha Tomer is currently working as an Assistant Professor in the Department of Horticulture and Food Science, VIT School of Agricultural Innovations and Advanced Learning, Vellore Institute of Technology, Vellore, Tamil Nadu, India, and has over seven years of experience in teaching and research. She has a MSc in Food Science and Technology with a certificate of merit with distinction (Vishishta Yogyata) for her outstanding academic performance. She completed her PhD in Food and Nutrition from Punjab Agricultural University (PAU), Ludhiana. Dr Tomer was awarded an INSPIRE Fellowship by the Department of Science and Technology, New Delhi, India, and a University Merit Scholarship by PAU, Ludhiana. She has also passed the National Eligibility Test and Senior Research Fellowship exam conducted by the Indian Council of Agricultural Research. She has authored 40 publications and 4 book chapters and is also an active reviewer of reputed journals. Dr Tomer is currently an active member of scientific societies including the Nutrition Society of India, the Indian Dietetic Association, and the American Society of Nutrition.

Dr Mukul Kumar is presently working as an Assistant Professor in the Department of Food Technology and Nutrition, School of Agriculture, Lovely Professional University, Phagwara, Punjab, India. He also served at the School of Bioengineering and Food Technology, Shoolini University, Solan, for about three years. He has seven years of experience in teaching and research. Dr Kumar is an alumnus of the Shoolini University, Solan, Himachal Pradesh, India, and an eminent researcher whose chief interests lie in herbal formulations to cure health problems, bioactive compounds utilizations, waste valorization, extruded products, beverage technology, *in-vitro* digestion, and packaging. He has authored 20 research publications, 6 book chapters, 13 patents, and 2 copyrights. He is guiding 4 PhD scholars and has guided 20 MSc students.

Dr Prince Chawla joined Lovely Professional University, Phagwara, Punjab, as an Assistant Professor in the Department of Food Technology and Nutrition (School of Agriculture). He is an alumnus of Chaudhary Devi Lal University, Sirsa, and Shoolini University, Solan. He has a chief interest in plant-based polysaccharides, mineral fortification, functional foods, protein modifications, food safety, and the detection of toxicants from foods using nanotechnology. Dr Chawla has worked on Department of Biotechnology- and Department of Science and Technology-funded research projects at the National Dairy Research Institute, Karnal, Haryana, and has eight years of research experience. He has 3 patents, 5 books, 82 international research articles, and 12 international book chapters. Dr Chawla is guiding 4 PhD students and has guided 22 MSc students. He is a life-time member of the Association of Food Scientists & Technologists (India). He is also the lead editor of a special issue of *Foods and Food Quality*. Dr Chawla is a recognized reviewer of more than 30 international journals and has reviewed several research and review articles.

Chapter 1

Origin, Taxonomy, Botanical Description, Biodiversity, Cultivation, Agrarian Requirements, and Production of Millets

1.1 INTRODUCTION

Millet is an umbrella term that represents cereals with small-seeded grains. This term is thought to have originated from the French word *"mille"* which means thousand. A handful of millet contains up to 1000 grains, which is how it got its name (Ramashia et al. 2019). Millets are one of the oldest foods known to humankind and were once the staple foods in many parts of Africa and Asia (Karuppasamy 2015). Worldwide, there are about 20 different species of millet that are cultivated today or were cultivated at different points in time (Habiyaremye et al. 2017). The major cultivated millets are pearl millet [*Pennisetum glaucum* (L.) R. Br., Syn. *Cenchrus americanus* (L.) Morron], foxtail millet (*Setaria italica*), proso millet (*Panicum miliaceum*), finger millet (*Eleusine coracana*), barnyard millet (*Echiconola esculenta, E. frumentacea*), kodo millet (*Paspalum scrobiculatum*), little millet (*Panicum sumatrense*), teff (*Eragrostis tef*), and fonio (*Digitaria exilis* Stapf, *D. iburua* Stapf). Among these, pearl millet is known as the major millet while all other millets are collectively represented by the term small millets. With progress in time millets were replaced with rice and wheat but in the past few decades millets have again attracted the attention of agriculturists and nutritionists. The key factors driving the interest of agriculturists are increasing global warming, irregular rains, limited capability to increase the irrigated lands, and the expansion of the drylands (a 10% increase in global drylands is expected by the end of the 21st century) (Calamai et al. 2020; Kumar et al. 2018). Due to the ability of millets to grow on low-fertility soils with limited water supply, they could be used to replace high-water demanding crops in the future. Being a rich source of dietary fibre, minerals, vitamins, and phytochemicals, millets are also a grain of interest for nutritionists (Kumar et al. 2021). They are also reported to reduce the glycaemic index of foods, control diabetes, obesity, cancer, and cardiovascular diseases, and build immunity (Kumar et al. 2021). The COVID-19 outbreak had a positive effect on millet consumption and the retail market, and people shifted from junk foods to nutrient-rich foods like millets. Millets have emerged as the most promising cereals in the changing context of agriculture and the demand for healthy foods. The global millet market is forecasted to grow at a compound annual growth rate of 4.5% from 2021 to 2026 (Research and Markets 2021) and is expected to reach 12 billion USD by 2025 (Pulidini and Pandey 2019). India alone has a 41.04% share in the global millet market. This chapter provides an overview of the origin, taxonomy, botanical description, agrarian requirements, germplasm accessions, regions of cultivation, and worldwide production of millets.

1.2 ORIGIN

Pearl millet, the most widely grown millet in the world, is believed to have originated in Africa about 4000–5000 years ago (Joshi et al. 2021; Muimba-Kankolongo 2018). Scholars disagree on the single centre of its origin, so the entire Sahel region from Mauritania to Western Sudan is considered as its native. The Sahara Desert is the ultimate source of the wild progenitors, *P. mondii* and *P. mollissimum*. The earliest clear evidence of the first domestication of pearl millet is found in the Sahel of Northern Mali where it was domesticated about 4500 years ago (Taylor 2019; Hu et al. 2015). From there, it was divided into two races, i.e. *globosum* and *typhoides*. The race *globosum* moved to the western side, and the race *typhoides* travelled to eastern Africa. From there it reached India and southern Africa about 2000–3000 years ago (Joshi et al. 2021). The genus *Setaria* has three gene pools, i.e. primary (*S. italica* and *S. viridis*), secondary (*S. adhaerans*, *S. verticillate*, and *S. faberii*), and tertiary (*S. glauca* and other wild species) (Jia et al. 2013). The cultivated foxtail, i.e. *S. italica*, is derived from its wild ancestor, *S. viridis* (green millet), which is interfertile. Early traces of foxtail millet domestication have been discovered in Thailand, where it was domesticated around 4300 years ago. However, it became popular in China and spread from there to India, Europe, and other parts of the world (Zohary and Hoff 2000). According to recent archaeological evidence, foxtail millet was domesticated from green foxtail around 9000–6000 years before present (YBP) in China (Hu et al. 2018; Diao and Jia 2017b). Several theories exist regarding the origin of proso millet. According to one theory, proso millet originated in East Asia, specifically Manchuria (Taylor and Emmambux 2008), whereas the Indian Council of Agricultural Research claims that proso millet originated in India and spread to other growing areas around the world (Prabhakar et al. 2017). According to other theories, it was domesticated in China between 8000 and 10,000 years ago (Das et al. 2019). Another theory holds that proso millet was domesticated in China and Europe, with evidence of proso millet in the form of charred grains and grain impressions in pottery discovered in Eastern Europe from around 7000 years ago (Vetriventhan et al. 2019). Although the precise time of the domestication of proso millet is unknown, archaeologists agree that it occurred in three different locations, namely northwest China, central China, and Inner Mongolia (Habiyaremye et al. 2017). Barnyard millet has two major types, i.e. Indian barnyard millet (*E. frumentacea*) and Japanese barnyard millet (*E. esculenta*). The Indian barnyard millet originated from the wild species *E. colona* (L.) which is also known as jungle rice. The exact history of the origin of the Indian barnyard millet is unclear. It is believed to have parallel lines of evolution in India and Africa. The Japanese species, *E. esculenta*, evolved from the wild species *E. crus-galli* and was domesticated around 4000 years ago (Renganathan et al. 2020; Sood et al. 2015). Finger millet is native to the Ethiopian highlands and was introduced to India about 4000 years ago. It has two subspecies, i.e. *africana* and *coracana*. Dida et al. (2008) have further classified the subspecies *coracana* into those originating from Africa and those originating from Asia based on phylogenetic and population structure analysis. Kodo millet is the coarsest of all the cereals. There are two schools of thought about its origin. Some researchers claim it is indigenous to India (Joshi et al. 2021; Taylor and Emmambux 2008), while other reports claim it originated in tropical Africa and was domesticated in India about 3000 years ago (Prabhakar et al. 2017). It is an important crop of the Deccan plateau of India (Kumar et al. 2016) and is known as a famine grain due to its ability to be preserved for a longer duration without any harm. The species grown in India is *Paspalum scrobiculatum* var. *scrobiculatum*, while the species grown in Africa is *Paspalum scrobiculatum* var. *commersonii* (Heuzé et al. 2015). Little millet is one of India's earliest cereal grains, having been domesticated in the Eastern Ghats around 5000 years ago (Johnson et al. 2019). Teff originated in eastern Africa and is believed to be first domesticated by pre-Semitic inhabitants of Ethiopia (Lee 2018). Fonio is the oldest indigenous cereal of West Africa. It has two major species, i.e. white fonio (*D. exilis* Stapf) and black fonio (*D. inurua* Stapf). It was domesticated about 7000 years ago in West Africa, in present-day Guinea and Mali (Abdul and Jideani 2019; Ibrhaim Bio Yerima et al. 2020).

1.3 TAXONOMY

Millets as a whole belong to the kingdom Plantae, the order Poales, and the family Poaceae or Gramineae. They are further divided into two subfamilies, i.e. Panicoideae or Chloridoideae. Pearl millet, foxtail millet, proso millet, barnyard millet, kodo millet, little millet, and fonio belong to the Paniceae tribe and Panicoideae subfamily, while finger millet and teff belong to Eragrostideae tribe and Chloridoideae subfamily. A detailed taxonomy of various millets is provided in Table 1.1.

1.4 GROWTH CYCLE

The growth cycle of millets can be divided into the vegetative stage, panicle development stage, and grain filling stage. The vegetative phase starts with the emergence of seedlings and continues up to the stage of panicle initiation. During the panicle initiation, the elongation of the panicle dome and the formation of a constriction at the base of the apex take place. The various other changes which take place during this stage are the development of spikelets, florets, glumes, stigmas, anthers, flowers, and pollination. After pollination, begins the next stage during which the fertilization of florets in the panicles of the main shoots takes place and continues up to the maturity of the plant. A substantial increase in the grain dry weight takes place during this stage. This begins the senescence and formation of a small dark layer of tissue in the hilar region of the grain that signals the end of grain filling (Maiti and Bidinger 1981).

1.5 BOTANICAL DESCRIPTION

Pearl millet is a high tillering, cross-pollinating C4 cereal that ranges from 6 to 14 feet in height. The stem is round to oval in shape with a diameter of about 0.5–1.5 cm. After 12 days of seedling emergence, the tiller leaf appears on both sides of the main stem. The leaves are 20–100 cm long and 0.5–5 cm wide and vary from densely hairy to glabrous. The panicle emerges after 35–70 days of sowing and may be cylindrical, conical, spindle, candle, lanceolate, dumb-bell, club, oblanceolate, or globose in shape. The most common shapes are cylindrical and conical. The inflorescence of pearl millet is a compound terminal spike called a panicle that has a length of 20–25 cm and a circumference of 7–9 cm (Anonymous 2021). The spikes are always surrounded by involucres, which are tightly filled with slender, glabrous to plumose bristles. The spikelets are either directly attached to the stalk (pedicellate) or are free from the stalk (sessile). The species are further classified based on duration (annual or perennial), ploidy level (diploid or octoploid), reproduction

TABLE 1.1 TAXONOMY AND THE LOCAL NAMES OF MILLETS

Millet type	Pearl millet	Foxtail millet	Proso millet	Barnyard millet	Finger millet	Kodo millet	Little millet	Teff	Fonio
Kingdom	Plantae	Plantae	Plantae	Plantae	Plantae	Plantae	Plantae	Plantae	Plantae
Tribe	Paniceae	Paniceae	Paniceae	Paniceae	Eragrostideae	Paniceae	Paniceae	Eragrostideae	Paniceae
Family	Poaceae	Poaceae	Poaceae	Poaceae	Poaceae	Poaceae	Poaceae	Poaceae	Poaceae
Subfamily	Panicoideae	Panicoideae	Panicoideae	Panicoideae	Chloridoideae	Panicoideae	Panicoideae	Chloridoideae	Panicoideae
Genus	Pennisetum/ Cenchrus	Setaria	Panicum	Echinochloa	Eleusine	Paspalum	Panicum	Eragrostis	Digitaria
Species	*P. glaucum/ C. americanus*	*S. italica*	*P. miliaceum*	*E. frumentacea* and *E. utilis*	*E. coracana*	*P. scorbiculatum*	*P. Sumatrense*	*E. tef*	*D. exilis* and *D. iburua*
Other English names	Bulrush millet, candle millet, dark millet, Indian millet, horse millet	Italian millet, German millet, Hungarian millet, dwarf setaria, giant setaria, green foxtail	Broomcorn millet, common millet, hog millet, Kashfi millet, red millet, panic millet	Cockspur milletgrass, Korean native millet, prickly millet	African millet, Caracan millet, Koracan	African bastard millet grass, cow grass, rice grass, ditch millet, native Paspalum, Indian crown grass	–	Teff, lovegrass, annual bunch grass, warm season bunch grass	Fonio, white fonio, hungry rice, hungry millet, hungry koos, fundi millet, acha grass
Country-specific names	Bajra (India), Gero (Nigeria), Hegni (Niger) Sanya (Mali) Dukhon (Sudan) Mahangu (Namibia) mil à chandelle or petit (French) Mhunga or mahango (Southern Africa)	Su, Guzi, Xiaomi (China), Kangani, Korra, Priyangu (India), Hirse (Germany), Ghomi (Georgia), Awa (Japan), Jo (Korea), Kaguno (Nepal),	Cheena, Cheno, Panivaragu, Baragu (India), mijo común (Spain); millet commun, millet blanc (France); kê Proso (Vietnam); Rispenhirse (Germany); Miglio (Italy)	Shayama, Sanwa, Swank, Khira (India), crête de coq, ergot de coq, millet du Japon, pied de coq (France), gewöhnliche Hühnerhirse, Hühnerhirse (Germany); Giavone comune (Italy), zacate de agua (Spain); Padi burung, Dwajan (Indonesia); Song chang (Vietnam)	Ragi, Nagli, Mandhal, Mandika, Nachni (India), Mijo africano (Spain), éleusine, coracan, millet africain (France), Fingerhirse (Germany)	Kodo, Kodra, Kodon, Kodra, Varagu, Arika (India)	Sama, Kutki, Same, Save, Samat (India)	Tef (Ethiopia), Tafi (Orimigna), Taf (Tigrigna), Mil ethiopien (France)	Findi, fonden, sangle (Senegal), Funde (Spain), Petit mil, millet digitaire (France), Fingi, Foyo (Niger), Atcha, Kabega (Ghana), Tschamma (Togo), Akka, Chyung (Nigeria)

method (sexual or asexual), somatic chromosome number (2n = 10–78), basic chromosome number (x = 5, 7, 8, or 9), chromosome size (92–395 Mbp), and genome size (ICx = 0.75–2.49 pg). The genus *Pennisetum* has approximately 140 species and nearly three-quarters of the species of the genus *Pennisetum* are polyploids. The cultivated species are diploid (2n = 2X = 14) and have a genome size of 2450 Mbp (Pattanshetti et al. 2016). Pearl millet seeds are oval and 3–4 mm long. These have the appearance of a pearl, which is where the name comes from (Dias-Martins et al. 2018).

The genus *Setaria* is widely distributed in warm and temperate regions of the world and has about 125 species. The cultivated foxtail millet is diploid (2n = 18) with a genome size of 423 Mbp. It has the smallest genome size among all the millets. The plant height ranges from 120 to 200 cm depending on the type of accession. The domesticated foxtail millet has a single stalk or a few tillers while the green foxtail has tillers and axillary branches that appear over the life of the plant (Diao and Jia 2017a). The stem is thin, erect, slender, and leafy with hollow internodes. The leaves are hairless, smooth, and arc-broad. The inflorescence is a contracted panicle that resembles a spike. It is about 5–30 cm long and gives the appearance of a fox's tail (Figure 1.1). The panicle type is also affected by environmental conditions and can be cylindrical, conical, highly branched, or like a long spike, spindle, palmate, cat foot, or hen beak. The anthers may be brownish-orange or white. Purple anthers are also found in some *Setaria* species like *S. gluaca*. The spikelets are closely packed and each spikelet contains only one flower with a yellow pistil. The domesticated species has large inflorescences that mature more or less at the same time. Each mature inflorescence may produce hundreds of small grains measuring about 2 mm in diameter. The domesticated species retain their grains on maturity while the inflorescence of wild species shatters readily on maturity to release the seeds (Singh et al. 2017; Doust et al. 2009). The grains may be red, orange, yellow, black, or brown.

Figure 1.1 Panicles of various millets. (a) Pearl millet, (b) proso millet, (c) foxtail millet, (d) finger millet, (e) barnyard millet, (f) kodo millet, (g) little millet.

The genus *Panicum* has more than 400 species. The two major species of this genus that are used for cultivation are *P. miliaceum* (proso milet) and *P. sumatrense* (little millet). The cultivated proso millet is a warm-season crop with an adventitious root system that varies in height from 20 to 130 cm. The stem is slender with distinctly swollen nodes. The stem diameter ranges from 2.5 to 8.8 mm and the stem node numbers are 3.7–9.1. The plant has few tillers and the average number of tillers is 2.33. The leaf area ranges from 39.94 to 317.76 cm^2 (Zhang et al. 2019) and there are six to seven linear, slender leaves per plant. The proso millet reaches 50% flowering 26–50 days after sowing (Vetriventhan et al. 2019; Habiyaremye et al. 2017). The panicle length ranges from 53 to 171 mm and the weight range is between 4.43 and 96.38 g. Depending on the panicle morphology and shape, proso millet is divided into five races, i.e. *miliaceum, patentissimum, contractum, compactum*, and *ovatum* (Vetriventhan et al. 2019). The inflorescence is an open or compact drooping panicle that ranges in length from 8 to 45 cm (Figure 1.1). It gives the appearance of an old broom which gives it its name, "broom corn". The spike length ranges from 12.8 to 48.5 cm and the spikelets are 0.5 cm long. The spikelets lack bristles and generally exist in isolation. The anthers are amber, dark brown, or black (Gupta et al. 2013). The seeds are round with a diameter of approx. 2–3 mm. They can be white, yellow, dark, or black and the seed yield per plant is 8.94–10.28 g (Calamai et al. 2020).

The genus *Echinochloa* consists of approximately 250 annual and perineal species. It is the fastest growing among all the millets and the average number of days to maturity is 45–60 depending on the accession and environmental conditions. The cultivated *Echinochloa* are annual, hexaploid, high tillering plants. It has a shallow fibrous root system and the plant height ranges from 60 to 122 cm in Japanese barnyard millet and up to 240 cm in Indian barnyard millet (Renganathan et al. 2020). The stem is flat near the base and the culms are slender to robust and vary in colour from green to purple. The leaves are green, 15–50 cm long, and 1–2.5 cm wide. The leaf blades are smooth and glabrous. It bears purplish flowers in summer or early fall. The inflorescence is a compact terminal panicle, i.e. usually erect, and is about 1–28 cm long. It is rarely drooping and can have a cylindrical, pyramidal, or globose to elliptic shape (Sood et al. 2015). The racemes are 0.5–3 cm long and range in number from 20 to 70 in Indian barnyard millet and 5 to 15 in Japanese barnyard millet. The spikelets are branching in Japanese barnyard millet and non-branching in Indian barnyard millet. The spikelets vary in length from 2 to 4 mm and the Japanese barnyard millet has smaller spikelets (Taylor 2019). The grain is caryopsis and ranges in colour from straw white, dark grey to brown in Indian barnyard millet, and pale yellow to light brown in Japanese barnyard millet (Renganathan et al. 2020). The geometric mean diameter and surface area of the barnyard millet whole grain at a moisture content of 0.5 kg/kg dry matter are between 1.2 and 1.4 mm and 5 and 6 mm,^{2} respectively (Singh et al. 2010).

Finger millet is a self-pollinating allotetraploid annual grass. It has two major subspecies, i.e. *africana* and *coracana*. Subspecies *africana* is a diploid (2n = 18) and subspecies *coracana* is a tetraploid (2n = 36) (Dida et al. 2008). The species used for cultivation is *E. coracana* (ICRISAT 2012). Subspecies *coracana* is classified into four races, i.e. *elongata, plana, compacta*, and *vulgaris* (Upadhyaya et al. 2007). The average plant height ranges between 96 and 115 cm and it can grow up to 170 cm. It has a shallow, branched, and fibrous root system with rooting at the lower nodes. The stem is green, slender, erect, compressed, and glabrous. It is free tillering and the average number of tillers is 4–18 per plant. The number of culm branches is between 2 and 4. The leaf blade is green, 30.3–36.4 cm long, 7.67–13.20 mm wide, and has a prominent midrib. The inflorescence or panicle grows at the end of the vegetative shoot and the length and width of the inflorescence are 82–181 mm and 69–112 mm, respectively. The panicles can be top curved, incurved, open, and fisty, and contains 3–20 spikes that are arranged like fingers on the hand. The rachis of the spikes is flat and contains alternate spikelets arranged on it. Flowering takes place between 63 and 84 days for different races and the florets open from bottom to top. Each spikelet contains four to seven seeds and the grain is globose to smooth. The naked grain is spherical with a diameter of 1–2 mm and can be black, brown, light brown, red, purple, or white (TNAU 2021; Taylor 2019). A pictorial representation of the various parts of finger millet is also provided in Figure 1.2.

The genus *Paspalum* contains about 400 species. The cultivated species of this genus is *P. scrobiculatum*, i.e. kodo millet. It is an annual self-pollinating high tillering cereal. The plant height ranges from 44 to 69 cm and the number of basal tillers ranges from 6 to 30. Culms branching may be present or absent. The stem may be pigmented (purple) or non-pigmented and the leaves are light green. There are five to seven slender leaves and the length and breadth of the flag leaf blade range from 15.6 to 22.6 cm and 5.9 to 8.4 mm, respectively (Upadhyaya et al. 2016). The average days for 50% flowering completion range between 80 and 90 days after sowing and the percentage of open flowers does not exceed 15–20%. The inflorescence is 2–12 cm long, may be semi-compact or open type, and contains two to six racemes. The raceme exertion may be complete or partial and varies in length from 3 to 15 cm. The rachis is 1.5–3 mm long and is usually sessile or on a short pedicel. The spikelets arrangement may be regular, irregular, or 2–3 rows irregular and are singly arranged in two rows on a flattened rachis. The crop matures in 4–6 months; it takes the longest time for maturity among all the millets (Taylor and Emmambux 2008). The seeds may have an orbiculate, elliptical, or oval shape and the diameter ranges between 2.15 and 2.31 mm. The colour of the grain may be golden to dark brown (Shrishat et al. 2008).

Little millet belongs to the genus *Panicum* which is same as that of proso millet. It is a self-pollinating, high tillering plant. Plant height ranges from 58 to 201 cm in different accessions and

Figure 1.2 Finger millet plant and its different parts. (a) A pictorial view of finger millet cultivated in fields of Himachal Pradesh, India, (b) a finger millet stalk with inflorescence, (c) inflorescence of finger millet, (d) spikelet of finger millet, (e) flowering bud, (f) finger millet grains.

the number of tillers per plant varies from 3 to 26 with a mean of 6.9. The initial growth behaviour in the majority of varieties is decumbent or erect. Prostrate growing behaviour is also reported in some of the varieties. The flag leaf is 18–41 cm long and 5–19 mm wide. Lead blade and leaf sheath pubescence are absent in most of the varieties. The days taken for 50% flowering vary with the accessions and it may take 30–139 days. The average number of days taken for flowering by most of the accessions is 65. The inflorescence is a contracted or thyrsiform panicle, i.e. about 19–34 cm long and 1–5 cm wide. The spikelets are produced on unequal pedicels and are 2.0–3.5 mm long. They are elliptical, dorsally compressed, acute, and grow singly at the end of branches. Each spikelet contains two minute flowers and three anthers about 1.5 mm long. The little millet grain is elliptical to oval and grey to straw white in colour. The diameter of little millet grain is about 1.90 mm (Rao et al. 2020). The grain yield per plant ranges between 2.9 and 50.1 g and the average grain yield per plant is 13.8 g.

Teff belongs to the genus *Eragrostis* and it has about 350 species. It is an annual self-pollinated allotetraploid ($2n = 4X = 40$). It has a slender stem with a massive shallow fibrous root system. The height of various teff cultivars ranges between 30–90 cm. It is high tillering and the number of tillers per plant is 4–22. In most of the species, the stem is erect; however, it may be bending or elbowing. The culm has a length of 11–82 cm and the diameter ranges between 0.9 and 2.4 cm. The leaf blade is 10–700 mm long and 3–6 mm wide. The average day of heading in most of the cultivars is 25–65 and the crop matures in 80–120 days. The inflorescence is an open panicle that may be compact or loose and varies in length from 10 to 65 cm. The peduncle length is 5.8–42.3 cm. The number of spikelets and florets per panicle may be 60–400 and 5–12, respectively. The anthers may be dark purple, deep red, or yellowish-white. The grains are ivory, medium brown to dark brown, dark reddish, purple, yellowish-white, or greyish-white. The diameter of grains is 0.5–1 mm (Assefa et al. 2017; Merchuk-Ovanut et al. 2020; Bedane et al. 2015; Arendt and Zannini 2013).

Fonio has four cultivated species around the world. These are *Digitaria exilis* (white fonio), *Digitaria iburua* (black fonio), *Diditaria sangwinalis* (European millet or red manna), and *Digitaria cruciata* (raishan). *D. exilis* and *D. iburua* are grown in West Africa, *D. sangwinalis* is grown in Europe, and *D. cruciata* is found in India and Vietnam (Kanlindogbe et al. 2020). The most cultivated varieties are *D. exilis* and *D. iburua*. The progenitor of cultivated fonio is most probably the annual wild weed *D. longiflora* that is distributed all across tropical Africa (Abrouk et al. 2020). The plant height of these species ranges from 60 to 86 cm and they have a fasciculated root system. The root diameter is about 1 mm and can grow up to 228 cm. The stubble is hollow, cylindrical, and erect. The size of stubble is 30–80 cm in white fonio and 45–140 cm in black fonio. There are two to eight tillers with five to nine nodes in white fonio and four to five nodes in black fonio. The number of internodes is 6–11 and the length of internodes is 18–25 cm. The number of leaves per plant is 79–212. The leaf length and width of the white variety are 5–15 cm and 0.3–0.9 cm, respectively, while those of the black variety are 30 cm and 1 cm. The flowering takes place between 68 and 92 days and 50% maturity is achieved in 85–108 days. The inflorescence has two to four racemes or spikes that are 5–12 cm long. The raceme bears spikelets (1.5–2 mm) that are attached through 1-mm-long slender pedicels. The spikelets contain sterile lower flowers and fertile upper bisexual flowers. Black fonio reproduces by absolute apomixes while white fonio is highly autogamous with a very low rate of cross-pollination. The number of grains per raceme varies from 70 to 120. The caryopsis is ovoid to ovo-ellipsoid in the white variety and ellipsoid in the black variety. The diameter of grain is 0.5–1 mm and the length varies from 0.75 to 2 mm. The grain colour is white to yellow in the white variety and purple wine red to blackish in the black variety (Kanlindogbe et al. 2020; Ibrhaim BioYerima et al. 2020).

1.6 GERMPLASM ACCESSIONS OF MILLETS

Germplasm is the genetic material of an organism that contains inherited characteristics that can be transmitted from one generation to another either sexually or somatically (Offord 2017). Accession is a uniquely identifiable sample of the seeds of a crop variety collected at a specific location and time for conservation and use (http://www.fao.org/wiews/glossary/en/). More than 164,777 germplasm accessions of millets are conserved worldwide. The conservation of germplasms helps in the conservation of biodiversity and biological integrity by preserving primitive, endangered, and extinct species. It also helps in the utilization of plant genetic resources throughout the world by providing the germplasms with validated phenotypic and genetic descriptions (Bhatia 2015). This section provides an insight into the germplasm accessions of various millets available in the various gene banks across the world.

Pearl millet has the highest number of germplasm accessions among millets and more than 66,682 wild and cultivated germplasm accessions of the *Pennisetum* spp. are conserved in the 97 gene banks across the 65 countries of the world. Nearly 37,580 germplasm accessions of its related genera *Cenhhrus* spp. are also conserved in the 50 gene banks of 32 countries (Pattanshetti et al. 2016). As of 2021, the largest collection of pearl millet germplasm is available at International Crops Research Institute for the Semi-Arid Tropics (ICRISAT), Patancheru, India, and National Bureau of Plant Genetic Resources (NBPGR), New Delhi, India. These institutions hold 24,390 and 8341 pearl millet germplasm accessions, respectively (ICRISAT 2021; NBPGR 2022). The other major gene banks that conserve pearl millet accessions are Embrapa Milho e Sorgo, Sete Lagoas, Brazil; Laboratoire des Ressources Génétiques et Amélioration des Plantes Tropicales, ORSTOM, Montpellier Cedex, France; Plant Gene Resources of Canada, Saskatoon Research Centre, Agriculture and Agri-Food Canada, Saskatoon, Canada; and International Crops Research Institute for Semi-Arid Tropics, Niamey, Niger, conserving more than 7225, 4418, 3806, and 2817 accessions, respectively. The national gene banks of Uganda, the USA, Niger, and Pakistan also hold 2144, 2063, 2052, and 1377 germplasm accessions, respectively (Pattanshetti et al. 2016).

Globally, more than 46,000 germplasm accessions of foxtail millet are conserved mainly in gene banks located in China, India, France, Japan, Korea, the USA, and Russia. The largest germplasm of foxtail millet is conserved at the Institute of Crop Science, Chinese Academy of Agricultural Sciences, China, having more than 27,069 germplasm accessions of cultivated and wild varieties. The other countries such as India, France, Japan, Korea, the USA, and Russia collectively contain about 15,000 germplasm accessions of foxtail millet (Diao and Jia 2017b). Among these, NBPGR, New Delhi, has the maximum number (4384) of germplasm accessions followed by ICRISAT, Hyderabad, India (1542) (Vetriventhan et al. 2020). The National Institute of Agrobiological Sciences, Tsukuba, Japan; the Department of Agriculture-Agricultural Research Service; the National Gene Bank of Korea; the Plant Genetic Resources Conservation Unit of the United States; and the National Gene Bank of Bangladesh contain about 1286, 960, 766, and 510 accessions, respectively.

Proso millet has more than 29,000 conserved germplasm accessions globally. N. I. Vavilov All-Russian Scientific Research Institute of Plant Industry, St. Petersburg, holds the largest (more than 8778) number of proso millet accessions followed by the Institute of Crop Germplasm Resources, Chinese Academy of Agricultural Sciences, Beijing, China, the Yuryev Plant Production Institute UAAS, Kharkiv, Ukraine, and the Ustymivka Experimental Station of Plant Production, S. Ustymivka, Ukraine, which hold more than 6517, 1046, and 976 germplasm accessions, respectively (Habiyaremye et al. 2017). As of 2021, ICRISAT, Patancheru, India, holds 849 germplasm accessions of proso millet (ICRISAT 2020). The remaining germplasm accessions of proso millet are conserved in the gene banks of the USA (1432), Poland (721), Mexico (400), and Japan (302).

Barnyard millet has about 8000 germplasm accessions conserved across the world. Japan and India are the largest contributors to the barnyard millet germplasms. The largest number of germplasm accessions (3671) is conserved at the Department of Genetic Resources, National Institute of Agrobiological Sciences, Japan (Vetriventhan et al. 2020). The Plant Germplasm Institute, Kyoto University, holds 65 germplasm accessions. In India, the NBPGR, the Indian Institute of Millet Research (IIMR), the University of Agricultural Sciences, Bangalore, the ICRISAT, and Vivekananda Parvatiya Krishi Anusandhan Sansthan, Almora, have conserved 1888, 1561, 985, 749, and 300 accessions of barnyard millet, respectively. Apart from these, the Institute of Crop Science, Chinese Academy of Agricultural Sciences, Bejing, China; the National Centre for Genetic Resources Conservation, Collins, USA; the North Central Regional Plant Introduction Station, Ames, USA; and the USDA Agricultural Research Service, Washington, USA, conserve 717, 306, 304, and 232 germplasm accessions, respectively. The gene banks of Kenya, Ethiopia, Australia, Pakistan, Norway, the United Kingdom, and Germany contain 208, 92, 66, 50, 44, 44, and 36 germplasm accessions, respectively (Renganathan et al. 2020).

Finger millet has approximately 36,873 accessions worldwide. The largest number of finger millet accessions, i.e. about 27% (10,507) of the world total, is maintained by the NBPGR, India (Vetriventhan et al. 2020). ICRISAT, India has 7519 germplasm accessions of finger millet as of 2021 (ICRISAT 2021). The All India Coordinated Minor Millet Project (AICMMP) holds 6257 finger millet accessions. Other institutes like the Kenya Agricultural Research Institute (KARI), Kenya; the Institute of Biodiversity Conservation (IBC), Ethiopia; and the USDA Agricultural Research Service (USDA-ARS), USA, have 2875, 2156, and 1452 germplasms of finger millet, respectively. The various institutions of other countries, i.e. Uganda, Zambia, Nepal, the USA, Japan, Zambia, and China, hold 1231, 1037, 809, 702, 565, 390, and 300 accessions of finger millet, respectively (Goron and Raizada 2015).

The studies on the germplasm accessions of kodo and little millet are very limited. The numbers of conserved germplasm accessions worldwide for kodo and little millet are 4780 and 3064, respectively. Among these, 3946 accessions of kodo millet are stored in India in various organizations like NBPGR (2,180), AICMMP (1,111), and ICRISAT (665) (ICRISAT 2021; Vetriventhan et al. 2020; Goron and Raizada 2015). The USDA-Agricultural Research Service (USDA-ARS) holds 336 accessions of kodo millet. The largest number of germplasm accessions (2830) of little millet are stored in Asia. NBPGR, India, holds the largest (1253) number of accessions followed by AICMMP (544), ICRISAT (473), and USDA-ARS (226) (Vetriventhan et al. 2020).

There are more than 8800 germplasm accessions of teff conserved worldwide. The highest number of teff germplasm accessions is stored at the Institute of Biodiversity Conservation, Addis Abada, Ethiopia. The other major gene banks with high numbers of teff germplasms are USDA-ARS, Washington State University, United States; the National Genebank of Kenya, Crop Plant Genetic Resources Centre, Muguga, Kenya; the Department of Genetic Resources, National Institute of Agrobiological Sciences, Japan; NBPGR, India; and Centro de Investigaciones Forestales y Agropecuarias, Instituto Nacional de Investigaciones Forestales, Agrícolas y Pecuarias, Mexico. These institutions hold 1302, 1051, 321, 269, and 258 germplasm accessions, respectively (FAO 2010).

Fonio has been given the least attention among all the millets and only a few germplasm accessions are conserved at the national research systems and the Consultative Group on International Agricultural Research (Ibrhaim BioYerima and Dako 2021). The first ever regional genetic resource collection of fonio was carried out in 1984 by Clement and Leblanc in six countries, namely Benin, Niger, Togo, Mali, Burkina-Faso, and Guinea. A total of 641 accessions of *D. exilis* were collected from these countries and were conserved at the Institut de Recherche pour le Développement (IRD), Marseilles, France (Ayenan et al. 2018). Germplasm accessions are also collected in various countries at the national level and some fonio accessions are maintained in different African and European agricultural research centres (Adoukonou-Sagbadja et al. 2004; Clottey et al. 2006). The conserved germplasm accessions of cultivated and wild fonio are distributed across France (>400), Nigeria (>257), Togo (>94), Burkina-Faso (>54), Benin (>50), Guinea (>45), Niger (>30), Ghana (>18), and Mali (>8) (Kanlindogbe et al. 2020).

1.7 CULTIVATION AND AGRARIAN REQUIREMENTS

Pearl millet is the sixth most cultivated cereal in the world and it is the major crop in arid and semiarid zones of India and Africa (Ausiku et al. 2020; Crookston et al. 2020). It can grow from sea level to a height of 2700 metres under a wide array of soils such as loamy, sandy loam, clay loam, clay, or shallow soils. It is highly drought-resistant and gives a good yield on soils with a limited number of nutrients and organic matter. Annual precipitation of 200–600 mm, pH of 6.0–7.0, and temperature of 30–34°C are optimal for its growth; however, it can tolerate a temperature as high as 46°C, rainfall of 150–800 mm, and soil pH of 8. It is also highly tolerant to soil salinity (11–12 dS/m) but economically viable yields are obtained only up to 8 dS/m. These attributes make it a good choice for regions where other cereal crops, such as wheat, rice, or maize, would not survive (Nambiar et al. 2010). It is highly susceptible to waterlogging conditions and standing water should be avoided in the crop (Lee et al. 2012). It is cultivated with the onset of monsoon, i.e. the first fortnight of July in North and Central India and in October in Tamil Nadu (Vikaspedia 2022a). The summer varieties of pearl millet are sown in the last week of January to the first week of February (Agropedia 2022).

Foxtail millet is the second most produced millet in the world. The major foxtail millet-producing countries are China, India, Russia, Afghanistan, Manchuria, Korea, Georgia, and the USA (Joshi et al. 2021). It is cultivated on sandy to loamy soils at an altitude of sea level to 2000 metres (Kumar et al. 2018). It grows rapidly in warm weather and can grow in semiarid conditions. It shows strong dehydration tolerance (Matsuura and An 2020); however, it has a shallow root system that does not easily recover from drought (Sheahan 2014). The maintenance of water-deficit conditions (20% soil water content) from 25 days after sowing decreased the yield to 62% of control (Matsuura and An 2020). The optimum temperature for its growth is 16–25°C; however, it can grow in a broad temperature range of 5–35°C. The desirable pH for its growth is 5.5–7.0 and it can tolerate soil salinity up to 6 dS/m. It requires an annual rainfall of 300–700 mm and is quite tolerant to waterlogging conditions. The maintenance of waterlogging conditions after 17 days of sowing to harvesting resulted in only a 4% reduction in grain yield (Matsuura et al. 2016). This millet can be cultivated in both kharif and rabi seasons. Rabi cultivation is done under irrigated conditions. Kharif sowing is done in the month of July–August and rabi sowing is done in the month of August–September or later (Hermuth et al. 2016).

Proso millet is well adapted to many soil and climatic conditions and altitudes. It is spread across the tropical, subtropical, and warmer temperate regions of the world. It is mainly cultivated in China, India, Russia, Iran, Iraq, Syria, Turkey, Afghanistan, and Romania (Saxena et al. 2018). It can be grown successfully on sandy loam, slightly acidic (5.5–6.5), saline, and low-fertility soils under hot dry weather conditions. The optimal soil temperature for seed germination is 20–30°C. It has the lowest water requirement among all the cereals and is very drought resistant. The low water requirement is not only due to its drought resistance but also due to its short maturing time. Average rainfall of 200–500 mm is required to obtain good yields; however, it can produce grains with annual precipitation of as low as 300–350 mm. In a study conducted by Matsuura and An (2020) the maintenance of water-deficit conditions (20% soil water content) from 25 days after sowing decreased the grain yield to 48% of the control yield. It is sensitive to frost and waterlogging conditions. In a study conducted by Matsuura et al. (2016) the maintenance of waterlogging conditions after 17 days of sowing to heading resulted in a 16% decrease in its yield. It is highly tolerant to salt and is considered an alternative crop to salt-affected areas. The salt tolerance also varies with the accessions. Liu et al. (2015) in his study on the effect of salt stress (120 mM/L solutions of NaCl and Na2SO4 mixed in 1:1 on a molar basis) on 155 proso millet accessions collected from China and other countries of the world found that 39 of the total accessions were highly tolerant [salt damage index (SDI) of 20% or less], 22 were salt-tolerant (SDI in between 20 and 40%), and 26 were intermediately salt-tolerant (SDI in between 40 and 60%). Only 30% of the total collected accessions had SDI of 60% and more and were intermediately sensitive or highly sensitive. Cultivation on coarse and sandy soils results in poor yields and plants grown on soils with pH above 7.8 show symptoms of iron chlorosis (Habiyaremye et al. 2017). It is cultivated with the onset of monsoon, preferably in July. The time of sowing might vary as per the region of cultivation.

Barnyard millet is the fastest growing among all the cereals and can grow on medium to heavy soils with a pH range of 4.6–7.4. It can grow in a temperature range of 15–33°C and the optimum temperature range to obtain good yields is 27–33°C. It is mainly cultivated as an alternative to rice in areas where the rice crop fails or the conditions are not suitable for the cultivation of rice. Indian barnyard millet is cultivated in India, Nepal, Pakistan, the Central African Republic, Tanzania, and Malawi and Japanese barnyard millet is mainly cultivated in Japan, Korea, China, Russia, and Germany (Saxena et al. 2018; Sood et al. 2015). The optimum annual rainfall is 60–80 cm (https://data-flair.training/blogs/millets-in-india/). It is more tolerant to soil salinity (3–5 dS/m) in comparison to rice that cannot grow above a soil salinity of 3 dS/m (Kumar et al. 2018). Barnyard millet can be cultivated in both the khariff and rabi seasons. Kharif sowing is generally done under rainfed conditions in the month of September–October and rabi sowing is done in the month of February–March under irrigated conditions. This crop is also considered a substitute for rice under water-stress conditions.

Finger millet is a crop of the tropics and subtropics and can be raised successfully from plains to hill slopes, from sea level to an altitude of 2300 metres. The major finger millet producers are India and East and South Africa. It grows best in a moist climate on rich loam to poor upland shallow soils with a pH of 4.5–7.5 and 500–600 mm of annual rainfall. A mean temperature of 26–29°C is optimum for its growth. Temperatures below 20°C result in poor yields. It also has high tolerance to salinity and some varieties perform better even under salinity as high as 12 dS/m (Shailaja and Thirumeni 2007). However, the response to salinity varies with accessions, and in a study conducted by Krishnamurthy et al. (2014) on the soil salinity tolerance of 68 accessions of finger millet, it was observed that 22 accessions were highly salt-tolerant, 20 were moderately salt-tolerant, and 25 were sensitive. Finger millet is generally cultivated in June–September under rainfed conditions; however, it can also be cultivated in the rabi season (December–January) under irrigated conditions.

Kodo millet is cultivated on fertile to marginal soils up to a height of 1500 metres above sea level. It can also grow on the gravelly and stony soils of hilly regions. It is cultivated in India, Pakistan, the Philippines, Indonesia, Vietnam, Thailand, and West Africa (Prabha et al. 2019). India is the major cultivator and consumer of kodo millet, where it is the major food crop of the

Deccan Plateau (Saxena et al. 2018). It has high drought tolerance and is eaten as a famine crop in West Africa. In contrast to this, wild scrobic (*Paspalum scrobiculatum* var. *commersonnii*) is grown in soils that remain wet and is well adapted to waterlogged and flooding conditions. The optimum rainfall required is 800–1200 mm but good yields are obtained even with an annual rainfall of 500–900 mm. It is frost-sensitive and prefers an optimum day temperature of 25–27°C. It is reported to tolerate a certain degree of alkalinity (Vikaspedia 2021), but the data on the optimum soil pH and salinity are scanty. The sowing of kodo millet is done with the onset of monsoon generally in the middle of June to the end of July. It can also be cultivated by the transplantation method. In the transplantation method, the nursery is raised in the month of May–July, and 3–4-week-old seedlings can be transplanted into the field (Vikaspedia 2022b).

Historically, little millet is cultivated mainly in India, China, the Caucasus, Nepal, Pakistan, Sri Lanka, Malaysia, and Myanmar. At present, its cultivation is mostly restricted to some hilly areas in India and it is an important crop in the Eastern Ghats of India (Kalaisekar et al. 2016; Gomez and Gupta 2003). This region is divided into three zones, i.e. the southern Orissa highlands, Chittor and Cuddapah districts of Rayalseema region in Andhra Pradesh, and the uplands and Nilgiri regions of Tamil Nadu, where the average annual rainfall is in the range of 2000–3000 mm, 685 mm, and 800–200 mm, respectively (http://www.rainwaterharvesting.org/eco/eg.htm). The average height in the Eastern Ghats is 600 metres (http://www.aees.gov.in/htmldocs/downloads/e-content_06_04_20/HANDOUT%203(3).pdf); however little millet can be grown up to a height of 2100 metres above sea level (Kumar et al. 2018). The average temperature in the Eastern Ghats is 20–25°C but never exceeds 41°C even during the hot seasons (https://kalpavriksh.org/wp-content/uploads/2018/12/Eastern-Ghats-Final-July-2004.pdf). Little millet is capable of tolerating both drought and waterlogging conditions but cannot withstand temperatures below 10°C (https://www.millets.res.in/technologies/brochures/Little_Millet_Brochure.pdf). The maintenance of water-deficit conditions (20% soil water content) in little millet from 25 days after sowing to harvest decreased the yield by 24% compared to the control (Matsuura and An 2020), while the maintenance of waterlogging conditions after 17 days of sowing to harvesting increased the yield by 210% (Matsuura et al. 2016). Little millet is sown in June–July. In some states of India, it is also cultivated in the rabi season. The sowing in the rabi season is done in September–October (Maitra and Shankar 2019).

Teff is cultivated in Ethiopia, Australia, Kenya, and South Africa, but is cultivated as a major staple crop only in Ethiopia (Gomez and Gupta 2003). Nowadays, it is also widely cultivated in India and Canada (Hayes and Jones 2016). Teff can be cultivated on waterlogged to well-drained heavy soils with an average annual rainfall of 1000 mm. It can be cultivated on heights up to 3000 metres above sea level; however, the optimum height for its cultivation is in the altitude range of 1100–2950 metres (Taylor and Emmambux 2008). Teff planting should be done when the soil temperature is 65°F or 18°C (https://www.agrifarming.in/teff-grain-farming-cultivation-practices). The desirable soil pH for teff cultivation is 6.0–6.5 (McIntosh 2020). It grows well below a salinity of 8 dS/m. Kinfemichael and Fisseha (2011) reported an increase in dry matter yield in some teff accessions/varieties at a salinity level of 2–4 dS/m; however, the increase of salinity to 8 dS/m reduced (7.1–63.6%) the yield significantly in most of the accessions/varieties. The sowing of teff is done in August.

The major fonio-cultivating countries are Guinea, Nigeria, and Mali. The other fonio-cultivating countries are Ghana, Niger, Gambia, Burkina Faso, Togo Benin, and Ivory Coast. White fonio is mainly cultivated from Senegal to Chad and black fonio is cultivated in Nigeria and the northern regions of Togo and Benin (Kalaisekar et al. 2016). It is a drought-tolerant crop that can be grown on low-fertility sandy soils. It is indigenous to West Africa and is considered one of the oldest cereals of this region. It can be cultivated in wide environmental conditions ranging from a tropical monsoon climate to a hot and desert climate (Abrouk et al. 2020). Fonio can be grown in plains as well as mountains up to a height of 1500 metres above sea level and an annual average precipitation of 400–3000 mm (Gomez and Gupta 2003). It requires cooler temperatures of 15–25°C during the growing season (Cruz et al. 2016). It is claimed to be resistant to acidic soils (Taylor and Emmambux 2008); however, the authors were not able to find literature on the optimum soil pH and salinity conditions for the cultivation of fonio. Its sowing is done in the last week of July.

Details on the agrarian conditions required for the cultivation of the various millets are also provided in Table 1.2.

1.8 PRODUCTION

Global millet production was 27.8 million tonnes in the year 2020 and increased to 29.79 million metric tonnes in the year 2021 (USDA 2021). The world's largest millet producers are Africa and Asia, which jointly account for 96.89% of the global millet production. Europe and the USA share only 3% and 1% of the global millet market share, respectively (Chandra et al. 2021). As a country, India is the largest producer (12,800 thousand metric tonnes, 41% of global production) of millets in the world. It is followed by Niger and China, which produce 3800 (12% of global production) and 2300 (7% of global production) thousand metric tonnes of millets, respectively. The top ten millet-producing countries in the world with their percent contribution to world millet production are provided in Table 1.3.

Pearl millet is the most produced millet in the world. It is cultivated in more than 30 countries of the world and covers about 31 million hectares worldwide, especially in the semiarid, tropical, and subtropical regions of Asia, Africa, and Latin America. The collective pearl millet production of the world is about 15 million tonnes and it accounts for almost half of global millet production (Chandra et al. 2021). India is the largest producer of pearl millet in the world. In 2018–2019, a total of 7.14 million hectares was under its cultivation in India and the production was about 9.93 million tonnes with average productivity of 1237 kg/hectare (ICAR-AICRP 2019). The remaining millets are represented with the collaborative term small millets. The exact data on the production of small millets is not available as some countries release the data collaboratively for small cereals or nutricereals which also include crops other than millets. According to some studies, India is the leading producer of small millets with an area of about 7.0 lakh hectares under their cultivation and a productivity of 4.41 lakh tonnes (Gowri and Shivakumar 2020). Among the small millets, the most produced millet is the foxtail millet. It has a global production of about 5 million tonnes. China is the leading producer of foxtail millet and it had an area of 0.72 million hectares under its cultivation in the year 2014 and the production was 1.81 million tonnes (Diao 2017). The production of foxtail millet in India is about 0.05 million tonnes (Bhat et al. 2018). The third most-produced millet is proso millet with a worldwide production of about 4 million tonnes. The area under proso millet cultivation in China, the USA, India, and Korea in the year 2016 was 0.32, 0.204, 0.041, and 0.002 million hectares, respectively (Prabhakar et al. 2017; Habiyaremye et al. 2017; Vetriventhan et al. 2019). The USA is the largest producer of proso millet. The production of proso millet was 16.6 million bushels (0.42 million metric tonnes approx. as the conversion factor of sorghum is used) in the year 2019 which was reduced to 9.21 million bushels (0.23 million metric tonnes approx.) in the year 2020 (Agricultural Marketing Resource Center 2018). It exports proso millet to more than 70 countries. The second-largest producer of

TABLE 1.2 OPTIMUM AGRARIAN CONDITIONS FOR THE CULTIVATION OF MILLETS

Crop	Scientific name	Cultivating countries	Optimum soil type	Height range	Temperature	pH	Soil salinity (dS/m)	Rainfall required
Pearl millet	*Pennisetum glacum*	India, Pakistan, Nigeria, Niger, Mali, Senegal, Burkina Faso, Sudan, Chad, Tanzania, Cameroon, Ghana, Yemen, Myanmar, Saudi Arabia	Loamy soils, shallow soils, soils with clay, clay loam and sandy loam texture	Sea level to 2700 metres	30–34°C *can grow up to 46°C	6.0–7.0 *can grow up to 8.0 pH	11–12 dS/m *yields are economically well up to ECe 8dS/m	20–60 cm
Foxtail millet	*Setaria italica*	China, India, Russia, Afghanistan, Manchuria, Korea, Georgia and the USA	Sandy to loamy soils	Sea level to 2000 metres	Range 5–35°C Average 16–25°C	5.5–7.0	6 dS/m	30–70 cm
Proso millet	*Panicum miliaceum*	China, India, Russia, Iran, Iraq, Syria, Turkey, Afghanistan and Romania	Sandy loam, slightly acidic, saline, low fertility soils	1200–3500 metres above sea level	20–30°C	5.5–6.5	–	20–50 cm
Barnyard millet	*Echinochloa, E. frumentacea* (Indian barnyard millet) and *E. esculenta* (Japanese barnyard millet)	Indian barnyard millet – India, Nepal, Pakistan, the Central African Republic, Tanzania and Malawi; Japanese barnyard millet – Japan, Korea, China, Russia and Germany	Medium to heavy soils	Sea level to 2000 metres	Range 15–33°C Average 27–33°C	4.6–7.4	3–5 dS/m	60–80 cm

(Continued)

TABLE 1.2 (CONTINUED) OPTIMUM AGRARIAN CONDITIONS FOR THE CULTIVATION OF MILLETS

Crop	Scientific name	Cultivating countries	Optimum soil type	Height range	Temperature	pH	Soil salinity (dS/m)	Rainfall required
Finger millet	*Eleusine coracana*	India, East and South Africa	Rich loam to poor upland shallow soils	Sea level to 2300 metres	26–29°C *lower productivity below 20°C	4.5 to 7.5	11–12 dS/m	50–60 cm
Kodo millet	*Paspalum scrobiculatum*	India, Pakistan, the Philippines, Indonesia, Vietnam, Thailand and West Africa	Fertile to marginal soils	Sea level to 1500 metres	25–27°C	–	–	80–120 cm
Little millet	*Panicum sumatrense*	India, China, the Caucasus, Nepal, Pakistan, Sri Lanka, Malaysia, and Myanmar	–	Sea level to 2100 metres	–	–	–	–
Teff	*Eragrostis tef*	India, China, the Caucasus, Nepal, Pakistan, Sri Lanka, Malaysia, and Myanmar	waterlogged to well-drained heavy soils	Sea level to 3000 metres	–	6.0–6.5	Up to 8 dS/m	Up to 100 cm
Fonio	*Digitaria exilis* Stapf; *D. iburua* Stapf	Guinea, Nigeria, and Mali	Low-fertility, sandy soils	Sea level to 3000 metres	15–25°C	–	–	40–300 cm

Source: Modified and adopted from Kumar et al. 2018.

TABLE 1.3   TOP MILLET-PRODUCING COUNTRIES OF THE WORLD

Country name	Area under cultivation (ha)	Productivity (hg/ha)	Production (tonnes)	Share in world production (%)
India	8,449,720	12,144	10,235,830	41
Niger	6,831,217	4788	3,270,453	12
China	900,249	25,553	2,300,379	7
Nigeria	2,778,395	7198	2,000,000	6
Mali	1,989,953	9440	1,878,527	6
Sudan	3,016,440	3756	1,133,000	5
Ethiopia	455,580	24,715	1,125,958	4
Burkina Faso	1,176,512	8246	970,176	3
Senegal	880,408	9167	807,044	3
Chad	1,180,431	6089	717,621	2

Source: http://www.fao.org/faostat/en/#data/QC.

this millet is China with production of 0.3 million tonnes (Diao 2017). Proso millet production in India in the year 2016 was 0.022 million tonnes (Prabhakar et al. 2017). Globally, barnyard millet is the fourth most produced millet with India being the largest producer with an area of 0.146 million hectares and production of 0.147 million metric tonnes (Renganathan et al. 2020). Global finger millet production is approximately 4.5–5 million tonnes and India is the largest producer of this millet. In 2015–2016, India had an area of 1138.2 hectares under its cultivation. The total production was 1821.9 thousand tonnes with a productivity of 1601 kg/hectare (Ramashia et al. 2019). The cultivation of kodo and little millet is mostly limited to India. In the year 2015–2016, the total area under kodo millet cultivation was 0.196 million hectares and the total production was 0.084 million tonnes (Prabhakar et al. 2017). Little millet was cultivated in the year 2018 on an area of 0.26 million hectares and the production was 0.12 million tonnes (ICRISAT 2021). The largest producer of teff is Ethiopia. It produces nearly 90% of global teff. In the year 2013–2014, it was cultivated on about 3 million hectares of Ethiopia and the total production was 4.4 million metric tonnes (Minten et al. 2018). An increasing trend (3.25% average annual growth rate) in the area under production and yield of fonio has been observed throughout the world. In the year 2020, it was cultivated on an area of 9.65 lakh hectares and its production was about 7.4 lakh tonnes (World Data Atlas 2022).

1.9 FUTURE PROSPECTS

Millets have reclaimed their lost identity, and because of their ability to grow under rainfed conditions on low-fertility soils, they have emerged as the most dependable solution for keeping the drylands productive. Millets are currently consumed primarily in developing regions of India and Africa, but their high nutritional value, gluten-free nature, and positive role in the prevention of diabetes, cardiovascular disease, and obesity have made them popular in developed countries as well. The consumption of millets is expected to increase at a compound annual growth rate of 4.5% from 2021 to 2026. The declaration of the General Assembly of the United Nations to observe the year 2023 as the "International Year of Millets" is further expected to enhance awareness of the health benefits of millets and the demand for millets. The COVID-19 outbreak could be another factor driving millet consumption, as there has been an increase in the consumption of healthy foods in the post-COVID era.

REFERENCES

Abdul, S. D., and A. I. Jideani. 2019. Fonio (Digitaria spp.) breeding. In *Advances in Plant Breeding Strategies: Cereals*, eds. J. M. Al-Khayri, S. M. Jain, and D. V. Johnson, 47–81. Gewerbestrasse, Cham Switzerland: Springer International Publishing.

Abrouk, M., H. I. Ahmed, P. Cubry, D. Simoníková, S. Cauet, Y. Pailles, J. Bettgenhaeuser, L. Gapa, N. Scarcelli, M. Couderc, L. Zekraoui, N. Kathiresan, J. Čížková, E. Hřibová, J. Doležel, S. Arribat, H. Bergès, J. J. Wieringa, M. Gueye, N. A. Kane, C. Leclerc, S. Causse, S. Vancoppenolle, C. Billot, T. Wicker, Y. Vigouroux, A. Barnaud, and S. G. Krattinger. 2020. Fonio millet genome unlocks African orphan crop diversity for agriculture in a changing climate. *Nature Communications* 11(1):4488.

Adoukonou-Sagbadja, H., A. Dansi, R. Vodouhè, and K. Akpagana. 2004. Collecting fonio (*Digitaria exilis* Kipp. Stapf, *D.iburua* Stapf) landraces in Togo. *Plant Genetic Resources Newsletter* 139:63–7.

Agricultural Marketing Resource Center. 2018. Proso millet. https://www.agmrc.org/commodities-products/grains-oilseeds/proso-millet. (accessed January 9, 2022).

Agropedia. 2022. Harvesting and storage of pearl millet. http://agropedia.iitk.ac.in/content/harvesting-and-storage-pearl-millet. (accessed September 2, 2022).

Vikaspedia. 2022a. Pearl millet. https://vikaspedia.in/agriculture/crop-production/package-of-practices/cereals-and-millets/bajra-1. (accessed May 21, 2022).

Anonymous. 2021. *Pearl Millet Biology*. Jodhpur: Project Coordinator, AICRP- Pearl millet (ICAR), ARS, Mandor. http://www.aicpmip.res.in/pmbiology.pdf. (accessed July 29, 2021).

Arendt, E. K., and E. Zannini. 2013. Teff. In *Cereal Grains for the Food and Beverage Industries*, es. E. K. Adrendt, and E. Zannini, 351–69. Cambridge: Woodhead Publishing Ltd.

Assefa, K., S. Chanyalew, and Z. Tadele. 2017. Tef, *Eragrostis tef* (Zucc.) Trotter. In *Millets and Sorghum: Biology and Genetic Improvement*, ed. V. S. Patil, 226–66. Hoboken: John Wiley & Sons, Inc.

Ausiku, A. P., J. G. Annandale, J. M. Steyn, and A. J. Sanewe. 2020. Improving pearl millet (*Pennisetum glaucum*) productivity through adaptive management of water and nitrogen. *Water* 12(2):422.

Ayenan, M. A. T., K. A. F. Sodedji, C. I. Nwankwo, K. F. Olodo, and M. E. B. Alladassi. 2018. Harnessing genetic resources and progress in plant genomics for fonio (*Digitaria spp.*) improvement. *Genetic Resources and Crop Evolution* 65(2):373–86.

Bedane, G. M., A. M. Saukuru, D. L. George, and M. L. Gupta. 2015. Evaluation of teff ('*Eragrostis tef*'[Zucc.] Trotter) lines for agronomic traits in Australia. *Australian Journal of Crop Science* 9(3):242–7.

Bhat, B. V., V. A. Tonapi, and B. D. Rao. 2018. Production and utilization of millets in India. In *International millet symposium on 3rd international symposium on broomcorn millet (3rd ISBM)*, eds. D. K. Santra and J. J. Johnson, 24–36. Mariott Inn, Fort Colins, USA.

Bhatia, S. 2015. Application of plant biotechnology. In *Modern Applications of Plant Biotechnology in Pharmaceutical Sciences*, eds. S. Bhatia, K. Sharma, R. Dahiya, and T. Bera, 157–207. Cambridge: Academic Press.

Calamai, A., A. Masoni, L. Marini, M. Dell'acqua, P. Ganugi, S. Boukail, S. Benedettelli, and E. Palchetti. 2020. Evaluation of the agronomic traits of 80 accessions of proso millet (*Panicum miliaceum* L.) under Mediterranean pedoclimatic conditions. *Agriculture* 10(12):578.

Chandra, A. K., R. Chandora, S. Sood, and N. Malhotra. 2021. Global production, demand, and supply. In *Millets and Pseudo Cereals*, eds. M. Singh and S. Sood, 7–18. Duxford, Unite Kingdom: Woodhead Publishing.

Clottey, V. A., W. A. Agyare, J. M. Kombiok, H. Abdulai, and A. H. Kaleem. 2006. Fonio (*Digitaria exilis* Stapf) germplasm assemblage for characterization, conservation and improvement in Ghana. *Plant Genetic Resources Newsletter* 146:24–7.

Crookston, B., B. Blaser, M. Darapuneni, and M. Rhoades. 2020. Pearl millet forage water use efficiency. *Agronomy* 10(11):1672.

Cruz, J. F., F. Béavogui, D. Dramé, and T. A. Diallo. 2016. *Fonio, an African Cereal*. Cirad, Montpellier, France, 154. https://fonio.cirad.fr/ (accessed August 02, 2021).

Das, S., R. Khound, M. Santra, and D. K. Santra. 2019. Beyond bird feed: Proso millet for human health and environment. *Agriculture* 9(3):64.

Diao, X. 2017. Production and genetic improvement of minor cereals in China. *The Crop Journal* 5(2):103–14.

Diao, X., and G. Jia. 2017a. Origin and domestication of foxtail millet. In *Genetics and Genomics of Setaria*, eds. A. Doust and X. Diao, 61–72. Gewerbestrasse Cham, Switzerland: Springer International Publishing.

Diao, X., and G. Jia. 2017b. Foxtail millet germplasm and inheritance of morphological characteristics. In *Genetics and Genomics of Setaria*, eds. A. Doust and X. Diao, 73–92. Gewerbestrasse Cham, Switzerland: Springer International Publishing.

Dias-Martins, A. M., K. L. F. Pessanha, S. Pacheco, J. A. S. Rodrigues, and C. W. P. Carvalho. 2018. Potential use of pearl millet (*Pennisetum glaucum* (L.) R. Br.) in Brazil: Food security, processing, health benefits and nutritional products. *Food Research International* 109:175–86.

Dida, M. M., N. Wanyera, M. L. H. Dunn, J. L. Bennetzen, and K. M. Devos. 2008. Population structure and diversity in finger millet (*Eleusine coracana*) germplasm. *Tropical Plant Biology* 1(2):131–41.

Doust, A. N., E. A. Kellogg, K. M. Devos, and J. L. Bennetzen. 2009. Foxtail millet: A sequence-driven grass model system. *Plant Physiology* 149(1):137–41.

FAO. 2010. *The Second Report on the State of the World's Plant Genetic Resources for Food and Agriculture*. Rome. http://www.fao.org/3/i1500e/i1500e.pdf (accessed August 01, 2020).

Gomez, M. I., and S. C. Gupta. 2003. Millets. In *Encyclopedia of Food Sciences and Nutrition*, eds. B. Caballero, P. Finglas, and F. Toldrá, 3974–9. Cambridge: Academic Press.

Goron, T. L., and M. N. Raizada. 2015. Genetic diversity and genomic resources available for the small millet crops to accelerate a new green revolution. *Frontiers in Plant Science* 6:157.

Gowri, M. U., and K. M. Shivkumar. 2020. Millet scenario in India. *Economic Affairs* 65(3):363–70.

Gupta, S., S. K. Shrivastava, and M. Shrivastava. 2013. Study of biological active factors in some new varieties of minor millet seeds. *International Journal of Innovations in Engineering and Technology* 3(2):115–7.

Habiyaremye, C., J. B. Matanguihan, J. D'Alpoim Guedes, G. M. Ganjyal, M. R. Whiteman, K. K. Kidwell, and K. M. Murphy. 2017. Proso millet (*Panicum miliaceum* L.) and its potential for cultivation in the Pacific Northwest, US: A review. *Frontiers in Plant Science* 7:1961.

Hayes, A. M. R., and J. M. Jones. 2016. Cultural differences in processing and consumption. In *Encyclopedia of Food Grains*, eds. C. Wrigley, H. Corke, K. Seetharaman, and J. Faubion, 35–42. Cambridge: Academic Press.

Hermuth, J., D. Janovská, P. H. Čepková, S. Usťak, Z. Strašil, and Z. Dvořáková. 2016. Sorghum and foxtail millet—Promising crops for the changing climate in central Europe. In *Alternative Crops and Cropping Systems*. London: IntechOpen.

Heuzé, V., G. Tran, and S. Giger-Reverdin. 2015. *Scrobic (Paspalum Scrobiculatum) Forage and Grain*. Feedipedia, a programme by INRAE, CIRAD. AFZ and FAO. https://www.feedipedia.org/node/401. (accessed August 09, 2021).

Hu, H., M. Mauro-Herrera, and A. N. Doust. 2018. Domestication and improvement in the model C4 grass, Setaria. *Frontiers in Plant Science* 9:719.

Hu, Z., B. Mbacké, R. Perumal, M. C. Guèye, O. Sy, S. Bouchet, V. P. Vara Prasad, and G. P. Morris. 2015. Population genomics of pearl millet (*Pennisetum glaucum* (L.) R. Br.): Comparative analysis of global accessions and Senegalese landraces. *BMC Genomics* 16:1048.

Ibrahim Bio Yerima, A. R., and E. G. Achigan-Dako. 2021. A review of the orphan small grain cereals improvement with a comprehensive plan for genomics-assisted breeding of fonio millet in West Africa. *Plant Breeding* 140(4):561–74.

Ibrahim Bio Yerima, A. R., E. G. Achigan-Dako, M. Aissata, E. Sekloka, C. Billot, C. O. A. Adje, A. Barnaud, and Y. Bakasso. 2020. Agromorphological characterization revealed three phenotypic groups in a region-wide germplasm of fonio (*Digitaria Exilis* (Kippist) Stapf) from West Africa. *Agronomy* 10(11):1653.

ICAR-AICRP. 2019. Project coordinator review. 54th Annual group meeting. ICAR- Indian Agricultural Research Institute, New Delhi. http://www.aicpmip.res.in/pcr2019.pdf. (accessed July 07, 2020).

ICRISAT. 2012. Global strategy for the ex-situ conservation of finger millet and its wild relatives. https://cdn.croptrust.org/wp/wp-content/uploads/2017/02/Finger-Millet-Strategy-FINAL-14May2012.pdf (accessed August 02, 2020).

ICRISAT. 2020. Proso millet passport. http://genebank.icrisat.org/IND/Passport?Crop=Proso%20millet (accessed July 30, 2020).

ICRISAT. 2021. Little millet that are big on nutrition and yield identified at ICRISAT. https://www.icrisat.org/little-millet-that-are-big-on-nutrition-and-yield-identified-at-icrisat/. (accessed July 07, 2021).

Jia, G., X. Huang, H. Zhi, Y. Zhao, Q. Zhao, W. Li, Y. Chai, L. Yang, K. Liu, H. Lu, C. Zhu, Y. Lu, C. Zhou, D. Fan, Q. Weng, Y. Guo, T. Huang, L. Zhang, T. Lu, Q. Feng, H. Hao, H. Liu, P. Lu, N. Zhang, Y. Li, E. Guo, E., S. Wang, S. Wang, J. Liu, W. Zhang, G. Chen, B. Zhang, W. Li, Y. Wang, H. Li, B. Zhao, J. Li, X. Diao, and B. Han. 2013. A haplotype map of genomic variations and genome-wide association studies of agronomic traits in foxtail millet (*Setaria italica*). *Nature Genetics* 15:957–61.

Johnson, M., S. Deshpande, M. Vetriventhan, H. D. Upadhyaya, and J. G. Wallace. 2019. Genome-Wide population structure analyses of three minor millets: Kodo millet, little millet, and proso millet. *Plant Genome-Us* 12(3):1–9.

Joshi, R. P., A. K. Jain, N. Malhotra, and M. Kumari. 2021. Origin, domestication, and spread. In *Millets and Pseudo Cereals: Genetic Resources and Breeding Advancements*, eds. M. Singh and S. Sood, 33–8. Duxford, United Kingdom: Woodhead Publishing.

Kalaisekar, A., P. G. Padmaja, V. R. Bhagwat, and J. V. Patil. 2016. *Insect Pests of Millets: Systematics, Bionomics, and Management*. London: Academic Press.

Kanlindogbe, C., E. Sekloka, and E. H. Kwon-Ndung. 2020. Genetic resources and varietal environment of grown fonio millets in West Africa: Challenges and perspectives. *Plant Breeding and Biotechnology* 8(2):77–88.

Karuppasamy, P. 2015. Overview on millets. *Trends in Biosciences* 8(13):3269–73.

Kinfemichael, G. A., and I. D. Fisseha. 2011. Response of dry matter production of tef [*Eragrostis tef* (Zucc.) Trotter] accessions and varieties to NaCl salinity. *Current Research Journal of Biological Sciences* 3(4):300–7.

Krishnamurthy, L., H. D. Upadhyaya, R. Purushothaman, C. Gowda, L. Lakkegowda, J. Kashiwagi, S. L. Dwivedi, S. Singh, and V. Vadez. 2014. The extent of variation in salinity tolerance of the minicore collection of finger millet (*Eleusine coracana* L. Gaertn.) germplasm. *Plant Science* 227:51–9.

Kumar, A., A. Kaur, K. Gupta, Y. Gat, and V. Kumar. 2021. Assessment of germination time of finger millet for value addition in functional foods. *Current Science* 120(2):406–13.

Kumar, A., V. Tomer, A. Kaur, V. Kumar, and K. Gupta. 2018. Millets: A solution to agrarian and nutritional challenges. *Agriculture and Food Security* 7(1):31.

Kumar, D., S. Patel, R. K. Naik, and N. K. Mishra. 2016. Study on physical properties of Indira Kodo-I (*Paspalum scrobiculatum* L.) millet. *International Journal of Engineering Research and Technology* 5(1):39–45.

Lee, D., W. Hanna, G. D. Buntin, W. Dozier, P. Timper, and J. P. Wilson. 2012. *Pearl Millet for Grain*. Bulletin #B 1216. College of Agriculture and Environmental Sciences, University of Georgia Cooperative Extension. Georgia: University of Georgia.

Lee, H. 2018. Teff, a rising global crop: Current status of teff production and value chain. *The Open Agriculture Journal* 12(1):185–93.

Liu, M., Z. Qiao, S. Zhang, Y. Wang, and P. Lu. 2015. Response of broomcorn millet (*Panicum miliaceum* L.) genotypes from semiarid regions of China to salt stress. *The Crop Journal* 3(1):57–66.

Maiti, R. K., and F. R. Bidinger. 1981. *Growth and Development of the Pearl Millet Plant*. Research Bulletin No 6. Patancheru: International Crops Research Institute for the Semi-Arid Tropics.

Maitra, S., and T. Shankar. 2019. Agronomic management in little millet (Panicum sumatrense L.) for enhancement of productivity and sustainability. *International Journal of Bioresource Science* 6:91–6.

Matsuura, A., and P. An. 2020. Factors related water and dry matter during pre-and post-heading in four millet species under severe water deficit. *Plant Production Science* 23(1):28–38.

Matsuura, A., P. An, K. Murata, and S. Inanaga. 2016. Effect of pre-and post-heading waterlogging on growth and grain yield of four millets. *Plant Production Science* 19(3):348–59.

McIntosh, D. 2020. https://utbeef.tennessee.edu/wp-content/uploads/sites/127/2020/11/Teff-Grass.pdf. (accessed August 10, 2021).

Merchuk-Ovnat, L., J. Bimro, N. Yaakov, Y. Kutsher, O. Amir-Segev, and M. Reuveni. 2020. In-depth field characterization of teff [*Eragrostis tef* (Zucc.) Trotter] variation: From agronomic to sensory traits. *Agronomy* 10(8):1107.

Minten, B., A. S. Taffesse, and P. Brown. 2018. *The Economics of Teff: Exploring Ethiopia's Biggest Cash Crop.* Washington, DC: International Food Policy Research Institute.

Muimba-Kankolongo, A. 2018. Cereal production. In *Food Crop Production by Smallholder Farmers in Southern Africa: Challenges and Opportunities for Improvement*, ed. A. Muimba-Kankolongo, 73–121. London: Academic Press.

Nambiar, V. S., J. J. Dhaduk, N. Sareen, T. Shahu, and R. Desai. 2010. Potential functional implications of pearl millet (*Pennise tumglaucum*) in health and disease. *Journal of Applied Pharmaceutical Science* 1(10):62–7.

NBPGR. 2022. *Status of Collections at National GeneBank (NGB), ICAR-NBPGR.* New Delhi. http://www.nbpgr.ernet.in/Research_Projects/Base_Collection_in_NGB.aspx. (accessed February 14, 2022).

Offord, C. A. 2017. Germplasm conservation. In *Encylopedia of Applied Plant Sciences*, eds. B. Thomas, B. G. Murray, and D. J. Murphy, 281–8. Cambridge: Academic Press.

Pattanashetti, S. K., H. D. Upadhyaya, S. L. Dwivedi, M. Vetriventhan, and K. N. Reddy. 2016. Pearl millet. In *Genetic and Genomic Resources for Grain Cereals Improvement*, eds. M. Singh and H. D. Upadhyaya, 253–89. San Diego: Academic Press, Elsevier.

Prabha, R., D. P. Singh, S. Gupta, V. K. Gupta, H. A. El-Enshasy, and M. K. Verma. 2019. Rhizosphere metagenomics of *Paspalum scrobiculatum* L. (kodo millet) reveals rhizobiome multifunctionalities. *Microorganisms* 7(12):608.

Prabhakar, C. G., B. Prabhu, B. Boraiah, C. Sujata, Kiran Nandini, V. Tippeswamy, and H. A. Manjunath. 2017. Improved production technology for proso millet. ICAR-AICRP on small millets. http://www.aicrpsm.res.in/Research/Package%20of%20Practices/Improved%20production%20technology%20for%20Proso%20millet.pdf. (accessed August 09, 2020).

Pulidini, K., and H. Pandey. 2019. Millets market size by product (organic [pearl, finger, proso, foxtail], Regular [pearl, finger, proso, foxtail]), by application (infant food, bakery products, beverages [alcoholic, non-alcoholic], breakfast food, fodder), by distribution channel (trade associations, supermarket, traditional grocery stores, online stores), industry analysis report regional outlook, growth potential, price trends, competitive market share & forecast, 2019–2025. https://www.gminsights.com/industry-analysis/millets-market. (accessed July 07, 2020).

Ramashia, S. E., T. A. Anyasi, E. T. Gwata, S. Meddows-Taylor, and A. I. O. Jideani. 2019. Processing, nutritional composition and health benefits of finger millet in sub-Saharan Africa. *Food Science and Technology* 39(2):253–66.

Rao, V. V., S. G. Swamy, D. S. Raja, and B. J. Wesley. 2020. Engineering properties of certain minor millet grains. *The Andhra Agricultural Journal* 67:89–92.

Renganathan, V. G., C. Vanniarajan, A. Karthikeyan, and J. Ramalingam. 2020. Barnyard millet for food and nutritional security: Current status and future research direction. *Frontiers in Genetics* 11:500.

Research and Markets. 2021. Millet market - Growth, trends, COVID-19 impact, and forecasts (2021–2026). Report ID: 4520082, p 82. https://www.researchandmarkets.com/reports/4520082/millet-market-growth-trends-covid-19-impact?utm_source=CI&utm_medium=PressRelease&utm_code=2lkpvm&utm_campaign=1278170+-+Global+Millet+Markets+2019-2024+-+India+Dominates+Global+Production+%26++Africa+Dominates+Global+Consumption&utm_exec=chdo54prd.

Saxena, R., S. K. Vanga, J. Wang, V. Orsat, and V. Raghavan. 2018. Millets for food security in the context of climate change: A review. *Sustainability* 10(7):2228.

Shailaja, H. B., and S. Thirumeni. 2007. Evaluation of salt-tolerance in finger millet (*Eleusine coracana*) genotypes at seedling stage. *The Indian Journal of Agricultural Sciences* 77(10):672–4.

Sheahan, C. M. 2014. *Plant Guide for Pearl Millet (Pennisetum Glaucum).* Cape May: USDA-Natural Resources Conservation Service, Cape May Plant Materials Center. Published 08/2014.

Shrishat, B., S. Patel, S. D. Kulkarni, P. H. Bakane, and M. Khedkar. 2008. Physical properties of kodo millet (*Pasplum scrobiculatum* L.). *International Journal of Agricultural Science* 4(2):580–7.

Singh, K. P., H. N. Mishra, and S. Saha. 2010. Moisture-dependent properties of barnyard millet grain and kernel. *Journal of Food Engineering* 96(4):598–606.

Singh, R. K., M. Muthamilarasan, and M. Prasad. 2017. Foxtail millet: An introduction. In *The Foxtail Millet Genome*, ed. M. Prasad, 1–9. Gewerbestrasse Cham, Switzerland: Springer international Publishing.

Sood, S., R. K. Khulbe, A. K. Gupta, P. K. Agrawal, H. D. Upadhyaya, and J. C. Bhatt. 2015. Barnyard millet–a potential food and feed crop of future. *Plant Breeding* 134(2):135–47.

Taylor, J. R. N. 2019. Sorghum and millets: Taxonomy, history, distribution, and production. In *Sorghum and Millets: Chemistry, Technology, and Nutritional Attributes*, eds. J. R. N. Taylor and K. G. Duodu, 1–21. Duxford, United Kingdom: Woodhead Publishing.

Taylor, J. R. N., and M. N. Emmambux. 2008. Gluten-free foods and beverages from millets. In *Gluten-Free Cereal Products and Beverages*, eds. E. K. Arendt and F. D. Bello, 119–48. Cambridge: Academic Press.

TNAU. 2021. http://www.agritech.tnau.ac.in/expert_system/ragi/botany.html. (accessed August 01, 2021).

Upadhyaya, H. D., R. P. S. Pundir, and C. L. L. Gowda. 2007. Genetic resources diversity of finger millet - A global perspective. In *Finger Millet Blast Management in East Africa: Creating Opportunities for Improving Production and Utilization of Finger Millet*, 90–101. Patancheru, India: ICRISAT. https://core.ac.uk/download/pdf/211012379.pdf.

Upadhyaya, H. D., M. Vetriventhan, S. L. Dwivedi, S. K. Pattanashetti, and S. K. Singh. 2016. Proso, barnyard, little, and kodo millets. In *Genetic and Genomic Resources for Grain Cereals Improvement*, eds. M. Singh and H. D. Upadhyaya, 321–43. Cambridge: Academic Press.

USDA. 2021. *Foreign Agricultural Service*. Millet: U. S. Department of Agriculture. https://ipad.fas.usda.gov/cropexplorer/cropview/commodityView.aspx?cropid=0459100. (accessed April 21, 2021).

Vetriventhan, M., V. C. Azevedo, H. D. Upadhyaya, and D. Naresh. 2019. Variability in the global proso millet (*Panicum miliaceum* L.) germplasm collection conserved at the ICRISAT GeneBank. *Agriculture* 9(5):112.

Vetriventhan, M., V. C. R. Azevedo, H. D. Upadhyaya, A. Nirmalakumari, J. Kane-Potaka, S. Anitha, S. A. Ceasar, M. Muthamilarasan, B. V. Bhat, K. Hariprasanna, A. Bellundagi, D. Cheruku, C. Backiyalakshmi, D. Santra, C. Vanniarajan, and V. A. Tonapi. 2020. Genetic and genomic resources, and breeding for accelerating improvement of small millets: Current status and future interventions. *The Nucleus* 63(3):217–39.

Vikaspedia. 2021. Kodo millet. https://vikaspedia.in/agriculture/crop-production/package-of-practices/cereals-and-millets/finger-millet-and-kodo-millet#:~:text=of%2050%2D60cm.-,Soil,a%20certain%20degree%20of%20alkalinity.&text=Kodo%20millet%20can%20be%20grown,as%20in%20the%20hilly%20region. (accessed August 9, 2021).

Vikaspedia. 2022b. *Aportal Govt of India, Ministry of Electronics and Information Technology, Govt.* Hyderabad: Centre for Development of Advanced Computing. https://vikaspedia.in/health/nutrition/nutritive-value-of-foods/nutritive-value-of-cereals-and-millets/milletsthe-nutricereals. (accessed August 12, 2021).

World Data Atlas. 2022. Fonio production in the world. https://knoema.com/data/agriculture-indicators-production+fonio. (accessed January 9, 2022).

Zhang, D. Z., R. B. Panhwar, J. J. Liu, X. W. Gong, J. B. Liang, M. Liu, P. Lu, X. L. Gao, and B. L. Feng. 2019. Morphological diversity and correlation analysis of phenotypes and quality traits of proso millet (*Panicum miliaceum* L.) core collections. *Journal of Integrative Agriculture* 18(5):958–69.

Zohary, D., and M. Hopf. 2000. *Domestication of Plants in the Old World*. New York: Oxford University Press.

Chapter 2

Harvesting, Threshing, and Storage of Millets

2.1 INTRODUCTION

Harvesting, threshing, and storage are inextricable operations required before any product development. The operation of gathering a ripe crop from the fields is known as harvesting. The major steps of harvesting are reaping/cutting and threshing, however, in general usage, the steps of handling, cleaning, sorting, packing, and cooling are also involved in harvesting. Millets are harvested at the fully ripe stage and harvesting can be carried out by traditional methods or modern automized methods. The traditional method of harvesting involves reaping the crop with a hand sickle and drying it in the sun. The harvested crop is threshed by tramping with bullocks, horses, etc., and cleaned by winnowing. The threshing operation separates the edible part of the grain from the straw. In modern methods of harvesting, the operations of reaping and threshing are carried out by specially designed machines. Modern-day threshers are equipped with a series of sieves and fans that clean the grains. Cleaned grains are then primarily processed by the operations of drying, cleaning, grading, tempering, and milling operations like hulling, pounding, grinding, sieving, etc. (Radhika et al. 2013; Prashant et al. 2015; Samuel et al. 2019; Balasubramanian et al. 2020; Karakannavar et al. 2021). Both traditional and modern methods can be used for the primary processing of grains. The traditional methods of processing involve cleaning the grains by winnowing, de-husking, and milling to flour by pounding in a pestle or mortar made of stone or wood. Alternatively, human-, water-, or air-driven stone mills can also be used (Powar et al. 2019). This results in the separation of the hull from the grain and the grinding of the edible part. Nowadays, machines with the dual ability of harvesting and threshing are also available. These machines give higher yields with less damage and increased shelf-life. Nowadays, machines known as "debranners" are also available. These machines can easily remove the bran and can increase the acceptability of millets. Debranning reduces the antinutritional factors present in the grain and improves nutritional value by increasing the biological availability of the nutrients (Pradhan et al. 2010; Hassan et al. 2021). The primary processing operations also make the grains ready for secondary processing where they can be processed into value-added products.

Harvesting and milling efficiency are affected by several factors such as the stage of maturity, timing of the harvest, environmental conditions (relative humidity and temperature during harvesting and processing), moisture content of grains, presence of impurities (stone, dirt, metal, plastic, glass, mud, grits), and the type of processing method (Birania et al. 2020). Modern machines can inspect the quality of millet grains at every step of processing and hence decrease the processing loss and help in maintaining the quality after processing. Modern machines are more work- and time-efficient and are hence cost-effective. The present chapter discusses the various traditional and modern methods of the primary processing of millet grains and the principles of their working.

2.2 HARVESTING OF THE MILLETS

The process of harvesting includes steps that are required to process the field crop into clean grains. This includes cutting/reaping the crop, threshing it to separate the grains, removing the straw and chaff from the grains, and cleaning the grains. Sorting, cooling, and packing are further extensions of the harvesting operations. The millets are generally cultivated on soils of low fertility in regions where there are limited irrigation facilities or non-uniform rain distribution with less overall precipitation. All these factors lead to lower production and productivity, limiting the mechanization of the millet harvesting operations. Hence, most of the postharvest operations of the millets are still done manually. However, the potential of millets as a nutrient-dense crop has been recognized in the past few years and much work has been done to promote the cultivation and processing of millets. This has improved mechanization in the cultivation, harvesting, and processing operations of millets. The modern machines also help to decrease processing losses which are otherwise higher in millets compared to other grains due to their smaller size. This section discusses both the traditional and modern methods of harvesting.

2.2.1 Cutting/Reaping of Millet

The cutting or reaping time varies with the time of sowing and the region of cultivation. Most of the millets are sown in the kharif season and are harvested in the month of September–October. However, some off-season varieties of millet are also available and their sowing and harvesting time can vary. The kharif crop of pearl millet is ready for harvest in the last week of September or the first fortnight of October. The right stage for the cutting of the crop or the separation of mature earheads can be judged from the colour of the leaves and the colour at the bottom of the grain in the hilar region. The leaves turn yellow and have a nearly dried-up appearance and a black spot is observed at the bottom of the grain in the hilar region. Another measure of the harvest of the grain may be seed moisture content. The ideal time for the harvesting of pearl millet grain is when the moisture content is below 20% (Khairwal et al. 2007). For grain production, pearl millet earheads are harvested first by cutting and the stalks are harvested later after approximately 1 week (Vikaspedia 2022). The foxtail millet is ready for harvest within 80–90 days of sowing. Harvesting is done in September–October for kharif crops and February–March for rabi crops. The plants of foxtail millet mature uniformly and almost lose the ability of natural seed dispersal, hence, the entire crop can be harvested at a single time (Hermuth et al. 2016). Harvesting can also be done by cutting the panicles. Proso millet is a short-duration crop that matures within 65–75 days of cultivation. The ripening is non-uniform as the seeds in the tip of the upper heads are ripe and shatter before the lower seeds. The panicles mature at the later stages. The crop is generally harvested when two-thirds of the seeds are ripe (Prabhakar et al. 2017). Barnyard millet crop reaches the reproductive phase within 3 weeks under suitable conditions and the seed matures about 20–40 days after the reproductive phase begins. The optimum time for harvesting is 30–35 days after heading. The harvesting is done by cutting the crop from ground level with the help of a sickle and the crop is heaped for about a week in the field before threshing (Malik et al. 2021). Finger millet crop is ready for harvesting within 3.5–5 months of cultivation. The entire crop does not mature at a single time; hence its harvesting is done in two stages. The first harvest is done when 50% of the earhead turn brown and the second harvest is performed 7 days after the first harvest; all the earheads including the green ones are harvested on this day (TNAU Agritech 2013). Kodo millet is ready for harvest after 100 days of sowing or when the heads turn from green to brown. The crop is harvested by cutting the plants close to ground level. Small bundles of the harvested crop are made and stacked in the field for at least 1 week before threshing (Chakrabarty et al., 2023). Little millet is ready for harvest after 80–85 days of cultivation (Kumar et al. 2018). The harvesting can be done by cutting the whole plant or by collecting only the panicle of the mature crop (Venkatesan and Sundaramari 2017). The harvesting of teff is done in January when the leaves turn from green to yellow. The harvesting is usually done with sickles (Lee 2018). Fonio reaches physiological maturity 75–98 days after sowing. The plants are cut using a sickle and gathered in sheaths for drying before threshing (Adai et al. 2022).

2.2.2 Threshing

Threshing is done to loosen the edible part of the crop (grains) from the straw. This operation threshes the grain out of the earhead, cleans the grain, and produces straw fodder suitable for animal feeding. The millets are generally threshed when the earheads/grain moisture reaches 15–17% (TNAU 2022). Threshing can be achieved by the action of rubbing, impact, and stripping. Traditionally, it is done by flat spreading the harvested crop in the open fields and beating it with sticks or feet, rolling over stone or wooden rollers, or trampling with animals (making farm animals like cattle or bullocks walk over it until the chaff of the crop breaks and the grains come out) (Figure 2.1) (Tomar and Kumar 2021). In some regions, the bundles of harvested crops are spread over a cement/*pakka* floor and a tractor is passed over them repeatedly. These operations are energy- and labour-intensive (Singh et al. 2015) as only the separation of the grains takes place in these methods and the farmers have many remaining operations to do like de-husking, cleaning, etc. Nowadays, mechanized methods of threshing are more popular, as the machines consume less time, and various operations of threshing, de-husking, and cleaning can be done

Figure 2.1 Traditional methods of grain processing. (A) Harvesting with sickles. (B) Threshing by beating with sticks. (C) Threshing by trampling under the feet of animals. (D) Winnowing process. (Pictures courtesy of Rohit, Uttar Pradesh; Amar Chandel, Himachal Pradesh; Narender Kumar, Himachal Pradesh).

with a single machine. Threshing efficiency is affected by feed rate, concave clearance, and moisture content.

2.3 CURING AND DRYING

Curing is the process of packing threshed millet grains in gunny or cloth bags and storing them before drying. The threshed grains may also be kept under steam for 3–4 days. This process is responsible for the maturity of millets and the development of brown colour (TNAU Agritech 2013). After curing, the millets are dried. Drying lowers the moisture content of foods (Ojediran et al. 2010). Mass and heat exchange is the basic phenomenon for the removal of moisture from millets during drying (Annavarapu 2018). In traditional systems, drying is achieved by spreading the grains in the open under the sun. In mechanized dehydration, several types of equipment, such as pneumatic conveyor dryers, shallow bed dryers, microwave drying, cabinet dryers, hot air dryers, tunnel dryers, and belt-trough dryers, are used (Radhika et al. 2013).

2.4 YIELD OF MILLETS

The yield of pearl millet is about 3.0–4.0 tons/ha in well-irrigated fields, while a yield of approximately 1.3–1.7 tons/ha is obtained under rainfed conditions. The yield of foxtail millet varies from 1.5 to 1.8 tons/ha (Feedipedia 2022). The average yield of proso millet is 2.5–3.0 tons/ha and a yield as high as 3.5 tons/ha can be obtained with improved varieties (Nandini et al. 2021). Barnyard millet has an average yield of 1.0–1.2 tons/ha (Sood et al. 2015; TNAU 2022). The yield of finger millet is reported to be in the range of 3.0–5.0 tons/ha (TNAU 2022). The yield of kodo millet varies from 1.2 to 3.8 tons/ha (Pali et al. 2019). The yield of little millet is 1.2–1.5 tons/ha. The grain yield varies from 0.6 to 1.5 tons/ha in teff (Lee, 2018; Mihretie et al. 2021) and 0.71 to 1.02 tons/ha in fonio (Adai et al. 2022).

2.5 PRIMARY PROCESSING OF MILLETS

The primary processing of millets includes the de-husking, cleaning, sorting/grading, and milling of grains to convert the grains to a ready-to-process form. The hull or the pericarp of many of the millets remains intact after harvesting. This is removed by the process of decortication. The threshed grains may also contain impurities in the form of stones, sand, and metal pieces (wearing losses from machines) that need to be removed before storage and milling. This section discusses the operations involved in preparing the grains for milling and the production of ready-to-process grains. The milling of the millet is discussed in the next chapter (Chapter 3).

2.5.1 De-Husking/Dehulling

De-husking/dehulling is the primary process used to remove the hull and outer pericarp of the grains (Kalse et al. 2022). Abrasion and attrition are the two major forces used for the removal of the outer layer of millet in a single pass (Balasubramanian et al. 2020). The efficiency of these processes depends upon the grain hardness, size, and moisture content. Both traditional and mechanical methods can be used to achieve de-husking. In the traditional processes, it is achieved by pounding grains in a wooden or stone pestle using a wooden or stone-made mortar. This method requires several repeated cycles of pounding and produces a large percentage of broken grains (Singh et al. 2018). The processing time might be as high as 1 hour for the de-husking of 1–2 kg of millet (Gyanwali et al. 2016). After the removal of the husk, the chaff is separated by winnowing and washing with water. In modern methods, various mechanized mills like Engle Berg hullers, under-runner disc hullers, rubber-roll shellers, husk separators, centrifugal dehullers, and abrasive dehullers are used for the removal of the outer layer (Patil et al. 2018). Several factors responsible for the selection of the huller are energy requirements, the efficiency of husk removal, percentage of damaged grains produced, capacity and durability of the machine, and maintenance cost of the machine. Separators and blowers are used to separate the husk after husking (Prashant et al. 2015).

2.5.2 Cleaning

Cleaning is the primary process for the removal of dirt, debris, ferrous metals, bolts, filings, grease, stones, soil, leaves, pods, twigs, skins, weeds, engine oil, excreta, larvae, fertilizers, herbicides, soft rots, yeasts, fungal growth, toxins, colours, and impurities from the grains (Kumar et al. 2011). The first step in cleaning is to remove the husk from the grains. This is traditionally done by winnowing, where the impurities are removed from the threshed grains by wind blowing. In addition to this, the grains are also cleaned by sieves or are separated based on density in the traditional types of equipment. In modern machines, this is achieved by aspirators. The lightweight impurities and unwanted dust particles are removed with the help of the aspirator (Fellow 2005). The remaining cleaning methods can be broadly classified into wet and dry methods. The wet method is the most commonly used method to remove impurities such as soil, dust, and pesticide residues from the grains. Traditionally, it can be achieved by simply soaking and washing the grains, while the modern methods of wet cleaning involve soaking, spraying, and mixing with agitators, ultrasonic vibrations, and washing with the flotation process, etc. (Singh et al. 2015). The use of warm water in washing the grains limits the growth of spoilage microorganisms. Spray washers, rod washers, and drum washers are types of washers used in wet cleaning for grains (Mohammed et al. 2021). In modern methods, dry cleaning can be achieved by multiple machines such as a scourer, destoner, magnetic separator, etc). The scourer removes impurities like soil, dust, and sand and improves the hygiene of grains during processing. The destoner removes heavy particles and stones based on the principle of density. The magnetic separator removes metal impurities such as broken parts of machines.

2.5.3 Sorting and Grading

Sorting and grading are the segregation and selection operations that are performed based on physical, chemical, and biological attributes. Various attributes for the sorting and grading of grains are size, shape, weight, density, composition, processing suitability, texture, colour, freshness, tradition, utility, variety, percentage of foreign matter, degree of insect infestation, microbial growth, and the ratio of damaged grains, etc. (Chhabra and Kaur 2017; Karakannavar et al. 2021). Sorting is generally done to remove undesirable/unwanted grains while grading is the selection of uniform-quality grains based on shape, size, colour, density, variety, composition, etc. Damaged grains and grains with high moisture content are more prone to spoilage by insect infestation and microbial growth (Rosentrater and Anthony 2017). The non-uniform size of grains is responsible for under- and over-processing. The composition of the grains also varies with their size and density and this affects their suitability for processing. The varietal purity is also important for processing as the texture and colour of the end product depend on the composition of the grains. Non-graded grains also fetch lower prices in the market. Hence, it is important to sort and grade the grains before sending them to the market. Sorting and grading can be achieved by different screens, i.e. flat, drum, and inclined screens (Birania et al. 2020). In addition to screens, vibratory and disc separators can also be used. The vibrator separator separates seeds by their differences in shape and surface texture. The disc separator removes oversized or undersized grains in one pass through the machine. These machines are provided with discs having indented pockets that lift the material which fits into the pockets and helps in the sizing of the grains (Singh et al. 2015). The grains can also be colour sorted to remove the impurities of similar shape, size, and density (e.g. the grains of some other varieties). The colour separator sorts materials according to colour or light-dark characteristics. The machines used for performing these operations are discussed in Chapter 4. The hectolitre weight of grains is also a potent tool for the grading of millets. This method measures the mass of the grains per litre volume. Hectolitre weight changes with the shape, size, and density of the grains as well as the percentage of foreign matter (Manley et al. 2009).

The de-husked, clean, sorted, and graded grains can be dried to a moisture content of below 10% for safe storage (TNAU 2022).

2.6 STORAGE OF MILLET GRAINS

Food grains are the most durable and most commonly stored commodities and are the pillars of the food security of a country. The efficient storage of grains can prevent postharvest losses and can ensure nutritional security and enough food for emergencies. The storage of grains also helps to maintain buffer stocks and ensure price stability and the supply of grains throughout the year. In India, the surplus of major grains like paddy, wheat, and maize is generally procured by government or private agencies and stored in large warehouses or silos but millet cultivation is practiced by a limited number of farmers, and that too in a limited region. Hence, most of the millet produce is stored at the household level using indigenous storage systems. The indigenous storage structures are more or less similar to the storage structures for other cereal grains and may differ according to their size (small or big storehouses), location (indoor or outdoor storage structures), type (temporary or permanent), and scale (individual or community level). The storage structures may also be open storage systems, semi-open storage systems, or closed systems. These structures are mostly prepared using the straw of paddy, wheat, or other local crops, the empty structures of gourds, a mixture of straw and mud, wood, bamboo, and metallic drums. Details on the major traditional storage structures are provided in Table 2.1 and some of the most commonly used storage structures are also discussed in the following sections.

2.6.1 Mud-Based Bins

Structures made of mud and straw are among the oldest traditional storage systems for the storage of dried grains (Figure 2.2). These structures are used for storing small quantities of grains. One such structure is *Bharola*, i.e. made using mud and straw with the appearance of an egg with an opening on the top. It can also be a rectangular box. It has a storage capacity of 40–80 kg and was used as a portable structure in Punjab, India (Dhaliwal and Singh 2010). One similar

TABLE 2.1   TRADITIONAL GRAIN STORAGE STRUCTURES USED IN INDIA

Type of structure	Name of structure	Place of storage	Shape	Storage capacity
Mud-based	*Bharola*	Inside the house	Oval/ rectangular	40–80 kg
	Ghadoli	Inside the house	Oval shape	Up to 100 kg
	Pitchers baked in fire	Inside the house	Pitcher shape	20–40 kg
	Mud-house	Outside the house	Rectangular or cylindrical	1-50 tonnes
Wood-based	Box	Inside the house	Rectangular	1–1.2 tonnes
		Outside the house	Cylindrical thatched roof	9–35 tonnes
Bamboo-based	*Peru/Kanaja/ Bharola*	Inside the house	Cylindrical	Different capacities according to the type of structure
Combined structures	*Morai*	Outside the house	Cylindrical thatched roof	3.5–18 tonnes
	Bukhari	Outside the house	Cylindrical or square thatched roof	3.5–18 tonnes

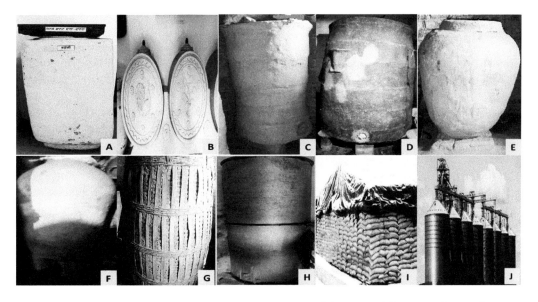

Figure 2.2 Traditional and modern grain storage structures. (A) *Bharola*. (B) *Ghadoli*. (C and D) *Kothila*. (E and F) *Kunna*. (G) *Peru/kannaja*. (H) Metal bin. (I) Cover and plinth storage system. (J) Silos. (Pictures courtesy of Amrit Kaur, Punjab; Manish Kumar, Bihar; Amar Chandel, Himachal Pradesh; Sunil Chandel, Himachal Pradesh).

structure is the mud *kothi* which has a rectangular (similar to a room) or cylindrical shape and a capacity of about 1–50 tonnes. They are equipped with a large door for the pouring of grains. This structure is also provided with a small outlet at the bottom for taking out the grains. Small-sized mud *kothies* are also available. Earthen pots made airtight by sun drying or kilning are also used for the storage of grains (TNAU 2022).

2.6.2 Wood-Based Storage Structures

Wood-based structures can be divided into small structures like s*andaka, peti,* s*andook,* etc., which are used for the storage of small quantities of grains for home consumption, and large structures like *Hapur kothi* which are used for the bulk storage of grains. "*Sandaka*" boxes have a storage capacity of 3 to 12 quintals. This structure is supported on wooden legs of about 1 foot height. The structure is also facilitated by a lid on the top for taking out the grains. This is polished at regular intervals to control insect infestation. Partitions can also be made inside this structured box to store two or more types of grains. Similar structures are also used in Himachal Pradesh, India, where they are called "*peti*" or "*sandook*". In Himachal, India, these wooden boxes are made using the termite- and insect-resistant wood of "*Tuni*" (*Cedrela toona* Roxb.) and "*Akhrot*" (*Juglans regia* Linn.) (Singh et al. 2017). *Hapur kothi* or the *Kothar* type of storage structures are outdoor structures with a capacity between 9 and 35 tonnes. These structures are built on raised pillars (1.5 meters above the ground) and both the floor and walls are made of wooden planks. The roof is a thatched or tiled roof type that is placed over it to protect the grains from the sun or rain. In the improved structures, the roof can be made of planks or corrugated metal sheets that sufficiently overhang on all sides. Rat-proofing cones are also provided on all pillars to avoid the entry of rats into the structure.

2.6.3 Bamboo Bins

Bamboo is one of the fastest-growing grasses and is found growing naturally in almost all parts of India except Kashmir (Bamboo Resources of the Country 2017). Split bamboo can be woven into different shapes. Utensils made from bamboo are used in almost all everyday household and agricultural practices, especially in hilly regions. Bamboo-based grain storage structures are woven from split bamboo in the shape of a cylinder or barrel with a wide base and narrow mouth as shown in Figure 2.2. The size of the structure may vary as per the requirement. These structures are plastered with a mud and cow dung mixture to prevent the spillage and pilferage of grains. After filling the grains, the top can be plastered with the mud and cow dung mixture. Alternatively, paddy straw and gunny bags can be used to cover it. These structures are not rat and moisture free and hence precautions need to be taken to prevent losses from rodents, insects, and moulds. This type of storage structure is known as "*Peru*" or "*Kannaja*" in Himachal Pradesh and "*Bhoral*" in Assam of India (Nagnur et al. 2006; Gogoi et al. 2017).

2.6.4 Composite Structures

Composite structures are made of a combination of two or more raw materials such as mud, wood, bamboo, metal, concrete, polyethylene sheet, etc. Such structures are used for the bulk storage of grains and provide better protection against insects, rodents, and moisture. *Morai* structures have a capacity of 3.5–18 tonnes and are used in the rural areas of eastern and southern India. These have the shape of an inverted cone. They are built on a raised platform supported by wooden or masonry pillars. The floor is a circular wooden plank supported on pillars using timber joints. All around the wooden floor, a 22-gauge corrugated metal cylinder of 90 cm height is nailed to it. The edge of the cylinder is flush with the bottom end of the floor. Inside the cylinder, 7.5 cm diameter ropes made of paddy straw or a similar material are placed, beginning from the floor level up to a height of 90 cm. Then bamboo splits are placed vertically along the inner surface without leaving any gap between them and, keeping the bamboo splits in position, the grain is poured into it up to the height of the metal cylinder. Following this the bamboo splits are held in an erect position and the winding of the rope as well as the pouring in of grain is done simultaneously till the required height is attained. The topmost ring of the rope is secured in position by tying it to the lower four rings. To provide a smooth surface, a layer of mud plaster about 1 cm thick is applied over the rope. A conical roof is placed on the top of the structure with an ample overhang all around. *Bukhari* structures are made with mud, bamboo, bricks, and cement. These structures have a square or cylindrical shape that is raised on timber or masonry pillars to a height of about 1.5 m from ground level. They have a capacity of about 3.5–18 tonnes. The floor of the bin is made either of timber planks or bamboo splits, plastered over with mud mixed with dung and paddy straw. In the improved structures, a cement base can also be built. The walls of the structure are made of a timber or bamboo framework and bamboo matting and are plastered with mud-straw plaster on both sides. These structures can also be lined or covered with a polyethylene sheet to protect against moisture. An overhanging cone-type roof is provided on the cylindrical structure. The roof is generally made of bamboo framework and straw. Rat-proofing cones are placed on all four pillars to prevent rats from entering the storage structure (Ashok and Sakunthla 2018).

2.6.5 Other Structures

The other structures are the "*Muda*" structures of Bihar which are built using "*Narai*" ropes and have a capacity of 1–3 tonnes. *Kuthla* storage structures are also found in rural areas of Bihar and Uttar Pradesh. These structures are kept inside and are made of burnt mud. An indigenous rat-proof grain storage system called "*Nahu*" is used by the "*Adi*" tribes of Arunachal Pradesh. This structure has a stone pad bottom, a wooden plate in the middle, and an airtight compartment at the top. Its storage capacity is 5.0–8.0 tonnes. "*Dhoosi*" is a traditional storage structure of the tribal people of Orissa that is made of long straw rope twined spirally. "*Garo*", a grain storage structure made of thatched grass, bamboo, and wooden poles, is used in Meghalaya. In addition, gunny bags are also used for the storage of grains.

2.6.6 Improved Structures for Household- and Farmhouse-Level Storage

Improved structures are made either by making suitable changes to the traditional structures or they are entirely new structures designed by various research institutes for the easy and safe storage of grains. In the improved bins, the mud-straw plaster is replaced with burnt bricks, concrete, coal-tar, and alkathene sheets that give them more strength and make them air- and moisture-proof. These changes make these bins suitable for the long-term storage of grains by reducing the qualitative losses. The Central Institute of Agricultural Engineering, India, developed a low-cost, small storage structure made of coal-tar (Singh et al. 2017). Punjab Agricultural University, Ludhiana, developed a galvanized metal iron structure with a capacity of 1.5–15 quintals for the storage of grains. It gives better protection against rodents and moisture. The Indian Agricultural Research Institute designed a mud, brick, and polythene-based modern small-scale storage structure. The structure is made of mud and bricks and the polythene sheets are placed between the layers of bricks and mud to provide proper sealing for a moisture-proof structure. A permanent type of rodent- and termite-proof storage structure is made from bricks, sand, cement, and stone. This structure is generally rectangular and the most widely adopted dimensions for length, breadth, and height are 2.43 × 1.82 × 2 ft. These structures are well plastered on the inside and outside to obtain smooth walls. Such structures are widely used in the southern part of India and are known as "*Kalangiyam*" in Tamil Nadu. The Indian Grain Storage Institute has also developed a cylindrical structure called "*Hapur Tekka*". This structure has a capacity of 2–10 tonnes and is made of galvanized iron or aluminium sheet and expandable cloth. It has a small circular or rectangular outlet at the bottom to take out the grains (Singh et al. 2017; Sharon et al. 2014).

2.6.7 Commercial Storage Structures

2.6.7.1 Cover and Plinth Storage System

This type of storage system is used by the Food Corporation of India for the bulk storage of bagged grains. This structure is constructed with the help of wooden crates that are fixed into the grooves of brick pillars at a height of nearly 1.17 feet above the ground. The bags are stacked vertically on the wooden crates one above another and the heap is covered with 250-micron low-density polyethylene (LDPE) sheets on all sides (Figure 2.2). This kind of storage structure is easy to build and cost-effective.

2.6.7.2 Silos

Silos are giant bulk storage structures made of concrete or metal that store up to 1000 tonnes of grains (Figure 2.2). They are generally cylindrical structures with conical roofs whose capacity depends upon their diameter and height (Singh et al. 2017). These structures are also provided with air intake vents at the bottom and silo vents at the top for the easy circulation of air. This stabilizes the grain temperature by reducing the internal silo temperature, humidity, and condensation build-up in silos. This also prevents the creation of hot spots and ensures that insects do not enjoy optimum storage conditions. They can also be equipped with silo fumigation units at the base. This facilitates the easy administration of chemical treatments and improves efficiency and safety (https://www.silovent.com/).

2.7 CONCLUSIONS

Millets are majorly sown as a kharif crop with the onset of monsoon and are available for harvest at the end of September or the first fortnight of October. The harvesting is done when the leaves turn yellow and give a nearly dried-up appearance. The ideal moisture content for the harvesting of millets is below 20% and for threshing 15–17%. The curing of grains before dehydration is essential in the case of millet where the grains do not mature uniformly and the crop is harvested before full maturity. The operation of curing helps in the uniform colour development of grains. The yield of millets range from 1.2 to 5 tonnes/hectare and the highest yield is obtained for finger millet. Being the crop of low fertile lands and rainfed regions, the yield of millets is comparatively low, and most of the harvesting, threshing, and postharvest processing is carried out using traditional methods. The major postharvest operations employed before the storage of grains are de-husking, cleaning, sorting and grading, etc. Machines are also available for millet processing but they can be used only for limited operations or specific millets. Millet grains are traditionally stored in structures made of wood, bamboo, or mud, or in composite structures. The bulk storage of millets can be carried out in large structures of steel known as silos and, alternatively, they can be packaged in plastic-lined gunny bags and stored under a cover and plinth system.

REFERENCES

Adai, W. A. 2022. Effects of alcoholic and alkaloid extract of Solanum nigrum in some aspect's life cycle of the green bottle fly (Lucilia sericata (Diptera: Calliphoridae). *Al-Qadisiyah Journal of Veterinary Medicine Sciences* 21(1):51–59.

Ashok, B. G., and N. M. Shakunthala. 2018. Different types of grain storage structures for the betterment of livelihood of Indian farmers. *International Journal of Pure and Applied Bioscience* 6(4):190–198.

Balasubramanian, S., S. D. Deshpande, and I. R. Bothe. 2020. Design, development and performance evaluation of CIAE-Millet mill. https://krishi.icar.gov.in/jspui/handle/123456789/47869. (accessed May 21, 2020).

Bamboo Resources of the Country. 2017. https://fsi.nic.in/isfr2017/isfr-bamboo-resource-of-the-country-2017.pdf. (accessed September 3, 2022).

Birania, S., P. Rohilla, R. Kumar, and N. Kumar. 2020. Post harvest processing of millets: A review on value added products. *International Journal of Chemical Studies* 8(1):1824–1829.

Chakrabarty, S. K., S. Basu, and W. Schipprach. 2023. Hybrid seed production technology. In M. Dadlani and D. K. Yadava *Seed Science and Technology: Biology, Production, Quality, eds.* 173–212. Singapore: Springer Nature.

Chhabra, N., and A. Kaur. 2017. Studies on physical and engineering characteristics of maize, pearl millet and soybean. *Journal of Pharmacognosy and Phytochemistry* 6(6):1–5.

Dhaliwal, R. K., and G. Singh. 2010. Traditional food grain storage practices of Punjab. https://nopr.niscpr.res.in/handle/123456789/9787. (accessed July 25, 2020).

Feedipedia. 2022. Seaweeds (marine macroalgae) content. Animal feed resources information system. Web. https://www.feedipedia.org/node/78. (accessed June 27, 2022).

Fellows, P., and A. Rottger. 2005. *Business Management for Small-Scale Agro-Processors*. AGSF Working Document 7. Food and Agriculture Organisation of the United Nations, Rome.

Gogoi, B., S. Bhagowati, and S. Das. 2017. Traditional crop management practices of central Brahmaputra valley zone of Assam, India. *International Journal of Current Microbiology and Applied Sciences* 6(7):2405–2407.

Gyanwali, K., B. Dhungana, D. Shrestha, G. K. Das, and M. Dhungana. 2016. Design, fabrication and testing of proso millet de-husking machine. In *Proceedings of IOE Graduate Conference* (pp. 123-131), Institute of Engineering, Tribhuvan University, Nepal).

Hassan, Z. M., N. A. Sebola, and M. Mabelebele. 2021. The nutritional use of millet grain for food and feed: A review. *Agriculture and Food Security* 10(1):1–14.

Hassan, M. M., H. M. Daffalla, H. I. Modwi, M. G. Osman, I. I. Ahmed, M. E. A. Gani, and A. G. E. Babiker. 2013. Effects of fungal strains on seeds germination of millet and Striga hermonthica. *Universal Journal of Agricultural Research* 2(2):83–88.

Hermuth, J., D. Janovská, P. H. Čepková, S. Usťak, Z. Strašil, and Z. Dvořáková. 2016. Sorghum and foxtail millet—Promising crops for the changing climate in central Europe. In *Alternative Crops and Cropping Systems*, ed. P. Konvalina, doi: 10.5772/62642, London: IntechOpen.

Kalse, S. B., S. B. Swami, and S. K. Jain. 2022. Millet: A review of its nutritional content, processing and machineries. *International Journal of Food and Fermentation Technology* 12(1):47–70.

Karakannavar, S. J., G. Nayak, and J. Kumar. 2021. Effect of different levels of polishing on physico-chemical characteristics of barnyard millet (Echinochloa frumentacea). *The Pharmaceutical Innovation* 10:50–54.

Khairwal, I. S., S. K. Yadav, K. N. Rai, H. D. Upadhyaya, D. Kachhawa, B. Nirwan, R. Bhattacharjee, B. S. Rajpurohit, and C. J. Dangaria. 2007. Evaluation and identification of promising pearl millet germplasm for grain and fodder traits. *Journal of SAT Agricultural Research* 5(1):1–6.

Kumar, A., V. Tomer, A. Kaur, V. Kumar, and K. Gupta. 2018. Millets: A solution to agrarian and nutritional challenges. *Agriculture and Food Security* 7(1):1–15.

Kumar, B., and J. Kumar. 2011. Management of blast disease of finger millet (Eleusine coracana) through fungicides, bioagents, and varietal mixture. *Indian Phytopathology* 64(3):272.

Kumar, K., G. Pal, A. Verma, and S. K. Verma. 2020. Seed inhabiting bacterial endophytes of finger millet (Eleusine coracana L.) promote seedling growth and development, and protect from fungal disease. *South African Journal of Botany* 134:91–98.

Lee, J. S., S. H. Choi, S. J. Yun, Y. I. Kim, S. Boandoh, J. H. Park, B. J. Shin. 2018. Wafer-scale single-crystal hexagonal boron nitride film via self-collimated grain formation. *Science* 362(6416):817–821.

Malik, S., N. Hussain, A. Singh, and M. I. Bhat. 2021. Finger millet pearling efficiency as affected by hydrothermal treatment. *International Journal of Current Microbiology and Applied Sciences* 10:1867–1874.

Manley, M., M. L. Engelbrecht, P. C. Williams, and M. Kidd. 2009. Assessment of variance in the measurement of hectolitre mass of wheat, using equipment from different grain producing and exporting countries. *Biosystems Engineering* 103(2):176–186.

Mihretie, F. A., A. Tsunekawa, N. Haregeweyn, E. Adgo, M. Tsubo, T. Masunaga, D. T. Meshesha, W. Tsuji, K. Ebabu, and A. Tassew. 2021. Tillage and sowing options for enhancing productivity and profitability of teff in a sub-tropical highland environment. *Field Crops Research* 263:108050.

Mohammed, A., and A. Aboshio. 2021. Durability characteristics of millet hush ash: A study on self-compacting concrete. *Teknika: Jurnal Sains Dan Teknologi* 17(1):106–112.

Nagnur, S., G. Channal, and N. Channamma. 2006. Indigenous grain structures and methods of storage. https://nopr.niscpr.res.in/handle/123456789/6811. (accessed January 17, 2022).

Nandini, B., H. Puttaswamy, R. K. Saini, H. S. Prakash, and N. Geetha. 2021. Trichovariability in rhizosphere soil samples and their biocontrol potential against downy mildew pathogen in pearl millet. *Scientific Reports* 11(1):9517.

Ojediran, J. O., M. A. Adamu, and D. L. Jim-George. 2010. Some physical properties of pearl millet (Pennisetum glaucum) seeds as a function of moisture content. *African Journal of General Agriculture* 6(1):39–46.

Pali, V., M. Ramani, H. E. Patil, S. Patil, and B. K. Patel. 2019. High yielding Kodo millet variety' GAK-3'for cultivation in hilly region of Gujarat. *Journal of Pharmacognosy and Phytochemistry* 8(6):533–537.

Powar, R., V. Aware, and P. Shahare. 2019. Modeling and optimization of finger millet pearling process by using RSM. *Journal of Food Science and Technology* 56(7):3272–3281.

Prabhakar, B., and D. R. More. 2017. Effect of addition of various proportion finger millet on chemical, sensory and microbial properties of sorghum papads. *Current Agriculture Research Journal* 5(2):191.

Pradhan, A., S. K. Nag, and S. K. Patil. 2010. Traditional technique of harvesting and processing for small millets in tribal region of Bastar. https://nopr.niscpr.res.in/handle/123456789/10318. (accessed October 18, 2021).

Prashant, S., and P. Rajashekar. 2015. Development of millet de-husking machine. *International Journal of Engineering Research & Technology (IJERT)*. ISSN:2278–0181.

Radhika, G. B., S. V. Satyanarayana, and D. G. Rao. 2013. An investigation on drying of millets in a microwave oven. *International Journal of Emerging Technology and Advanced Engineering* 3:583–589.

Rosentrater, K. A., and A. D. Evers. 2017. *Kent's Technology of Cereals: An Introduction for Sudents of Food Science and Agriculture*. Woodhead Publishing.

Samuel, A. A., K. A. Alice, S. Mawuli, and A. O. Adetunla. 2019. Development of a threshing device for pearl millet. *Current Journal of Applied Science and Technology* 33(1):1–12.

Sharon, M., C. V. Abirami, and K. Alagusundaram. 2014. Grain storage management in India. *Journal of Postharvest Technology* 2(1):12–24.

Singh, S. N., S. Srivastav, and F. M. Sahu. 2018. Milling technology of grains. In *Fundamentals of Food Engineering and Application*, eds. S. Shrivastava and P. M. Ganorkar (pp. 89–136). Brillion Publishing.

Singh, K. P., K. N. Poddar, K. N. Agrawal, S. Hota, and M. K. Singh. 2015. Development and evaluation of multi-millet thresher. *Journal of Applied and Natural Science* 7(2):939–948.

Singh, V., D. K. Verma, and P. P. Srivastav. 2017. Food grain storage structures: Introduction and applications. In *Engineering Interventions in Foods and Plants*, eds. D. K. Verma and M. K. Goyal (pp. 247–284). Apple Academic Press.

Sood, S., R. K. Khulbe, A. K. Gupta, P. K. Agrawal, H. D. Upadhyaya, and J. C. Bhatt. 2015. Barnyard millet–a potential food and feed crop of future. *Plant Breeding* 134(2):135–147.
Tomar, A., and V. Kumar. 2021. Truthful label seed production techniques of Kodo millets under seed hub on millets in Bundelkhand regions. *The Pharma Innovation Journal* 10:560–563.
TNAU. 2022. Crop production: Millets. http://www.agritech.tnau.ac.in/agriculture/millets_index.html. (accessed May 12, 2022).
TNAU Agritech. 2013. Post harvest technology. http://www.agritech.tnau.ac.in/postharvest/postharvest.html. (accessed May 12, 2022).
Venkatesan, P., and M. Sundaramari. 2017. Indigenous technical knowledge in little millet cultivation among Malayali tribes of India. https://krishi.icar.gov.in/jspui/handle/123456789/14770. (accessed May 11, 2021).
Vikaspedia. 2022. *Aportal Govt of India, Ministry of Electronics and Information Technology, Govt.* Centre for Development of Advanced Computing. https://vikaspedia.in/health/nutrition/nutritive-value-of-foods/nutritive-value-of-cereals-and-millets/milletsthe-nutricereals. (accessed March 12, 2021).

Chapter 3

Milling and Secondary Processing of Millets

3.1 INTRODUCTION

Milling is a size reduction and separation process that pulverizes grains to flour or divides them into various fractions like seed coat, germ, and endosperm (Kulkarni et al. 2018). Milling operations are classified into two types, i.e. dry milling and wet milling. Dry milling separates the outer fibrous layer and germ using an abrasive method that gradually removes the seed coat, aleurone layer, sub-aleurone layer, and germ to produce polished grain. The major dry milling operations are debranning, polishing, tempering, grinding/pulverization, sieving, etc. In this type of milling the endosperm is simply broken into smaller fragments with no conscious separation of starch and protein (Rosentrater and Evers 2018). Starch granules in the endosperm are released from the protein network by wet milling and this milling operation results in the recovery of separate fragments like germ, bran, starch, and protein. The major operations of wet milling cereal grains are the soaking or maceration of the grains in water followed by milling. Soaking softens the grain and affects the chemical and physical properties of the bran and endosperm. The bran becomes more elastic and the possibility of contamination of flour with bran is reduced.

The operations of primary processing and milling produce the basic elements for secondary processing. These secondary operations convert the end products of primary processing and milling into edible products. The various secondary processing techniques are baking, cooking, flaking, frying, puffing, popping, extrusion, fermentation, roasting, malting, etc. In-depth information on millet milling and the procedures involved in millet secondary processing is provided in this chapter.

3.2 MILLING

The milling process separates the grain into three components, i.e. seed coat, germ, and endosperm. Dry milling and wet milling are the two major processes used for grain milling. Dry milling includes conditioning/tempering, debranning, polishing, grinding/pulverization, and so on, whereas wet milling involves soaking the grains in water for 24–48 hours followed by milling using abrasive mills, disc mills, or pin mills (Microtek Engineering Group 2022). Mechanical force of different types such as shear, friction, compression, impact, hammering, collisions, etc., is used to perform milling. This section discusses various milling procedures.

3.2.1 Conditioning/Tempering

"Conditioning" is the operation of the addition of small quantities of water to the cleaned grains before milling. After the addition of water, the grains are allowed to rest at a particular temperature for a particular time and that process is known as "tempering". Tempering time depends on kernel hardness, kernel temperature, initial moisture content of the grain, and the dampening method. Tempering time is also affected by mixing and it can be as high as 72 hours at low-speed mixing and as low as 7–12 hours at high-speed mixing or in intensive dampening systems. The tempering treatment helps in the uniform distribution of moisture throughout the grain. The combined operation of conditioning and tempering toughens the bran coat and makes it more elastic, prevents its fine grinding, and avoids the contamination of flour with bran (Alfin 2020; Perdon et al. 2020). It also makes the endosperm more friable and helps in its milling. The target tempering moisture level of different grains can be quite variable depending on the grain type and the time of the year. The hydrothermal treatment of millet grains also aids in their decortication by hardening the endosperm (Shobana and Malleshi 2007). The elevated pressure and temperature of hydrothermal treatment have been reported to increase the milling yield. Dharmaraj et al. (2013) reported a pressure of 313.8 kPa and a steaming time of 17.5 minutes as the optimum condition for the hydrothermal treatment of finger millet grains. The tempering (the addition of 5% water at stage 1 of milling and 4% water at stage 2 of milling) of the hydrothermally processed grains with 16% moisture resulted in a 64.4% yield of decorticated grains. The hydrothermal treatment of millet grains followed by drying to a moisture content of 12–13% (w.b.) improves the dehulling of pearl millet (Amadou et al. 2014). The equilabration of the moisture content of finger millet to a moisture content of $33 \pm 2\%$, steaming for 20 minutes at atmospheric pressure, and drying to a final moisture content of $12 \pm 2\%$ is reported to enhance the hardness (1.1–7.1 kg/cm) of the millet kernel and enable its decortication (Shobana and Malleshi 2007). The conditioning of pearl millet and finger millet grains by increasing their moisture content to 15% and tempering them for 10 minutes has been reported to reduce the milling of the seed coat of millets (Malleshi et al. 2004). For puffing, pearl millet grains are conditioned at a moisture content of 18% and are tempered for 6 hours while finger millet grains are conditioned by adding 3–5% moisture and are tempered for 2–4 hours (Selladurai et al. 2022). The literature on the conditioning of other millets is limited.

3.2.2 Debranning

The bran of millets is rich in polyphenols and anti-nutritional factors. The milling of whole millet produces dark flour with a poor aesthetic appearance. The increased level of bran may also turn the flour bitter (especially in finger millet and kodo millet) and produces bitter bread that has lower overall acceptability. Therefore it is important to remove bran in such millets. This process of the removal of the outer layer/bran of the grains is known as debranning. In this process, the seed coat, which is tightly bound with the grain endosperm, is loosened with the help of hydrothermal treatment and abrasive and frictional force (Malik et al. 2021). Debranning also helps in reducing the anti-nutritional factors (Kate and Singh 2021), as the most of antinutrients in millets are localized in the bran layer (Kumar et al. 2020). In a nutshell, debranning improves the biological availability of nutrients and consumer acceptance (Samtiya et al. 2020). The major mills used for debranning are abrasive decorticators and metal friction machines (Saleh et al. 2013).

3.2.3 Polishing/Pearling

The process of debranning results in nonuniform surfaces with less acceptability among consumers. So, to make the product more attractive, the grains are polished or pearled. This process makes the grain surface smooth and shiny with the abrasive and mild friction developed by the rubbering process or chemical treatment (Karakannavar et al. 2021). The traditional methods of pearling are stone grinding (*Jatta*), the rubbing of grains in gunny bags, and leg pounding (Powar et al. 2019). Millet polishers are also commercially available in the market (see Chapter 4). Chemical methods using sodium bicarbonate solution are used as a quick method of pearling to obtain average polishing. Pearling can also be achieved by an enzymatic process. In this process cell wall-degrading enzymes such as cellulose, xylanases, and β-glucanases are used (Malavika et al. 2020). Xylanases break down xylans in the millet's cell wall and enhance the cooking quality of millets. Enzymatic methods also result in minimum loss of nutrients and are eco-friendly methods. The fibre and crude oil content of grains also decreases with enzymatic treatment (Das et al. 2019). The polishing process increases the strength of polished products by reducing the stress concentration. This process also improves the taste and appearance of the products (Malik et al. 2021).

3.2.4 Grinding/Pulverization

Grinding/pulverization is a process in which grains are put under mechanical forces to reduce the size and increase the surface area of the grains (Skovgaard et al. 2010). The various forces that can be applied to achieve grinding are compression, shearing, and impact (Seo et al. 2015). The force applied to the grains is absorbed and deforms the grain. When the deformation reaches a critical point, it creates fissures and cracks in the grain that cause its breakdown. The factors that affect the rate of size reduction are the shape, size, hardness, composition, presence or absence of crease, moisture content, type of machine, the working principle of the machine, etc. (Gowda et al. 2022). Different machines used for size reduction are attrition mills, ball mills, pin mills, hammer mills, disc mills, roller mills, disc-beater mills, vertical toothed mills, etc. (Embashu et al. 2019, Kate and Singh 2021). The grinding operation may be achieved in a single-stage, double-stage, or multi-stage mill (Ferreira et al. 2013). The single-stage mill grinds the millets to a coarse flour. A double-stage grinder pulverizes the material into a powder to produce fine flour (Seo et al. 2015). The multi-stage mill contains two different chambers for the feeding of grains and converting them into flour. Grinding machines can be broadly classified into rough grinding machines and precision grinding machines. Rough grinding machines are used for the removal of stock with no change in the property of the product whereas precision grinding machines are used for the accurate dimension of the final product (García-Granero et al. 2017). The endosperm can be ground into various sizes such as semolina (300–750 µm), middlings (125–300 µm), and flour (less than 125 µm) (Lorusso et al. 2017). The coarse ground grains, larger in size than semolina, are used for porridge preparation. Details on the working principles and structures of these mills are provided in Chapter 4. The millet milling process has a significant effect on the quality of final products (Beukes 2021).

3.2.5 Packaging of Flour

The bulk storage of flour can be done in steel containers or gunny bags lined with low-density polyethylene (LDPE). Retail packages of 500 g or 1 kg are also available in laminate packaging or polypropylene bags. The millet grains or flour may be passed through an impact entoleter before packaging to improve their hygiene and shelf life (Shaaya et al. 2005). The impact entoleter is a high-speed/rpm machine that works on the principle of striking the feed material against the outer wall. When the material enters the rotor, it is thrown outward due to the centrifugal force of the spinning rotor. The high impact removes damaged or unmatured grains, kills insects, and destroys their eggs. This operation increases the shelf life of flour. This machine is also used to mix micronutrients and other additives into the flour before packaging (Udawat 2022).

3.3 SECONDARY PROCESSING OF MILLETS

The secondary processing of grains is the process of converting the end-products of primary processing and milling into an edible form. The edible products of the millet can be prepared using cleaned unmilled or milled grains (flour). The unmilled grains are generally processed by puffing/popping, flaking, roasting, and malting. The milled grains or flour are generally made into dough or slurry and are then processed by baking, cooking, or the combined operations of fermentation and cooking. These operations increase the nutritional and organoleptic properties (appearance, taste, texture, and aroma) of foods. The shelf life and safety of the food products are also enhanced as the high temperatures of cooking, baking, popping, and roasting kill the microorganisms. This section discusses various methods used for the secondary processing of millets.

3.3.1 Roasting

Roasting is a traditional process of cooking foods by employing dry or diffused heat. The food can be roasted in an open flame, oven, or microwave or heat may be applied indirectly from other sources like in sand roasting (Singh et al. 2018). The basic mechanism of the roasting process is to control the temperature and time of cooking which helps in making the product edible for human consumption. Roasting takes place in two phases, i.e. the first phase is responsible for the drying of grains and the second phase is responsible for the development of texture and flavour (Sudha et al. 2021). There is an ample supply of air in this type of cooking and this leads to the development of aroma and colour. The major reactions responsible for the development of aroma and colour are Maillard browning and caramelization. Roasting also results in the denaturation of proteins, inactivation of enzymes, and change in the texture of the product. The anti-nutritional factors are reduced and the viscosity of roasted flour is reduced compared to the native flour (Ranganathan et al. 2014). The water solubility index, oil absorption capacity, and iron content of the pearl millet flour increase with roasting. Potassium was found to be the dominant mineral in the roasted samples. The highest iron content was obtained when roasting was carried out at 180°C for 10 minutes (Obadina et al. 2017). The roasting of finger millet has been reported to decrease moisture, fat, protein, phenols, and antioxidant activity and increase total carbohydrate, ash, and fibre content. An increase in the bioavailability of minerals like calcium and iron is also reported (Singh et al. 2018). Roasting of proso millet flour increases its polydispersity (the weight average molecular weight divided by the number average molecular weight), whiteness, and viscosity (peak, trough, final, breakdown, and setback) (Shinoda et al. 2002). The roasting of barnyard millet significantly increases the nutrients and lowers the viscosity (Anbalagan and Nazni 2020). Kalam Azad et al. (2019) reported increased nutraceutical compounds and improved functionality in proso millet on roasting. The proso millet samples roasted at 110°C were reported to have a significantly higher content of total phenolic acids (670 mg/100 g of ferulic acid equivalent) and total flavonoids (391 mg/100 g of rutin equivalent) compared to control (untreated whole proso millet) (295 mg/100 g, 183 mg/100 g, respectively). The reason for an increase in phenolic compounds has been described by Hu et al. (2009) as the hydrolysis of C-glycosyl flavones during roasting. Their hydrolysis promotes the release of polyphenols (Hu et al. 2009). An increase in the total phenolic, flavonoid, and tannin contents of roasted little millet has been also reported by Pradeep and Guha (2011). The roasted little millet samples were also reported to have high 2,2-diphenylpicrylhydrazyl(DPPH) radical scavenging activity and iron-reducing power. The roasting of teff grains increased their oil content, carotenoid content, total flavonoid content, and antioxidant activity. The content of these nutrients was highest in microwave-roasted grains compared to oven-roasted grains (Ahmed et al. 2021). Ballogou et al. (2015) roasted dried grains of two fonio landraces, i.e. Iporhouwan and Namba, in a pot at about 300°C for 30 to 35 minutes and studied the effect of roasting on their colour and nutritional value. A decrease in brightness (L*) and an increase in redness (a*) and yellowness (b*) were recorded in roasted samples of both varieties compared to their native samples. Roasting was also reported to reduce the particle size distribution as the roasted fonio showed a higher percentage of particles with sizes smaller than 710 μm. The roasting of fonio was also reported to decrease crude protein, crude fat, crude fibre, total ash, iron, and zinc content. A small increase in total carbohydrates was also reported. However, among the two studied landraces, the Namba landrace reported an increase in crude lipid content on roasting. Roasting kodo millet for 15 minutes before milling has been reported to increase the whiteness index and decrease the yellow index (Bhawna et al. 2009).

The selection of the method of roasting is directly dependent on the type of food matrix. The various types of machines used for roasting grains are horizontal rotating drums, vertical rotating bowls, vertical static drums with blades, fluidized beds, etc. The horizontal rotating drum is of two major types, i.e. perforated wall type and solid wall type (Asghari et al. 2013). The temperature and time of the roasting should be controlled carefully as excessively high temperatures and roasting time can degrade the sensory and nutritional components. The roasting process is also used for the popping or puffing of millet grains.

3.3.2 Popping/Puffing

Popping and puffing are dry heat-based processing methods. In these methods, the application of heat produces superheated vapours that cook the grains, break the outer skin, and expand and gelatinize the endosperm suddenly (Gowda et al. 2020). Small angular fragments are observed on the popped and puffed product. Air bubbles are also formed in the kernels that are responsible for the expansion of the endosperm. The cell wall of protein comes below the aleurone layer (Mohapatra and Das 2011). Popping and puffing are similar in many aspects, the only difference is that expansion and gelatinization take place simultaneously in popping, while in puffing the controlled expansion of pre-gelatinized kernels takes place (Sharma et al. 2014). The various methods used for popping/puffing are the dry heat method, salt and sand roasting method, hot oil frying method, hot air popping method (gun popping), fluidized bed puffing, microwave heating method, etc. The volume expansion of popped grains is affected by their variety, moisture content, shape of grains, pericarp thickness, chemical composition, amylose content, grain hardness, protein solubility, temperature of processing, and method of processing. In sand puffing, the grains come in contact with hot sand at a temperature of about 250°C, and in oil roasting, the temperature of the oil is about 200–220°C (Mishra et al. 2015). In gun puffing, preheated grains are added to a high-pressure chamber where they come in contact with superheated steam. In this puffing, the gun is uniformly rotated to make the product puffed. The texture of the puffed product directly depends upon the pressure of the steam. In high-temperature short-time fluidized bed puffing the product surface is uniformly exposed to a heating medium and hence there is increased heat and mass transfer. The popping/puffing yields are influenced by the moisture content of grains as well as the moisture in the heating media (Mishra et al. 2014). The optimum conditions for the puffing can vary according to the design of the machine. In general, grain puffing at 240–270°C for 7–9 seconds gives a good yield. High-temperature short-time fluidized bed puffing is more efficient than hot air or conduction puffing and can be run in continuous mode which results in a higher puffing yield per hour. This method also produces a product with good colour (Farahnaky et al. 2013). Microwave popping/puffing is the most common method for the preparation of popped and puffed products worldwide (Mishra et al. 2015). The microwave heating of food products depends on their electrical conductivity, water content, and dielectric properties (Dogan Halkman et al. 2014). The microwaves cause the agitation of water molecules and charged ions which converts electromagnetic energy to thermal energy. The factors that affect popping in a microwave are the residence time of the sample, microwave power density, and the level of microwaves (Mishra et al. 2015, Sharma et al. 2014). In a study conducted by Garud et al. (2022), the highest popping yield (72.83%) and expansion ratio (6.15) for pearl millet variety 92901 were obtained using conventional salt popping at 260°C. Chauhan et al. (2015) screened eight germplasm varieties of pearl millet for their popping characteristics and optimum processing conditions. It was found in the study that the maximum popping yield (62%) was obtained at a moisture content of 15.92%, tempering time of 7.16 hours, and temperature of 282°C. In a study conducted by Mirza et al. (2015) on the popping characteristics of brown and white finger millet genotypes, it was found that the brown genotype PRM-6107 with the highest grain hardness and protein solubility with a moisture content of 10.65%, crude protein content of 9.92%, and crude fibre content of 3.6% had the highest popping ability.

Popping is also reported to affect the nutritional properties of the millets. It reduces the antinutritional factors such as phytic acid, oxalic acid, tannins, etc., and increases the level of some minerals. A decrease in tannins, phytic acid, oxalic acid, and trypsin inhibition activity, i.e. of 29.88%, 60.86%, 29.6%, and 26.2%, respectively, was observed on the popping of finger millet (Chauhan and Sarita 2018). Choudhury et al. (2011) reported the highest popping yield in yellow (30%) and purple (26.3%) varieties of foxtail millet at 230°C. A temperature above 250°C was reported to decrease the popping yield. A decrease in crude fat (32.18–36.04%), crude fibre (54.87–64.6%), and minerals (18.1–37.8%), and an increase in the protein (8.1–9.3%) and carbohydrates (10.6–13.4%) content of both yellow and purple varieties of foxtail millet were found compared to the non-treated foxtail millet grains. The popping of the grains removed the seed coat, decreased fat, crude fibre, and minerals content, and increased the protein and carbohydrate content. This is due to the fact that in cereals most of the fibre, fat, and minerals are present in the seed coat (Kumar et al. 2020; Chaudhury et al. 2011). The popping percentage of proso millet grains is about 62% (Ranganna et al. 2012). Delost-Lewis and Tribelhorn (1992) studied the puffing quality of eight proso millet varieties and the best puffing characteristics were found in Variety 2027. Further, the effects of moisture (12, 15, and 18%) and puffing gun pressure (120, 140, and 160 psi) were also studied. Improved puffing characteristics were obtained at the moisture content of 15 and 18% and puffing gun pressure of 140 and 160 psi. The puffing was also reported to increase the protein content and starch digestibility, while the ash and dietary fibre content was lowered. The popping percentage of kodo millet is 32–36% (Ranganna et al. 2012). Patel et al.

(2018) studied the puffing characteristics of two kodo millet varieties JK48 and JK155. The puffing of both kodo millet varieties decreased moisture (7.92–7.94% to 3.35–3.36%), fat (1.43–1.44% to 1.38–1.41%), ash (3.98 to 3.84–3.92%), and fibre (9.83–10.83 to 8.92–9.10%) content. The content of protein and carbohydrates was increased from 7.92–7.94% to 8.02–8.12%, and 68.35–69.48% to 74.20–74.38%, respectively. Jaybhaye et al. (2011) optimized the high-temperature short-time puffing process for the development of barnyard millet-based ready-to-eat (RTE) puffed products. To prepare the RTE product, at first, a cold extruded dough sheet was prepared from barnyard millet flour, potato mash, and tapioca powder in the ratios 60:37:3, followed by steam cooking and puffing in a hot air puffing setup. The process parameters were optimized using a central composite rotational design of response surface methodology. It was found in the study that the texture of the product was mainly dependent on moisture content, while volume expansion was highly dependent on steaming pressure and steaming time. The optimized puffing conditions were a steaming pressure of 0.18 kg/cm^2, a steaming time of 10 minutes, a hot air temperature of 234°C, and a puffing time of 39 seconds. The optimum popping grain moisture and particulate medium temperature for little millet are reported to be 16% and 260°C, respectively (Kapoor 2013). Studies are not available on the optimization of popping conditions for teff and fonio.

3.3.3 Flaking

Flaking is the process of flattening the partially cooked grains with the help of a roller. The flakes of desired thickness are produced by passing the pellets or grits of grain through large counter-rotating metal rollers that are internally cooled by passing water. The flakes are removed from the rollers with the help of a scraper knife. The rolled flakes are then passed through a hot air blast to remove the moisture and develop desirable colour and flavour (Taylor et al. 2017). The flakes are generally dried to a moisture content of below 7% (Gowda et al. 2020). A variety of flaked products including millet-based flaked products (see detail in Chapter 14) are available in the market. The flakes are consumed as ready-to-eat foods or quick-cooking foods.

3.3.4 Frying

Frying is generally the cooking of foods in hot oil; however, nowadays, air frying processes are also available. Oil frying can be further categorized as deep-fat frying and shallow frying. In the deep-fat frying process, the food is completely immersed in cooking oil, while, in shallow frying, the food is cooked in a thin layer of oil in which the food is not completely dipped. The high temperature of the oil results in instant cooking and moisture removal. In this cooking method, the outer layer of the food becomes crisp while the inside of the product is still soft and contains reasonable moisture. Several changes that occur during this processing are hydrolytic cleavages, oxidation reactions, protein denaturation, and starch gelatinization (Choe and Min 2007). This process also helps in the development of a uniform colour and unique taste of the product. Plenty of Indian fritters like vegetable *pakoras*, bread *pakoras*, *besan* egg, *bonda*, etc., are prepared using this method. Crisp products like Indian *papads* are also prepared using this method. The higher temperature of frying kills the microorganisms, deactivates the enzymes, and enhances the shelf life of the product. The high temperature also degrades the nutrients, especially vitamins (Jiménez-Monreal et al. 2009). The sensory quality and shelf life of the product directly depend on the food material to be fried and the medium and condition of frying (Oke et al. 2018).

3.3.5 Malting

Malting is a three-step process that consists of the soaking, germination, and kilning of the grains. The millet grains are soaked overnight (Kumar et al. 2020) and germinated under controlled conditions to develop the enzymes that degrade the cell wall and hydrolyze the starch, protein, and nucleic acid molecules into smaller molecules that can be easily utilized by yeast during brewing processes. High α-amylase activity has been reported in pearl millet and finger millet malt within 2–3 days of germination (Malleshi and Desikachar 1986). The malting process is arrested by kilning (drying the grains at $50 \pm 2°C$) (Baranwal and Sankhla 2019; Von et al. 2003). A characteristic malt flavour and colour can be produced by heating the malted grains at high temperatures. Malting also plays a crucial role in the modification of the physical structure and nutritional composition of the grains. It drastically lowers the peak viscosity of millets (Malleshi and Desikachar 1986). The malting of pearl millet decreases the bulk density of flour and increases the swelling capacity and oil and water absorption capacity. A slight increase in the crystallinity of the fermented and malted millet flours has been also observed on X-ray diffraction (Adebiyi et al. 2016). Zarnkow et al. (2014) reported 25–30°C germination temperature and 3–5 days germination time as optimal conditions for the malting of pearl millet. These conditions were reported to have high diastatic power, α- and β-amylase activity, good free α-amino nitrogen content, and moderate malting loss. A slight increase in the protein content of malted grains is also observed due to the respiration of carbohydrates. However, the removal of roots and shoots of germinated grains decreases the protein content. An increase in protein digestibility is also reported (Taylor and Taylor 2017). An 11.86–12.6% decrease in protein content, 31.39–54.02% decrease in fat content, 10–21.6% decrease in minerals, and 9.7–12.5% increase in crude fibre was observed on the malting of yellow and purple varieties of foxtail millet (Chaudhury et al. 2011). Sharma et al. (2015) optimized the germination conditions for the foxtail millet based on its phytochemical content. In the study an increase in the antioxidant activity, total phenolics, and flavonoid content was observed with an increase in soaking time, germination time, and germination temperature. The highest antioxidant activity (90.5%), total phenolics content (45.67 mg gallic acid equivalent/100 g sample), and total flavonoid content (43.9 mg rutin equivalent/g sample) were obtained at a soaking time of 15.84 hours, germination temperature of 25°C, and germination time of 40 hours. The optimized malting for proso millet is achieved on the fifth day of germination, at 44% degree of steeping and a steeping and germination temperature of 22°C (Zarnkow et al. 2012). Chandraprabha and Sharon (2021) reported soaking of 10 hours followed by 24 hours of germination as best for the malting of barnyard millet. Kumar et al. (2021) soaked the finger millet overnight and carried out its germination at $25 \pm 2°C$ and 95% relative humidity for varying (12–96 hours) time intervals. The germinated millet samples were dried at $50 \pm 2°C$ and were analyzed for physiochemical changes and their suitability for the development of value-added products. An increase in malting loss, reducing sugars, total sugars, and crude fibre and a decrease in proteins, ash, and fat content were observed with an increase in germination time. The anti-nutritional factors showed a decreasing trend in the initial hours of germination (up to 48 hours) followed by an increasing trend afterwards. The water solubility index was highest (15.3%) after 96 hours of germination, while water absorption (1.45 ml/g) and oil absorption (2.52 ml/g) were highest in the samples germinated for 48 and 60 hours, respectively. Foaming capacity (15.33%) and foam stability (72.02%) were highest after 24 hours of germination. Based on the obtained results, a germination time below 48 hours was recommended to reduce the anti-nutritional factors, while a prolonged temperature of 72 hours was recommended to prepare malt-based beverages. The impact of soaking and sprouting on antioxidant and anti-nutritional factors in little millet, foxtail millet, pearl millet, finger millet, and kodo millet was studied by Bhuvaneshwari et al. (2020). It was found in the study that a soaking period of 24 hours followed by germination for 24 hours was best for the production of nutritionally enriched products. Germination was also reported to reduce the level of antinutrients and increase antioxidant activity. Sharma et al. (2017) optimized the malting conditions for kodo millet using response surface methodology and reported a soaking time of 13.81 hours, germination temperature of 38.75°C, and germination time of 35.82 hours as optimum for the production of germinated kodo millet flour. The optimized germination conditions reduced the phytates and tannins by 25.8 and 85.4%, respectively. An increase in protein content (17.9%), dietary fibre (8.6%), minerals (8.1%), and gamma-aminobutyric acid (406%) was also observed under these germination conditions. The flour germinated under these conditions also had high antioxidant activity (91.34%).

Malted cereals are also used in many food preparations. This process has been used since ancient times for the production of beer, whisky, malt vinegar, and malted milk. It also helps in the conversion of starch to sugar molecules. Malted millets are also used for the production of high-value-added products such as instant beverages, drink mixes, malted flour, and malted enzyme-rich products. Finger millet is the most widely used millet for the malting process due to its high amylase activity compared to other millets (Birania et al. 2020). Some studies reported that the enzymes present in millets are responsible for enhancing the flavour of the product (Desai et al. 2010). Another study reported that millet malt is also used for the preparation of beverages with a blend of lukewarm water, milk, sugar, and infant foods. It also increases the digestibility and nutritional value of the food (Birania et al. 2020). However, the malting process shows the lowering activity of anti-nutritional factors (Desai et al. 2010). The malting process's main aim is to increase the nutritive value, vitamins, enzymatic activity, and flavour. It also reduces the phytate level and increases the stability of zinc and iron. The germination process is reported to increase the level of vitamins B and C (Coulibaly and Chen 2011). In another study, scientists revealed that copper, zinc, niacin, riboflavin, calcium, ascorbic acid, and magnesium content improve in germinated grains (Nkhata et al. 2018). Ohtsubo et al. (2005) reported a change in texture during the malting process. Malting is responsible for the decomposition of polymers of higher molecular weight. This process also enhances the organoleptic properties.

3.3.6 Baking

Baking is a dry heat process used to cook mixtures of flour, water, sugar, egg, leavening agents, fat, and additives to produce edible products like bread, bread rolls, puddings, doughnuts, cakes, cookies, and biscuits. The basic phenomena involved in this process are heat and mass transfer. This type of cooking is carried out in specially designed baking ovens where the heat transfers to the food by the mechanisms of convection, conduction, and radiation (Stear 1990). The baking process changes the chemical, biochemical, and physical properties of the product. The major changes are the evaporation of water, expansion of product volume, gelatinization of starch, development of porous structure, browning reaction, and development of crust. The loss of moisture leads to weight loss and the development of a hard and dry texture (Kotoki and Deka 2010). This also enhances the quality and shelf life of the product. The expansion of the product volume takes place due to the production and expansion of gases. The leavening agents (yeast/chemical agents) are responsible for the gas production and the major produced gas is carbon dioxide. The gases get trapped inside cells of the dough during fermentation and expand on heating which in turn increases the volume of the product. The porous structure is the result of the escape of gas on baking. The control of the escape rate of gas is important to obtain quality baked products as a high gas escape rate results in the production of heavy/bulky baked products with poor texture. The rate of gas retention depends on the protein/gluten network. High-protein flours are recommended for products like bread and low-protein flours are used for biscuits. The coagulation of proteins takes place above 165°F/75°C which also determines the structure of the baked goods. Inadequate baking temperatures result in improper structure and volume (Mior Zakuan Azmi et al. 2019). In gelatinization, the starch granules swell up on heating in the presence of moisture, become soft, and burst due to a pressure build-up on heating. It changes the thickness of the dough and batter and also helps in maintaining the rigid structure of the baked goods (size and shape). The gelatinization temperatures of various millets are provided in Table 3.1.

Baking also results in the loss of some aromatic compounds and the development of colour and new flavouring compounds. The new flavouring compounds are produced due to caramelization and the Maillard reaction. The high temperature of baking decreases the microbial load and enzymatic activity. The baking process also changes the nutritional value of the product by destroying the vitamins such as vitamin C and thiamine. It also improves the digestion of proteins and starches.

The various types of ovens used for baking are convection ovens, rack ovens, microwave ovens, and Dutch ovens. Other types of equipment used for baking are the bread toaster, double boiler, gas range, heat gun, blow torch/gun, etc. (Sharma et al. 2022). Recipes and the process for the development of millet-based baked products are discussed in Chapter 14.

3.3.7 Extrusion

Extrusion is a process of pushing the mixed ingredients out through small openings of different shapes under high pressure with the help of a single screw or set of screws (Gu Bon-Jae et al. 2017). The extrusion process can be divided into cold extrusion and hot extrusion. The cold extrusion process is carried out at a low temperature, high pressure, and high shear, while in hot extrusion the ingredients are subjected to high temperature, high pressure, and high shear. The various steps of extrusion processing are the mixing or preconditioning of ingredients, cooking, shaping, and post-cooking operations like cutting, drying, and packaging. The mixing of ingredients

TABLE 3.1   GELATINIZATION TEMPERATURE OF VARIOUS MILLET STARCHES

Millet type	T_o °C	T_p °C	T_c °C	References
Pearl millet	64.76	69.78	74.63	Punia et al. 2021
Foxtail millet	69.65	73.47	77.68	Kim et al. 2009
Proso millet	68.56	74.53	82.43	Singh and Adedeji 2017
Barnyard millet	64.66	69.65	74.71	Sharma et al. 2021
Finger millet	63.41	68.64	74.00	Roopa and Premavalli 2008
Kodo millet	76.60	81.63	89.60	Deshpande et al. 2015
Little millet	73.41	79.47	85.00	Shinoj et al. 2006
Teff	63.8–65.4	70.2–71.3	81.3–81.5	Bultosa and Taylor 2003
White fonio	60	72	–	Ubwa et al. 2012
Black fonio	60	74	–	Ubwa et al. 2012

Where T_o – onset temperature; T_p – peak temperature; and T_c – concentration temperature.

TABLE 3.2 EFFECT OF EXTRUSION PARAMETERS ON THE PRODUCTION AND NUTRITIONAL QUALITY OF EXTRUDED FOOD PRODUCTS

S. no.	Parameter	Effect
1.	Die diameter	Retention time decreases with an increase in die diameter
2.	Feed rate	An increase in feed rate decreases oil holding capacity and increases the expansion ratio, bulk density, product throughput rate, and steam evaporation rate
3.	Feed moisture	An increase in feed moisture increases the expansion ratio and steam expansion rate, decreases product throughput rate, and has a non-significant effect on oil holding capacity and bulk density Protein and lysine digestibility is also decreased with an increase in moisture content from 93 to 167°C
4.	Speed of screw	Digestibility increases with the increase in speed of the screw due to the denaturation of proteins under high shear force
5.	Temperature	Extrusion cooking at high temperatures reduces the protein quality compared to low-temperature extrusion due to a reduction in amino acid concentration

Source: Adelye et al. 2020; Faridah et al. 2020; Wanjala et al. 2020.

is usually done with ribbon blenders or preconditioners. This step ensures the uniformity of the blend. The thermal energy of the cooking operation converts the raw ingredients/solid powder into a melted state. The molten ingredients are forced through a die that gives the ingredients a unique shape and introduces them to atmospheric pressure. The instant drop in pressure results in a rapid expansion of ingredients that gives the extruded products their unique texture. This is followed by cutting them into desirable sizes and drying them to the optimum moisture level. The cooled products are then packaged. The various chemical changes that take place during extrusion are the gelatinization of starch, the denaturation of proteins, an increase in the solubility of dietary fibre, an increase in mineral bioavailability, and the development of flavour. It also results in enhanced lipid oxidation and degradation of vitamins (Chandraprabha et al. 2021). The various extruders used in the food industry are direct type, indirect type, and combination type. The extruders can be further classified as single screw type, double screw type, roller type, hydraulic type, and pneumatic type. The screw type extruders produce a high shear rate, while the hydraulic, pneumatic, and roller type of extruders produce a low shear rate (Singh et al. 2018; Chandraprabha et al. 2021).

The nutritional value and the quality of the extruded products are affected by feed rate, feed ratio, feed moisture, speed of the screw, the temperature of extrusion, and die diameter (Table 3.2). The various products produced using the extrusion process are pasta, snack foods, noodles, breakfast cereals, and ready-to-eat foods. Recipes for millet-based extruded products are discussed in Chapter 14.

3.3.8 Fermentation

Fermentation is the process in which complex molecules are converted into simpler ones by the action of microorganisms and their enzymes. It also reduces the anti-nutritional factors of cereal grains and produces many essential vitamins. The process of fermentation can be carried out in the presence or absence of oxygen and the major end products of fermentation are alcohol, acids, and carbon dioxide (Wang et al. 2022). Depending on the end product of the process, the fermentation is known as alcoholic fermentation, lactic acid fermentation, acetic acid fermentation, propionic acid fermentation, or butyric acid fermentation. The major fermented products prepared from millet are alcoholic beverages and lactic acid fermented products (see Chapters 13 and 14). Alcoholic fermentation is carried out by yeast. This fermentation can be divided into two major steps. In the first step, the glucose molecules are converted into pyruvic acid, and in the second step, the pyruvic acid is converted into acetaldehyde and carbon dioxide in the presence of the enzyme pyruvate decarboxylase. Further, acetaldehyde is converted into alcohol by the enzyme alcohol dehydrogenase (Birania et al. 2020). In lactic acid fermentation glucose is first converted to pyruvate and is then converted to lactate by the enzyme lactate dehydrogenase (Wang et al. 2021). The fermentation of millets has been also reported to improve their nutritional value. The fermentation of foxtail millet bran degrades its cellulose and hemicellulose content and enhances the solubility of dietary fibre by 10.9% (Chu et al. 2019). Akinola et al. (2017) reported an increase in protein (10.99–13.65%) and crude fat (1.83–3.7%) in the fermentation of millets. Nour et al. (2018) reported an increase in the protein content of pearl millet flour on fermentation. Pushparaj and Urooj (2017) reported a decrease in the phytic acid content of the pearl millet. The protein digestibility (45.75–64.38%) was also improved on fermentation (Nour et al. 2018).

3.4 CONCLUSIONS

Milling is a size-reduction process that removes the non-edible portions (hull, bran, etc.) of the grains and produces fractions (flour, middlings, semolina, etc.) that can be readily processed into secondary products. The dehulled and debranned cereal grains can be roasted, popped, and flaked for consumption without pulverization. Milling and secondary processing also help to reduce the anti-nutritional factors of the grains and improve their nutritional availability. However, the loss of some minerals and vitamins is also reported. The secondary processing operations like the roasting and popping of millets develop new flavours and increase their consumer acceptability. The flakes can be used as ready-to-eat foods or quick-eating foods. The operation of frying increases the sensory acceptability and shelf life of the product. The malting operations like soaking and germination increase enzymatic activity, reduce anti-nutritional factors, and increase the overall nutritional value. The millet malt can be used for the production of specialized food products like malt drinks, malt foods, weaning foods, etc. Baking is the most common cooking method for the preparation of bread, biscuits, *chappatis*, cookies, etc. The fermentation of millets increases the protein content and protein digestibility. Fermentation can also be used to produce alcoholic beverages and lactic acid-fermented products.

REFERENCES

Adeleye, O. O., S. T. Awodiran, A. O. Ajayi, and T. F. Ogunmoyela. 2020. Effect of high-temperature, short-time cooking conditions on in vitro protein digestibility, enzyme inhibitor activity and amino acid profile of selected legume grains. *Heliyon* 6(11):e05419.

Adebiyi, J. A., A. O. Obadina, A. F. Mulaba-Bafubiandi, O. A. Adebo, and E. Kayitesi. 2016. Effect of fermentation and malting on the microstructure and selected physicochemical properties of pearl millet (Pennisetum glaucum) flour and biscuit. *Journal of Cereal Science* 70:132–139.

Ahmed, I. A. M., N. Uslu, F. Al Juhaimi, M. M. Özcan, M. A. Osman, H. A. Alqah, E. E. Babiker, and K. Ghafoor. 2021. Effect of roasting treatments on total phenol, antioxidant activity, fatty acid compositions, and phenolic compounds of teff grains. *Cereal Chemistry* 98(5):1027–1037.

Akinola, S. A., A. A. Badejo, O. F. Osundahunsi, and M. O. Edema. 2017. Effect of preprocessing techniques on pearl millet flour and changes in technological properties. *International Journal of Food Science and Technology* 52(4):992–999.

Alfin, F. 2020. Wheat conditioning and tempering. *Miller: World Milling and Pulses Technologies Referred Magazine* 14:58–60.

Amadou, I., M. E. Gounga, Y. H. Shi, and G. W. Le. 2014. Fermentation and heat-moisture treatment induced changes on the physicochemical properties of foxtail millet (Setaria italica) flour. *Food and Bioproducts Processing* 92(1):38–45.

Anbalagan, S., and P. Nazni. 2020. Effect of processing techniques on nutritional, viscosity and osmolarity of barnyard millet based diarrheal replacement fluids. *Current Research in Nutrition and Food Science* 8(1): 164–173.

Asghari, I., S. M. Mousavi, F. Amiri, and S. Tavassoli. 2013. Bioleaching of spent refinery catalysts: A review. *Journal of Industrial and Engineering Chemistry* 19(4):1069–1081.

Ballogou, V. Y., F. S. Sagbo, M. M. Soumanou, J. T. Manful, F. Toukourou, and J. D. Hounhouigan. 2015. Effect of processing method on physico-chemical and functional properties of two fonio (Digitaria exilis) landraces. *Journal of Food Science and Technology* 52(3):1570–1577.

Baranwal, D., and A. Sankhla. 2019. Physical and functional properties of malted composite flour for biscuit production. *Journal of Pharmacognosy and Phytochemistry* 8(2):959–965.

Bhawna, S., S. D. Kulkarni, and S. Patel. 2009. Effect of premilling treatments on colour analysis of kodo (Paspalum scrobiculatum L.) millet. *International Journal of Agricultural Engineering* 2(1):153–156.

Beukes, M. 2021. *Effect of Stone and Roller Milling on Physicochemical, Functional and Structural Properties of Sifted Wheat Flour* (Doctoral dissertation. Stellenbosch: Stellenbosch University).

Bhavana, J., R. Krishnan, P. M. Arthanari, R. Kalpana, C. N. Chandrasekhar, and R. Ravikesavan. 2023. Screening of herbicides in transplanted finger millet [Eleusine coracana (L.) gaertn]. *Indian Journal of Agricultural Research* 57(2):194–197.

Bhuvaneshwari, G., A. Nirmalakumari, and S. Kalaiselvi. 2020. Impact of soaking, sprouting on antioxidant and anti-nutritional factors in millet grains. *Journal of Phytology* 12:62–66.

Birania, S., P. Rohilla, R. Kumar, and N. Kumar. 2020. Post harvest processing of millets: A review on value added products. *International Journal of Chemical Studies* 8(1):1824–1829.

Bultosa, G., and J. R. Taylor. 2003. Chemical and physical characterisation of grain tef [Eragrostis tef (Zucc.) Trotter] starch granule composition. *Starch-Stärke* 55(7):304–312.

Chandraprabha, S., and C. L. Sharon. 2021. Optimisation of conditions for barnyard millet germination. *Plant Archives* 21(1):1676–1680.

Sarita, E. S. C. 2018. Effects of processing (germination and popping) on the nutritional and anti-nutritional properties of finger millet (Eleusine coracana). *Current Research in Nutrition and Food Science Journal* 6(2):566–572.

Chauhan, S. S., S. K. Jha, G. K. Jha, D. K. Sharma, T. Satyavathi, and J. Kumari. 2015. Germplasm screening of pearl millet (Pennisetum glaucum) for popping characteristics. *Indian Journal of Agricultural Sciences* 85(3):344–348.

Choe, E., and D. B. Min. 2007. Chemistry of deep-fat frying oils. *Journal of Food Science* 72(5):R77–R86.

Choudhury, M., P. Das, and B. Baroova. 2011. Nutritional evaluation of popped and malted indigenous millet of Assam. *Journal of Food Science and Technology* 48:706–711.

Chu, J., H. Zhao, Z. Lu, F. Lu, X. Bie, and C. Zhang. 2019. Improved physicochemical and functional properties of dietary fiber from millet bran fermented by Bacillus natto. *Food Chemistry* 294:79–86.

Coulibaly, A., and J. Chen. 2011. Evolution of energetic compounds, antioxidant capacity, some vitamins and minerals, phytase and amylase activity during the germination of foxtail millet. *American Journal of Food Technology* 6(1):40–51.

Das, S., R. Khound, M. Santra, and D. K. Santra. 2019. Beyond bird feed: Proso millet for human health and environment. *Agriculture* 9(3):64.

Delost-Lewis, K., K. Lorenz, and R. Tribelhorn. 1992. Puffing quality of experimental varieties of proso millets (Panicum miliaceum). *Cereal Chemistry* 69(4):359–365.

Deshpande, S. S., D. Mohapatra, M. K. Tripathi, and R. H. Sadvatha. 2015. Kodo millet-nutritional value and utilization in Indian foods. *Journal of Grain Processing and Storage* 2(2):16–23.

Desai, A. D., S. S. Kulkarni, A. K. Sahoo, R. C. Ranveer, and P. B. Dandge. 2010. Effect of supplementation of malted ragi flour on the nutritional and sensorial quality characteristics of cake. *Advance Journal of Food Science and Technology* 2(1):67–71.

Devisetti, R., S. N. Yadahally, and S. Bhattacharya. 2014. Nutrients and antinutrients in foxtail and proso millet milled fractions: Evaluation of their flour functionality. *LWT-Food Science and Technology* 59(2):889–895.

Dharmaraj, U., R. Ravi, and N. G. Malleshi, 2013. Optimization of process parameters for decortication of finger millet through response surface methodology. *Food and Bioprocess Technology* 6:207–216.

Dogan Halkman, H. B., P. K. Yücel, and A. K. Halkman. 2014. Non-thermal processing. Microwave. In: *Encyclopedia of Food Microbiology*, eds. C. A. Batt, and M. L. Tortorello. Academic Press, 962–965.

Embashu, W., and K. M. N. Komeine. 2019. Malts: Quality and phenolic content of pearl millet and sorghum varieties for brewing nonalcoholic beverages and opaque beers. *Cereal Chemistry* 96(4):765–774.

Faridah, H. S., Y. M. Goh, M. M. Noordin, and J. B. Liang. 2020. Extrusion enhances apparent metabolizable energy, ileal protein and amino acid digestibility of palm kernel cake in broilers. *Asian-Australasian Journal of Animal Sciences* 33(12):1965–1974.

Farahnaky A., M. Aliopour, and M. Mazoobi. 2013. Popping properties of corn grains of two different varieties at different moistures. *Journal of Agricultural Science and Technology* 15: 771–780.

Ferreira, L. C., A. Donoso-Bravo, P. J. Nilsen, F. Fdz-Polanco, and S. I. Pérez-Elvira. 2013. Influence of thermal pretreatment on the biochemical methane potential of wheat straw. *Bioresource Technology* 143:251–257.

García-Granero, J. J., C. Lancelotti, and M. Madella. 2017. A methodological approach to the study of microbotanical remains from grinding stones: A case study in northern Gujarat (India). *Vegetation History and Archaeobotany* 26(1):43–57.

Garud, S. R., A. G. Lamdande, H. A. Tavanandi, N. K. Mohite, and U. Nidoni. 2022. Effect of physicochemical properties on popping characteristics of selected pearl millet varieties. *Journal of the Science of Food and Agriculture* 102(15):7370–7378.

Gowda A., D. Pange, and S. Shewalkar. 2020. Processing of cereals, pulses, and millets, the stages and more. FnBnews.com. http://www.fnbnews.com/FB-Specials/processing-of-cereals-pulses-and-millets-the-stages-and-more-62597. (accessed October 25, 2022).

Gowda, N. N., K. Siliveru, P. V. Prasad, Y. Bhatt, B. P. Netravati, and C. Gurikar. 2022. Modern processing of Indian millets: A perspective on changes in nutritional properties. *Foods* 11(4):499.

Bon-Jae, G., R. J. Kowalski, and G. M. Ganjyal. 2017. Food extrusion processing: An overview. Fact sheet (Washington State University. Extension), Washington State University Extension; Pullman, Washington. https://hdl.handle.net/2376/12875. (accessed October 23, 2022).

Hu, X., J. Wang, P. Lu, and H. Zhang. 2009. Assessment of genetic diversity in broomcorn millet (Panicum miliaceum L.) using SSR markers. *Journal of Genetics and Genomics* 36(8):491–500.

Jaybhaye, R. V., D. N. Kshirsagar, and P. P. Srivastav. 2011. Development of barnyard millet puffed product using hot air puffing and optimization of process parameters. *International Journal of Food Engineering* 1:1–25.

Jiménez-Monreal, A. M., L. García-Diz, M. Martínez-Tomé, M. M. M. A. Mariscal, and M. A. Murcia. 2009. Influence of cooking methods on antioxidant activity of vegetables. *Journal of Food Science* 74(3):H97–H103.

Kalam Azad, M. O., D. I. Jeong, M. Adnan, T. Salitxay, J. W. Heo, M. T. Naznin, J. D. Lim, D. H. Cho, B. J. Park, and C. H. Park. 2019. Effect of different processing methods on the accumulation of the phenolic compounds and antioxidant profile of broomcorn millet (Panicum miliaceum L.) flour. *Foods* 8:230.

Kapoor, P. 2013. *Nutritional and Functional Properties of Popped Little Millet (Panicum Sumatrense)*. (Ph.D. dissertation. Quebec, Canada: Dept. Bioresource. Eng., McGill University Libraries).

Karakannavar, S. J., G. Nayak, and J. S. Hilli. 2021. Effect of different levels of polishing on physico-chemical characteristics of barnyard millet (Echinochloa frumentacea). *The Pharma Innovation Journal* 10:50–54.

Kate, A., and A. Singh. 2021. Processing technology for value addition in millets. In: *Millets and Millet Technology,* eds. A. Kumar, M. K. Tripathi, D. Joshi, V. Kumar, 239–254. Singapore: Springer.

Kim, S. K., E. Y. Sohn, and I. J. Lee. 2009. Starch properties of native foxtail millet, Setaria italica Beauv. *Journal of Crop Science and Biotechnology* 12(1):59–62.

Kotoki, D., and S. C. Deka. 2010. Baking loss of bread with special emphasis on increasing water holding capacity. *Journal of Food Science and Technology* 47:128–131.

Kulkarni, V. N., V. N. Gaitonde, V. Hadimani, and V. Aiholi. 2018. Analysis of wire EDM process parameters in machining of NiTi superelastic alloy. *Materials Today: Proceedings* 5(9):19303–19312.

Kumar, M., K. Rani, B. C. Ajay, M. S. Patel, K. D. Mungra, and M. P. Patel. 2020. Multivariate diversity analysis for grain micronutrients concentration, yield and agro-morphological traits in pearl millet (Pennisetum glaucum (L) R. Br.). *International Journal of Current Microbiology and Applied Sciences* 9(3):2209–2226.

Kumar, A., A. Kaur, K. Gupta, Y. Gat, and V. Kumar. 2021. Assessment of germination time of finger millet for value addition in functional foods. *Current Science* 120(2):406–413.

Lorusso, A., M. Verni, M. Montemurro, R. Coda, M. Gobbetti, and C. G. Rizzello. 2017. Use of fermented quinoa flour for pasta making and evaluation of the technological and nutritional features. *LWT - Food Science and Technology* 78:215–221.

Malavika, M., S. Shobana, P. Vijayalakshmi, R. Ganeshjeevan, R. Gayathri, V. Kavitha, N. Gayathri, R. Savitha, R. Unnikrishnan, R. M. Anjana, N. G. Malleshi. 2020. Assessment of quality of minor millets available in the south Indian market & glycaemic index of cooked unpolished little & foxtail millet. *The Indian Journal of Medical Research* 152(4):401.

Malik, S., K. Krishnaswamy, and A. Mustapha. 2021. Physical properties of complementary food powder obtained from upcycling of Greek yogurt acid whey with kodo and proso millets. *Journal of Food Process Engineering* 44(11):13878.

Malleshi, N. G., and H. S. R. Desikachar. 1986. Studies on comparative malting characteristics of some tropical cereals and millets. *Journal of the Institute of Brewing* 92(2):174–176.

Malleshi, N. G., P. V. Reddy, and C. F. Klopfenstein. 2004. Milling trials of sorghum, pearl millet and finger millet in quadrumat junior mill and experimental roll stands and the nutrient composition of milling fractions. *Journal of Food Science and Technology* 41(6):618–622.

Microtek Engineering Group. 2022. Corn milling machine. https://microtecco.com/products/corn-millingmachine/#:~:text=For%20these%20 two%20milling%2C%20two,of%20germ%20from%20corn%20kernel. (accessed 09 December, 2022).

Mior Zakuan Azmi, M., F. S. Taip, S. M. M. Kamal, and N. L. Chin. 2019. Effects of temperature and time on the physical characteristics of moist cakes baked in air fryer. *Journal of Food Science and Technology* 56:4616–4624.

Mirza, N., N. Sharma, S. Srivastava, and A. Kumar. 2015. Variation in popping quality related to physical, biochemical and nutritional properties of finger millet genotypes. *Proceedings of the National Academy of Sciences, India Section B: (Biological Sciences)* 85(2):507–515.

Mishra, G., D. C. Joshi, and B. K. Panda. 2014. Popping and puffing of cereal grains: A review. *Journal of Grain Processing and Storage* 1(2):34–46.

Mishra, G., D. C. Joshi, D. Mohapatra, and V. B. Babu. 2015. Varietal influence on the microwave popping characteristics of sorghum. *Journal of Cereal Science* 65:19–24.

Mohapatra, M., and S. K. Das. 2011. Mechanizing the conditioning process of rice before puffing. *ORYZA-An International Journal on Rice* 48(2):114–118.

Nkhata, S. G., E. Ayua, E. H. Kamau, and J. B. Shingiro. 2018. Fermentation and germination improve nutritional value of cereals and legumes through activation of endogenous enzymes. *Food Science and Nutrition* 6(8):2446–2458.

Obadina, A. O., C. A. Arogbokun, A. O. Soares, C. W. P. de Carvalho, H. T. Barboza, and I. O. Adekoya. 2017. Changes in nutritional and physico-chemical properties of pearl millet (Pennisetum glaucum) ex-Borno variety flour as a result of malting. *Journal of Food Science and Technology* 54(13):4442–4451.

Ohtsubo, K. I., K. Suzuki, Y. Yasui, and T. Kasumi. 2005. Bio-functional components in the processed pre-germinated brown rice by a twin-screw extruder. *Journal of Food Composition and Analysis* 18(4):303–316.

Oke, E. K., M. A. Idowu, O. P. Sobukola, S. A. O. Adeyeye, and A. O. Akinsola. 2018. Frying of food: A critical review. *Journal of Culinary Science and Technology* 16(2):107–127.

Patel, A., P. Parihar, and K. Dhumketi. 2018. Nutritional evaluation of kodo millet and puffed kodo. *International Journal of Chemical Studies* 6(2):1639–1642.

Perdon, A. A., S. L. Schonauer, and K. Poutanen. 2020. *Breakfast Cereals and How They Are Made: Raw Materials, Processing, and Production* 459 p. Duxford, United Kingdom: Woodhead Publishing and AACC International Press.

Powar, R. V., V. V. Aware, and P. U. Shahare. 2019. Optimizing operational parameters of finger millet threshing drum using RSM. *Journal of Food Science and Technology* 56(7):3481–3491.

Pradeep, S. R., and M. Guha. 2011. Effect of processing methods on the nutraceutical and antioxidant properties of little millet (Panicum sumatrense) extracts. *Food Chemistry* 126(4):1643–1647.

Punia, S., M. Kumar, A. K. Siroha, J. F. Kennedy, S. B. Dhull, and W. S. Whiteside. 2021. Pearl millet grain as an emerging source of starch: A review on its structure, physicochemical properties, functionalization, and industrial applications. *Carbohydrate Polymers* 260:117776.

Pushparaj, S. F., and Urooj, A. 2014. Antioxidant activity in two pearl millet (Pennisetum typhoideum) cultivars as influenced by processing. *Antioxidants* 3(1):55–66.

Ranganathan, V., I. T. Nunjundiah, and S. Bhattacharya. 2014. Effect of roasting on rheological and functional properties of sorghum flour. *Food Science and Technology International* 20(8):579–589.

Ranganna, B., K. G. Ramya, B. Kalpana, and P. Arunkumar. 2012. Nutri products from popped kodo and proso millet grains. *Mysore Journal of Agricultural Sciences* 46(3):500–504.

Roopa, S., and K. S. Premavalli. 2008. Effect of processing on starch fractions in different varieties of finger millet. *Food Chemistry* 106(3):875–882.

Rosentrater, K. A., and A. D. Evers. 2018. Wet milling. *Kent's Technology of Cereals*. Elsevier, 839–860.

Saleh, A. S., Q. Zhang, J. Chen, and Q. Shen. 2013. Millet grains: Nutritional quality, processing, and potential health benefits. *Comprehensive Reviews in Food Science and Food Safety* 12(3):281–295.

Samtiya, M., R. E. Aluko, and T. Dhewa. 2020. Plant food anti-nutritional factors and their reduction strategies: An overview. *Food Production, Processing and Nutrition* 2:1–14.

Seo, K. H., J. E. Ra, S. J. Lee, J. H. Lee, S. R. Kim, J. H. Lee, and W. D. Seo. 2015. Anti-hyperglycemic activity of polyphenols isolated from barnyard millet (Echinochloa utilis L.) and their role inhibiting α-glucosidase. *Journal of the Korean Society for Applied Biological Chemistry* 58(4):571–579.

Selladurai, M., M. K. Pulivarthi, A. S. Raj, M. Iftikhar, P. V. Prasad, and K. Siliveru. 2022. Considerations for gluten free foods-pearl and finger millet processing and market demand. *Grain & Oil Science and Technology* 6(2):59–70.

Shaaya, E., R. Mailer, M. Kostyukovsky, and L. Mailer. 2005. Improving the control of insects in food processing. In: *Handbook of Hygiene Control in the Food Industry*, eds. H. Lelieveld, J. Holah, and D. Gabric, 407–424. Dusford, United Kingdom: Woodhead Publishing.

Sharma, P. C., D. Singh, D. Sehgal, G. Singh, C. T. Hash, and R. S. Yadav. 2014. Further evidence that a terminal drought tolerance QTL of pearl millet is associated with reduced salt uptake. *Environmental and Experimental Botany* 102(100):48–57.

Sharma, S., D. C. Saxena, and C. S. Riar. 2015. Antioxidant activity, total phenolics, flavonoids and antinutritional characteristics of germinated foxtail millet (Setaria italica). *Cogent Food and Agriculture* 1(1):1081728.

Sharma, S., D. C. Saxena, and C. S. Riar. 2017. Using combined optimization, GC–MS and analytical technique to analyze the germination effect on phenolics, dietary fibers, minerals and GABA contents of Kodo millet (Paspalum scrobiculatum). *Food Chemistry* 233:20–28.

Sharma, R., S. Sharma, B. N. Dar, and B. Singh. 2021. Millets as potential nutri-cereals: A review of nutrient composition, phytochemical profile and techno-functionality. *International Journal of Food Science and Technology* 56(8):3703–3718.

Shinoda, K., T. Takahashi, Z. Jin, M. Miura, and S. Kobayashi. 2002. Processing properties of modified proso millet flour, 1: Effect of roasting and heat-moisture treatments on physical properties of proso millet flour. *Journal of the Japanese Society for Food Science and Technology (Japan)* 49(7):491–497.

Shinoj, S., R. Viswanathan, M. S. Sajeev, and S. N. Moorthy. 2006. Gelatinisation and rheological characteristics of minor millet flours. *Biosystems Engineering* 95(1):51–59.

Shobana, S., and N. G. Malleshi. 2007. Preparation and functional properties of decorticated finger millet (Eleusine coracana). *Journal of Food Engineering* 79(2):529–538.

Singh, M., and A. A. Adedeji. 2017. Characterization of hydrothermal and acid modified proso millet starch. *LWT - Food Science and Technology* 79:21–26.

Singh, N., D. John, D. K. Thompkinson, B. S. Seelam, H. Rajput, and S. Morya. 2018. Effect of roasting on functional and phytochemical constituents of finger millet (Eleusine coracana L.). *The Pharma Innovation Journal* 7(4):414–418.

Skovgaard, M., A. Ahniyaz, B. F. Sørensen, K. Almdal, and A. V. Lelieveld. 2010. Effect of microscale shear stresses on the martensitic phase transformation of nanocrystalline tetragonal zirconia powders. *Journal of the European Ceramic Society* 30(13):2749–2755.

Stear, C. A. 1990. Formulation and processing techniques for specialty-breads. In: *Handbook of BreadmakingTtechnology*, ed. C. A. Stear. Springer US, 522–535.

Sudha, K. V., S. J. Karakannavar, N. B. Yenagi, and B. Inamdar. 2021. Effect of roasting on the physicochemical and nutritional properties of foxtail millet (Setaria italica) and bengal gram dhal flours. *The Pharma Innovation Journal* 10(5):1543–1547.

Taylor, J. R. N., and J. Taylor. 2017. Proteins from sorghum and millets. In *Sustainable Protein Sources*, eds. S. R. Nadathur, J. P. D. Wanasundara, and L. Scanlin, 79-104. Cambridge, United States: Academic Press..

Ubwa, S. T., J. Abah, K. Asemave, and T. Shambe. 2012. Studies on the gelatinization temperature of some cereal starches. *International Journal of Chemistry* 4(6):22–28.

Udawat, A. S., and P. Singh. 2022. An automated detection of atrial fibrillation from singlelead ECG using HRV features and machine learning. *Journal of Electrocardiology* 75:70–81.

Von, B. R. 2003. Genetic diversity for quantitatively inherited agronomic and malting quality traits. *Diversity in Barley (Hordeum vulgare)* 7:201.

Wang, H., Y. Fu, Q. Zhao, D. Hou, X. Yang, S. Bai, X. Diao, Y. Xue, and Q. Shen. 2022. Effect of different processing methods on the millet polyphenols and their anti-diabetic potential. *Frontiers in Nutrition* 9:101.

Wang, Y., J. Wu, M. Lv, Z. Shao, M. Hungwe, J. Wang, and W. B. Geng. 2021. Metabolism characteristics of lactic acid bacteria and the expanding applications in food industry. *Frontiers in Bioengineering and Biotechnology* 9:612285.

Wanjala, W. N., O. Mary, and M. M. Symon. 2020. Influence of feed rate, moisture and mixture composition from composites containing rice (Oryza sativa), Sorghum [Sorghum bicolor (L.) Moench] and Bamboo (Yushania alpine) shoots on physical properties of extruded flour and mass transfer. *Food and Nutrition Sciences* 11(8):807–823.

Zarnkow, M. 2014. Fermented foods/beverages from sorghum and millet. In: *Encyclopedia of Food Microbiology*, Second Edition, eds. C. A. Batt, R. K. Robinson, and C. A. Batt. Elsevier, 839–845.

Zarnkow, M., M. Keßler, F. Burberg, W. Back, E. K. Arendt, and S. Kreisz. 2007. The use of response surface methodology to optimise malting conditions of proso millet (Panicum miliaceum L.) as a raw material for gluten-free foods. *Journal of the Institute of Brewing* 113(3):280–292.

Chapter 4
Machines for Processing Millets

4.1 INTRODUCTION

Primary processing is the conversion of cereal crops into a form that can be readily used for the development of an edible product. The major operations involved in the primary processing of millets are harvesting, threshing, cleaning, dehulling, pearling, sorting, grading, pulverization, etc. In secondary processing, the products obtained from the primary processing are processed into edible products. The various secondary processing operations for millets are baking, cooking, extrusion, fermentation, flaking, frying, malting, puffing, etc. A combination of two or more of these operations can also be used for the production of a single product. The history of the tools used for millet processing is as old as their cultivation. Traditionally, the harvesting of millets was done manually with sickles, and the threshing was done by beating the millets with sticks or by the repetitive walking of oxen. The grains obtained from threshing were cleaned by winnowing and further the husk was removed by pounding. The pounded grains were again winnowed and then milled with the help of hand-operated stone mills, water-operated mills, or oxen-driven mills. The traditional processing of millets was labour-intensive, time-consuming, and involved a lot of drudgeries. Nowadays, several machines have been developed to ease these operations. Mechanization has speeded up the processing of grains and the quality of grains is also improved. The machines also enable us to mill the millet grains in various shapes and sizes, and hence a variety of food products can be produced from them. The use of extrusion, puffing, and flaking machines in the secondary processing of millets has also enabled us to produce products that possess a different taste, texture, and flavour compared to the products produced using traditional methods. In this chapter, we discuss both traditional tools and modern machines involved in the primary processing of millets. The chapter also highlights the major research work carried out on the development of millet processing machines in recent years.

4.2 TRADITIONAL EQUIPMENT

Humans are among the most curious species on the earth; they have developed a rational thinking ability over the years and have made many inventions to make their lives easier. Many methods, processes, and machines have been developed by them to ease their work since prehistoric times. In this section, we discuss some of the instruments, equipment, and machines that were used for the cleaning and milling of grains in the pre-electric period. The harvesting of cereal crops was carried out manually with sickles (Figure 2.1a, Chapter 2). The threshing was carried out by spreading the loosened sheaves on hard ground and beating them with sticks or oxen, horses, and other animals were allowed to walk on them repeatedly to tread out the grain (Figure 2.1b, c, Chapter 2). The cleaning of the grains was carried out by winnowing. Winnowing is the process of removing light particles like chaff from heavier grains with the help of air or wind (Figure 2.1d, Chapter 2). The left-over impurities like stone and residual husk were removed by using the "*chazz*" or "*supa*". The *chazz* was traditionally made from bamboo and could be a circular or curved trapezoid structure (Figure 4.1a) with a broad open side and a narrow side closed with a small raised wall of a few centimetres. The *chazz* is moved to and fro manually which causes a vibratory effect and results in the separation of particles based on density. Milling in the pre-electric era was carried out using stone- or wood-based pestles and mortar or stone mills. These were operated either by humans, animals, or the power of water or wind. The pestle and mortar are manually operated instruments that help to convert the solid grains into a fine flour (McDonough et al. 2000). These structures can be made up of stone, wood, metal, or a combination of all of these. The degree of grinding and fineness of the flour depend on the impact and

Figure 4.1 Traditional equipment used for the processing of millets. (A) *Chazz/supa*, (B and C) *khori and battu*, (D and E) stone-based *okhal and moosal* for wet milling, (F) stone *okhal* and wood *moosal* for bulk grinding, (G and H) stone mill (*Chakki*), (I and J) "*gharat*". (Pictures courtesy of Ashwani Kumar, Sarita Thakur, Kiran Bhunal, Himachal Pradesh).

abrasion forces. Pestles and mortars of many designs and capacities were available in the past and many are still in use. These structures can be used for both the dry and wet milling of grains. The dry milling of grains was carried out in wood- or stone-based *khori-battu* or *okhal-moosal*. In the "*khori* and *battu*" system (Figure 4.1b, c) the *khori* is a stoned-based structure with a hole carved into it that contains the grains. It is dug into the ground leaving the hole open. The grains to be milled are filled into this hole and are beaten with a *battu* made of stone and attached to a wooden handle to operate it. *Okhal-moosal* made of stone, wood, or a combination of stone and wood is available (Figure 4.1d, e, f). The stone-based *okhal-moosal* is generally used for the wet milling of grains. The grains to be milled are filled into the *okhal* (a round structure made of wood, stone, or metal) and are ground using *moosal*. The stone-based *okhal* and *moosal* shown in Figure 4.1d are operated by two persons sitting adjacent to each other. The stone and wood-based *okhal-moosal* can be operated by a single person. The grains to be milled are filled into the *okhal* and beaten with the *moosal* (a long wooden rod/bar) (Figure 4.1f). A typical stone mill (*chakki*) consists of two stones in a horizontal position (Figure 4.1h), where the bottom stone is stationary and the upper stone rotates. The upper stone can be moved manually or using the force generated from the flow of the water stream, or a windmill, or it can be driven by animals. The basic principle of the stone mills is rotational force, impact force, and abrasion forces. In India, the typical water-operated mill is known as "*Gharat*" (Kashyap et al. 2023) (Figure 4.1i, j). In this, the water is allowed to pass through a narrow drain and is allowed to fall with pressure on a wooden wheel from a certain height. The pressure of the water causes the wheel to rotate. This wheel is attached to the upper stone disc of the mill through a shaft and this causes it to rotate. This results in the milling of the grains. Similar animal- or wind-operated mills are also used in other parts of the world.

4.3 MACHINES USED IN PRIMARY PROCESSING OF MILLETS

The various primary processing operations are harvesting, threshing, cleaning, dehulling, pearling, sorting, grading, pulverization, etc. These operations can be achieved by employing machines such as harvesters, threshers, cleaners, dehullers, sorters, graders, pearlers, pulverizers, etc. This section discusses in detail the functioning of these machines. Details on various machines used in the processing of millets and information on their availability in the Indian market are also provided in Table 4.1.

4.3.1 Harvesters

Windrowers as well as combines can be used for the harvesting of millets depending on millet type and maturity level. The windrower is a self-propelled or tractor-drawn harvester that lays the stalks in windrows for later threshing and cleaning. The windrower consists of a cutter bar that is driven by an engine or a shaft that derives its power from a tractor, a reel to sweep the grain onto a platform, and a canvas conveyor to carry it to one side and deposit it in a windrow for drying. The crop harvested with the windrower needs to be threshed separately, however; the windrowers are more helpful when the moisture content of grains is high (Britannica 2020). Millet-specific harvesters have also been developed. Nisha et al. (2022) developed a tractor-mounted harvester for finger millet. The harvesting using this machine can be divided into two stages, i.e. an upper cutting unit that harvests the ear heads at a height of 460 mm, and a lower cutting unit that cuts 100 mm above the ground. In this machine the minimum header

TABLE 4.1 VARIOUS MILLET-PROCESSING MACHINES AVAILABLE IN THE MARKET

	Machine	Purpose	Power/travel speed	Productivity	Supplier name	Approximate price	Link for the information
Millet harvesters	Windrowers	Cuts grain and lays the stalks in windrows for later threshing and cleaning	15 km/h	2.4–3.9 ha/h	Gomselmash India Private Limited	Rs. 12.50 lakh INR	https://www.indiamart.com/proddetail/combine-harvester-windrower-zhvz-5-for-grain-20605573430.html
	Combines	Cuts, threshes, and cleans the grains	102 hp	1.30 ha/h	Jiangsu World Agriculture Machinery Co., Ltd	Rs. 15 lakh INR	https://jiangsuworld.en.made-in-china.com/product/ReIjMCDZYHW/China-Ruilong-4lz-6-0p-Millet-Combine-Harvester-with-Good-Quality.html
Threshers	Drummy thresher	Separates grains from the harvested crop and provides clean grains	15 hp	1500–2000 kg/hr	Unnati Threshers	Rs. 6.05 lakh INR	https://www.indiamart.com/proddetail/ground-nut-thresher-12578749130.html
	Axial flow thresher	Separates the seeds from the stalks and husks	5.45 kW	1500 kg/hr	Bharat Industries	Rs. 1.60 lakh INR	https://www.indiamart.com/proddetail/axial-flow-paddy-thresher-4900691062.html
	Regular-type thresher	Separates the seeds from the stalks and husks	4kW	200–500 kg/hr	Landforce	Rs. 1.88 lakh INR	https://dir.indiamart.com/impcat/thresher.html
	High impact thresher	Separates the grains from stalks and husks	5–7 hp	1000–1500 kg/hr	Amar Agricultural Implements Works	Rs. 10 lakh INR	https://www.indiamart.com/proddetail/amarshakti-thresher-multicrop-high-capacity-thresher-balwan-6372950812.html?pos=1&kwd=high%20impact%20thresher&tags=A|||8190.9683||product|||SS|c|type=attr=1|attrS
Cleaning	Semi-automatic millet cleaner	Removes stones and sand	2–3 kW	250 kg/hr	Mr. S. Nallasivam	Rs. 1.05 lakh INR	https://www.nndestoner.com/millet-cleaning-machine-6539485.html
	Air screen cleaner	Removes the lighter impurities	380 V	1000–1200 kg/hr	V-Tech	Rs. 16,500 INR	https://www.indiamart.com/proddetail/seed-grader-air-screen-cleaner-type-19736241962.html

(Continued)

Machines for Processing Millets

TABLE 4.1 (CONTINUED) VARIOUS MILLET-PROCESSING MACHINES AVAILABLE IN THE MARKET

	Machine	Purpose	Power/travel speed	Productivity	Supplier name	Approximate price	Link for the information
	Vertical cleaning machine	Removal of impurities such as weeds, dirt, and stalks from the grains	380 V	1000–1200 kg/hr	NN Engineering products	Rs. 1.05 lakh INR	https://www.indiamart.com/proddetail/nn-millet-cleaning-machinery-22569329148.html
	Aspirator	Removes lightweight impurities like dusk, husk, and leaves	4.75 kW	50–100 kg/hr	Mr. Udhay Gopal	Rs. 1.25 lakh INR	https://www.tradeindia.com/products/destoner-cum-grader-cum-aspirator-for-millet-c5261716.html
Separators/sorters	Specific gravity separators	Separates products of the same size but with a difference in specific weight	380V	300–500 kg/hr	Goldin Engineering Co.	Rs. 9 lakh INR	https://www.indiamart.com/proddetail/gravity-separators-vacuum-3492441073.html
	Magnetic separators	Removes ferrous tramp metal contaminants from dry particulate, liquids, and slurries in the process of grain, feed	1.5–11 kW	10–280 ton/hr	Magna tronix	Rs. 1.44 lakh INR	https://www.indiamart.com/proddetail/industrial-magnetic-separator-14603695230.html
	Color separators	Sorting of foreign grains	3.5 kW	370 ton/hr	Promech Industries Private Limited	Rs. 15 lakh INR	https://www.indiamart.com/proddetail/khas-khas-beej-color-sorting-machine-t20-3-chute-23832078333.html
	Spiral separators	Removal of weeds from grains	5–6 kW	250 ton/hr	Star trace Private Limited	Rs. 1.10 lakh INR	https://www.indiamart.com/proddetail/spiral-separators-2411440030.html
	Dehusker	Removes the husk from the grains without damaging it	1 hp	100 kg/hr	Thomas International	Rs. 65,101 INR	https://www.indiamart.com/proddetail/sunflower-seed-sheller-de-husker-machine-11891150155.html?pos=1&kwd=dehusker&tags=A||||8132.985|Price|proxy|||SSlc
	Scalper	Remove large impurities	100 kW	6–10 tonnes/day	Mr. S.B. Singh	Rs. 40 lakh INR	https://www.srishtechmetalcasting.com/scalping-machine.html
	Screens	Separates and transports granulated materials	1.5 kW	3–5 tonnes/day	PK Machinery	Rs. 1 lakh INR	https://www.pkmachinery.com/vibrating-screen/rotary-vibrating-screen.html

loss (1.25%) was achieved at 30 rpm rotational speed of the reel, 40 cm mounting height of the reel, and 2.68 km/h forward speed. A portable finger millet harvesting machine that can cut ear heads and straw separately was also designed by Ghimire et al. (2019). The machine had a good harvesting efficiency at a speed of 1800 rpm and power of 0.9 kW.

Combine harvesters are gigantic multi-operation machines that cut, thresh, and clean the grains. The crops to be harvested are gathered at the front by a hydraulic powered header and then a reel (a slow rotating wheel with horizontal bars and vertical teeth to grip the plants) pushes the crop towards the cutter. The cutter bar is present underneath the reel and covers the entire length of the header. The cut crops are fed inside the main part of the combine by the spinning augers where a threshing drum beats the cut crops to separate the grains from stalks. The separated grains pass through a sieve and are collected in a tank below. The crop residue, i.e. chaffs and stalks, is transported through the conveyors called straw walkers to the back of the machine where it is discharged into the field (Woodford 2021). Necessary arrangements in the cylinder speed, chaffer height, sieve dimensions, aspirator speed, etc., need to be maintained. For pearl millet harvesting a cylinder speed of 700 rpm, concave of 5/16 inch, chaffer at 1/2 inch, sieve at 3/16 inch, and cleaning fan speed of 700 rpm are recommended (Lee et al. 2012).

4.3.2 Threshers

Threshing is the process of detaching the kernels from the plant, ear heads, panicles, pods, and cobs and threshers are the machines that are used to perform this operation. The basic principle followed in this machine is rubbing, impact, and wearing force to separate the grains from the stalks. This machine consists of different components such as a machine holder, motor, hopper, feed chute, frame, cylinder, shaft, spikes, concave, aspirator, vibrating screens, suspension lever, sample outlet, etc., as shown in Figure 4.2a, b. The threshers can be classified as drummy, axial flow, or regular type according to their functional components. The drummy threshers

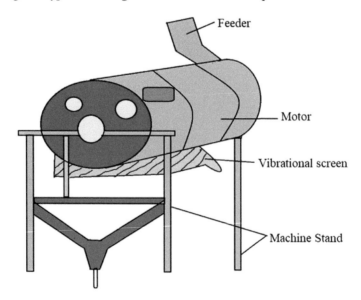

Figure 4.2a A drummy-type thresher.

(Figure 4.2a) are low-cost threshers that are easy to operate and are known for their simple structures. In these threshers, the stainless-steel beaters are radially arranged on the shaft that rotates inside an enclosed casing. The inside upper half of the cover is provided with ribs that give better threshing. The lower half of the casing known as concave is provided with rectangular openings. The crop to be harvested is received from the feeding chute and the impact of rotating beaters reduces it in size till it can pass through the concave. These threshers are not facilitated with provisions for the separation and cleaning of grains and the threshed material is separated and cleaned by small pedal-type blowers in a separate operation. In the axial type of threshers, the crop is fed into the cylinder through a feeding chute located at one end of the threshing drum. The thresher axially moves the crop several times around the threshing cylinder drum and louvers, that cleanly separate the grain from the stalk. The axial movement with proper clearance between the cylinder and concave provides adjustable speed and long threshing exposure. The straw is removed from the machine using paddles and the grains are further cleaned by a set of oscillating sieves and an aspirator. This type of thresher requires high power input and gives good results even with wet crops. A regular-type thresher mainly consists of a revolving cylinder and concaves. The crop flows tangentially through the gap between the drum and the concave. The crop is fed through a feeder from where it is taken to the cylinder through a conveying belt where the separation of grains takes place by the impact of rotating pegs mounted on the cylinder. The threshing cylinders may be spike tooth type/peg type, rasp bar type, angled bar type, wire loop type, cutter blade/syndicator type, or hammer mill type (Powar et al. 2019). A rasp-bar type cylinder is common in most threshers. The speed arrangements in the cylinder are made as per the crop type. The grains fall through the concave and are further cleaned using the aspirators and a series of screens (Chaturvedi and Rathore 2018). The threshers can be a horizontal

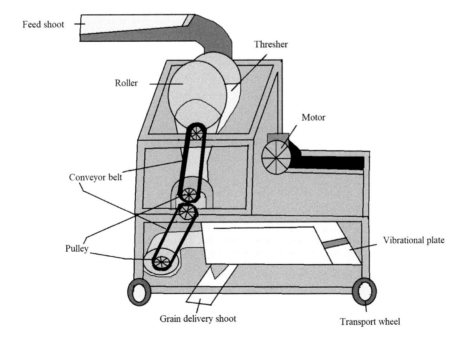

Figure 4.2b A horizontal-type thresher.

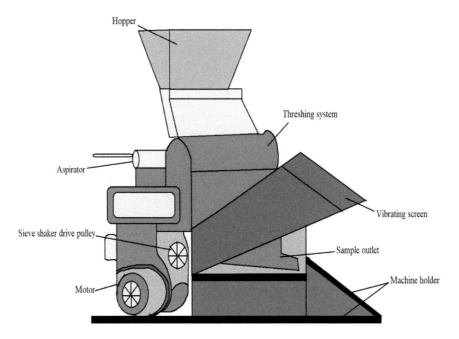

Figure 4.2c A vertical-type thresher.

type (Figure 4.2b) or a vertical type (Figure 4.2c) depending on their flow type. The horizontal machines are also known as feed-in type threshers. In these threshers, the whole crop is fed into the machine. The horizontal machines require more power and give high throughput. The vertical-type threshers are also known as head feed or hold-on threshers. In these machines, only the panicle is fed to the machine. These machines are complex and have low throughput.

Hanumantharaju (2017) developed a whole crop finger millet thresher and evaluated its performance at three different cylinder speeds (720, 800, 1050 rpm), concave clearances (4, 6, 8 mm), and seed moisture content (12.06, 13.39, and 14.50%). The best cleaning and threshing efficiencies, i.e. 94.15 and 93.25%, respectively, were achieved at the cylinder speed of 1050 rpm, concave clearance of 4 mm, and moisture content of 12.06–13.39%. A multi-millet thresher was developed by Singh et al. (2015) and its operating parameters were optimized for little millet using response surface methodology. Maximum threshing (95.13%) and cleaning efficiency (94.12%) for little millet were obtained at a feed rate of 105 kg/h, cylinder speed of 625 rpm, and threshing sieve size of 5 mm. On making the necessary adjustments in the machine the threshing capacities obtained for little, kodo, foxtail, proso, barnyard, and finger millet were 89, 137, 140, 91, 88, and 99 kg/h, respectively. The threshing efficiency obtained for all the millets was higher than 96% with less than 2% broken. Kamble et al. (2003) developed a pearl millet thresher and evaluated it for threshing efficiency, grain damage, seed germination, and energy consumption at three levels of cylinder speed (450, 700, 850 rpm), concave clearance (5, 10, 15 mm), grain moisture (10.2, 14.5, 17.2%), and feed rate (300, 400, 500 kg/h). The highest threshing efficiency and the maximum seed germination under these operating conditions were 96.8% and 87%, respectively. The minimum grain damage and grain loss achieved were 2.75 and 2.10%, respectively. The minimum energy consumption was 0.8 MJ per quintal of input crop. Chaturvedi et al. (2019) developed a thresher cylinder for millet crops and tested it at three speeds (580, 712, and 1068 rpm), two seed moisture contents (13 and 14.2%), and two concave clearances (2 mm and 5 mm). The highest threshing efficiency (99.6%), cleaning efficiency (98.6%), and output capacity (19.68 kg/h) were obtained at 13% moisture content, cylinder speed of 1068 rpm, and concave clearance of 5 mm. In a study conducted by Kamble et al. (2003), the threshing efficiency of pearl millet was found to increase from 83.73% to 92.53% with an increase in peripheral speed from 450–800 m/min. The increase in concave clearance, moisture content, and feed rate was found to decrease threshing efficiency due to a reduction in impact action, higher moisture content, and increased grain handling. Powar et al. (2019) developed a finger millet threshing drum and optimized the effect of operational parameters such as feed rate, drum speed, and concave clearance on its threshing efficiency, pearling efficiency, and grain damage using response surface methodology. An increase in threshing efficiency was observed in this study with a decrease in concave clearance and an increase in drum speed and feed rate. The maximum threshing efficiency (98.7%) was obtained at 3 mm conclave clearance, 10 m/s drum speed, and 40 kg/h feed rate. The pearling efficiency increased with an increase in drum speed and feed rate and a decrease in concave clearance. The highest pearling efficiency (84%) was obtained at 3 mm concave clearance, 10 m/s drum speed, and 40 kg/h feed rate.

4.3.3 Cleaners

Cleaning is the process of the removal of impurities such as dust, straw, chaff, parts of the stem, insects, soil, immature grains, small stones, animal excreta, etc., from the grains (Yang et al. 2012). The cleaning process can be divided into wet methods or dry methods (Chaturvedi and Rathore 2018). The wet methods help remove soil, pesticides, and dust using water or other solvents. The washed grains are then appropriately dried to remove the adhering moisture. The dry method cleans the grains based on shape, size, density, aerodynamic properties, magnetic properties, friction coefficient, suspended speed, colour, etc. The cleaning based on size and shape is done using sifters or screen cleaners. The grains with impurities are dropped on a screening surface which is vibrated. The screen divides the entire mass into two fractions depending on the shape and size of the apertures. The screening efficiency can be increased by using two or more screens. An improvement in screen cleaning is the use of air screen cleaners. In the air screen cleaners the cleaning screens are equipped with aspirators and the air-blast from the aspirators performs the satisfactory cleaning and separation of lighter impurities. Air screen cleaners may be vibratory or rotary depending on the movement of the screen. The vibratory air screen cleaners consist of two to eight screens that are tightened together and suspended by hangers. Such cleaners have a dual horizontal oscillating and slightly vertical motion that results in tossing as well the downward movement of grains from the screen. The slope of the screen can be adjusted as per the grain type and it also controls the rate of downward travel of the grains. The rotary screen cleaners have a circular deck with a circular motion in the horizontal plane. In such cleaners, the grain mixture is fed into the hopper from which it travels to the rotary screen through gravity. The grains pass through the screen while the oversized impurities are retained above and pass out through an outlet. The sound grains then come out at the centre of the screen drum and fall onto a vibratory screen. The vibratory screen removes the dirt particles and the light impurities are removed by the aspirator. The clean grains are then discharged through a discharge chute. The gravity graders are used to remove impurities based on density, friction coefficient, and suspended speed. These cleaners use terminal velocity to clean the grains. The basic assembly of gravity graders consists of a sloped perforated deck through which air is blown using a fan. The air lifts the grains from the deck surface and the materials are stratified as per their densities or specific gravities in the lower chamber and upper chamber. The deck also moves in

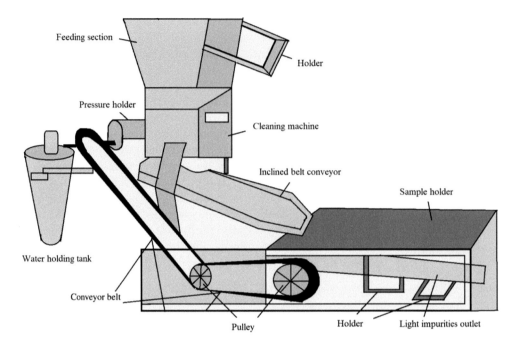

Figure 4.3 A vertical cleaning machine.

an oscillatory motion that moves the stratified mass along the direction of conveyance and discharges at the right edge of the deck (Hemasankari et al. 2023). The light impurities are removed using aspirators or pneumatic separators. These separators use terminal velocity to separate different fractions. The particles with low terminal velocities are lifted with the air current while the particles with high terminal velocities fall down. Broken parts of machines or other ferromagnetic impurities are removed using magnetic separators. The magnetic separators consist of a magnetic drum that attracts the ferromagnetic impurities. A diagrammatic representation of a millet cleaning machine is given in Figure 4.3.

Tejaswini et al. (2018) developed a cleaner cum pearler for finger millet and evaluated it for its cleaning and pearling efficiency. The machine was operated on a 2.5 hp electric motor and consisted of a hopper, outer cylinder, inner cylinder [with 12 cotton felts ($10 \times 10 \times 1$ cm^3) with 1 end of them bolted on its surface], mainframe, and aspirator. The cleaning frame was a stainless-steel plate with circular perforations of 2 mm in diameter. The screen was vibrated by the motor and the cleaned grains from this screen travel to the pearling unit which consists of inner and outer cylinders. The clearance between the cylinders was 2 cm and 1 cm at chamber for the maximum compression and shearing of grains so that the grains are pearled. The husk was removed by using an aspirator and the cleaned grains were collected at other outlets. The optimum value for pearling efficiency was 80.1% at 10% seed moisture, 900 rpm speed, and 150 kg/h feed rate.

4.3.4 Sorting/Grading Machines

Sorting and grading machines are used to remove damaged, diseased, or foreign grains. These machines use vibratory screens, aspirators, or sensors to perform this operation and can be run in semi-automatic (Figure 4.4a) or automatic (Figure 4.4b) mode.

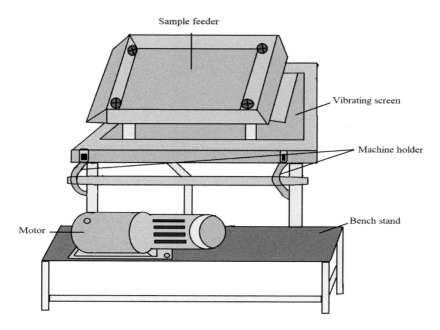

Figure 4.4a A semi-sorting machine.

A semi-automatic sorting machine consists of a sample feeder, vibrating screen, machine holder, motor, and bench stand as shown in Figure 4.4a. In this machine, grains enter from the top through the feeder. Then the grains move towards the chamber with the help of a chute where the different sizes of screens are inserted. The vibration of these screens segregates the grains based on their size and shape and the lightweight impurities are removed from the chamber with the help of an aspirator (Maitra et al. 2023). The fully automatic grading machines are also equipped with colour sorters. These machines contain sensors [charge coupled device (CCD) chips] in addition to the basic parts like the sample feeder, vibrating screen, machine holder, motor and bench stand, etc. (Lu 2022). A screen segregates the grains based on size. The screens have variable sizes and they may be fixed, moveable, and vibrating. These can segregate the grains based on size into two or more fractions. The capacity of screening depends on the number of grains passed through per square metre per second. The parameters that control the rate of separation are the size and shape of grains, feed rate, type of screen, arrangement of screens, aperture size, opening area, shaking frequency and amplitude of the machine, the nature of the sieve material, and the blocking prevention of the screen (Mustač et al. 2020; Gbabo et al. 2013). Details on the screens available in the market for the sorting of grains are provided in Table 4.2.

The various size sieves used in grain separation are HS32.35 (200 mm), HS32.40 (300 mm), HS32.45 (300 mm), and HS32.50 (450 mm) (Gbabo et al. 2013; Jun et al. 2015).

4.3.4.1 Colour Sorting Machine

Colour sorting machines are used to remove any impurity that has a similar shape, size, density, and terminal velocity as that of the main grain. Colour sorting machines work on photoelectric detection technology to separate the materials based on their optical properties (Miedaner and Geiger 2015). The various parts of the machines are the feeder, the detecting system, the signal processing module, and the rejecting system as shown in Figure 4.5. The grains enter the

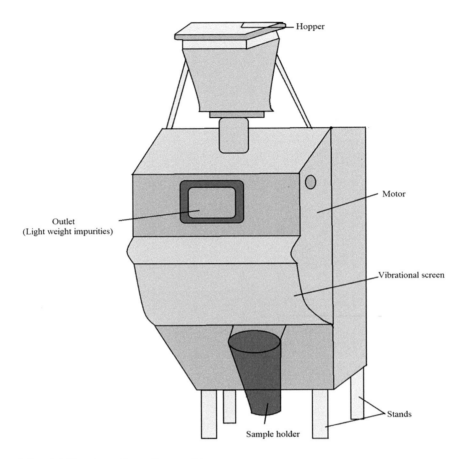

Figure 4.4b A fully automatic sorting machine.

machine through a feeding system or hopper from which they are transported to a chute using a vibration system. The material slides along the chute and enters the observation area where it passes through a light-emitting diode (LED) light and a detecting system like a high-definition camera scans the material in real-time and sends the pictures to the signal processing and control system, where the pictures are analyzed and the material to be rejected is marked and the control module sends the rejection data to the rejection system. The rejection system blows the different coloured material (material having colour other than main crop) by opening the designated nozzle of the spray valve into the defect cavity of the receiving hopper. The good quality grains are obtained from other outlets (Metra 6as 2022; Grotech 2021).

The first generation of colour sorters employed fluorescent light as the lighting system, a photodiode as the acquisition module, and a microcontroller or complex programming logic device as the core processing unit. They had a lower resolution and were replaced by second-generation devices that worked on field programmable gate arrays (FPGA), a digital signal processor (DSP), or a hybrid of FPGA and DSP. The second-generation colour sorters were also the first to use LED lights. The limitation of these sorters was their monochromatic nature and they were able to capture only in grayscale. The third-generation colour sorters use complementary metal-oxide-semiconductor (CMOS) technology and can identify various colour differences in materials (Grotech 2021).

TABLE 4.2 DIFFERENT TYPE OF SCREENS AVAILABLE IN THE MARKET FOR THE SORTING OF THE GRAINS

Screen	Angle	Shape
Micro-perforated holes	45, 60, and 90	Round
Oblongs/slots	–	Oblong
Square holes	–	Square
Hexagonal holes	–	Hexagonal
Flared and milled	–	Polygonal
Zone perforations	–	Round and oblong
Special holes	–	Round
Sectors and sieves	–	Square and round
Perforated disk	–	Round
Embossed sheets	–	Round

Source: Li et al. 2021; Duodu et al. 2019; Gomashe et al. 2017; Hua et al. 2017; Paulsen et al. 2015; Weisskopf et al. 2015.

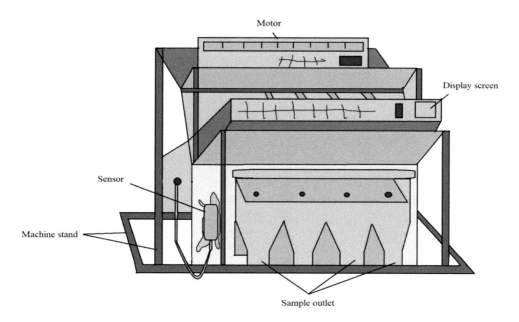

Figure 4.5 A colour sorting machine.

4.3.4.2 Destoner cum Grader cum Aspirator

A destoner cum grader cum aspirator, as is clear from the name, performs three functions, i.e. the removal of stones, the grading of grains, and the removal of light impurities. The machine is made up of heavy metal steel and the various components of this machine are a hopper, motor,

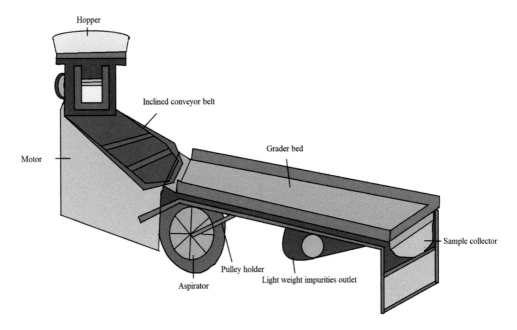

Figure 4.6 Destoner cum grader cum aspirator machine.

inclined conveyor belt, aspirator, pulley holder, grader bed, lightweight impurities outlet, and sample collector as shown in Figure 4.6. The millets are fed through the hopper from which they reach the grader bed with the help of an inclined conveyor belt. The eccentric shaft at the base of the destoner causes vibration in the screens that removes the stones from the grains based on the principle of specific gravity. The openings in the screens also help to separate the grains based on their size and shape. The aspirator removes the lighter impurities from the grains (Srinivas 2022).

4.3.5 Milling

Millets are categorized as naked grains (sorghum, pearl millet, finger millet, teff) and husked grains (foxtail millet, barnyard millet, proso millet, kodo millet, little millet, fonio). The husk is composed of non-digestible components like cellulose, hemicelluloses, and lignin and hence it needs to be removed before the food application. The process of the removal of the husk is known as dehusking or dehulling and the machines used for this purpose are known as dehullers. The dehulled grains are then pulverized into flour, grits (semolina/*rawa*), or middlings before their food application.

4.3.5.1 Dehuskers/Dehullers

Dehullers work on the principle of gentle abrasion and attrition. Dehullers based on centrifugal or impact forces have also been developed. The husk detached from the grains is then removed using aspirators or cyclone separators. The dehuller can be a single-stage or a double-stage machine. A diagrammatical representation of millet hullers is provided in Figure 4.7a and b. A description of millet dehuskers available in the market is provided in Table 4.1.

Krishnappa et al. (2021) developed a double-stage foxtail millet dehuller and evaluated it for its dehulling capacity. The major components of the dehuller were a hopper, two knurled rolls

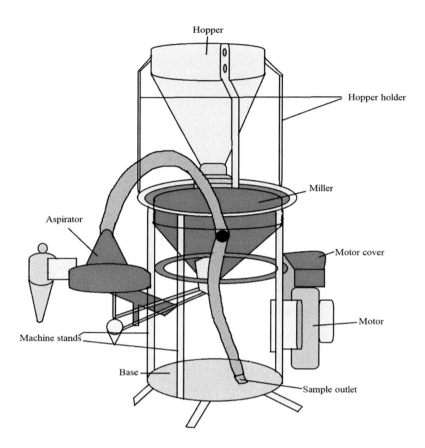

Figure 4.7a A dehusker.

(diamond angular and straight knurled), two rubber rolls, an air blower, and a grader. The maximum dehulling (82.24%) and cleaning (82.47%) efficiencies were observed using a diamond-type knurled roll at a moisture content of 12% wet basis and feed rate of 100 kg/h. The highest yield for head grain (97.82%) was again obtained with a diamond-type knurled roll at a feed rate of 300 kg/h. It was concluded in this study that optimum dehulling conditions that provided good dehulling capacity (82.24%), cleaning efficiency (82.47%), head grain yield (95.84%), and broken (4.16%) were obtained using a diamond-type knurled roll at a moisture content of 12% wet basis and feed rate of 100 kg/h. Ullegaddi et al. (2021) developed a 25 kg batch-type abrasive foxtail deshelling/dehulling machine of dimensions 700 mm × 320 mm × 450 mm. The machine was operated using a motor of 0.5 hp and the operating speed was between 1000 and 2000 rpm, the distance between the discs was 18 mm, and the torque was 4.5 Nm. The machine could handle 2.5 kg of grains per batch. The machine produced 77.91% of fully deshelled millets, 13.81% of shelled millets, and 8.47% of damaged millets. Gouda et al. (2019) developed a table-top centrifugal dehuller with a dehulling capacity of 6–8 kg/h at the University of Agricultural Sciences, Bengaluru, India. The components of the dehuller were a hopper, feed controller, dehulling drum, centrifugal impeller, husk separating chamber, centrifugal blower, and airflow rate control valve. The machine was operated using a 1 hp motor with a speed of 2880 rpm and a V-belt driving mechanism. The machine was also evaluated for the dehulling of five millets, i.e. foxtail,

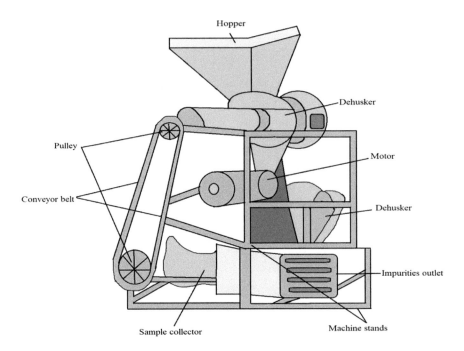

Figure 4.7b A double-stage dehusker.

little, proso, kodo, and barnyard millets, at three moisture contents, i.e. 11.1, 13.6, and 16.2% dry basis, at three dehulling speeds (5000, 5500, and 6000 rpm). The highest dehulling efficiency for foxtail (92.7%), little (91%), and proso (93.5%) millets was obtained at a moisture content of 13.6% dry basis and dehulling speed of 5500 rpm. The highest dehulling efficiency for kodo (80.9%) and barnyard (74.4%) millet was obtained at a dehulling speed of 6000 rpm and 11.1% db moisture content. The highest percentage of broken was found in barnyard millet (9.2%), followed by kodo millet (5.7%). The percentage of broken in all other studied millets was less than 5%. Durairaj et al. (2019) designed a double-chamber centrifugal dehuller for millet dehulling. The dehuller's components included a feeding hopper, impellers with curved vanes, two cast iron centrifugal chambers, a blower/aspirator, and outlets for dehusked grains and husks. The grains enter the impeller from the hopper via the feed housing, gaining momentum and being thrown at high velocity against the cast iron chamber. The high impact causes husk separation with little damage to the bran. The resulting mixture of husk and dehulled grains is separated using a blower. On the comparison of the dehulling results of this machine with the abrasive type of machine, it was discovered that the kernel recovery was approximately 10% higher than the abrasive huller, regardless of millet type. Breakage was also lower (4–5%) than abrasive milling (20%). Shirsat et al. (2008) milled kodo millet using a laboratory model rice polisher. Before milling the sample was soaked for 60–300 minutes at 60, 70, and 80°C, steamed for 30 minutes, and roasted for 15 minutes. The maximum head yield (61.52%) was obtained for the samples soaked for 30 minutes. Samples soaked at 80°C for 3 hours produced minimum broken (5.89%) and head yield (58.19%). Chin et al. (2018) developed a composite rubber roller and centrifugal mill for the dehulling of little millet. The major components of the mill were a feeding hopper, a rubber roll for dehulling, and a centrifugal mill for detached hull removal. The machine was operated on a 1 hp motor with a speed of 1100 rpm. The speed differential of the rolls was 2:3, i.e. 1100:1640 rpm, and the spacing between the two rolls was 0.55 mm. The mill achieved 96.52% milling efficiency and the broken were less than 1.7%.

4.3.5.2 Pulverization

The term "pulverization" refers to the grinding of grains primarily for maximizing the separation of the endosperm, bran, and germ, lowering particle size, and allowing the creation of refined flour. The millets can be pulverized/ground by both dry milling and wet milling processes. The selection of the process for millet pulverization is mostly determined by the end-use (Taylor and Kruger 2019). The dry milling process normally has two primary goals: particle size reduction and grain tissue fractionation. Whole millet flour can be produced by electrically operated attrition mills or plate mills. In the attrition mills, grinding is done between two discs or two plates equipped with replaceable wearing surfaces. In this machine, generally, one wheel/plate is stationary and another is revolving. However, both wheels can revolve and, in that case, they revolve in opposite directions. The grain size is reduced by the shearing and cutting action caused by the rotation of the wheel. The rotating wheel also causes the impact action. This mill is generally used for the smooth grinding of materials that have been previously ground by other mills. These mills can also be used for the wet milling of mixtures containing liquids (Shankar et al. 2013; Sadler et al. 1975). Modern technologies can produce flour, i.e. free from bran, and are available in various fractions. This involves a two-step process involving mechanical abrasive decortication (often referred to as dehulling) (Taylor and Kruger 2019; Mensah and Tomkins 2003), followed by particle size reduction by hammer milling, pin milling, or roller milling. The hammer mill is a communiting mill that works on the principle of impact and is hence also known as an impact mill. Hammer mills are steel drums containing a horizontal or vertical shaft on which hammers are mounted (Figure 4.8a). The hammers are fixed on the central rotor and are free to

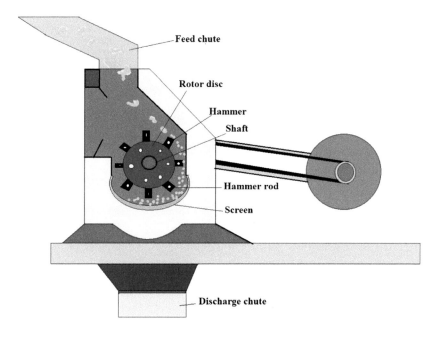

Figure 4.8a A hammer mill.

swing. The grains enter the drum from the feeding hopper and are comminuted by the hammers rotating at high speed until the grains are reduced to a size smaller than the openings of the screen (Ibrahim et al. 2019; Ebunilo et al. 2010).

The pin mills are vertical shaft impact mills that consist of two rotating discs with pins embedded on the surface of one or both of the discs (Figure 4.8b). The pins repeatedly move past each other and this action communicates the material through repeated impact (Shelake et al. 2017).

The bran can also be removed by conditioning and tempering. Conditioning is the addition of small quantities of water to grains and tempering is the holding of moist grains at constant temperature for a defined time interval (discussed in detail in Chapter 3). These processes make the bran elastic and the endosperm friable. The entire or part of the outer bran (pericarp and germ tissues) is removed during milling while retaining the endosperm. Modern milling machines use a series of metal cylinders/rollers (break rolls and reduction rolls) and sieves to perform the milling and obtain various fractions (Kebakile et al. 2007). The machines used for the milling of millets are common to that of other cereals, however the necessary arrangements need to be made in the machine as per the millet type to be milled. Palaniswamy (2018) studied the effect of milling parameters on the quality and nutritive value of flour obtained from two finger millet varieties, i.e. GPU-28 and L-15. The fineness modulus increased from 2.04 to 3.44 in variety GPU-28 and 2.05 to 3.45 in variety L-15 with an increase in plate clearance from 0.3 to 0.7 mm. A decrease in fineness

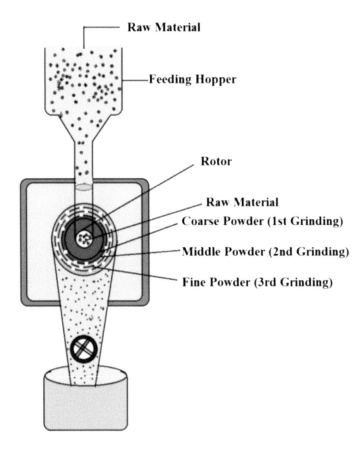

Figure 4.8b A rotary spinning mill.

modulus was also observed with an increase in feed rate and plate speed. The fineness modulus decreased from 2.18 to 2.04 and 2.21 to 2.05 in both varieties with an increase in feed rate. The lowest fineness modulus was obtained at a feed rate of 100 kg/h. The fineness modulus decreased from 2.36 to 2.04 and 2.37 to 2.05 on increasing the plate speed from 450 rpm to 600 rpm. A loss in calcium (37.73%), phosphorus (12%), and protein (9.6%) was also observed when the plate clearance was reduced from 0.7 to 0.3 mm. The flour quality is also affected by the type of mill used. In a comparative study on the flour quality obtained by finger millet milling using different mills, it was found that the flour milled by roller, emery, and plate mills was finer compared to that obtained from pin or hammer mills. The highest damaged starch content was found in roller mills (22.8%) followed by emery (10.6%), plate (9.4%), pin (7.2%), and hammer mills (5%), respectively. The flour obtained from the pin and hammer mill had a higher content of intact starch granules while the flour from the plate, emery, and roller mills had flour particles with a large amount of exposed starch (Smitha et al. 2008). Abdelrahman et al. (1983) optimized the milling process to produce low-fat grits from pearl millet. Two millet types, i.e. decorticated and whole grain, with a moisture content of 12% were used in this study. Both millet types were tempered to 22% moisture content and held for about 6 hours before milling. The milling was carried out using two different systems. In the first milling system, the grains were passed only once through a corrugated roll mill with 22 corrugations at a 0.015-inch roll gap. The milled grains were then sifted into three fractions, i.e. 16–20 wire, 20 wire, and over 20 wire. In the second milling system, the grains were passed through a three-break system where the first roll had 12 corrugations at a roll gap of 0.02 inches, the second roll had 22 corrugations at a 0.013-inch roll gap, and the third roll had 22 corrugations at a 0.009-inch roll gap. The effect of roll differential (1:1.30, 1:2.12, and 1:4.24) and corrugation was also studied on the decorticated samples conditioned to 22% moisture and tempered for 6 hours. It was found in the study that tempering helped to reduce the contamination of grits with pieces of pericarp and also helped to separate the germ. Tempering the decorticated grains up to a moisture content of 22% for 6 hours increased the yield of low-fat grits. The grits produced after decortication in a single-pass system were also low in fat or germ fraction. However, decortication had no advantage over the whole grain system in producing low-fat grits in a three-break system. The yield of grits was increased by using fine corrugation rolls. A roll speed differential above 1:2.5 did not affect particle size distribution.

4.4 SECONDARY PROCESSING OF MILLETS

Millets and millet flours can be processed into various traditional as well as modern products such as *roti*, *halwa*, *idli*, *dosa*, *uthpam*, *injera*, alcoholic beverages, puffed products like popcorns, roasted products, extruded products, bakery products, baby health foods, gluten-free products, etc. The machines used for the preparation of these products are similar to the machines used for making products from any other flour so such machines are not discussed in this book. The process, recipes, and scientific studies on the production of millet-based traditional and modern products are discussed in Chapters 13 and 14 of this book.

4.5 CONCLUSIONS

The first step in the processing of millets is harvesting. Harvesting can be carried out manually with sickles or with modern machines known as windrowers and combines. Harvesting is followed by threshing. Traditionally, threshing was done by beating the millets with sticks or trampling them under the feet of animals. Nowadays, a variety of threshers like drummy threshers, axial flow threshers, or regular-type threshers are used. These threshers can be further categorized into horizontal threshers and vertical threshers. The horizontal threshers are used for the threshing of the whole crop and the vertical threshers are used for the threshing of panicles. Further processing of millets depends upon the type of millet, i.e. whether it belongs to the naked or husked category of grains. The major steps in the processing of naked grains are cleaning, sorting, grading, milling, etc. The husked grains need to be dehusked before the operations of cleaning, sorting, grading, and milling. The cleaning can be done by wet or dry cleaning methods. The wet cleaning methods use water for cleaning while the dry cleaning is performed based on the principles of rubbing, vibration, and aspiration. Traditionally, dry cleaning is done by winnowing using *chazz* or *supa*. Various modern machines are also available for the cleaning of millets. Modern processing methods also facilitate the sorting of damaged and under-mature grains. Foreign impurities of similar shape and size can be removed using colour sorters. The grains are milled using the principle of shearing, impact, or abrasion using traditional as well as modern mills. The end products of milling are further processed into edible products in secondary processing.

REFERENCES

Abdelrahman, A., R. C. Hoseney, and E. Varriano-Marston. 1983. Milling process to produce low-fat grits from pearl millet. *Cereal Chemistry* 60(3):189–191.

Britannica, T. 2020. Editors of encyclopaedia. *Windrower. Encyclopedia Britannica.* https://www.britannica.com/technology/windrower. (accessed February 28, 2023).

Chaturvedi, S., F. Rathore, and S. Pandey. 2019. Performance evaluation of developed thresher cylinder on millet crop. *International Journal of Current Microbiology and Applied Sciences* 8:2319–7706.

Chin, R., N. Matlashewski, and K. Swan. 2012. Development of a little millet mill. https://www.mcgill.ca/bioeng/files/bioeng/nick_rebecca_kris_2012.pdf. (accessed January 23, 2023).

Chaturvedi, S. and F. Rathore. 2018. Development of threshing cylinder for small millets. *Agricultural Science Digest-A Research Journal* 38(3):178–182.

Duodu, K. G., and F. E. Dowell. 2019. Sorghum and millets: Quality management systems. In *Sorghum and Millets*, eds. J. R. N. Taylor and K. G. Duodu, 421–442. Cambridge: AACC International Press.

Durairaj, M., G. Gurumurthy, V. Nachimuthu, K. Muniappan, and S. Balasubramanian. 2019. Dehulled small millets: The promising nutricereals for improving the nutrition of children. *Maternal and Child Nutrition* 15:e12791.

Ebunilo, P. O., A. I. Obanor, and G. O. Ariavie. 2010. Design and preliminary testing of a hammer mill with end suction lift capability suitable for commercial processing of grains and solid minerals in Nigeria. *International Journal of Engineering, Science and Technology* 2(6):1581–1593.

Gbabo, A., I. M. Gana, and S. A. Matthew. 2013. Design, fabrication and testing of a millet thresher. *Net Journal of Agricultural Science* 1(4):100–106.

Ghimire, S. K., R. Awal, P. Paneru, P. Bokati, and S. Wasti. 2019. Design, fabrication and testing of finger-millet harvesting machine. *Journal of the Institute of Engineering* 15(1):71–76.

Gomashe, S. S. 2017. Proso Millet, Panicum miliaceum (L.): Genetic improvement and research needs. In *Millets and Sorghum: Biology and Genetic Improvement*, ed. J. V. Patil, 150–169. New Jersey, United States: Wiley, Publishing company.

Gouda, G. P., H. Sharanagouda, U. Kumar, N. C. Ramachandra, N. J. Naik, N. Ananada, and G. Ganjyal. 2019. Studies on engineering properties of foxtail millet [Setaria italica (L.) Beauv.]. *Journal of Farm Sciences* 32(3):340–345.

Grotech. 2021. Grotech multifunction color sorter. https://www.grotechcolorsorter.com/blog/four-technical-upgrades-of sorting-machine-from-monochrome-recognition-to-artificial intelligence_b96. (accessed March 12, 2022).

Hanumantharaju, K. N., L. Vikas, and P. Kumar. 2017. Comparison study of prototype thresher with different methods of threshing whole crop finger millet. *International Journal of Science, Environment and Technology* 6(1):391–398.

Hemasankari, P., B. D. Rao, V. M. Malathi, E. Kiranmai, and C. V. Rathnavathi. 2023. Value addition in millets through agricultural processing techniques. *Indian Farming* 73:107–109.

Hua, L., W. Jinshuang, Y. Jianbo, Y. Wenqing, W. Zhiming, and Q. Youzhang. 2017. Analysis of threshed rice mixture separation through vibration screen using discrete element method. *International Journal of Agricultural and Biological Engineering* 10(6):231–239.

Ibrahim, M., M. Omran, and E. Abd EL-Rhman. 2019. Design and evaluation of crushing hammer mill. *Misr Journal of Agricultural Engineering* 36(1):1–24.

Ikubanni, P. P., O. O. Komolafe, O. O. Agboola, and C. O. Osueke. 2017. Moringa seed dehulling machine: A new conceptual design. *Journal of Production Engineering* 20(2):73–78.

Jun, H. J., T. G. Kang, D. K. Choi, Y. Choi, D. K. Choi, and C. K. Lee. 2015. Study on performance improvement of a head-feeding rice combine for foxtail millet harvesting 40(1):10–18.

Kamble, H. G., A. P. Srivastava, and J. S. Panwar. 2003. Development and evaluation of a pearl-millet thresher. *Journal of Agricultural Engineering* 40(1):18–25.

Kashyap, T., R. Thakur, M. Sethi, R. K. Tripathi, A. Thakur, S. Kumar, S. Chand, B. Goel, and S. Chand. 2023. The influence of reflector on the thermal performance of single and double pass solar air heater: A comprehensive review. *Materials Today: Proceedings* 72:1260–1269.

Kebakile, M. M., L. W. Rooney, and J. R. Taylor. 2007. Effects of hand pounding, abrasive decortication-hammer milling, roller milling, and sorghum type on sorghum meal extraction and quality. *Cereal Foods World* 52(3):129–137.

Krishnappa, G., S. Savadi, B. S. Tyagi, S. K. Singh, H. M. Mamrutha, S. Kumar, C. N. Mishra, H. Khan, K. Gangadhara, G. Uday, G. Singh, and G. P. Singh. 2021. Integrated genomic selection for rapid improvement of crops. *Genomics* 113(3):1070–1086.

Lee, D., W. Hanna, G. D. Buntin, W. Dozier, P. Timper, and J. P. Wilson. 2012. *Pearl Millet for Grain*. Bulletin 1216. The University of Georgia Cooperative Extension. College of Agricultural and Environmental Sciences and College of Family and Consumer Sciences. https://secure.caes.uga.edu/extension/publications/files/pdf/B%20 1216_3.PDF (accessed February 28, 2022).

Li, X., G. Zhao, W. Wang, Y. Huang, and J. Ji. 2021. Design and experiment of vibrating screen millet cleaning devise with double-fan. *Applied Engineering in Agriculture* 37(2):319–331.

Lu, H. 2022. Design of margin detection algorithm for uneven illumination picture based on SVM model. In *2nd International Conference on Networking, Communications and Information Technology (NetCIT)*, 627–630.

Maitra, S., T. S. S. K. Patro, A. Reddy, A. Hossain, B. Pramanick, K. Brahmachari, and M. Sairam. 2023. Brown top millet (Brachiaria ramosa L. Stapf; Panicum ramosum L.)—A neglected and smart crop in fighting against hunger and malnutrition. In *Neglected and Underutilized Crops*, eds. F. Muhammad and K.H.M. Siddique, 221–245. Cambridge, United States: Academic Press.

McDonough, C. M., L. W. Rooney, and S. O. Serna-Saldivar. 2000. The millets. In *Handbook of Cereal Science and Technology*, eds. K. Kulp, and J. G. Ponte Jr., 177–201. Florida, United States: CRC Press.

Mensah, P., and A. Tomkins. 2003. Household-level technologies to improve the availability and preparation of adequate and safe complementary foods. *Food and Nutrition Bulletin* 24(1):104–125.

Metra 6as. 6AS-68 Grain color sorter users manual. https://graincleaner.com/ web/manual/web-instr-metra-6as.pdf (accessed March 12, 2022).

Miedaner, T., and H. H. Geiger. 2015. Biology, genetics, and management of ergot (Claviceps spp.) in rye, sorghum, and pearl millet. *Toxins* 7(3):659–678.

Mustač, N. Č., D. Novotni, M. Habuš, S. Drakula, L. Nanjara, B. Voučko, M. Benković, and D. Ćurić. 2020. Storage stability, micronisation, and application of nutrient-dense fraction of proso millet bran in gluten-free bread. *Journal of Cereal Science* 91:102864.

Nisha, N., M. Saravanakumar, and B. Shridar. 2022. Development of tractor front mounted harvester for finger millet. *Madras Agricultural Journal* 109:1.

Palaniswamy, S. 2018. *Development of a Millet Dehuller (Hand-Operated) to Reduce Drudgery in Processing and Utilization of Millet Waste (Hulls) in Antioxidant Extraction*. McGill University. https://www.

proquest.com/openview/821636654ae8df212646185f9bd84cce/1?pq-origsite=gscholar&cbl=18750&diss=y (accessed April 2018).

Paulsen, M. R., P. K. Kalita, and K. D. Rausch. 2015. Postharvest losses due to harvesting operations in developing countries: A review. In *ASABE Annual International Meeting*, American Society of Agricultural and Biological Engineers, ASABE Annual International Meeting 152176663. doi:10.13031/aim.20152176663

Powar, R. V., V. V. Aware, and P. U. Shahare. 2019. Optimizing operational parameters of finger millet threshing drum using RSM. *Journal of Food Science and Technology* 56(7):3481–3491.

Sadler, L. Y., D. A. Stanley, and D. R. Brooks. 1975. Attrition mill operating characteristics. *Powder Technology* 12(1):19–28.

Shankar, M., G. M. Chowde, R. Manikandan, U. Ravindra, and Honabyraiah. 2013. Performance evaluation of attrition mill used in the finger millet processing industries. *International Journal of Technical Research and Applications* 1:59–62.

Shelake, P. S., M. N. Dabhi, R. D. Nalawade, and M. L. Jadhav. 2017. Design and development of pin mill for size reduction of turmeric (Curcuma longa) rhizome. *International Journal of Current Microbiology and Applied Sciences* 6(10):2102–2107.

Shirsat B. W., S. D. Kulkarni, S. Patel, and S. P. Singh. 2008. Milling characteristics of Kodo (Paspalum scrobiculatum L.) millet. *International Journal of Agricultural Sciences* 4(2):712–718.

Singh, K. P., R. R. Poddar, K. N. Agrawal, S. Hota, and M. K. Singh. 2015. Development and evaluation of multi-millet thresher. *Journal of Applied and Natural Science* 7(2):939–948.

Smitha, V. K., K. Sourav, and N. G. Malleshi. 2008. Studies on the effect of milling finger millet in different pulverisers on physico-chemical properties of the flour. *Journal of Food Science and Technology* 45(5):398–405.

Srinivas, A. 2022. Millet milling technologies. In *Handbook of Millets-Processing, Quality, and Nutrition Status*, eds. C. Anandharamakrishnan, A. Rawson, C. K. Sunil, 173–203. Singapore: Springer Nature.

Taylor, J. R., and J. Kruger. 2019. Sorghum and millets: Food and beverage nutritional attributes. In *Sorghum and Millets*, eds. J. R. N. Taylor and K. G. Duodu, 171–224. Cambridge, United States: AACC International Press.

Tejaswini, V. V., R. D. Bhaskara, R. Lakshmipathy, and S. Kumar. 2018. Development and evaluation of cleaner cum pearler for finger millet. *International Journal of Current Microbiology and Applied Sciences* 7(11):1819–1830.

Ullegaddi, M. M., N. M. Babu, A. R. Faisal, M. Mohammad, M. S. Shreenidhi, and S. Anjum. 2021. Design and development of compact Foxtail millet deshelling machine. *Materials Today: Proceedings* 42:781–785.

Weisskopf, A., Z. Deng, L. Qin, and Q. Dorian. 2015. The interplay of millets and rice in Neolithic central China: Integrating phytoliths into the archaeobotany of Baligang. *Archaeological Research in Asia* 4:36–45.

Woodford, C. 2021. Combine harvesters. https://www.explainthatstuff.com/.howcombineharvesterswork.html. (accessed February 28, 2022).

Yang, X., Z. Wan, L. Perry, H. Lu, Q. Wang, C. Zhao, J. Li, F. Xie, J. Yu, T. Cui, T. Wang, M. Li, and Q. Ge. 2012. Early millet use in northern China. *Proceedings of the National Academy of Sciences of the United States of America* 109(10):3726–3730.

Chapter 5

Structural Composition and Engineering Properties of Millets

5.1 INTRODUCTION

The structural as well as engineering properties of grains play an important role in their handling, processing, and safe storage. The physical and dimensional properties like shape, length, breadth, thickness, diameter, sphericity, and surface area are required to design the equipment for the seeding (cultivation), threshing, processing, and storage of grains. Gravimetric properties like thousand kernel weight, bulk density, true density, porosity, and frictional properties like static and dynamic friction, angle of repose, etc., affect structure loads (vertical and horizontal loads acting on silo walls) (Kumar et al. 2020). These properties are also important in designing drying, aeration, and transportation structures. The coefficient of friction of the grain against various surfaces is also necessary for designing conveying, grain flow, and storage structures. Knowledge of the structures and composition of grains is a must for their optimum utilization. The composition of the grain, components of grain tissue, and end-use properties of the grain are also affected by the grain structure. It greatly impacts the processing, utilization, and nutritive value of the grains. The amount and thickness of the hull can determine the resistance of the grain against pest infestation. The seed coat of the millet plays a crucial role in protecting the embryo from the external environment, i.e. microbial contamination, mechanical stress, fluctuations in temperature, and storage conditions (Radchuk and Borisjuk 2014). It can also provide information regarding the force required for the milling of grains. The size and the location of the germ in the grain affect the retention of the germ during processing and milling. The hardness of the endosperm determines the breakage of grains during milling (Pomeranz 1982). This chapter discusses in detail the structural composition and engineering properties of millets.

5.2 TYPE OF MILLET SEEDS

Millets are small-seeded grasses divided into two types of seeds *viz.*, caryopsis, and utricles. The caryopsis seeds are those whose pericarp is completely fused in the seeds. Examples are pearl millet, teff, and fonio. In utricles, the pericarp is present in the sac that surrounds the seeds, and only one end of the pericarp is attached to the seeds. Examples are foxtail millet, finger millet, and proso millet.

5.3 SHAPE OF MILLET GRAINS

Millet grains may be oval, spherical, elliptical, polyhedral, globose, or hexagonal (Figure 5.1). Among these the largest variations are found in pearl millet grains which may be obovate, globose, lanceolate, or hexagonal. The grains of finger millet are spherical to globose; foxtail millet grains are spheroid or spherical to polygonal; proso millet grains are oval and spherical (Salahudeen and Bosede 2021); barnyard millet grains are round (Singh et al., 2018); and kodo and little millet are elliptical to oval (Heiru et al. 2019).

5.4 STRUCTURE AND COMPOSITION OF MILLET GRAINS

Grains are made up of three major parts, i.e. bran, endosperm, and germ. Figure 5.2 represents the different components of millet grains.

Detailed information on the parts of millet grains is given in the following sections. Information on the structural and compositional characteristics of different millets is also provided in Table 5.1.

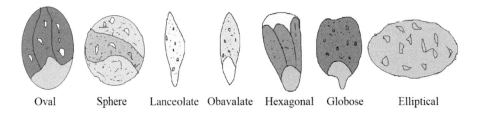

Figure 5.1 Different shapes of millet grains.

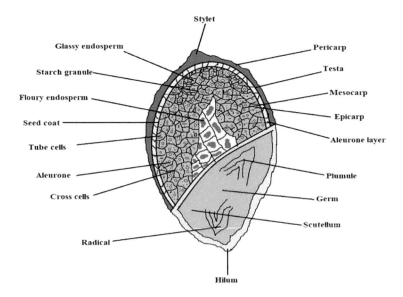

Figure 5.2a Structure of a pearl millet grain.

Structural Composition and Properties 77

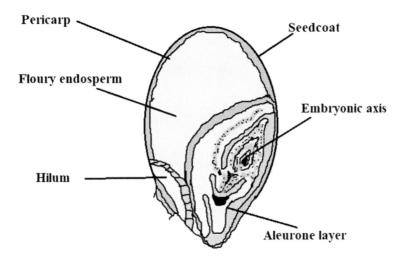

Figure 5.2b Structure of a foxtail millet grain.

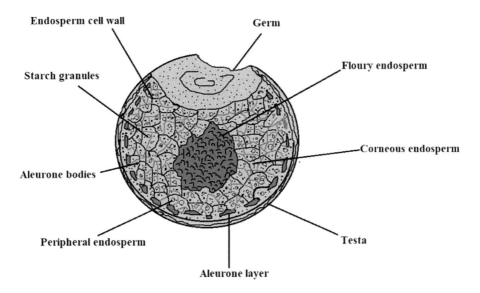

Figure 5.2c Structure of a finger millet grain.

5.4.1 Bran

The bran is a multilayered outer skin that is composed of a pericarp, seed coat, and aleurone layer. The pericarp is also known as the fruit coat and it surrounds the entire seed. It is further divided into the outer pericarp and inner pericarp. The outer pericarp consists of the epidermis/epicarp, hypodermis, and the innermost layer called the remnants of thin-walled cells. The inner pericarp is present adjacent to the remnants and is composed of a single layer of cross cells and tube cells. The cross cells are long, cylindrical, have no intercellular space, are tightly bound, and the long axis of the cross cells is perpendicular to the long axis of the grain. The tube cells

78 Millets

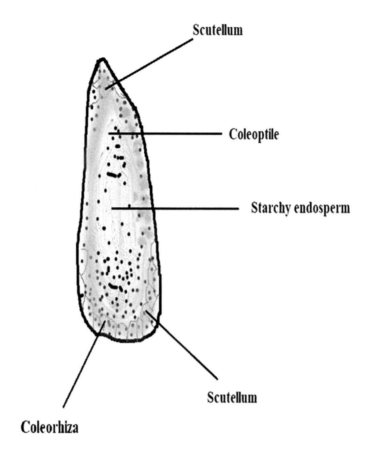

Figure 5.2d Structure of a fonio grain.

are also long and cylindrical but are loosely packed and their long axis is parallel to the long axis of the grain. The hypodermis layer gives support to the kernel and protects the epidermis from external factors (Hassan 2021). The removal of the outer pericarp helps in the movement of water to the inside of the grain. Next to this is the seed coat which is the outermost covering of the seed and is firmly joined to the tube cells on the outside and the nucellar epidermis on the inside. The seed coat has further two layers, i.e. the outer testa and inner tegmen. The seed coat may be pigmented or non-pigmented depending on the millet type. The innermost layer of bran is the aleurone layer that surrounds the kernel. It covers all the starchy endosperm and the majority of the germ except for that adjacent to the scutellum (Grundas 2003). This layer is rich in minerals and vitamins. All these layers of the bran collectively protect the kernel from pests, sunlight, water, and diseases.

The percentage of pericarp, endosperm, and germ in pearl millet grain is 8%, 73%, and 17.4%, respectively. The pericarp of the pearl millet is thick and is entirely attached to the endosperm (Hassan et al. 2021). In foxtail millet, the pericarp is thin with two epidermal layers firmly compressed to each other. The foxtail pericarp also contains large numbers of starch granules that disappear after maturation (Serna-Saldivar et al. 2019). The pericarp of proso and foxtail millet is loosely attached to the kernel and can be easily removed by rubbing or soaking. The pericarp of the finger millet, also known as a glume, is thinner. Finger millet contains three layers of pericarp,

Structural Composition and Properties 79

TABLE 5.1 STRUCTURAL AND COMPOSITIONAL CHARACTERISTICS OF DIFFERENT TYPES OF MILLET SEEDS

	Parameters		Pearl millet	Foxtail millet	Proso millet	Barnyard millet	Finger millet	Kodo millet	Little millet	Teff	Fonio
Bran	Pericarp										
	Seed coat	Number of layers	1	1	–	–	5	–	–	1	1
		Pigment	Sometimes in the aleurone layer and endosperm	–	No	–	Yes	–	–	No	No
		Thickness (Pm)	0.4	–	0.2–0.4	–	10.8–24.2	–	–	1.5–2.0	0.5–1.5
	Aleurone layer	Number of layers	1	1	1	1	1	1	–	1	–
		Cell size (pm)	16–30 × 5–15	–	12 × 6	–	18 × 7.6	–	–	12 × 4	12 × 4
		Starch type	Simple	Simple	Simple	–	Compound	Simple	–	–	–
		Starch granule (μm)	6.4–12.0	5–25	1.3–17	2.5–20	8.0–21.0	7–15	–	–	–
Endosperm	Starch granules	Diameter (μm)	101–2	0.8–9.6	2–10	1.3–8.0	3–21	1.2–9.5	1.0–9.0	–	7.0–8.0
		Peripheral zone (μm)	6.4–6.8	–	3.9	–	8–16.5	–	–	5.6–7.2	7.2
		Corneous zone (μm)	6.4–6.5	–	4.1	–	3–19	–	–	15.1–22.4	7.8
		Floury zone (μm)	7.6–7.8	–	4.1	–	11–21	–	–	10.2–15.8	6.5
	Protein bodies	Type	Simple	Spherical	Simple	Spherical and Polygonal	Simple	Spherical and Polygonal	Spherical	Simple/compound	Simple
		Size (μm)	0.6–0.7	1–2	0.5–1.7	–	0.5–1.7	–	–	1.2–2.4	1.2–2.4
		Location	All area	Peripheral cell	Peripheral	–	Peripheral	–	–	All areas	Peripheral

(*Continued*)

TABLE 5.1 (CONTINUED) STRUCTURAL AND COMPOSITIONAL CHARACTERISTICS OF DIFFERENT TYPES OF MILLET SEEDS

	Parameters	Pearl millet	Foxtail millet	Proso millet	Barnyard millet	Finger millet	Kodo millet	Little millet	Teff	Fonio
Germ	Size (L × W) (μm)	1420 × 620	270 × 980	100 × 310	–	980 × 270	–	–	160 × 725	517 × 977
	Shape	Obovate, globose, lanceolate, hexagonal	Spherical, polygonal	Spherical, oval	Oval, concave	Spherical, globose	Elliptical to oval	Elliptical to oval	Oval	Polyhedral, globose ellipsoids, oval
	Colour	Yellow, white, grey, brown, purple	Red, black, white, brown, yellow	Yellow, red, orange, white cream, brown to black	White or yellow	Violet, white, red, brown, yellow	Blackish brown to dark brown	Grey to straw white	Pale white to ivory white	White to pale brown or purplish
	Pericarp	Thin and thick	Thin	–	–	Thin	–	–	–	Thin
	Texture	Very floury, very corneous, intermediate	–	–	–	–	–	–	Smooth and gelatinous	–
	Germ type	Caryopsis	Caryopsis	Utricle	Caryopsis	Utricle, caryopsis	Caryopsis	–	Caryopsis	Caryopsis
	Protein in body size (mm)	0.6–1.2	1–2	0.5–2.5	–	2.0	–	–	–	–
	Wax type	No	Yes	Yes	No	Yes	No	–	No	No
	Endosperm germ ratio	4.5:1	12:1	12:1	–	11:1	–	–	3:1	6:1

Source: Barretto et al. 2021; Salahudeen and Bosede 2021; Heiru et al. 2019; Pawase et al. 2019; Singh et al. 2018; Berwal et al. 2017; Dayakar et al. 2017; Bultosu 2016; Guha et al. 2015; Upadhyaya et al. 2014; Saleh et al. 2013; Chukwu and Abdul-kadir 2008; Kulp 2000; Kumari and Thayumanavan 1998.

namely endocarp, mesocarp, and epicarp (Ramashia 2018). The pericarp of barnyard millet consists of two epidermal layers and the cells of the inner epidermis are closely compressed against the outer part. The pericarp of the fonio is 1.4–1.5 mm long, 0.8–0.9 mm wide, and 0.6 mm thick. The outer pericarp of the teff is thin and membranous and wears a slime layer rich in pectin beneath the cuticle toward the nucellar epidermis. In the inner surface of the pericarp, the mesocarp and endocarp are fused and appear as a single layer. In this fused layer, starch granules are also present (Bultosa 2016). Fonio grain has a shiny pericarp that varies in colour from white to yellow to violet depending on the variety. The data on the anatomy of the kodo and little millet is scarce, however, the bran and husk of kodo constitute about 37% of the total grain (Taylor and Emmambux 2008). The seed coat or testa in the pearl millet is present in a thin layer and also contains the pigment in aleurone and endosperm. The seed coat (testa) of the finger millet contains five layers which appear red to purple (Kumar et al. 2016). The seed coat of pearl millet, foxtail millet, proso millet, teff, and fonio millet contains a single layer (Guha et al. 2015; Dayakar et al. 2017). The seed coat thickness of the finger millet is the highest (10.8–24.2 picometre) (Pm) amongst all the millets. The seed coat thickness of pearl millet, proso millet, teff, and fonio is 0.4 Pm, 0.2–0.4 Pm, 1.5–2.0 Pm, and 0.5–1.5 Pm, respectively (Kulp 2000). The aleurone layer of millets consists of a single layer of living cells that is present on the outer side of the endosperm (Guha et al. 2015). It plays an important role in germination process due to the presence of a higher content of protein. Pearl millet (16–30 × 5–15 Pm) and finger millet (18 × 7.6 Pm) have the largest aleurone layer cell size compared to proso millet (12 × 6 Pm), teff (12 × 4 Pm), and fonio (12 × 4 Pm) (Prakash et al. 2019; Powar et al. 2018). The lipid droplets and protein bodies are also present in aleurone cells (Portères 2011). The starch granules in the aleurone layer range from 8.0 to 21.0 μm for finger millet, 7 to 15 μm for kodo millet, 6.4 to 12 μm for pearl millet, 5 to 25 μm for foxtail millet, 1.3 to 17 μm for proso millet, and 2.5 to 20 μm for barnyard millet (Taylor and Duodu 2018; Saleh et al. 2013; Guha et al. 2015).

5.4.2 Endosperm

The endosperm is the largest part of the grains and is composed mainly of starch and proteins. The size of the endosperm in the pearl millet has been reported in the range of 73–75% (Satyavati et al. 2017). Foxtail millet (12:1), proso millet (12:1), and finger millet (11:1) have the highest endosperm-to-germ ratios, while fonio (6:1), pearl millet (4.5:1), and teff (3:1) have the smallest endosperms and endosperm to germ ratios (Berwal et al. 2017; Singh et al. 2018; Asha 2021). The starch content of the endosperm varies according to the millet type. The highest (80%) starch content is found in the endosperm of fonio millet followed by teff (73%), kodo millet (72%), pearl millet (71.6%), finger millet (59%), proso millet (57.1%), foxtail millet (55.1%), barnyard millet (19.2%), and little millet (18.8%) (Sharma and Chauhan 2018; McDonough et al. 2000). The highest protein content is reported for the endosperm of barnyard millet (14.1%) and pearl millet (14%) followed by proso millet (13.4%), foxtail millet (11.7%), teff millet (10.9%), kodo millet (10.4%), fonio (8.7 %), and finger millet (8%).

Structurally, the endosperm is divided into the aleurone layer, the peripheral area, the corneous (hard) area, and the floury (soft) area (Taylor and Anyango 2011). The aleurone layer is the innermost layer of the bran and the outermost layer of the endosperm. The composition of this layer has already been discussed under the bran section. The starch granules of higher size are found in finger millet (8.0–21.0 μm), kodo millet (7–15 μm), pearl millet (6.4–12 μm), foxtail millet (5–25 μm), proso millet (1.3–17 μm), and barnyard millet (2.5–20 μm) (Guha et al. 2015; Saleh et al. 2013).

The next part is the peripheral endosperm which contains several layers of dense cells containing small, compact, and angular starch (6.4–6.8 μm) units along with high protein content. The protein matrix consists of the glutelin and protein bodies. The glutelin is alkali-soluble and protein bodies are alcohol-soluble prolamins (Berwal et al. 2017). The corneous and the floury areas comprise the largest portions of the endosperm and they also control the texture of the millet kernel. The soft textured kernels have more floury endosperm, while solid kernels have more thickly filled corneous endosperm (Hassan et al. 2021). The cells of the corneous endosperm contain a continuous matrix of protein bodies. The floury endosperm is the innermost part of the cereal grain and consists of relatively larger and less angular starch granules. It is also associated with thinner walls and less protein content is present in some of the millets such as foxtail millet, barnyard millet, teff, fonio, and finger millet. The floury appearance is due to the presence of minuscule air spaces in the grain. The pearl millet kernel may contain floury, very soft, corneous, and very hard endosperm. The endosperm of pearl millet may vary from yellow to grey as per the variety. The corneous endosperm of pearl millet is yellow. A grey endosperm indicates the presence of pigments. The pearl millet endosperm may also contain β-carotene which is responsible for the external colour of the kernel. The outer part of the endosperm is vitreous and hard, whereas the inner part consists of the soft endosperm with an opaque region. The hard endosperm does not contain air spaces in a tightly packed cellular structure. The starch is polygonal and uniform in shape and a small number of protein bodies are embedded in the matrix. The size of protein bodies is 1.5 μm and the starch granules are about 10 μm. The protein is present in the hard endosperm, whereas the soft endosperm does not contain protein bodies (Gomez and Gupta 2003). In pearl millet, the endosperm protein bodies are present mostly in the floury zone compared to the peripheral and corneous zone (Ojediran et al. 2010). It also contains magnesium, calcium, phosphorous, and potassium. The starch granules in the floury endosperm are big and spherical (Pawase et al. 2019).

The nature of foxtail millet endosperm is waxy (high amylopectin content), non-waxy (high amylose content), and normal (low amylose content) (Sharma et al. 2021). The starch granules have a size of approximately 7.6 μm (Zhu 2014) and may be polygonal, round, or irregular in shape with the attachment of protein bodies (Babu et al. 2019). The angular and polygonal starch granules are found in the flinty endosperm and the spherical granules are found in the mealy endosperm. The protein bodies in the endosperm of foxtail millet are spherical with a diameter of 1–2 μm. The size, shape, and packaging of the starch granules in proso millet are different at different endosperm sites and are classified into monomodal and bimodal. The bimodal is further categorized into A-granules with a size >10 μm and B-granules with a size <10 μm. Starch granules with a diameter <2.5 μm may be categorized as C-granules (Zarnkow 2007).

The starch granules in barnyard millet are simple, they do not contain the waxy part, and vary in size from 2.5 to 20 μm (Gaurav et al. 2021, Rao et al. 2020). Barnyard millet also contains the testa layer and the starch hilum in the centre (Paschapur et al. 2021). In finger millet, the endosperm is attached to the seed coat and endosperm cell walls show strong fluorescence which tells the presence of phenolic compounds. The finger millet endosperm contains equal proportions of corneous and floury areas. The starch granules in the endosperm are rhombic, spherical, and polygonal and vary in size from 5 to 15 μm (McDonough et al. 1986). Finger millet also contains protein bodies (the protein is stored in starchy endosperm) with a diameter of 2–5 μm (Shobana et al. 2013). The endosperm of kodo millet may have polygonal, spherical, or large polygonal starch granules which vary in diameter from 1.2 to 9.5 μm. The protein bodies of endosperm are spherical and polygonal (Kumari and Thayumanavan 1998). The little millet endosperm contains small, spherical starch granules of 1.0–9.0 μm and the shape of the protein bodies is spherical and polygonal (Bean et al. 2019). Teff contains a floury endosperm with a huge number of starch granules. The starch granules are polygonal in shape, 2–6 μm in diameter, and are devoid of surface pores and channels (Emmambux and Taylor 2013). The endosperm of fonio consists of a starchy endosperm and an aleurone layer. This layer contains lipids and proteins at the periphery in large amounts. The starch granules are polyhedral in shape, have a diameter of 2–13 μm, and are embedded with a protein matrix of the continuous phase and protein bodies (Babu et al. 2019).

5.4.3 Germ

The embryonic axis, scutellum, plumule, and primary root are the major portion of the germ. The embryonic axis is responsible for the split of immature plants and embryos. This axis terminates at the part where the root develops and the plant matures (Ramakrishnan et al. 2019). The scutellum is the modified cotyledon which is the part between the embryo and endosperm. It appears in wing-like expansions, contains cells of irregular shape (Serna-Saldivar and Ramirez 2019), and helps to transfer nutrients from the endosperm to the developing embryo. The plumule is the plant embryo's younger shoot above the cotyledons which contain the immature leaves and epicotyl. The primary root plays a crucial role in the growth of seedlings and the survival of root systems (Qin et al. 2019). There are two common germ types, i.e. caryopsis and utricle. Pearl millet, foxtail millet, barnyard millet, kodo millet, teff, and fonio contain the caryopsis germ type, whereas proso millet contains the utricle germ type (Singh et al. 2018a). The pearl millet contains the largest (1420 × 620 μm) germ compared to finger millet (980 × 270 μm), fonio (517 × 977 μm), foxtail millet (270 × 980 μm), teff (160 × 725 μm), and proso millet (100 × 310 μm) (Bultosu 2016; Berwal et al. 2017; Singh et al. 2018; Crop 2020; Barretto et al. 2021). The germ may also have different shapes such as obovate, globose, lanceolate, hexagonal, spherical, polygonal, elliptical, oval, polyhedral, and globose ellipsoids. The germ of pearl millet contains obovate, globose, lanceolate, or hexagonal shapes, whereas foxtail millet, proso millet, and finger millet germs are spherical, polygonal, and oval. The germ of kodo millet and little millet is elliptical to oval. The germ of fonio millet is polyhedral, globose ellipsoids, and oval-shaped (Singh et al. 2018; Ramashia 2018; Barretto et al. 2021). The germ of pearl millet, foxtail millet, proso millet, and finger millet contains protein bodies of the size 0.5–2.5 mm, 1–2 mm, 0.6–1.2 mm, and 2.0–2.33 mm, respectively. It is rich in lipids, proteins, minerals, vitamins, and enzymes.

5.5 PHYSICAL CHARACTERISTICS OF MILLETS

5.5.1 Colour

The colour of millets is an indication of their nutritional and phytochemical composition. It also determines their functional value and is a key component in determining the acceptance of a food (Dey and Nagababu 2022). The pearl millet has white, yellow, and brown grains. It is generally believed that the dark grey varieties of pearl millet are high in iron; however, a study concluded that the grain colour had a non-significant effect on the iron and zinc content of pearl millet varieties (Govindraj et al. 2017). The foxtail millet grains are white, red, or black. Mikulajova et al. (2017) compared the yellow and red genotypes of foxtail millet for the proximate composition and antioxidant content. It was found in the study that the protein (15.32 g/100g) and fat (6.6 g/100g) content of the Slovensky yellow genotype was higher than other studied varieties. Red genotype 00002 had the highest content of individual phenolics. Slovensky had the higher total antioxidant capacity as assessed by DPPH in comparison to Red 0011. The red genotype 0011 was found to have a higher content of p-coumaric acid, ferulic acid, and cinnamic acid derivatives than the yellow genotype Slovensky. Two brown-coloured grain genotypes (PS-4 and SIA-3126) had a fat content of 4.8 and 4.9 g/100 g and protein content of 11.3 and 13.4 g/100 g), respectively (Devisetti et al. 2014).

Proso millet grains are white, black, brown, yellow, red, and orange, and barnyard millet grains have a white to yellow colour. Literature is scanty regarding the impact of colour on the nutritional and phytochemical profile of proso and barnyard millet and therefore further research is warranted in this area. Finger millet grains vary from white to dark brown and the most commonly found colours are white, yellow, red, and dark brown. Higher protein content is reported in brown-seeded varieties in comparison to white-seeded varieties. In contrast to this study, Rao et al. (1994) reported higher protein content in the white variety of finger millet than in the brown variety. However, no significant difference in the amino acid composition of brown and white varieties of finger millet was reported. Seetharam (2001) reported a higher content of calcium in the white varieties than in the brown varieties. The brown varieties of finger millet were also reported to have more total phenols, flavonoids, and tannins compared to the white varieties (Xiang et al. 2019, Rao et al. 1994). Kodo millet grains vary from brick red to dark brown, little millet grains vary from straw white to grey, teff grains are ivory white to pale yellow, and fonio grains are white to pale brown (Table 5.1) (Berwal et al. 2017; Singh et al. 2018; Pawase et al. 2019; Barretto et al. 2021). Enough literature is not available on the effect of colour on the nutritional, phytochemical, and functional properties of kodo millet, little millet, teff, and fonio and hence more studies are suggested for these millets.

5.5.2 Hardness

Hardness is an important quality attribute of grains that gives an idea of the soundness and health of grains. It also determines the grains' resistance to scratching and deformation and is one of the most important factors to determine the milling quality of flour (Hourston et al. 2017). A higher hardness value results in higher milling yields compared to soft grains. Among millets, kodo millet grains have the highest hardness value (52.63 N) followed by pearl millet (49.4 N), barnyard millet (31.48 N), finger millet (37.51 N), little millet (32.85 N), foxtail millet (29.62 N), proso millet (20.99 N), and teff (5.99 N) (Rajsekhar et al. 2018; Chaturvedi et al. 2019; Rao et al. 2020; Gaurav et al. 2021; Pulivarthi et al. 2022). A decrease in the hardness value of millet grains is observed with an increase in moisture content and vice-versa. Balasubramanian and Viswanathan (2010) reported a decrease in grain hardness from 30.7 to 12.4 N for all minor millets when the moisture was increased from 11.1% to 25% on a dry basis. Sunil et al. (2016) reported a hardness value of 24.11 N in foxtail millet at a moisture content of 11.67%.

5.6 ENGINEERING PROPERTIES

The principle axial dimensions, i.e., length, width, and thickness, play an important role in designing machines for the threshing, cleaning, handling, and milling of grains. These properties also play an important role in designing storage structures known as "silos". This section discusses the engineering properties of major millet species grown for food purposes. Details on the engineering properties of various millet varieties are provided in Table 5.2.

5.6.1 Principle Axial Dimensions: Length, Width, and Thickness

Millet grains are generally spherical and have a tiny size. Hence, it is very tedious to measure their principal axial dimensions such as length, width, and thickness. These parameters can be measured by selecting ten random grains and taking the average of their length, width, and thickness,

TABLE 5.2 ENGINEERING PROPERTIES OF DIFFERENT MILLETS

Properties	Pearl millet	Foxtail millet	Proso millet	Barnyard millet	Finger millet	Kodo millet	Little millet	Teff	Fonio
Weight (mg)	3–15	3–4	5–9	3–4	2–3	3–6	3–4	0.2–0.4	0.5–0.6
Length (mm)	1.20–2.00	2.12–2.22	2.27–2.41	2–3	1.52–1.85	2.61–2.74	2.32–3.01	0.9–1.7	1.4–1.8
Width (mm)	0.60–2.50	1.20–1.76	2.21–2.31	–	1.42–1.56	1.96–2.23	–	–	–
Thickness (mm)	0.61–1.10	1.06–1.78	1.59–1.89	–	1.44–1.51	1.33–1.45	1.85–1.90	1.50–2.00	0.50–1.50
1000 kernel weight (g)	2.5–14	2.0–2.6	6.1–7.1	3–4	2.3–2.6	5–6	4–5	–	0.5
Geometric mean diameter (mm)	0.07–2.12	1.42–1.89	1.96–2.17	–	1.44–1.59	–	–	–	–
Sphericity (%)	58.1–95.3	65.9–88.1	86.1–92.3	89.1–96.3	86.1–97.6	–	–	–	–
Surface area (mm²)	–	–	12.12–14.47	9.32–12.16	6.65–8.35	9.06–11.53	–	–	–
Bulk density (kgm^{-3})	780–833	737.12–756.21	782.56–812.34	–	790–842	620–670	–	–	–
True density (kgm^{-3})	1110–1280	1260.132–1267.65	–	1225.50–1308.1	1	1200.12–1242.33	–	–	–
Porosity (%)	36.20–49.02	41.47–48.01	42.87–48.78	43.00–57.47	–	43.99–50.27	–	–	–
Angle of response (°)	22.68–25.11	26.78–26.86	21.95–26.32	24.47–39.13	24.93–32.81	25–26	–	–	–
Terminal velocity (ms^{-1})	–	–	–	4.22–6.59	–	–	–	–	–
Hardness (N)	–	24.21–39.83	31.38–37.21	–	–	–	–	–	–
Coefficient of static friction (g/g)	0.5–06	0.3057–0.4478	–	0.56–0.84	–	–	–	–	–

Source: Kumar et al. 2020; Singh et al. 2018; Chhabra and Kaur 2017; Kumar et al. 2016; Bhise et al. 2014; Singh et al. 2010; Bargeh 2002; Jain and Bal 1997.

or the size of the grains can be zoomed to 10X with the help of a projector, and this image can be drawn on white paper, and further measurements can be made from this paper (Kumar et al. 2020). The average of three or more grain samples gives more reliable results. In general, the length of pearl millet, finger millet, foxtail millet, proso millet, barnyard millet, kodo millet, little millet, teff, and fonio grains varies in the range of 2–4 mm, 1–3 mm, 2–3 mm, 2–3 mm, 3 mm, 3–4 mm, 2–3 mm, 0.9–1.7 mm, and 1.4–1.8 mm, respectively (Kumar et al. 2016, Singh et al. 2017, Singh et al. 2018, Sunil et al. 2016). The maximum width is reported for pearl millet (2.60 mm), followed by proso millet (2.01 mm), little millet (1.83 mm), foxtail millet (1.74 mm), and finger millet (1.58 mm) (Prakash 2020, Rao et al. 2020, Singh et al. 2018a, Ojediran et al. 2010). The thickness values are highest for pearl millet (2.20 mm), followed by foxtail millet (2.12 mm), teff (2.00 mm), proso millet (1.52 mm), little millet (1.52 mm), fonio (1.50 mm), and barnyard millet (1.37 mm) (Rao et al. 2020). The average grain weight for pearl millet, foxtial millet, barnyard millet, proso millet, finger millet, kodo millet, little millet, teff, and fonio varies from 3–15 mg, 2–3 mg, 3–4 mg, 5–9 mg, 3–4 mg, 3–6 mg, 2–3 mg, 0.2–0.4 mg, and 0.5–0.6 mg, respectively. (Crop 2020; Upadhya et al. 2014). The grain size is controlled and affected by genetic makeup, environmental factors, proteome, and metabolic factors (Kesavan et al. 2013). These properties are affected by factors like moisture content and storage temperature. An increase in millets' length, width, and thickness has been observed with an increase in moisture content. The increase in length can be attributed to the hygroscopic nature of millet grains and the swelling property of millet starch (Balasubramanian and Viswanathan 2010). Pearl millet's grain length, width, and thickness increased from 2.997 mm to 3.172 mm, 2.433 mm to 2.609 mm, and 2.036 mm to 2.209 mm, respectively, with an increase in moisture content from 10 to 30% on a dry basis (Prakash 2020). In another study, the length, width, and thickness of the pearl millet grain increased from 3.16 mm to 4.46 mm, 2.30 mm to 3.39 mm, and 1.54 mm to 2.51 mm, respectively, with an increase in moisture content from 10 to 20% on a wet basis (Ojediran et al. 2010). Sunil et al. 2016) reported similar results for the principle axial dimensions of foxtail millet, where the length, width, and thickness increased from 2.353 mm to 2.493 mm, 1.423 mm to 1.627 mm, and 1.06 mm to 1.78 mm, with an increase in moisture content from 9.8 to 21% on a wet basis. Increases in the length (2.27 to 2.37 mm), width (2.08 to 2.29 mm), and thickness (1.59 to 1.84 mm) of proso millet were observed with an increase in moisture content from 0.065 to 0.265 kg kg^{-1} (Singh et al. 2010). In barnyard millet, the length (2.43 to 2.57 mm), breadth (1.94 to 2.01 mm), and thickness (1.26 to 1.30 mm) increased with an increase in moisture content from 8 to 14% on a dry basis (Lohani and Pandey 2008). In finger millet, grain length (1.61 to 1.81 mm) and width (1.58 to 1.84 mm) increased with an increase in moisture from 0.065 to 0.265 kg kg^{-1} (Singh et al. 2010a). Singh et al. (2018) reported an increase in the thickness (1.41 to 1.61 mm) of finger millet with an increase in moisture content. Kumar et al. (2016) also reported an increase in length (2.62 to 2.74 mm) and thickness (1.33 to 1.45 mm) with an increase in moisture content from 8.19 to 12.71%. However, a decrease in length breadth ratio (1.33 mm to 1.23 mm) and an increase in width (1.96 to 2.23 mm) were observed with an increase in moisture content (Kumar et al. 2016). In teff millet, the length (101 to 127 mm) and width (59 to 68 mm) were increased with an increase in moisture content from 5.6 to 29.6% on a wet basis (Zewdu and Solomon 2007).

5.6.2 Diameter

The shape of cereal grains can be considered as a prolate sphere, from which the diameter (D_p) can be calculated by using the equation,

$$D_p \text{ (Diameter)} = (L((W + T)^2/4))^{1/3}$$

Where L is the length of the grain, W is the width of the grain, and T is the thickness of the grain (Varnamkhasti et al. 2008; Kumar et al. 2020).

Pearl millet has a significantly higher (2.50 mm) diameter compared to proso millet (2.05 mm), little millet (1.90), barnyard millet (1.90 mm), foxtail millet (1.80 mm), and finger millet (1.70 mm) (Ramashia et al. 2018, Sunil et al. 2016, Prakash 2019, Rao et al. 2020). Finger millet's diameter and moisture content showed a linear relationship. Ramashia et al. (2018) reported an increase in the diameter of finger millet from 1.41 mm to 1.58 mm with an increase in moisture content from 7 to 25% on a dry basis. An increase in the diameter of finger millet grain from 1.53 mm to 1.75 mm was also reported with an increase in the moisture content from 0.065 to 0.265 kg kg^{-1} (Singh et al. 2017). The diameter of finger millet decreases slightly during hydrothermal treatment and this might be due to the shrinkage of grain. The longitudinal and lateral diameter of finger millet grains decreased from 1.68 mm to 1.63 mm and 1.60 mm to 1.46 mm respectively on dehydration (Rajasekhar et al. 2018). In foxtail millet, the range of geometric mean diameter varies from 1.38 mm to 1.75 mm (Sunil et al. 2016). Sunil et al. 2016) reported an increase in the geometric mean diameter of barnyard millet from 1.68 mm to 1.90 mm with an increase in moisture content. The geometric mean diameter of pearl millet increased from 2.45 mm to 2.62 mm, with an increase in moisture content from 10 to 30% on a dry basis (Prakash 2019). The increase in moisture content from 0.065 to 0.265 kg kg^{-1} in proso millet is responsible for increasing the geometric mean diameter from 1.96 mm to 2.14 mm (Singh et al. 2010). The geometric mean diameter of barnyard millet increased from 1.3 mm to 1.5 mm due to an increase in moisture content from 9.12 to 17.06% on a dry basis (Singh et al. 2010).

5.6.3 Sphericity

Sphericity indicates how closely the shape of the grain resembles the shape of a sphere. It is defined as the ratio of the surface area of a sphere to the surface area of the sample grain. The sphericity of the grains can be calculated using the following formula.

$$\text{Sphericity } (\phi) = (LWT)^{1/3}/L$$

Where L is the length of the grain, W is the width of the grain, and T is the thickness of the grain (Varnamkhasti et al. 2008).

The sphericity of millet is influenced by the soaking temperature, steaming pressure, and moisture content (Ushakumari et al. 2007). Finger millet grain has higher sphericity (0.910) (Sunil and Venkatachalapathy 2016) compared to pearl millet (0.820) (Prakash et al. 2019), proso millet (0.728) (Ojediran et al. 2010), foxtail millet (0.678), barnyard millet (0.655), and little millet (0.584) (Singh et al. 2018a, Chaturvedi et al. 2019). The sphericity of finger millet increased from 0.91 to 0.92 with an increase in moisture content from 7 to 25% (Powar et al. 2018). Singh et al. (2018) reported that the sphericity of finger millet grain increased from 0.95 to 0.96 with an increase in moisture content from 0.065 to 0.265 kg kg^{-1}. In foxtail millet, grain sphericity increased from 0.714 to 0.762 with an increase in moisture content from 9.8 to 21% on a wet basis (Sunil et al. 2016). Sunil et al. (2016) also reported an increase in sphericity (0.659 to 0.881) of foxtail millet with an increase in moisture content. The sphericity of the pearl millet varies from 0.821 to 0.833, with a change in the moisture content from 10 to 30% on a dry basis (Prakash et al. 2019). For pearl millet, it increased from 0.667 to 0.744 with an increase in moisture content (Ojediran et al. 2010). The sphericity of proso millet increased from 0.86 to 0.91 with an increase in moisture content from 0.045 to 1.45 kg kg^{-1} (Singh et al. 2010). The sphericity of proso millet

(0.96 to 0.97) significantly increased with an increase in moisture content from 6.5 to 26.5% on a dry basis (Singh et al. 2018).

5.6.4 Thousand Kernel Weight

The thousand kernel weight for cereal grains can be determined by counting a thousand kernels and weighing them on an electrical balance (Mariotti et al. 2006). Thousand kernel weight of teff (0.2–0.4 g), and fonio (0.5–0.6 g) is very less compared to other millets like pearl millet (9.47–15 g), proso millet (5–7 g), barnyard millet (3–4 g), finger millet (2.89–3.182 g), foxtail millet (3.44–4.0 g), etc. (Rao et al. 2020; Rajasekhar et al. 2018; Singh et al. 2017; Ojediran et al. 2010). The thousand kernel weight of millet is influenced by various parameters such as moisture content, soaking, and steaming pressure. Powar et al. (2018) reported an increase in the thousand kernel weight of finger millet (2.297 to 2.918) with an increase in moisture content from 7 to 25% on a dry basis. Singh et al. (2018) also reported an increase in the thousand kernel weight (2.527 to 2.987 g) of finger millet with an increase in moisture content from 0.065 to 0.265 kg kg^{-1} (Singh et al. 2010). The hysteresis effect of foxtail millet is responsible for an increase in thousand kernel weight from 2.436 g to 2.837 g (Sunil et al. 2016). The thousand kernel weight of pearl millet increased from 9.11 to 10.12 g on a 7.97% increase in moisture content (Chhabra and Kaur 2017). Another study reported that the thousand kernel weight of pearl millet varied from 7.3 to 11.95 g with an increase in moisture from 10 to 20% on a wet basis (Ojediran et al. 2010). The thousand-grain weight of proso millet increased from 4.69 to 6.19 g with an increase in moisture content from 0.065 to 0.265 kg kg^{-1} (Singh et al. 2010). The thousand-grain weight of proso and barnyard millet varied from 5.25 to 6.70 g and 4.17 to 5.45 g due to an increase in moisture content from 6.5 to 26.5% (Singh et al. 2018).

5.6.5 Bulk Density

The bulk density is the relationship between the mass and volume of the grain and can be measured by weighing a known volume of cereal grains (Kumar et al. 2020).

$$\text{Bulk density} = \text{Weight of grain}/\text{Volume}$$

Pearl millet has a bulk density of 0.852 g/mL. Among the minor millets, finger millet has a significantly higher (0.660 g/mL) bulk density compared to kodo millet (0.620 g/mL), proso millet (0.587 g/mL), barnyard millet (0.552 g/mL), and little millet (0.496 g/mL) (Singh et al. 2018; Sunil and Venkatachalapathy 2017; Kumar et al. 2016; Upadhyaya et al. 2014; Guha et al. 2015; Ramashia et al. 2018; Prakash et al. 2019; Singh et al. 2010; Panda et al. 2021). Several studies reported a decrease in bulk density with an increase in moisture content due to the rise in the kernel's volume and mass (Al-Mohasnen and Rababah 2007). Rao et al. (2020) reported a higher bulk density (0.773 g/mL) for foxtail millet grains at a moisture content of 11.06% compared to 12.19%. The increase in moisture content (0.065–0.265 kg kg^{-1}) of finger millet is responsible for a decrease in bulk density (788.1 to 627.6 kgm^{-3}) (Singh et al. 2017). Powar et al. (2018) reported a decrease in bulk density of finger millet from 749 to 665 mm^3 with an increase in moisture content from 7 to 25%. In foxtail millet, the increase in moisture content from 9.8 to 21% on a wet basis was responsible for decreased bulk density (749.5 to 729.94 kg/m^3) (Sunil and Venkatachalapathy 2017). Similar results were obtained for kodo millet where the bulk density decreased from 0.67–0.63 g mL^{-1} with an increase in moisture content from 8.19 to 12.71% on a dry basis (Kumar et al. 2016). A study reported decrease in bulk density of pearl millet from 827.5 to 711.9 kg/m^3 with increase in moisture content from 10 to 30% on a dry basis (Prakash et al. 2019). Ojediran et al. (2010) reported the bulk density of pearl millet decreased from 817.64 to 646.4 kg/m^3 with an increase in moisture content from 10 to 20% on a wet basis. In proso millet, the bulk density varies from 774.30 to 809.67 kg/m^3 due to an increase in moisture content from 0.045 to 0.145 kg/kg (Singh et al. 2010). Another study observed a decrease in the bulk density of proso millet from 765.5 to 697.57 kg/m^3 with an increase in moisture content from 6.5 to 26.5% on a dry basis (Singh et al. 2018).

5.6.6 True Density

True density is the measure of the mass per unit volume of the grains excluding the pore spaces in between the samples. It can be determined by the displacement method (using a non-absorbing solvent like toluene, kerosene oil, etc.) in which the weighed amount of cereal is added to a pre-recorded volume of non-absorbing liquid, and the volume change caused by the added grains is used to calculate the true density (Kumar et al. 2020).

$$\text{True density} = \text{Weight of grain}/\text{Volume of liquid displaced}$$

The true density of the pearl millet is higher (1.531 g/mL) compared to kodo millet (1.230 g/mL), foxtail millet (1.190 g/mL), proso millet (1.136 g/mL), and little millet (1.027 g/mL) (Sunil et al. 2016; Kumar et al. 2016; Ramashia et al. 2018; Rao et al. 2020; Barretto et al. 2021; Pawase et al. 2019; Heiru et al. 2019). According to the toluene displacement method, the true density and proximate composition of millets are affected by an increase in moisture content. Finger millet and barnyard millet show a similar true density (1.000 g/mL) with an increase in moisture content from 11.06 to 12.19% on a wet basis (Rao et al. 2020). The increase in the moisture content of finger millet from 0.065 to 0.265 kg/kg is responsible for an increase in the true density from 1336.3 to 1438.9 kg/m^3 (Singh et al. 2017). Powar et al. (2018) reported a decrease in true density (1515 to 1000 kg/m^3) of finger millet with an increase in moisture content from 7 to 25% on a dry basis. The true density of foxtail millet decreased from 1315.71 to 1204.83 kg/m^3 with an increase in moisture from 8 to 21% on a wet basis (Sunil and Venkatachalapathy 2017). In contrary to this, the true density of kodo millet increased from 1.20 to 1.24 g/ml with an increase in moisture content from 8.19 to 12.71% (Kumar et al. 2016). In pearl millet, the true density decreased from 1314.17 to 1216.97 kg/m^3 with an increase in moisture from 15 to 20% (Prakash et al. 2019). Singh et al. (2018) reported that the true density of proso millet and barnyard millet increased from 1354.02 to 1467.37 kg/m^3 and 1225.5 to 1308.1 kg/m^3, respectively, with an increase in water content from 0.065 to 0.265 kg/kg (Singh et al. 2010).

5.6.7 Volume

Volume is the amount of space covered by an object within the boundaries. The volume of cereal grain can be calculated by using the equation

$$\text{Volume (V)} = 0.25((\pi/6)L(W + T)^2)$$

Where L is the length of the grain, W is the width of the grain, and T is the thickness of the grain (Kumar et al. 2020).

The volume range of pearl millet is significantly higher (6.324–7.833 mm^3) compared to kodo millet (2.59–3.46 mm^3), foxtail millet (1.785–2.696 mm^3), and finger millet (1.62–2.20 mm^3) (Singh et al. 2010; Sunil and Venkatachalapathy 2017; Ramashia et al. 201). Powar et al. (2018) reported an increase in the volume (1.40–2.10 mm³) of finger millet kernel with an increase in moisture content from 7 to 25% on a dry basis (Powar et al. 2018). Singh et al. (2018) reported an increase in the volume (1.75 to 2.69 mm³) of finger millet with an increase in moisture content from 6.5 to 26.5%. The increase in steaming pressure from 1 to 2 kg/cm² increased the grain volume from 6.324 to 7.833 mm³ on a wet basis (Prakash et al. 2019).

5.6.8 Interstice

An interstice is a small narrow space or interval between things or parts. It strongly relates to the moisture content in finger millet, showing the kernels' decrease in intermolecular space. The increase in moisture content (0.065 to 0.265 kg/kg) reduces the interstice property of finger millet from 0.26 to 0.01. A decrease in the interstice of barnyard millet is also reported with an increase in moisture content (Singh et al. 2010).

5.6.9 Surface Area

Surface area is the region occupied by any object's surface. The surface area of cereal grains can be calculated by using the following equation,

$$S = \pi BL^2/(2L - B)$$

Where

$$B = \sqrt{WT}$$

S is the surface area (mm²), B is the breadth of the grain (mm), L is the length of the grain (mm), W is the width of the grain (mm), and T is the thickness of the grain (mm) (Varnamkhasti et al. 2008).

The foxtail millet surface area significantly increased from 7.53 to 9.75 mm², while the finger millet surface area decreased from 6.78 to 6.25 mm² due to changes in storage conditions and a reduction of moisture content by 3.2% in hydrothermally treated millet at 14% moisture content on a dry basis (Sunil and Venkatachalapathy 2017; Rajasekhar et al. 2018). The surface area of finger millet decreases from 6.782 to 6.246 mm² when steaming pressure is increased from 1 to 2 kg/cm² (Prakash et al. 2019). However, the surface area of finger millet increased from 6.31 to 7.88 mm² due to an increase in moisture content (Barretto et al. 2021). Another study reported an increase in the surface area of millets from 7.05 to 9.35 mm² with an increase in moisture content from 9.8 to 21% on a wet basis (Sunil and Venkatachalapathy 2017).

5.6.10 Porosity

Porosity is the object's pores volume ratio. It is expressed as a percentage (%). Porosity for cereal grains can be calculated by

$$\text{Porosity } (\varepsilon) = (\rho_t - \rho_b)/\rho_t \times 100$$

Where ρ_t is the true density, and ρ_b is the bulk density (Kumar et al. 2020).

The porosity of foxtail millet is 35% (Rao et al. 2020), pearl millet is 76.83% (Ramashia et al. 2018), proso millet is 48.32% (Rao et al. 2020), little millet is 51.74%, finger millet is 51% (Powar et al. 2018), and barnyard millet is 44.80% (Sunil et al. 2016). In foxtail millet and kodo millet, the porosity decreased from 49.58 to 38.89% and 43.99 to 50.27% due to an increase in moisture content from 9.8 to 21% on a wet basis (Sunil and Venkatachalapathy 2017). In pearl millet, an increase in moisture content from 10 to 30% increased the porosity from 37.02 to 41.56% (Parkash et al. 2019). Powar et al. (2018) observed a decrease in porosity (51 to 34%) and mass (47 to 30%) with an increase in moisture content from 7 to 25% on a dry basis in finger millet. Singh et al. (2018) also reported an increase in the porosity of finger millet with an increase in moisture content from 41.02 to 55.43%. Sunil et al. (2016) reported an increase in foxtail millet porosity (40.08 to 43.92%) with an increase in moisture content from 0.065 to 0.265 kg/kg. The porosity of pearl millet decreases with an increase in bulk density (Ojediran et al. 2010). The porosity of proso millet and barnyard millet increased from 42.87 to 44.59% and 34.2 to 45.8%, with an increase in moisture content from 0.065 to 0.265 kg/kg (Singh et al. 2010). Germination has been found to significantly reduce the porosity of finger millet during the initial 72 hours due to nutrient migration to the surface and filling of pores and increase in the later hours potentially due to shrinking and crack formation on the grain (Kumar et al. 2020).

5.6.11 Coefficient of Friction

The coefficient of friction is the ratio of the frictional force resisting the motion of two surfaces in contact to the normal force pressing the two surfaces together. It is a dimensionless quantity as both frictional force and normal force are measured in Newtons or pounds (Brittanica 2020). It is calculated as:

$$\mu = F/N$$

Where μ is the static friction coefficient, F is the frictional force, and N is the normal force.

The coefficient of friction varies for static friction and kinetic friction. Static friction is the frictional force that resists the force applied to an object. This force is required to change the resting stage of an object. The static friction coefficient is the ratio of maximum static force between the surfaces in contact before the movement commences to the normal force (Behera and Hari 2010). In kinetic friction, the frictional force resists the motion of an object (Brittanica 2020). It is the ratio of the friction force to the normal force experienced by a body moving on a dry, non-smooth surface.

The friction coefficient of millets is affected by different factors, such as moisture content, the surface area and surface properties of the grains, force applied, state of matter, movement, and temperature (Nwakuba et al. 2019). An increase in static friction coefficients of foxtail millet was reported on glass (0.315 to 0.438), galvanized iron sheets (0.428 to 0.487), and acrylic sheets (0.295 to 0.375) with an increase in moisture content from 9.8 to 21% (Sunil and Venkatachalapathy 2017). Subramanian and Viswanathan (2007), reported that millet flours' static friction on a stainless-steel surface increased from 0.62 to 0.88 as their moisture content increased from 11.11 to 42.86% dry basis. Jain and Bal (1997) observed that at 7.4% moisture content, the average coefficient of static friction for three kinds of pearl millet (*P. gambiense*) was 0.25 on galvanized iron and 0.26 on mild steel plates. Baryeh (2002) reported an increase in the coefficient of friction with an increase in the grain moisture content.

5.6.12 Coefficient of Internal Friction

The coefficient of internal friction is the friction between solid particles or grain mass moving against each other (Neikov and Yefimov 2019). The coefficient of internal friction can be measured using an apparatus consisting of two wooden rectangular boxes, with one being stationary and the other one being free to slide on the stationary one. In this measurement, the top box is made to slide on the stationary box through a pulley-rope arrangement with a loading pan. The outer box is placed in position on the sliding surface with some grains filled into it. This is followed by placing the small box in its centre and filling it with sample. The grain mass sample contained in the internal box acts on the layer of the same material present in the bottom stationary box. An incremental load is applied on the loading pan to slide the internal box. The force required to slide the empty internal box is subtracted from the frictional force required to get actual frictional force to overcome the friction due to the material. The coefficient of internal friction is calculated as the ratio of weights added (frictional force) and grain mass (normal force) as given below.

$$\mu_i = W_2 - W_1/W$$

Where μ_i is the coefficient of internal friction, W_2 is the weight required to cause the sliding of a filled smaller box, W_1 is the weight to cause the sliding of an empty smaller box, and W is the weight of the material inside the smaller box.

Research revealed that the coefficient of internal friction of finger millet is directly related to the moisture content. Singh et al. (2018) reported an increase in the internal friction coefficient of finger millet from 0.536 to 0.774 with an increase in moisture from 6.5 to 26.5% on a dry basis (Singh et al. 2018). The increase in the coefficient of internal friction of barnyard millet (0.66 to 0.91) was reported with an increase in moisture content from 0.065 to 0.265 kg/kg (Singh et al. 2010). The values of the coefficient of static friction and the coefficient of internal friction showed a linear connection with the amount of moisture present, with greater coefficients of fit being associated with linear relationships (Subramanian and Viswanathan 2007).

5.6.13 Coefficient of External Friction

The coefficient of external friction is the coefficient of sliding friction. It is defined as the sliding stress between the grain and the horizontal plane against the wall (Bai et al. 2012). The coefficient of external friction is determined using an apparatus consisting of a frictionless pulley fitted on a frame, rectangular wooden boxes with open ends, a loading pan, and a test surface. Firstly, a blank is run to determine the weight required to slide the empty container. Following this, the container placed on the wooden surface is filled with a known quantity of material, and weights are added to the loading pan until the container begins to slide. The coefficient of external friction is calculated as:

$$\mu_e = W_2 - W_1/W$$

Where μ_e is the coefficient of external friction, W_2 is the weight required to cause the sliding of the filled box, W_1 is the weight required to cause the sliding of the empty box, and W is the weight of the material inside the small box.

The coefficient of external friction for finger millet varies from 0.44 to 0.50 (Bai et al. 2012). The coefficient of external friction for unhusked foxtail millet is 0.40 and it decreases to 0.27 on dehusking (Joshi et al. 2018). In barnyard millet, the coefficient of external friction was reported to increase from 0.26 to 0.62 with an increase in moisture content from 11.11 to 42.86% on a dry basis (Subramanian and Viswanathan 2007). A higher coefficient of sliding friction means that steeper hopper walls are required for the flow (Neikov and Yefimov 2019). Chhabra and Kaur (2017) reported a coefficient of external friction of 0.45 for foxtail millet on galvanized iron sheet and 0.52 on cardboard. The increase in moisture content of flour increases the coefficient of external friction.

5.6.14 Angle of Repose

This is the angle formed between the heap of grains and the flat horizontal surface and can be calculated by using the base and height of the heap formed.

$$\text{Angle of repose } (\theta) = \tan^{-1} 2h/d$$

Where h is the height, and d is the diameter (Kumar et al. 2020).

The angle of repose of different millets varies due to variations in size, shape, moisture content, internal friction, velocity, and surface area (Rajasekhar et al. 2018). The angle of repose for pearl millet, foxtail millet, proso millet, barnyard millet, finger millet, kodo millet, little millet, and teff varies from 23 to 25°, 25.87 to 27.47°, 24 to 36.25°, 23 to 37.10°, 21 to 25°, 19.3 to 23.9°, and 20.1 to 23.6°, respectively (Rao et al. 2020; Tsegaye and Abera 2020; Rajasekhar et al. 2018; Kumar et al. 2016; Sunil et al. 2016; Ojediran et al. 2010; Jain and Bal 1997). The angle of repose is affected by the moisture content. The angle of repose of finger millet grains linearly increases from 21 to 30° with an increase in moisture from 7 to 25% on a dry basis (Coracana and Ragi 2018). An increase in the angle of repose of foxtail millet from 24.63 to 28.96° was also reported with an increase in moisture content from 9.8 to 21% on a wet basis (Sunil and Venkatachalapathy 2017). Rao et al. (2020) reported an angle of repose of 33.75° for foxtail millet at a moisture content of 0.265 kg/kg. The angle of repose of kodo millet increases from 8.19 to 12.71° with an increase in moisture content from 8.19 to 12.71% (Kumar et al. 2016). Prakash et al. (2019) reported an increase in the angle of repose of pearl millet from 28.83 to 33.57° with an increase in moisture content. The angle of repose of pearl millet increased from 29.33 to 40.00° with an increase in moisture content from 10 to 20% (Ojediran et al. 2010). The angle of repose of proso millet increased from 21.95 to 26.68° with an increase in moisture content from 6.5 to 26.5%, on a dry basis (Singh et al. 2018).

5.6.15 Terminal Velocity

Terminal velocity is the highest velocity attained by an object falling through a gas or liquid. The instrument used to measure it is known as a hot wire anemometer. The terminal velocity of cereal grains can be calculated by placing 20 g of sample on a fan screen and increasing its speed gradually until the particles begin to float in the vessel. The air velocity at that level is measured by a hot wire anemometer (Rajabipour et al. 2006). Another method to measure the terminal velocity is using an air column experimental device (Chavoshgoli et al. 2014, Singh et al. 2010). In this device, a hollow cylindrical air column made of transparent plastic is fed with a 100 g sample for 30 seconds under controlled air flow. The terminal velocity of finger millet (6.88 ms^{-1}) is higher compared to barnyard millet (5.78 ms^{-1}) and proso millet (2.01 ms^{-1}) due to the change in seed mass. The terminal velocity of grains increases with an increase in moisture. An increase in the terminal velocity of finger millet (4.1 to 5.9 ms^{-1}) and proso millet (1.68 to 2.77 ms^{-1}) was reported by Singh et al. (2018) with an increase in moisture content from 6.5 to 26.5% on a dry basis. An increase in the terminal velocity of finger millet (3.5 to 3.7 ms^{-1}) was also reported by Singh et al. (2018) with an increase in surface area from 12.73 to 15.64 mm^2. In barnyard millet, the increase in moisture content from 0.065 to 0.265 kg/kg was reported to increase the terminal velocity from 3.82 to 5.96 ms^{-1} (Singh et al. 2010) (Table 5.2).

5.7 CONCLUSIONS

Millets can be divided into caryopsis (pericarp is completely fused in the seeds) and utricles (pericarp is completely fused in the seeds) types based on their structure. Their shape may be oval, spherical, elliptical, polyhedral, globose, or hexagonal. The major components of their structure are bran, endosperm, and starch. Bran is composed of the pericarp, seed coat, and aleurone layer. The endosperm contains starch and protein matrix. The size of starch granules in the millets varies from 1.3 to 21.0 μm. The largest starch granules (8.0–21 μm) are found in finger millet. The largest germ (1420 × 620 μm) is found in pearl millet. The colour and hardness of the grain also affect the nutritional composition and milling yield. The coloured varieties have high phenolic contents. A higher hardness value results in higher milling yields compared to soft grains. Amongst the millets, kodo millet has the highest hardness value. The weight of millet grains ranges from 0.2 to 15 mg and the highest weight per grain (3–15 mg) is found for pearl millet. The pearl millet grain also has the highest thousand kernel weight (9.47–11.45 g), bulk density (0.852 g/mL), true density (1.531 g/mL), volume (6.324–7.833 mm^3), and porosity (76.83%). The sphericity (0.910) is highest for finger millet. The dimensional properties of millets are affected by moisture, temperature, and processing methods. Grain length, width, thickness, diameter, sphericity, thousand kernel weight, porosity, true density, volume, static friction coefficient, coefficient of internal friction, coefficient of external friction, angle of repose, and terminal velocity of millet grains increase with increased moisture content, while decreases in bulk density and interstice are observed. The surface area decreases with an increase in steaming pressure.

REFERENCES

Asha, A. B. 2021. A review on nutritional values and health benefits of teff (Eragrostis tef) and its product (injera): Evidence from Ethiopian context. *International Journal of Academic Health and Medical Research* 5(6): 9–16.

Babu, A. S., J. M. Rangarajan, and R. Parimalavalli. 2019. Effect of single and dual-modifications on stability and structural characteristics of foxtail millet starch. *Food Chemistry* 271: 457–465.

Bai, R. S. R., H. Sharanagouda, and N. Udaykumar. 2012. Engineering properties of finger millet (Eleusine Coracana L.) Grains. *International Journal of Agricultural Engineering* 5(2): 178–181.

Balasubramanian, S., and R. Viswanathan. 2010. Influence of moisture content on physical properties of minor millets. *Journal of Food Science and Technology* 47(3): 279–284.

Barretto, R., R. M. Buenavista, J. L. Rivera, S. Wang, P. V. V. Prasad, and K. Siliveru. 2021. Teff (Eragrostis tef) processing, utilization and future opportunities: A review. *International Journal of Food Science and Technology* 56(7): 3125–3137.

Baryeh, E. A. 2002. Physical properties of millet. *Journal of Food Engineering* 51(1): 39–46.

Bean, S. R., L. Zhu, B. M. Smith, J. D. Wilson, B. P. Ioerger, and M. Tilley. 2019. Starch and protein chemistry and functional properties. In John R. N. and Duodu K. G. eds., *Sorghum and Millets* (Second Edition), 131–170. AACC International Press, Cambridge.

Behera, B. K., and P. K. Hari. 2010. Friction and other aspects of the surface behavior of woven fabrics. In *Woven Textile Structure*, 230–242. Woodhead Publishing, Sawston.

Berwal, M. K., L. K. Chugh, P. Goyal, and D. Vart. 2017. Protein, micronutrient, antioxidant potential and phytate content of Pearl Millet hybrids and composites adopted for cultivation by Farmers of Haryana, India. *International Journal of Current Microbiology and Applied Sciences* 6(3): 376–386.

Britannica, the Editors of Encyclopaedia. 2020, June 24. Coefficient of friction. *Encyclopedia Britannica*. https://www.britannica.com/science/coefficient-of-friction (accessed 25 May 2023).

Bultosa, G. 2016. Teff: Overview. Encyclopedia of food grains. https://scitechconnect.elsevier.com/encyclopedia-food-grains/ (accessed December 2015).

Chaturvedi, S., A. K. Shrivastava, G. K. Koutu, S. Ramakrishnan, and S. Singh. 2019. Study of physical property of Kodo (Paspalum scrobiculatum L.) millet. *International Journal of Current Microbiology and Applied Sciences* 8(12): 1503–1510.

Chavoshgoli, E., S. Abdollahpour, R. Abdi and A. Babaie. 2014. Aerodynamic and some physical properties of sunflower seeds as affected by moisture content. *Agricultural Engineering International: CIGR Journal* 16(2): 136–142.

Chhabra, N., and A. Kaur. 2017. Studies on physical and engineering characteristics of maize, pearl millet and soybean. *Journal of Pharmacognosy and Phytochemistry* 6(6): 01–05.

Chukwu, O., and A. J. A. Kadir. 2008. Proximate chemical composition of acha (Digitaria exilis and Digitaria iburua) grains. *Journal of Food Technology* 6(5): 214–216.

Coracana, E., and G. Ragi. 2018. Moisture-dependent physical properties of finger Millet Grain and Kernel ((L.)) Eleusine coracana Gaertn. *Journal of the Indian Society of Coastal Agricultural Research* 36(1): 48–56.

Dayakar, R. B., K. Bhaskarachary, A. G. D. Christina, G. S. Devi, A. T. Vilas, and A. Tonapi. 2017. Nutritional and health benefits of millets. *ICAR_Indian Institute of Millets Research (IIMR). Rajendranagar, Hyderabad* 112: 33.

Dey, S. and B. H. Nagababu. 2022. Applications of food color and bio-preservatives in the food and its effect on the human health. *Food Chemistry Advances* 1: 100019.

Devisetti, R., S. N. Yadahally and S. Bhattacharya. 2014. Nutrients and antinutrients in foxtail and proso millet milled fractions: Evaluation of their flour functionality. *LWT-Food Science and Technology* 59(2): 889–895.

Emmambux, M. N., and J. R. N. Taylor. 2013. Morphology, physical, chemical, and functional properties of starches from cereals, legumes, and tubers cultivated in Africa: A review. *Starch-Stärke* 65(9–10): 715–729.

Gaurav, A., R. C. Pradhan, and S. Mishra. 2021. Comparative study of physical properties of whole and hulled minor millets for equipment designing. *Journal of Scientific and Industrial Research* 80(08): 658–667.

Gomez, M. I., and S. C. Gupta. 2003. Millets. In L. Trugo and P. M. Finglas (eds.), *Encyclopedia of Food Sciences and Nutrition*, 3974–3979. Elsevier.

Govindaraj, M., K. N. Rai, A. Kanatti, H. D. Upadhyaya, H. Shivade, and A. S. Rao. 2020. Exploring the genetic variability and diversity of pearl millet core collection germplasm for grain nutritional traits improvement. *Scientific Reports*. 10(1): 21177.

Grundas, S. T. 2003. Wheat grain structure of wheat and wheat-based products. *Encyclopedia of Food Sciences and Nutrition* 1: 6137–6146.

Guha, M., Y. N. Sreerama, and N. G. Malleshi. 2015. Influence of processing on nutraceuticals of little millet (Panicum sumatrense). In V. Preedy (ed.), *Processing and Impact on Active Components in Food*, 353–360. Academic Press, Cambridge.

Hassan, Z. M., N. A. Sebola, and M. Mabelebele. 2021. The nutritional use of millet grain for food and feed: A review. *Agriculture & Food Security* 10(1): 1–14.

Heiru, M., G. Bultosa, and N. Busa. 2019. Effect of grain teff, finger millet and peanut blending ratio and processing condition on weaning food quality. *Cogent Food & Agriculture* 5(1): 1671116.

Hourston, J. E., M. Ignatz, M. Reith, G. L. Metzger, and T. Steinbrecher. 2017. Biomechanical properties of wheat grains: The implications on milling. *Journal of the Royal Society. Interface / The Royal Society* 14(126): 20160828.

Jain, R., and S. Bal. 1997. Properties of pearl millet. *Journal of Agricultural Engineering Research* 66(2): 85–91.

Joshi, P. P., R. Biradar, P. Mathad and V. Jha. 2018. Effects of different retail packaging materials on the Shelflife of Dehusked foxtail millet. *Journal of Applied Packaging Research* 10(3): 5.

Kesavan, M., J. T. Song, and H. S. Seo. 2013. Seed size: A priority trait in cereal crops. *Physiologia Plantarum* 147(2): 113–120.

Kulp, K. 2000. *Handbook of Cereal Science and Technology, Revised and Expanded*. CRC Press.

Kumar, A., A. Kaur, V. Kumar, and Y. Gat. 2020. Effects of soaking and germination time on the engineering properties of finger millet (Eleusine coracana). *Carpathian Journal of Food Science and Technology* 12(1).

Kumar, A., M. Metwal, S. Kaur, A. K. Gupta, S. Puranik, S. Singh, and M. Singh. 2016. Nutraceutical value of finger millet [Eleusine coracana (L.) Gaertn.], and their improvement using omics approaches. *Frontiers in Plant Science* 7: 934.

Kumar, D., S. Patel, R. K. Naik, and N. K. Mishra. 2016. Study on physical properties of indira kodo-I (Paspalum scrobiculatum L.) millet. *International Journal of Engineering Research and Technology* 5(01): 39–45.

Kumari, K. S., and B. Thayumanavan. 1998. Characterization of starches of proso, foxtail, barnyard, kodo, and little millets. *Plant Foods for Human Nutrition* 53: 47–56.

Mariotti, M., C. Alamprese, M. A. Pagani, and M. Lucisano. 2006. Effect of puffing on ultrastructure and physical characteristics of cereal grains and flours. *Journal of Cereal Science* 43(1): 47–56.

McDonough, C. M., and L. W. Rooney. 1986. Structural characteristics of Pennisetum americanum (Pearl Millet) using scanning electron and fluorescence microscopy. *Food Structure* 8(1): 16.

McDonough, C. M., L. W. Ronney, and S. O. Serna-Saldivar. 2000. The millets. In K. Kulp and J. G. Ponte (eds), *Handbook of Cereal Science and Technology*, 177–202. CRC Press, Florida, USA.

Mikulajova, A., Šedivá, D., Čertík, M., Gereková, P., Németh, K., & Hybenová, E. (2017). Genotypic variation in nutritive and bioactive composition of foxtail millet. *Cereal Research Communications* 45, 442–455.

Neikov, O. D., and N. A. Yefimov. 2019. Powder characterization and testing. *Handbook of Non-ferrous Metal Powders (Second Edition)*, 3–62. Elesevier, Amsterdam.

Nwakuba, N. R., O. C. Chukwuezie, F. C. Uzoigwe, and P. Chukwu. 2019. Friction coefficients of local food grains on different structural surfaces. *Journal of Engineering Research and Reports* 6(3): 1–9.

Ojediran, J. O., M. A. Adamu, and D. L. Jim-George. 2010. Some physical properties of pearl millet (Pennisetum glaucum) seeds as a function of moisture content. *African Journal of General Agriculture* 6(1).

Panda, D., N. H. Sailaja, P. K. Behera, K. Lenka, S. S. Sharma, and S. K. Lenka. 2021. Genetic diversity of under-utilized indigenous finger millet genotypes from Koraput, India for crop improvement. *Journal of Plant Biochemistry and Biotechnology* 30(1): 99–116.

Paschapur, A. U., D. Joshi, K. K. Mishra, L. Kant, V. Kumar, and A. Kumar. 2021. Millets for life: A brief introduction. In A. Kumar, M. K. Tripathi, D. Joshi, and V. Kumar (eds), Millets and Millet Technology. Springer, Singapore, 1–32.

Pawase, P., U. Chavan, and S. Lande. 2019. Pearl millet processing and its effect on antinational factors. *International Journal of Food Science and Nutrition* 4: 10–18.

Pomeranz, Y. 1982. Grain structure and end-use properties. *Food Structure* 1: 2.

Portères, R. 2011. African cereals: Eleusine, Fonio, Black Fonio, Tefi, Brachiaria. Paspalum, Pennisetum, and African Rice. In J. R. Harlan, J. M. J. E. WET, and A. B. L. Stemler, (eds), *Origins of African Plant Domestication*, 409–453. Mouton Publishers, Paris.

Powar, R. V., V. V. Aware, P. U. Shahre, S. P. Sonawane, and K. G. Dhande. 2018. Moisture-dependent physical properties of finger Millet Grain and Kernel (L.) Eleusine Coracana Gaertn. *Journal of the Indian Society of Coastal Agricultural Research* 36(1): 48–56.

Prakash, J., V. Waisundara, and V. Prakash. 2020. Nutritional and Health Aspects of Food in South Asian Countries. Academic Press.

Prakash, O. M., S. K. Jha, A. Kar, J. P. Sinha, C. T. Satyavathi, and M. A. Iquebal. 2019. Physical properties of Pearl Millet Grain. http://krishi.icar.gov.in/jspui/handle/123456789/70231 (accessed August 2019).

Pulivarthi, M. K., M. Selladurai, E. Nkurikiye, Y. Li and K. Siliveru. 2022. Significance of milling methods on brown teff flour, dough, and bread properties. *Journal of Texture Studies* 53(4): 478–489.

Qin, L., Y. Zhang, E. Chen, Y. Yang, F. Li, and Y. Guan. 2019. Screening for germplasms tolerant to salt at germination stage and response of protective enzymes to salt stress in foxtail millet. *Scientia Agricultura Sinica* 52(22): 4027–4038.

Radchuk, V., and L. Borisjuk. 2014. Physical, metabolic and developmental functions of the seed coat. *Frontiers in Plant Science* 5: 510.

Rajabipour, A., A. H. M. A. D. Tabatabaeefar, and M. E. H. D. I. Farahani. 2006. Effect of moisture on terminal velocity of wheat varieties. *International Journal of Agriculture and Biology* 8(1): 10–13.

Rajasekhar, M., L. Edukondalu, D. D. Smith, and G. Veeraprasad. 2018. Changes in engineering properties of finger millet (PPR-2700, Vakula) on hydrothermal treatment. *The Andhra Agricutlural Journal* 65(2): 420–429.

Ramakrishnan, S. R., K. Ravichandran, and U. Antony. 2019. Millets. In S. A. Mir, A. Manickavasagan, and M. A. Shah (eds.), *Whole Grains*, 103–127. CRC Press.

Ramashia, S. E., E. T. Gwata, S. Meddows-Taylor, T. A. Anyasi, and A. I. O. Jideani. 2018. Some physical and functional properties of finger millet (Eleusine coracana) obtained in sub-Saharan Africa. *Food Research International* 104: 110–118.

Rao, P. U. 1994. Evaluation of protein quality of brown and white ragi (Eleusine coracana) before and after malting. *Food chemistry* 51(4): 433–436.

Rao, V. V., S. V. S. G. Swamy, D. S. Raja, and B. J. Wesley. 2020. Engineering properties of certain minor millet grains. *The Andhra Agricultural Journal* 67: 89–92.

Salahudeen, H. O., and B. A. Orhevba. 2021. Nutritional composition, health benefits and utilization of Fonio (Digitaria exilis) grains: A review. Pan African Society for Agricultural Engineering & The Nigerian Institution of Agricultural Engineers (A Division of the Nigerian Society of Engineers). 2021 Virtual International Conference., 2021. http://repository.futminna.edu.ng:8080/jspui/handle/123456789/6899 (accessed 2021).

Saleh, A. S., Q. Zhang, J. Chen, and Q. Shen. 2013. Millet grains: Nutritional quality, processing, and potential health benefits. *Comprehensive Reviews in Food Science and Food Safety* 12(3): 281–295.

Satyavathi, T. C., S. Praveen, S. Mazumdar, L. K. Chugh, and A. Kawatra. 2017. Enhancing Demand of Pearl millet as Super grain - Current Status and Way Forward. ICAR All India Coordinated Research Project on Pearl millet, Jodhpur.

Seetharam, A. 2001. Annual Report 2000–01 All India Coordinated Small Millets Improvement Project, Bangalore, pp. 1–28.

Serna-Saldivar, S. O., and J. E. Ramírez. 2019. Grain structure and grain chemical composition. In R. N. John, K. Taylor, and G. Duodu (eds.), *Sorghum and Millets*, 85–129. AACC International Press.

Sharma, K., and E. S. Chauhan. 2018. Nutritional composition, physical characteristics and health benefits of teff grain for human consumption: A review. *The Pharma Innovation Journal* 7(10): 3–7.

Sharma, R., S. Sharma, B. N. Dar, and B. Singh. 2021. Millets as potential nutri-cereals: A review of nutrient composition, phytochemical profile and techno-functionality. *International Journal of Food Science and Technology* 56(8): 3703–3718.

Shobana, S., K. Krishnaswamy, V. Sudha, N. G. Malleshi, R. M. Anjana, L. Palaniappan, and V. Mohan. 2013. Finger millet (Ragi, Eleusine coracana L.): A review of its nutritional properties, processing, and plausible health benefits. *Advances in Food and Nutrition Research* 69: 1–39.

Singh, K. P., H. N. Mishra, and S. Saha. 2010. Moisture-dependent properties of barnyard millet grain and kernel. *Journal of Food Engineering* 96(4): 598–606.

Singh, K. P., N. S. Chandel, R. R. Potdar, D. Jat, K. N. Agrawal, and S. Hota. 2018a. Assessment of engineering properties of proso millet (Panicum miliaceum). *Journal of Agricultural Engineering* 55(2): 42–51.

Singh, K. P., R. R. Potdar, K. N. Agrawal, P. S. Tiwari, and S. Hota. 2017. Effect of moisture content on physical properties of finger (Eleusine coracana) millet. *Agricultural Mechanization in Asia, Africa and Latin America* 48: 24–32.

Singh, K. P., S. Saha, and H. N. Mishra. 2010a. Optimization of machine parameters of finger millet thresher-cum-pearler. *Ama, Agricultural Mechanization in Asia, Africa & Latin America* 41(1): 60.

Singh, M., A. Adedeji, and D. Santra. 2018. Physico-chemical and functional properties of nine proso millet cultivars. *Transactions of the ASABE* 61(3): 1165–1174.

Subramanian, S., and R. Viswanathan. 2007. Bulk density and friction coefficients of selected minor millet grains and flours. *Journal of Food Engineering* 81(1): 118–126.

Sunil, C. K., and N. Venkatachalapathy. 2016. Engineering properties of foxtail Millet (Setaria italic L) as a function of moisture content. *Trends in Biosciences* 10(20): 3990–3996.

Sunil, C. K., N. Venkatachalapathy, S. Shanmugasundaram, and M. Loganathan. 2016. Engineering properties of foxtail millet (Setaria italic L): Variety-HMT 1001. *International Journal of Science, Environment and Technology* 5(2): 632–637.

Taylor, J., and K. G. Duodu, (eds.). 2018. *Sorghum and Millets: Chemistry, Technology, and Nutritional Attributes*. Elsevier.

Taylor, J. R. N., and J. O. Anyango. 2011. Sorghum flour and flour products: Production, nutritional quality, and fortification. In V. R. Preedy and R. R. Watson (eds.), *Flour and Breads and their Fortification in Health and Disease Prevention*, 127–139. Academic Press, Cambridge.

Taylor, J. R., and M. N. Emmambux. 2008. Gluten-free foods and beverages from millets. In *Gluten-free cereal products and beverages*, eds. E. K. Arendt and F. D. Bello, 119-V. Academic Press.

Tsegaye, G. A., and S. Abera. 2020. The study of some engineering properties of teff [Eragrostis teff (ZUCC.) Trotter] GrainVarieties. *International Journal of Food Engineering and Technology* 4(1): 9.

Upadhyaya, H. D., K. N. Reddy, S. Singh, C. L. L. Gowda, M. I. Ahmed, and S. Ramachandran. 2014. Latitudinal patterns of diversity in the world collection of pearl millet landraces at the ICRISAT GeneBank. *Plant Genetic Resources* 12(1): 91–102.

Ushakumari, S. R., N. K. Rastogi, and N. G. Malleshi. 2007. Optimization of process variables for the preparation of expanded finger millet using response surface methodology. *Journal of Food Engineering* 82(1): 35–42.

Vadivoo, A. S., R. Joseph, and N. M. Garesan. 1998. Genetic variability and calcium contents in finger millet (*Eleusine coracana* L. Gaertn) in relation to grain colour. *Plant Foods for Human Nutrition* 52(4): 353–364.

Varnamkhasti, M. G., H. Mobli, A. Jafari, A. R. Keyhani, M. H. Soltanabadi, S. Rafiee, and K. Kheiralipour. 2008. Some physical properties of rough rice (Oryza sativa L.) grain. *Journal of Cereal Science* 47(3): 496–501.

Xiang, J., F. B. Apea-Bah, V. U. Ndolo, M. C. Katundu, and T. Beta. 2019. Profile of phenolic compounds and antioxidant activity of finger millet varieties. *Food Chemistry* 275: 361–368. https://doi.org/10.1016/j.foodchem.2018.09.120.

Zarnkow, M., A. Mauch, W. Back, E. K. Arendt, and S. Kreisz. 2007. Proso millet (Panicum miliaceum L.): An evaluation of the microstructural changes in the endosperm during the malting process by using scanning-electron and confocal laser microscopy. *Journal of the Institute of Brewing* 113(4): 355–364.

Zewdu, A. D., and W. K. Solomon. 2007. Moisture-dependent physical properties of tef seed. *Biosystems Engineering* 96(1): 57–63.

Zhu, F. 2014. Structure, physicochemical properties, and uses of millet starch. *Food Research International* 64: 200–211.

Chapter 6

Nutritional and Functional Composition of Millets

6.1 INTRODUCTION

Millets are highly rich in nutrients and their nutritional profile is considered better than staple cereals like rice, and wheat. These contain macronutrients, micronutrients, and phytochemicals with bioactive properties. The composition of millets varies with their type and variety. Apart from this, several other factors like genetics and environment play an important role in determining the final composition of the grain. This chapter discusses the nutritional composition of millets in detail.

6.2 MACRONUTRIENTS IN MILLETS

Macronutrients are energy-giving compounds which are important for the body and include carbohydrates, proteins, and fats. Macronutrients are also important in metabolic activities, muscle building, and various other vital functions. Millets are known to have various health-benefitting properties that can partly be attributed to the type of macronutrients. For example, the presence of complex carbohydrates, amino acid sequences, and unsaturated fatty acids gives millet grains a low glycaemic index, antiobesity effects, and heart-benefitting properties. In general, millets mainly contain carbohydrates, in the range of 60–75% and proteins (7–12%) and have low fat (2–5%) content. All millets are rich in dietary fibre (15–20%). The current section explains the characteristics of the three macronutrients in millets.

6.2.1 Carbohydrates

Millet grain contains approximately 60–75% carbohydrates. Carbohydrates in millets are usually simple sugars, starch, non-starch polysaccharides, and dietary fibre. Unlike rice and other refined products, the millet endosperm contains a considerable quantity of dietary fibre. Also, millets contain other carbohydrates which are complex in nature and are not easily digestible. These components make millets potentially effective against many diseases and disorders from as miniscule as constipation to as complex as diabetes and cardiovascular diseases. Low incidences of such disorders have been reported in millet-eating populations. The details of health benefits of millets will be discussed in this chapter. Many studies show varied compositions in the carbohydrate content of the same millet type as well.

Yankah et al. (2020) compared the carbohydrate composition of millets with brown rice and maize for varietal complementary feeding. The analysis revealed that millet had the lowest amount of carbohydrate (70.41%) in comparison to brown rice (77.94%) and maize (73.94%). The highest total carbohydrates were reported in proso millet and are given in Table 6.1. Malleshi et al. (2021) reported proso millet to have a content of 70.4 g/100 g while the lowest was reported for foxtail millet (60.9 g/100 g) (Malleshi et al., 2021). The carbohydrate content of finger millet and kodo millet was reported to be 66.8 and 66.2 g/100 g respectively. Similar carbohydrate content was reported for barnyard and little millet (65.5 and 65.6 g/100 g respectively).

Millets vary largely not just in the content but in the composition of carbohydrates. On average, starch occupies from 50 to 75% of the total weight of the millet grain. In fonio millet, starch is the major component of carbohydrates (up to 80% of total starch on dry basis) (Zhu 2020). Similar observations were made for teff (Baye 2014). Teff contains starch content of more than 70% (Bultosa 2007). The dry weight yields of starches of different millets were assessed by Annor et al. (2014). The starch yield using the alkaline extraction method was highest for proso millet (93.7%) and lowest for finger millet (63.4%). Foxtail and pearl millet had similar yields of starch (69.1 and 70.4% respectively) with starch damage between 0.8 and 1.2% (Annor et al. 2014).

The granule size of the millet starch generally lies between 2.5 to 24 μm with foxtail millet having the biggest followed by pearl millet with a peak size of 7.6 μm (Sharma et al. 2017). Proso millet and finger millet have comparatively smaller granules with a similar size distribution and smaller peak size (5.7 μm) (Annor et al. 2014). The arrangement is highly compact and organized with a distribution of some spherical granules distributed between the major polygonal granules. The hydrogen bonds hold both amylose and amylopectin and provide the polygonal or spherical arrangement. Analysis of finger millet starch did not reveal any spherical granules (Annor et al. 2014). Varietal variations were also observed in millet starch granule size. Hadimani et al. (2001) isolated starch from three pearl millet varieties (MBH157, CO6, and TV) with a yield of 65.8–75.3%. The MBH157 variety of pearl millet was found to have larger starch granules (7.5–16 μm) in comparison to CO6 and Suma and Urooj (2015) isolated starch from pearl millet cultivars kalukombu and Maharashtra Rabi Bajra observed a low yield for both varieties (34.5 and 39.4 g/100 g respectively). The low amylose content of the starch indicated its non-waxy nature. Such starches are fragile, disintegrate easily, and have low water- and oil-holding capacity. An A-type starch pattern was observed in these varieties. Such varieties are more prone to enzymatic hydrolysis. Fonio starch is also A-type and polygonal in shape with an approximate diameter of 8 micrometres. The white variety has more branched amylose than black varieties. On average, the amylopectin chain has a chain length of approximately twenty glucosyl residue. The swelling capacity and water solubility of both white and black fonio varieties were reported to be similar (Zhu 2020).

Amylose content is inversely related to the glycaemic index. High amylose content is associated with poor starch digestibility. The amylose content of starch in four millet varieties, pearl, finger, proso, and foxtail millet, ranged from 32.5 to 28.6% in descending order. Also, the study revealed that the low molecular weight amylopectin fractions with fraction number >40 were found in higher quantities in finger and pearl millets in comparison to other millet types (Annor et al. 2014). The amylose component of the starch of the millet grain is utilized first during the process of germination as the grain contains 23–30% amylase. The red, pink, yellow, or green discolouration of the endosperm is due to the pericarp pigments. The mean amylose content of

TABLE 6.1 CARBOHYDRATE AND SUGAR PROFILE OF MILLETS (G/100 G)

Millet type	Total carbohydrates	Total available carbohydrates	Total starch	Glucose	Sucrose	Total free sugars	Crude fibre	Dietary fibre (total)	Dietary fibre (soluble)
Pearl millet	61.78 ± 0.85	56.02 ± 2.57	55.21 ± 2.57	6.83 ± 0.02		0.81 ± 0.01	2–2.5	11.49 ± 0.62	2.34 ± 0.42
Finger millet	66.82 ± 0.73	62.47 ± 1.24	62.13 ± 1.13	0.25 ± 0.06	0.12 ± 0.02	0.34 ± 0.01	3.6–4.2	11.18 ± 1.14	1.67 ± 0.55
Proso millet	70.2 ± 0.3		73.71 ± 3.65				2–9	39.85 ± 1.1	51.44 ± 2.84
Foxtail millet	71 ± 11	64.4 ± 0.4	18.5	13	33	–	4.25 ± 0.15	11.5	
Barnyard millet	64.1 ± 9.5		12.6				2.98–6.6	21.65	4.2
Kodo millet	66.19 ± 1.19	66.25 ± 2.90	64.96 ± 2.93	0.89 ± 0.11	0.40 ± 0.02	1.29 ± 0.10	5–9	6.39 ± 0.6	2.11 ± 0.34
Little millet	65.55 ± 1.29	56.43 ± 4.09	56.07 ± 4.12	0.24 ± 0.10	0.13 ± 0.01	0.37 ± 0.09	4–8	6.39 ± 0.60	2.27 ± 0.52
Teff	85.6%		74%				1–2	4.5	0.9
Fonio White	74.3						2–3.5	2.2	
Black	78							0.5	

Source: Longvah et al. 2017; Osman Magdi 2011; Himanshu et al. 2018; Jideani & Akingbala 1993; Irving & Jideani 1997; Gebru et al. 2020; Verma et al. 2015, Sharma et al. 2016; Ugare et al. 2014; Nazni & Devi 2016; Malavika et al. 2020; Wankhede et al. 1978; Shen et al. 2018; Yang et al. 2018.

nine Chinese varieties of proso millet was 10.33 ± 13.21. Non-waxy varieties contained high levels of amylose (22.95%) in comparison to waxy varieties (0.23%). Resistant starch was also high in non-waxy varieties (2.02%) in comparison to waxy ones (0.78%) (Shen et al. 2018). The amylose content of fonio millet varies between 23 and 26%, whereas that reported for teff is 20–26%. The functional differences between fonio and other cereals can be attributed to amylopectin structure (Bultosa 2007; Zhu 2020). The ratio of amylose and amylopectin in the starch component is important for the functional properties of millets. The ratio of amylose to amylopectin showed wide variation ranging from 16–28 to 72–84 in terms of percentage.

Varietal variations were also observed in millet amylose content. Hadimani et al. (2001) isolated starch from three pearl millet varieties (MBH157, CO6, and TV) and found that the total amylose content was highest for the MBH157 variety. The study also revealed that an increase in temperature from 60 to 80°C enhanced the swelling capacity and solubility of the starch. The non-starch polysaccharide content for the TV variety was found to be 10.3%, the highest among the three (Hadimani et al. 2001).

The gelatinization temperature is higher for starches obtained from corneous endosperm as compared to those compared to floury endosperm. In general, the gelatinization temperature of the starch is 70–80°C. The thermal properties and pasting behaviour indicate the shortness and cohesiveness of the paste. Also, the water-binding capacity of millet starch is lower in comparison to maize starch. The swelling capacity (at 90°C) of millet starch is higher than maize starch but the solubility is lower. Waxy millet varieties are known to have shorter cooking time, poor cooking stability and paste clarity, and high viscosity, water-binding capacity, and retrogradation.

Resistant starch is highly important in determination of the glycaemic index of the grain and is also responsible for increasing bulk and preventing constipation. The resistant starch of millet varieties was found to be between 1.4 and 5.1 g/100 g. Resistant starch content of 10–20 g/100 g is generally considered a requirement to show any visible effects to the physiological quality of the food products. Nutritionally, starch can be differentiated into resistant dietary starch, soluble dietary starch, and resistant starch. Pearl millet flour showed values of 10.2 and 11.4 g/100 g, 6 and 7.6 g/100 g, and 5.1 and 4.0 g/100 g for two different varieties (K and MRB, respectively). The starch digestion index measures the relative rate at which starch is digested. The starch digestion index was reported to be 49–57% (for K and MRB starch, respectively) (Suma and Urooj 2015). These values were similar to those of freshly cooked spaghetti (52%) and lentils (44%). The physical inaccessibility of the pearl millet starch to human digestion enzymes can be attributed to its low starch digestion index (Alsaffar 2011).

The non-starch polysaccharide fractions included arabinose, xylose, glucose, and uronic acid as the main monosaccharides. The good popping and milling varieties were identified with higher proportions of water-soluble non-starch polysaccharides containing 2-6-fold higher pectic polysaccharides than the other varieties. These components have good prebiotic potential and are good for gut health. Hadimani et al. (2001) characterized the carbohydrate composition of pearl millet varieties exhibiting different milling and popping abilities. Three pearl millet cultivars, CO6, TV, and MBH 157, were assessed for their carbohydrate composition. The endosperm texture of good milling varieties (CO6, TV) was corneous. The starch content was highest for the CO6 (72.9%) and MBH (75.3%) varieties. The non-starch polysaccharide fraction was highest in the good milling but poor popping variety (TV–15.7%) followed by MBH and CO6 (8.5% and 6.5%) respectively. It was also observed that the non-starch polysaccharide fractions were composed of mainly glucose, arabinose, and xylose sugars. The amylose:amylopectin ratio of eight varieties of proso millet contained 28–38% amylose of total starch. Processing methods like decortication were found to make no difference to the final ratio (Bagdi et al. 2011).

Dietary fibre is generally classified into soluble (oligosaccharides, pectins, b-glucans, and galactomannan gums alginate, psyllium fibre) and insoluble fractions (cellulose, hemicellulose, and lignin) based on their solubility in water. The total dietary fibre content of finger millet and proso millet is reported to be 11.2 g. Foxtail and barnyard millet also have similar total dietary fibre content (10.5 and 9.9 g respectively). Little millet and kodo millet have comparatively low total dietary fibre content (7.7g and 6.4 g respectively) (Sharma et al 2017a). Sharma et al. (2016) estimated the dietary fibre content in barnyard millet and reported it to be 21.65%. The dietary fibre content in fonio millet (white) ranges from 0.5 and 18.2% whereas the black varieties of fonio millet had 0.5–4.8% dietary fibre (Zhu 2020). The insoluble dietary fibre content of finger millet, proso millet, barnyard millet, and foxtail millet (9.5, 9.3, 8.8, and 8.7 g per 100 g respectively) is high in comparison to little millet and kodo millet (5.5 and 4.3 g per 100 g respectively). The content of soluble dietary fibre is low in finger millet, proso millet, barnyard millet, and foxtail millet (1.7, 1.9, 1.1, and 1.8 g/100 g respectively). Little millet and kodo millet have comparatively high soluble dietary fibre content (2.1 and 2.3 g/100 g respectively) (Malleshi et al. 2021).

The process of germination increased the percentage of dietary fibre in the grain to 23.74%. With germination the cell wall polysaccharide structure is modified due to the disruption of the protein carbohydrate interaction (Sharma 2017). An increase in dietary fibre on germination was also reported in foxtail millet by Duodu and Awika (2019). The dietary fibre content of little millet was reported to be 5 g/100 g.

Foxtail millet bran (60%) contains approximately three times more dietary fibre than rice (27%). This makes it an excellent fibre source. Mohamed et al. (2009) extracted dietary fibre from foxtail millet bran with a purity of 79.22%. It included a high amount of insoluble fraction (77.57%) and a very low soluble fraction (1.75%). Even such low concentrations of soluble dietary fibre can actually have a positive health impact due to its high power of regulating cholesterol and glucose levels in the body (Galisteo et al. 2008). Similar results were obtained by Zhu et al. (2018), who also identified that foxtail millet bran dietary fibre mainly consisted of insoluble dietary fibre (77.57%). Hemicellulose and celluloses accounted for the major portions of the foxtail millet bran dietary fibre and could yield hexoses and pentoses like xylose, galactose, and arabinose on hydrolysis. The major hemicellulose identified was arabinoxylan which made the flour highly viscous. Hemicellulose could be hydrolyzed into hexose and pentose such as xylose, mannose, galactose, and arabinose by the use of acid. In the insoluble dietary fibre fraction, the major monosaccharides reported were xylose and arabinose. The soluble fraction of dietary fibre of foxtail millet bran consisted of glucose, galactose, and arabinose (68.23, 12.99, and 7.74% respectively) as the major monosaccharides in descending order (Zhu et al. 2018). Glucose is the main soluble sugar in pearl millet.

6.2.2 Fats

Lipids are relatively minor constituents in millets. Most of the lipids are located in the scutellar area of the germ, the pericarp, and the aleurone layer. Thus, lipid content is significantly reduced when the germ is removed during decortication or degermination. The content of lipid in any millet variety depends upon the variety, soil, temperature, harvesting time, extraction method, etc.

Ether extract from the different millet varieties analysed revealed the lipid content to be in the range of 3.2 to 6.2% on dry weight basis. The fat content of fonio millet ranged from 1.1 to 4.7% for both white and black varieties and had no relationship with the colour of the grain (Zhu 2020). The lipid content of foxtail millet bran was found to be 9.39% which is much lower than rice bran (Liang and Liang 2019; Liang et al. 2010). The fat content of seven varieties of teff was observed to be in the range of 2.92–3.34% (Amare et al. 2021). The lipid composition of proso millet (*Panicum miliaceum*), studied by Lorenz and Hwang (1984), revealed information about the free and bound fractions of the lipids. The free lipid fraction contains hydrocarbons triacylglycerols, diacylglycerols, and free fatty acids mainly comprising of palmitic, linoleic, and oleic acids. The bound lipids contained mono and digalactosyl, diacylglycerols, phosphatidyl serine, choline, and ethanolamine. The lipid content of different millets is given in Table 6.2.

Lipids can be subdivided into polar, nonpolar, and nonsaponifiable lipids. The most abundant fats in millets are the nonpolar lipids, 70–80%. The composition of the nonpolar lipids was clearly dominated by triglycerides, 85%, followed by sterols, 4.1%, and diglycerides, 4.0%. Triglycerides serve as a reserve material for germination. The less abundant polar lipids (i.e. glycolipids, 2.5 to 6.2%, and phospholipids, 17 to 25%) have an important biochemical function. Nonsaponifiable compounds, 3 to 5%, include carotenoids, phytosterols, and tocopherols. Ghodsizad and Safekordi (2012) assessed the composition of cold-pressed millet oil. The oil yield was 4.7% and linoleic acid was the predominant fatty acid. The oil is composed of polyunsaturated (65%), monounsaturated (25.2%), and saturated (8.1%) fatty acids. The iodine value of millet oil is equivalent to soyabean oil and higher than other vegetable oils like sunflower and canola. The major sterols found in millet oil were beta-sitosterol, campesterol, citostanol, and stigmasterol in descending order. Millet oil has high antioxidant activity due to the presence of α, β, ϒ, and δ tocopherols. ϒ tocopherol was found in the highest quantities. In white-coloured fonio varieties, the major lipids were found to be linoleic, oleic, and palmitic acid in decreasing order (Ballogou et al. 2013; Glew et al. 2013).

Finger millet fixed oil composition was assessed by Poonia et al. (2012). Petroleum ether extract of dry powder finger millet was 0.74% and 13 fatty acids were identified from it. The major fatty acids identified were palmitic (23.06%), oleic (47.17%), and linoleic acid (24.78%). The fatty acids present in minor amounts were stearic (0.58%), linolenic (2.26%), and arachidonic acid

TABLE 6.2   FATS PROFILE AND ENERGY VALUE OF MILLETS (MG/100 G)

Millet type	Total fat	Total saturated fatty acid (mg)	Total unsaturated fatty acid	Total mono unsaturated	Oleic acid	Alpha linoleic acid	Energy
Pearl millet	5.43 ± 0.64	875 ± 34.5		1047 ± 39.9	1040 ± 39.8	1844 ± 56.7	363–412
Finger millet	1.92 ± 0.14	317 ± 17.0		585 ± 36.3	585 ± 36.3	362 ± 15.3	328–336
Proso millet	1–3.5						330–340
Foxtail millet	4–7				13.0	66.50	330–350
Barnyard millet	2–4				53.80	34.90	300–310
Kodo millet	1.4–3.6	246 ± 2.3		297 ± 6.8	291 ± 7.2	576 ± 17.8	309–353
Little millet	5–6	589 ± 31.9		868 ± 24.2	868 ± 24.2	1230 ± 42.9	329–341
Teff	2.5	23.5%	76.2%	25.86%	29%	0.9	330--340
Fonio	0.5–2						360–370

Source: Gebru et al. 2020; Baye 2014; Himanshu et al. 2018; Longvah et al. 2017; Gopalan 1996; Rao et al. 2017; Amare et al. 2021.

(0.27%). Similar results were obtained in other studies as well (Serna-Saldivar 2010; Ramashia et al. 2019). Finger millet oil also contains phospholipids mainly comprised of both cephalins and lecithins. Teff grains contain good quantities of oleic and linoleic acid and the ratio of linoleic acid to alpha-linolenic acid is 7:1 which meets the standard range as per the codex standards (Koletzko et al. 2005).

Millet small bran, a by-product of millet processing, is rich in fat in comparison to endosperm where the fat content may even reach 21%. Li et al. (2009) identified over 90% unsaturated fatty acids of which 72% were essential fatty acids. Similar results were obtained by Guifeng et al. (2018) who obtained 23% crude fat and a very high phospholipid content (4.83%). Zhao Chenyong et al. (2011) assessed the fatty acid composition of millet bran oil and compared the composition of oils obtained by different methods of extraction. The study compared the fatty acid profile of the oils obtained by supercritical carbon dioxide extraction and isopropanol solvent extraction. The study revealed that the content of linoleic acid in millet oil was higher than other vegetable oils (71.8%). Supercritical carbon dioxide extraction showed a decrease in unsaturated fatty acid (0.11%) and phospholipid (99.6%) but an increase in saturated fatty acids (0.9%), vitamin E (20.6%), and sterols (34.8%) as compared to the isopropanol solvent extraction method.

Ji et al. (2019) studied the oleochemical properties of foxtail millet bran obtained from various fractions of the grain. The authors studied the coarse bran, skin bran, polished bran, and mixed bran. Foxtail bran oil is a high-value plant oil with high amounts of unsaturated acid, the main fatty acids being linoleic (65–69%) and oleic (12–17%) acid. The main triglyceride obtained were trilinolein and oleodilinolein. There was no significant difference observed in the oil composition obtained from different bran fractions. However, a difference in the physical characteristics was observed in oil from different fractions. Other components present in the lipid fraction of the millets are tocopherols. Crude foxtail millet bran oil contains α- (15.53 mg/100 g), γ- (48.79 mg/100 g). The level of γ-tocopherol in foxtail millet bran is much higher than in rice bran oil. The total unsaturated fat in foxtail millet bran was reported to be up to 83% with a ratio of unsaturated to saturated fatty acid of 5. The major fatty acids reported in this oil were palmitate (20%), oleic acid (42%), and linoleic acid (32%). Linoleic and oleic acids were positioned on the sn-2 position of the triacylglycerol molecule. Like other oils, the sn-2 position of the triglyceride was occupied mainly (89%) by the unsaturated fatty acids. Amare et al. (2021) assessed seven varieties of teff for their lipid content. Linoleic, oleic, and palmitic acid were the major fatty acids identified (89% of the total fatty acid content). The ratio of saturated to unsaturated fatty acid was found in the range of 0.3–0.32. The study also compared the thrombogenicity and atherogenicity index of teff (0.38–0.41 and 0.23–0.25 respectively) with that of other grains like amaranth, quinoa, olive oil, oats, etc. Teff outperformed all other grains on these indices (Amare et al. 2021). The fat content of white fonio (3%) was reported to be lower than black fonio varieties of Nigeria (Sadiq et al. 2015).

6.2.3 Proteins

Proteins are one of the essential and most diverse functional nutrients in the human body. They are not only essential for structural integrity but also useful for many other functions. Millets typically contain higher quantities of essential amino acids and are higher in fat content than maize, rice, and sorghum. The protein content of millets is highly variable based on type and variety. The proteins present in millet grain are distributed across different parts of the grain, mainly (80%) concentrated in the endosperm. The germ part of the seed contains 16% and approximately 3% is found in the pericarp. The amount of protein depends on various factors like soil characteristics viz., nitrogen %, water content, temperatures, yield, etc. (Qaisrani et al. 2019). Pearl millet (10.49%) has more than double the protein of brown rice (4.67%) and maize (8.9%) (Yankah et al. 2020). The protein content of fonio millet for both white and black varieties ranged from 4.4 to 10% (Zhu 2020). Among all millets, the highest amount of protein is present in pearl millet while the lowest is reported in fonio and finger millet. The protein contents of other millet varieties are mentioned in Table 6.3.

Millets contain prolamins or kafirins as the major proteins. Prolamines are known to contain lower amounts of lysine, arginine, and glycine and higher quantities of alanine, methionine, and leucine in comparison to the albumin/globulin protein. Proso millet contains prolamins with a molecular weight in the range of 20–28 kDa with high content of nonpolar amino acids. Hence, proso millet protein is soluble in aqueous ethanol but insoluble in water. This property makes proso millet protein a good surface active agent in an oil–water interface. The proso millet protein structure is very similar to zein protein which forms characteristic spherical and brick-like shapes in an aqueous medium. Proso millet finds applications for encapsulating several bioactive materials (Wang et al. 2018). The germ portion of the grain is rich in albumin and globulin proteins while the endosperm portion contains prolamines and glutelins. The bran portion of the grain together with the germ is relatively poor in essential amino acids phenylalanine, leucine, and isoleucine but contains high amount of lysine, arginine, and glycine in comparison to the whole grain. Like other millets, foxtail millet protein is composed of 60–65% prolamins and minor amounts of albumins and globulins (11.6 to 29.6%). The content of protein is positively correlated with the prolamin content and negatively correlated with lysine. In pearl millet both prolamins and glutelins were present in equal quantities as storage protein. In millet grains like barnyard and kodo millet, the major protein found was glutelin as in rice (Sachdev et al. 2021). White-coloured fonio contains more prolamins and less glutelins and amino acids in comparison to black-coloured varieties. In teff, glutelins and albumins are the major (45 and 37% respectively) storage proteins whereas prolamines are minor protein types. However, recent contradictory studies show prolamines to be the major protein type in teff grains (Adebowale et al. 2011). Kalinova and Moundry (2006) analysed 13 varieties of proso millet over a span of 3 years. The crude protein content in different varieties was in the range of 11.41 to 5.62% dry matter. One significant observation in this study was that both protein content and quality were affected by weather. Dry weather enhanced protein quantity but diminished the quality of protein.

The prolamine fraction for all millets is distributed with 3–4 bands usually within the range of 27 kDa to 13 kDa. Among these, a band with molecular weight of 20 kDa is found in most concentration in all millets, proso millet being an exception. Kumar and Parameswaran (1998) reported the albumin/globulin fraction across a wide range of molecular weights, such as 12, 22, 34, and 54 kDa for foxtail millet. The prolamine fractions of kodo millet and barnyard millet are of comparatively low molecular weight. SDS-PAGE analysis revealed that foxtail millet had only four major polypeptide bands which is in contrast to the maize prolamins with at least ten polypeptide prolamines. The glutelin fraction was indicated by a large number of bands. Protein extracted from enzymatically produced foxtail millet protein concentrate was found to have a protein composition similar to original foxtail millet flour (Sachdev et al. 2021).

The protein content of pearl millet variety CO6 was the highest (13.6%) followed by TV (13%) and MBH157 (10%). Prolamine constituted about 45% of the total proteins in all the varieties analysed in pearl millet. Albumin and globulin together accounted for 25% of the proteins. The continuity of the protein matrix and physical contact between starch and storage protein result in a harder texture of the endosperm. The higher matrix protein observed in CO6 and TV varieties may also contribute to their corneous endosperm textures which further facilitate their good

TABLE 6.3 PROTEIN AND AMINO ACID PROFILE OF MILLETS (% PER GRAM PROTEIN)

Millet type	Total protein	Histidine	Isoleucine	Leucine	Lysine	Methionine	Cystine	Phenylalanine	Threonine	Tryptophan	Valine
Pearl millet	10.96 ± 0.26	2.15 ± 0.37	3.45 ± 0.74	8.52 ± 0.49	3.19 ± 0.49	2.11 ± 0.50	1.23 ± 0.33	4.82 ± 1.18	3.55 ± 0.4	1.33 ± 0.30	4.79 ± 1.04
Finger millet	7.16 ± 0.63	2.37 ± 0.46	3.7 ± 0.44	8.86 ± 0.54	2.83 ± 0.34	2.74 ± 0.27	1.48 ± 0.23	5.7 ± 1.27	3.84 ± 0.45	0.91 ± 0.3	5.65 ± 0.44
Proso millet	12.5	2.1	5.34	13	3.3	2.22	-	6.01	4.08	4.25	6.01
Foxtail millet	12.3	2.11	4.59	13.60	1.59	3.06	0.45	6.27	3.68	-	5.81
Barnyard millet	11.6	1.8–2.0	4.5–4.6	11.4–11.7	1.6–1.8	1.6–2.0		5.5–6.3	3.6–3.7	1.0	6.1–6.2
Kodo	8.3	2.14 ± 0.07	4.55 ± 0.22	11.98 ± 1.65	1.42 ± 0.17	2.69 ± 0.16	1.92 ± 0.05	6.27 ± 0.34	3.89 ± 0.16	1.32 ± 0.19	5.49 ± 0.23
Little millet	8.92 ± 1.09	2.35 ± 0.18	4.14 ± 0.08	8.08 ± 0.06	2.42 ± 0.10	2.21 ± 0.1	1.85 ± 0.14	6.14 ± 0.1	4.24 ± 0.12	1.35 ± 0.1	5.31 ± 0.16
Teff	11	3.2	4.1	8.5	3.7	4.1	2.5	5.7	4.3	1.3	5.5
Fonio	7.2	1.3	1.37	4.4	1.9	2.98	3.1	2.37	1.89	0.95	2.34

Source: Longvah et al. 2017; Sharma & Niranjan 2017; Kalinová 2007; Kalinova & Moudry 2006, Amadou et al. 2013, Nassarawa & Sulaiman 2019; Gopalan 1996; Rao et al. 2017; Renganathan et al. 2020.

milling properties (Hadimani et al. 2001). In another study, Anitha et al. (2019) determined the protein content of pearl millet (Dhanshakthi, Proagro9444) and foxtail millet (VR847, GPU28). The total protein content was 10.53% for Dhanshakthi and 9.06% for Proagro9444 varieties of pearl millet. The protein contents of the two finger millet varieties were similar (6.34% for VR847 and 6.31% for GPU28).

Bagdi et al. (2011) characterized the protein of eight varieties (GK alba, GK Piroska, Fertodi-2, Biserka, GyGyrka, G, Rumenka, Biopont, Natura) of kodo millet. The varieties had a protein content in the range of 11.58–14.8%. The molecular masses of similar polypeptides obtained from all the varieties were 58, 50, 40, 18, 17, 12, 7, and 5 kDa. However, there were at least 18 different polypeptides identified in the varieties in varied proportions. Processing methods like decortication had no impact on these kodo millet varieties. Similar results were obtained for pearl millet (Lestienne et al. 2007). However, finger millet was observed to have lower protein (22%) following decortication (Shobana & Malleshi 2007). The biological value of millet protein was between 48.3 and 56.5 (Geervani & Eggum 1989). Kulkarni and Naik (2000) assessed three varieties (PSC-1, PSC-3, and local) of kodo millet for their protein quality. The protein content was higher for varieties PSC-1 (9.92 g/100 g) and PSC-3 (9.26 g/100 g). In another study, Srilekha et al. (2019) assessed the protein quantity and in-vitro protein digestibility of kodo millet flour. The protein content of kodo millet flour was found to be 7.60 ± 0.1% and in-vitro protein digestibility was found to be 74.58 ± 0.35%, which was similar to the digestibility of the yellow and white varieties of foxtail millet (Mohamed et al. 2009). Fonio millet has poor solubility in comparison to wheat protein, probably due to the larger number of sulphide bonds.

Protein concentrates extracted from white and yellow varieties of foxtail millet were assessed for their digestibility and functional properties. The protein content of the millet flour was in the range of 11.39 to 11.92 g/100 g. An important observation in the study was that the protein contents of whole millet flour and defatted flour were relatively close and did not have any significant difference. On processing, the concentrates were found to have a protein content of 80.04 and 75.69% respectively (Mohamed et al. 2009).

6.2.3.1 Amino Acids

It has been observed that as the overall protein content of the grain increased, all amino acids increased in the fractions of the grain but the relative distribution of amino acids in the protein varied in different fractions. Like all other cereals, millets are also deficient in lysine. The protein of the bran-germ fraction contained approximately four times as much lysine and two times as much arginine and glycine as the protein of endosperm fractions.

Mohamed et al. (2009) analysed the amino acid composition of foxtail millet protein concentrates. The millet possesses all essential amino acids and the proportion of essential amino acids in foxtail millet protein concentrate was comparable to soy protein concentrate. Similar to cereals, the amount of lysine in foxtail millet is also limited. The essential amino acid content of the defatted samples was found to be higher than the whole millet protein concentrates. There was not much difference observed in the white and yellow varieties. The isoleucine and leucine contents of the varieties were 3.82–3.91 and 8.58–8.72 g/100 g of protein respectively. The lysine content was between 5.17 and 6.07 g/100 g of protein. The content of methionine and cysteine in total was reported to be 3.15–3.79 g/100 g of protein. The content of phenylalanine and tyrosine together ranged from 9 to 9.33 g/100 g of protein. The protein concentrates contained good quantities of other essential amino acids like threonine, valine, histidine, and tryptophan. As analysed in the study, the content of all the individual essential amino acids was higher in the foxtail millet protein concentrates in comparison to the soy protein concentrate, hence proving the better protein quality of millet protein. The non-essential amino acids reported in foxtail millet concentrate are alanine, arginine, aspartic acid, cysteine, glutamic acid, glycine, serine, tyrosine, and proline. Like other millets, foxtail millet also lacks lysine which is a characteristic feature of prolamins. Kamara et al. (2009) showed that the alkaline treatment caused reduced levels of threonine and arginine followed by tryptophan loss due to the lowering of pH. The amino acid profile of pearl millet is very close to rice and wheat but better than sorghum and maize. There is a difference in the amino acid content across the seed as well. The germ fraction of pearl millet contains approximately four times more lysine than the endosperm. Finger millet contains a higher amount of lysine in comparison to other cereals. White-coloured fonio millet has high content of amino acid methionine in comparison to most other cereals (Glew et al. 2013).

The amino acid composition of finger millet is different from foxtail millet with lower contents of phenylalanine (5.2 g/100 g protein) and histidine. The content of limiting amino acid lysine in finger millet is only 2.2 g/100 g. The contents of isoleucine and leucine are 6.6–9.5 and 4.3 g/100 g protein respectively. Other essential amino acids present are histidine, methionine, threonine, tryptophan, and valine (Rao et al. 2017). The content of total essential amino acids in proso millet was between 33.95 to 42.27 g/100 g. Red varieties had higher amounts of tyrosine and phenylalanine as compared to other varieties. Proso millet contains 3.3 g/kg of lysine (Kalinova and Moundry 2006). The content of these is given in Table 6.3.

6.2.3.2 Protein Digestibility

Protein digestibility determines the final availability of amino acids in the body and acts as an indicator of protein bioavailability. It indicates the susceptibility of protein to hydrolyse. Proteins with high digestibility are considered far superior as they provide more amino acids available for digestion. Millet proteins are poorly digestible. The type of protein and the arrangement of the molecules with starch affect protein digestibility. Proteins with high proline content pose higher resistance against proteolytic hydrolysis. Apart from this, digestibility is influenced by the degree of folding and crosslinking between proteins. The tighter the folding, the lower the access of the enzyme to the substrate. However, apparent quality and digestibility can be improved through various processing technologies. Protein digestibility can be determined by the degree of hydrolysis of proteins by different proteolytic enzymes like pepsin, trypsin, and papain. Fourteen foxtail varieties were assessed for their protein digestibility using different proteolytic enzyme systems and presented different values with pepsin (95.5%), trypsin (26.8%), and papain (92.95%) (Sachdev et al. 2021).

The in-vitro trypsin digestibility of foxtail millet protein concentrate is lower than that of soyabean protein concentrate but the solubility is higher. Maximum solubility of the foxtail millet protein concentrate was found to be in alkaline conditions. The emulsification property of foxtail millet protein is higher than that of soyabean protein concentrate meaning it has potential for industrial applications (Mohamed et al. 2009). The essential amino acid index in proso millet which indicates the quality of protein was found to be 51% higher than wheat (Kalinova and Moundry 2006). The amino acid contents of different millets reported in various studies are given in Table 6.3.

6.3 MICRONUTRIENTS

Millets are also rich in micronutrients. Micronutrients perform various important functions in the body like cell growth, macronutrient metabolism, cell synthesis, division and differentiation, vision, and many more. Pearl millet is known for its high iron content, finger millet for high calcium, other small millets for phosphorus and iron, etc. These also contain good amounts of vitamins except vitamin C which is generally not found in millets. The current section deals with the micronutrients present in millets.

6.3.1 Vitamins

Millets contain both water- and fat-soluble vitamins. Water-soluble B vitamins of finger millet are concentrated in the aleurone layer and germ, while lipid-soluble vitamins are mainly in the germ (Gull et al. 2016).

6.3.1.1 Water-Soluble Vitamins

All millets are good sources of B-complex vitamins with vitamin B12 an exception. Although total niacin content has been reported up to 10.88 mg, only 13% was found to be cold water extractable. Also, studies indicate that the total ascorbic acid content is comparatively low in matured millet grains. Ochanda et al. (2010) analysed the content of some water-soluble vitamins in pearl millet. The content of folic acid was found to be 0.02 mg/100 g. The contents of other B vitamins like niacin, pyridoxine, and riboflavin were found to be 4.32 mg/100 g, 0.35 mg/100 g, and 0.21 mg/100 g respectively. In another study, the thiamine content of pearl millet and finger millet was 289 and 192 μg % respectively. The riboflavin content of pearl millet and finger millet was found to be 835.67 and 245 μg % respectively. The niacin (4.61 mg % and 2.89 mg % respectively) and ascorbic acid (2.78 mg % and 3.19 mg % respectively) content of pearl millet and finger millet has been reported. Malting is said to improve the B vitamin content in all millets especially thiamine and ascorbic acid (Barikmo et al. 2004, Barikmo et al. 2007). The thiamine content of finger millet, little millet, kodo millet, foxtail millet, and barnyard millet was found to be 0.26, 0.29, 0.59, 0.33, and 0.41 mg/100 g respectively. The riboflavin content of finger millet, little millet, kodo millet, foxtail millet, barnyard millet, and proso millet was reported to be 0.17, 0.05, 0.20, 0.11, 0.10, and 0.28 mg/100 g. The thiamine content of fonio millet was measured to be 0.17 ± 0.09 mg/100 g (Table 6.4). The niacin content of finger millet, little millet, kodo millet, foxtail millet, barnyard millet, and proso millet was reported to be 1.34, 1.29, 1.49, 3.2, 4.2, and 4.5 mg/100 g. The total folate content of pearl millet is 36.11 mg/100 g. The total folate content of finger millet, little millet, and kodo millet is found to be 34.66, 36.2, and 39.49 mg/100 g respectively. The content of riboflavin and niacin in fonio millet was reported 0.22 ± 0.04 mg/100 g and 1.15 ± 0.26 mg/100 g respectively.

6.3.1.2 Fat-Soluble Vitamins

The mature kernels of pearl millet are rich in vitamin A. The total carotenoid in pearl millet is 293 mg/100 g. Finger millet and little millet are rich in total carotenoids. The total carotenoid contents of finger millet and little millet are 154 and 120 mg/100 g respectively.

The vitamin E (tocopherol) content of millets has been found to be comparatively lower than that of soyabean and corn oil. Millets contain mainly alpha tocopherol which has high potential activity in comparison to other tocopherols viz., the efficacy of gamma tocopherol is less than 10% of the alpha form. Details regarding tocopherols have already been discussed under the subheading "fat". The unrefined fat from common millet contained vitamin A equivalent (8.3–10.5 mg) and vitamin E (87–96 mg) per 100 g. The alpha tocopherol contents of pearl millet, ragi, and little millet were reported to be 0.24, 0.16, and 0.55 mg/100 g respectively.

TABLE 6.4 FAT- AND WATER-SOLUBLE VITAMIN IN MILLETS

Millet type	Thiamine (mg)	Riboflavin (mg)	Niacin (mg)	Pantothenic (mg)	Total B6 (mg)	Total folate (µg)	Ergocalciferol (µg)	Alpha-tocopherols (mg)	Phylloquinones (µg)
Pearl millet	0.25 ± 0.044	0.20 ± 0.038	0.86 ± 0.10	0.50 ± 0.05	0.27 ± 0.09	36.11 ± 5.05	5.65 ± 0.27	0.24 ± 0.02	2.85 ± 0.6
Finger millet	0.37 ± 0.041	0.17 ± 0.008	1.34 ± 0.03	0.29 ± 0.19	0.05 ± 0.007	34.66 ± 4.97	41.46 ± 3.12	0.16 ± 0.01	3 ± 0.4
Proso millet	0.2	0.18	2.3	-	-	-	-	-	-
Foxtail millet	0.59	0.11	3.2	-	-	15.0	-	-	-
Barnyard millet	0.4	0.1	4.2	-	-	-	-	-	-
Kodo millet	0.269 ± 0.054	0.2 ± 0.018	1.49 ± 0.08	0.63 ± 0.07	0.07 ± 0.017	39.49 ± 4.52	-	0.07 ± 0.02	3.75 ± 0.6
Little millet	0.26 ± 0.042	0.06 ± 0.008	1.29 ± 0.02	0.6 ± 0.07	0.04 ± 0.006	36.20 ± 4.97	3.75 ± 0.80	0.55 ± 0.16	4.47 ± 0.3
Teff	0.39	0.27	3.363		0.482			0.08	1.9
Fonio	0.17	0.22	1.15						

Source: Barikmo et al. (2004; Barikmo et al. 2007; Gebru et al. 2020; Longvah et al. 2017; Gopalan 1996; Rao et al. 2017; Renganathan et al. 2020.

Finger millet has high quantities of ergocalciferol in comparison to other millets. The ergocalciferol content of finger millet was 41.46 microgram/100 g which is far higher than pearl millet and little millet (5.65 microgram and 3.75 microgram per 100 g) respectively (Table 6.4). The phylloquinone content of millets is quite low as compared to sorghum (43.82 microgram/100 g). The phylloquinone content of pearl millet, finger millet, and little millet is 2.85, 0, and 4.47 microgram/100 g respectively (Table 6.4).

6.3.2 Minerals

Millets act as a major mineral source in the diet. The mineral content of millets is similar to sorghum, and they even have higher contents of calcium and magnesium. Some high-yielding and high-protein varieties of finger millet have calcium content in the range of 290–390 mg/100 g. Rao et al. (2006) identified the intake from sorghum and pearl millet varieties from different states of India and analysed iron and zinc content. The intake of iron from sorghum and pearl millet was highest in the western region of Rajasthan and the dry region of Gujrat had the highest iron content (20.2 and 15.3 mg/person per day). The highest intake of zinc (7.8 and 5.9 mg/person per day respectively) was also obtained from the same regions. The replacement of a rice-based diet with finger millet for girls showed not only better calcium retention but also better nitrogen balance. Although most millets are high in one or another mineral, the majority of it is concentrated in the hull. Dehulling is a major step in the primary processing of millets. Dehulling has been reported to decrease the mineral content significantly, though this varies with the species (Himanshu et al. 2018). The content of various minerals for different millets in given in Table 6.5.

Among millets, finger millet has the highest proportion of minerals (332%) (Nazni and Bhuvaneswari 2015). Anitha et al. (2019) analysed the mineral content of two varieties of pearl millet (Dhanshakti and Proagro9444) and finger millet (VR847 and GPU28). Pearl millet varieties had higher content of iron and zinc than finger millet. The calcium content of finger millet (VR847 – 359.79 mg/100 g and GPU28 – 450.33 mg/100 g) was much higher than pearl millet (Dhanshakthi – 30.04 mg/100 g and Proagro 9444 – 22.44 mg/100 g). The iron content of pearl millet in different studies was reported in between 4 and 11.2 mg/100 g (Velu et al. 2008, Abdelrahman et al. 2005, Abdalla et al. 1998).

As given in Table 6.5, the highest content of phosphorus was present in pearl millet followed by foxtail, barnyard, and finger millet. Wide variation was observed in the case of finger millet ranging from 130 to 250 mg/100 g. Kodo millet had the lowest phosphorus content (188 mg/100 g) (Ramashia et al., 2018). Millets are a good source of potassium. Finger millet had the highest content of potassium among millets (430–490 mg/100 g). Pearl millet was reported to have 307 mg/100 g of potassium followed by foxtail and kodo millet (250 and 144 mg/100 g respectively) (Ramashia et al. 2019). The potassium content of fonio millet is between 2200 and 3070 microgram/gram (Glew et al. 2013). The magnesium content of foxtail and barnyard millet is 81–82 mg/100 g. Finger millet was reported to have a content of 78–201 mg/100 g. The magnesium content of kodo and barnyard millet was reported to be approximately 228 and 137 mg/100 g respectively. Finger millet had a magnesium content of 78–201 mg/100 g (Ramashia et al. 2018). The magnesium content of fonio millet was reported to be 43 ppm (Annongu et al. 2019).

The calcium content of millets is quite varied. Pearl millet (398 mg/100 g) was reported to have the highest calcium content among all millets. Foxtail, kodo, barnyard, and pearl millet were reported to have a calcium content of 31, 27, 22, and 42 mg/100 g (Ramashia et al. 2018). The

TABLE 6.5 MINERALS AND TRACE ELEMENTS IN MILLETS (MG/100 G)

Millet type	Aluminium	Calcium	Chromium	Cobalt	Copper	Iron	Magnesium	Phosphorus	Zinc	Sodium	Potassium
Pearl millet	2.21 ± 0.78	27.35 ± 2.16	0.025 ± 0.006	0.03 ± 0.015	0.54 ± 0.11	6.42 ± 1.04	130–137	350–379	43.7 5.2	10–12	440–442
Finger millet	3.64 ± 0.69	364 ± 58	0.032 ± 0.019	0.02	0.4–4	4.62 ± 0.36	11–137	240–320	2–2.3	7–11	408–570
Proso millet		14			0.83–5.8	0.8	153	206	1.4	8.2	113
Foxtail millet		31			1–3	2.8	81	290	2.4	4.6	250
Barnyard millet		21.66			0.6	37.68	38.19			22.07	
Kodo millet	1.07 ± 0.83	15.27 ± 1.28	0.021 ± 0.027	1.6–5.8	0.26 ± 0.05	2.34 ± 0.46	130–166	215–310	0.7–1.5	4.6–10	144–170
Little millet	-	16.06 ± 1.54	0.016 ± 0.006	0.001 ± 0	0.34 ± 0.08	1.26 ± 0.44	120–133	251–260	3.5–11	6–8.1	129–370
Teff		78.8–147			1.6	11.5–>150	184	429	3.8–3.9	12	427
Fonio		30				12	70		1.5	30	12

Source: Longvah et al. 2017; Gopalan 1996; Rao et al. 2017; Bayeetal et al. 2014; Jideani & Akingbala 1993; Irving & Jideani 1997; Gebru et al. 2020; Sharma et al. 2016; Himanshu et al. 2018; Renganathan et al. 2020.

calcium content of fonio millet is reported to be 27 ppm. There is no difference in the white and black varieties of fonio (Annongu et al. 2019). Red varieties of teff have higher calcium content in comparison to the white varieties.

Sodium is an important electrolyte in the human body. If present in high quantities it can be injurious to health. The sodium content of foxtail and kodo millet is approximately 4.6 mg/100 g. Finger millet has high sodium content (Ramashia et al. 2018). The sodium content for the white and black varieties was reported to be in the range of 20–30 mg/g (Sadiq et al. 2015). Zinc content is high in pearl millet (3.1 mg/100 g). Barnyard, foxtail, and finger millet also are reported to have appreciable amounts of zinc (3.0, 2.4, and 2.3 mg/100 g respectively). The lowest zinc content was reported in kodo millet. Due to its high zinc content, pearl millet has been used in zinc fortification programmes. The zinc content of fonio millet was reported to be 3.2 ppm (Annongu et al. 2019). Pearl millet has high iron content (8 mg/100 g). Several iron-rich varieties have been developed with iron content as high as 78 mg/100 g. Even some varieties of finger millet were reported to have iron content as high as 14.89 mg/100 g. Foxtail and kodo millet have relatively lower iron content. The iron content of fonio millet is 1.2–2.5 ppm (Annongu et al. 2019). Red varieties of teff have higher calcium content in comparison to the white varieties. However, the copper content of white teff varieties is better than red (Baye 2014).

6.4 CONCLUSION

Millets are rich sources of nutrients and bioactive compounds. They contain approximately 60–75% carbohydrates. Carbohydrates in millets are usually simple sugars, starch, non-starch polysaccharides, and dietary fibre. They have high amounts of dietary fibre contributing to various health benefits. Lipids are present in relatively minor quantities in millets (1–7%). The proteins present in millet grain are distributed across different parts of the grain, mainly (80%) concentrating in the endosperm. The germ part of the seed contains 16% protein and approximately 3% is found in the pericarp. Apart from macronutrients, millets are also rich in micronutrients. However, more research needs to be conducted on the nutritional aspects of millets and changes in the quantity and properties on processing. The consumption of a combination of millets is recommended for better nutritional benefits.

REFERENCES

Abdalla, A. A., A. H. E. L. Tinay, B. E. Mohamed, and A. H. Abdalla. 1998. Effect of traditional processes on phytate and mineral content of pearl millet. *Food Chemistry* 63(1): 79–84.

Abdelrahman, S. M., H. B. Elmaki, W. H. Idris, E. E. Babiker, and A. H. E. L. Tinay. 2005. Antinutritional factors content and minerals availability of pearl millet (*Pennisetum glaucum*) as influenced by domestic processing methods and cultivation. *Journal of Food Technology* 3: 397–403.

Adebowale, A. R. A., M. N. Emmambux, M. Beukes, and J. R. N. Taylor. 2011. Fractionation and characterization of teff proteins. *Journal of Cereal Science* 54(3): 380–386. https://doi.org/10.1016/J.JCS.2011.08.002.

Alsaffar, A. A. 2011. Effect of food processing on the resistant starch content of cereals and cereal products—A review. *International Journal of Food Science and Technology* 46(3): 455–462.

Amadou, I., M. E. Gounga, and G. W. Le. 2013. Millets: Nutritional composition, some health benefits and processing - A review. *Emirates Journal of Food and Agriculture* 25(7): 501–508. https://doi.org/10.9755/EJFA.V25I7.12045.

Amare, E., L. Grigoletto, V. Corich, A. Giacomini, and A. Lante. 2021. Fatty acid profile, lipid quality and squalene content of teff (*Eragrostis teff* (*Zucc.*) *Trotter*) and amaranth (*Amaranthus caudatus* L.) varieties from Ethiopia. *Applied Sciences* 11(8): 3590.

Anitha, S., M. Govindaraj, and J. Kane-Potaka. 2019. Balanced amino acid and higher micronutrients in millets complements legumes for improved human dietary nutrition. *Cereal Chemistry* 97(1): 74–84. https://doi.org/10.1002/cche.10227.

Annongu, A. A., J. H. Edoh, R. M. O. Kayode, A. O. Gawati, et al. 2019. Phytochemicals and nutrients constituents of varieties of hungry rice and their significance in nutrition. *International Journal of Research Publications* 41: 1–10.

Annor, G. A., M. Marcone, E. Bertoft, and K. Seetharaman. 2014. Physical and molecular characterization of millet starches. *Cereal Chemistry Journal* 91(3): 286–292. https://doi.org/10.1094/CCHEM-08-13-0155-R.

Bagdi, A., G. Balázs, J. Schmidt, M. Szatmári, R. Schoenlechner, E. Berghofer, and S. Tömöskőzia. 2011. Protein characterization and nutrient composition of Hungarian proso millet varieties and the effect of decortication. *Acta Alimentaria* 40(1): 128–141. https://doi.org/10.1556/AAlim.40.2011.1.15.

Ballogou, V. Y., M. M. Soumanou, F. Toukourou, and J. D. Hounhouigan. 2013. Structure and nutritional composition of fonio (*Digitaria exilis*) grains: A review. *International Research Journal of Biological Sciences* 2: 73–79.

Barikmo, I., F. Ouattara, and A. Oshaug. 2004. Protein, carbohydrate and fibre in cereals from Mali-how to fit the results in a food composition table and database. *Journal of Food Composition and Analysis* 17(3–4): 291–300.

Barikmo, I., F. Ouattara, and A. Oshaug. 2007. Differences in micronutrients content found in cereals from various parts of Mali. *Journal of Food Composition and Analysis* 20(8): 681–687.

Baye, K. 2014. *Teff: Nutrient Composition and Health Benefits* (Vol. 67). International Food Policy Research Institute.

Bultosa, G. 2007. Physicochemical characteristics of grain and flour in 13 tef [*Eragrostis Tef* (*Zucc.*) *Trotter*] grain varieties. *Journal of Applied Science Research* 3(12): 2042–2051.

Chen-yong, Z., W. Chang-Qing, X. Jie, F. Ying, and F. Wang. 2011. Effects of extraction methods on nutritional components in millet bran oil. *Food Science* 14. https://en.cnki.com.cn/Article_en/CJFDTotal-SPKX201114064.htm.

Duodu, K. G., and J. M. Awika. 2019. Phytochemical-related health-promoting attributes of sorghum and millets. In *Sorghum and Millets: Chemistry, Technology, and Nutritional Attributes*, 225–258. https://doi.org/10.1016/B978-0-12-811527-5.00008-3.

Galisteo, M., J. Duarte, and A. Zarzuelo. 2008. Effects of dietary fibers on disturbances clustered in the metabolic syndrome. *The Journal of Nutritional Biochemistry* 19(2): 71–84. https://doi.org/10.1016/J.JNUTBIO.2007.02.009.

Gebru, Y. A., D. B. Sbhatu, and K. P. Kim. 2020. Nutritional composition and health benefits of teff (*Eragrostis tef (Zucc.) Trotter*). *Journal of Food Quality*. https://doi.org/10.1155/2020/9595086.

Geervani, P., and B. O. Eggum. 1989. Nutrient composition and protein quality of minor millets. *Plant Foods for Human Nutrition* 39(2): 201–208. https://doi.org/10.1007/BF01091900.

Ghodsizad, G., and A. A. Safekordi. 2012. Oil extraction from millet seed –Chemical evaluation of extracted oil. In: *Journal of Food Biosciences and Technology* 2(2): 71–76.

Glew, R. H., E. P. Laabes, J. M. Presley, J. Schulze, R. Andrews, Y. C. Wang, Y. C. Chang, and L. T. Chuang. 2013. Fatty acid, amino acid, mineral and antioxidant contents of acha (*Digitaria exilis*) grown on the Jos Plateau, Nigeria. *International Journal of Nutrition and Metabolism* 5(1): 1–8.

Gopalan, C., B. V. Ramasastri, and S. C. Balasubramanian. 1996. *Nutritive Value of Indian Foods*. National Institute of Nutrition, Indian Council of Medical Research.

Guifeng, L., W. Jianhu, B. Huijuan, and Z. Lei. 2018. Process optimization for extraction of millet small bran oil by aqueous ethanol. *IOP Conference Series: Material Science and Engineering* 392(5): 052023. https://doi.org/10.1088/1757-899X/392/5/052023.

Gull, A., N. G. Ahmad, K. Prasad, and P. Kumar. 2016. Technological, processing and nutritional approach of finger millet (*Eleusine coracana*)-a mini review. *Journal of Food Processing and Technology* 7(593): 2.

Hadimani, N. A., G. Muralikrishna, R. N. Tharanathan, and N. G. Malleshi. 2001. Nature of carbohydrates and proteins in three pearl millet varieties varying in processing characteristics and kernel texture. *Journal of Cereal Science* 33(1): 17–25. https://doi.org/10.1006/jcrs.2000.0342.

Himanshu, C. M., Sachin, K., S. S. Sonawane, and S. Arya. 2018. Nutritional and Nutraceutical properties of millets: A review. *Clinical Journal of Nutrition and Dietetics* 1(1): 1–8.

Irving, D. W., and I. A. Jideani. 1997. Microstructure and composition of *Digitaria exilis Stapf* (acha): A potential crop. *Cereal Chemistry* 74(3): 224–228.

Ji, J., Y. Liu, Z. Ge, Y. Zhang, and X. Wang. 2019. Oleochemical properties for different fractions of foxtail millet bran. *Journal of Oleo Science* 68(8): 709–718. https://doi.org/10.5650/jos.ess19063.

Jideani, A. I., and J. O. Akingbala. 1993. Some physicochemical properties of acha (*Digitaria exilis Stapf*) and Iburu (*Digitaria iburua Stapf*) grains. *Journal of the Science of Food and Agriculture* 63(3): 369–374.

Kalinová, J. 2007. Nutritionally Important Components of Proso Millet (*Panicum miliaceum L.*). *Food Global Science Books* 1: 91–100.

Kalinova, J., and J. Moudry. 2006. Content and quality of protein in proso millet (*Panicum miliaceum* L.) varieties. *Plant Foods for Human Nutrition* 61(1): 43. https://doi.org/10.1007/s11130-006-0013-9.

Kamara, M. T., Z. Huiming, Z. Kexue, I. Amadou, and F. Tarawalie. 2009. Comparative study of chemical composition and physicochemical properties of two varieties of defatted foxtail millet flour grown in China. *American Journal of Food Technology* 4(6): 255–267.

Koletzko, B., S. Baker, G. Cleghorn, U. Fagundes Neto, S. Gopalan, O. Hernell, Q. S. Hock, P. Jirapinyo, B. Lonnerdal, et al. 2005. Global standard for the composition of infant formula: Recommendations of an Espghan coordinated international expert group. *Journal of Pediatric Gastroenterology and Nutrition* 41(5): 584–599.

Kulkarni, L. R., and R. Naik 2000. Nutritive value, protein quality and organoleptic quality of kodo millet (*Paspalum scrobiculatum*). *Karnataka Journal of Agricultural Sciences* 13(1): 125–129. https://www.cabdirect.org/cabdirect/abstract/20013057019.

Kumar, K. K., and K. P. Parameswaran. 1998. Characterisation of storage protein from selected varieties of foxtail millet (*Setaria italica* (L) Beauv). *Journal of the Science of Food and Agriculture* 77(4): 535–542.

Lestienne, I., M. Buisson, V. Lullien-Pellerin, C. Picq, and S. Trèche. 2007. Losses of nutrients and anti-nutritional factors during abrasive decortication of two pearl millet cultivars (*Pennisetum glaucum*). *Food Chemistry* 100(4): 1316–1323.

Li, L., C. Y. O. Chen, H. K. Chun, S. M. Cho, K. M. Park, Y. C. Lee-Kim, J. B. Blumberg, R. M. Russell, and K. J. Yeum. 2009. A fluorometric assay to determine antioxidant activity of both hydrophilic and lipophilic components in plant foods. *Journal of Nutritional Biochemistry* 20(3): 219–226.

Liang, S., G. Yang, and Y. Ma. 2010. Chemical characteristics and fatty acid profile of foxtail millet bran oil. *Journal of the American Oil Chemists' Society* 87(1): 63–67. https://doi.org/10.1007/S11746-009-1475-3.

Liang, S., and K. Liang. 2019. Millet grain as a candidate antioxidant food resource: A review. *Https://Doi.Org/10.1080/10942912.2019.1668406* 22(1): 1652–1661. https://doi.org/10.1080/10942912.2019.1668406.

Longvah, T., I. An_antan_, K. Bhaskarachary, K. Venkaiah, and T. Longvah. 2017. *Indian Food Composition Tables*. National Institute of Nutrition, Indian Council of Medical Research, 2–58.

Lorenz, K., and Y. S. Hwang. 1984. Lipids in proso millet (*Panicum miliaceum*) flours and brans. *Cereal Chemistry* 63: 387–390.

Malavika, M., S. Shobana, P. Vijayalakshmi, R. Ganeshjeevan, et al. 2020. Assessment of quality of minor millets available in the south Indian market & glycaemic index of cooked unpolished little & foxtail millet. *Indian Journal of Medical Research* 152(4): 401–409. https://doi.org/10.4103/ijmr.IJMR_2309_18.

Malleshi, N. G., A. Agarwal, A. Tiwari, and S. Sood. 2021. Nutritional quality and health benefits. *Millets and Pseudo Cereals*: 159–168. https://doi.org/10.1016/b978-0-12-820089-6.00009-4.

Mohamed, T. K., K. Zhu, A. Issoufou, T. Fatmata, and H. Zhou. 2009. Functionality, in vitro digestibility and physicochemical properties of two varieties of defatted foxtail millet protein concentrates. *International Journal of Molecular Sciences* 10(12): 5224. https://doi.org/10.3390/IJMS10125224.

Nassarawa, S. S., and S. A. Sulaiman. 2019. Comparative analyses on the chemical compostion, phytochemcimal and antioxidant properties of selected milled varieties (finger and pearl millet). *International Journal of Food Sciences* 2(1): 25–42.

Nazni, P. S., and J. Bhuvaneshwari. 2015. Analysis of physico chemical and functional characteristics of finger millet (*Eleusine corcana* L.) and little millet (*Panicum sumantarnce*). *International Journal of Food and Nutritional Sciences* 4(3): 109–114.

Nazni, P. S., and R. Devi. 2016. Effect of processing on the characteristics changes in barnyard and foxtail millet. *Journal of Food Processing & Technology* 07(03). https://doi.org/10.4172/2157-7110.1000566.

Ochanda, S. O., C. A. Onyango, M. A. Mwasaru, J. K. Ochieng, and F. M. Mathooko. 2010. Effects of alkali treatment on tannins and phytates in red sorghum, white sorghum and pearl millet. *Journal of Applied Biologicalsciences* 36: 2409–2418.

Osman Magdi, A. 2011. Effect of traditional fermentation process on the nutrient and antinutrient contents of pearl millet during preparation of Lohoh. *Journal of the Saudi Society of Agricultural Sciences* 10(1): 1–6.

Poonia, K., S. Chavan, and M. Daniel. 2012. Fixed oil composition, polyphenols and medical approach of finger millet (Eleusine coracona L.). *Journal of Biological Forum* 4(1): 45–47.

Qaisrani, S. N., S. Murtaza, A. H. Khan, F. Bibi, S. M. J. Iqbal, F. Azam, I. Hussain, and T. N. Pasha. 2019. Variability in millet: Factors influencing its nutritional profile and zootechnical performance in poultry. *Journal of Applied Poultry Research* 28(2): 242–252. https://doi.org/10.3382/JAPR/PFY073.

Ramashia, S. E. 2018. *Physical, Functional and Nutritional Properties of Flours from Finger Millet (Eleusine coracana) Varieties Fortified with Vitamin B2 and Zinc Oxide* (PhD thesis). http://hdl.handle.net/11602/1245.

Ramashia, S. E., T. A. Anyasi, E. T. Gwata, S. Meddows-Taylor, and A. I. O. Jideani. 2019. Processing, nutritional composition and health benefits of finger millet in sub-Saharan Africa. *Food Science and Technology* 39(2): 253–266. https://doi.org/10.1590/FST.25017.

Rao, D. B., K. Bhaskarachary, G. D. Arlene Christina, G. Sudha Devi, A. T. Vilas, and A. Tonapi. 2017. *Nutritional and Health Benefits of Millets*. ICAR_Indian Institute of Millets Research (IIMR) Rajendranagar, 112.

Rao, P. P., P. S. Birthal, B. V. Reddy, K. N. Rai, and S. Ramesh. 2006. Diagnostics of sorghum and pearl millet grains-based nutrition in India. *International Sorghum and Millets Newsletter* 47: 93–96.

Renganathan, V. G., C. Vanniarajan, A. Karthikeyan, and J. Ramalingam. 2020. Barnyard millet for food and nutritional security: Current status and future research direction. *Frontiers in Genetics* 11: 500. https://doi.org/10.3389/fgene.2020.00500.

Sachdev, N., S. Goomer, and L. R. Singh. 2021. Foxtail millet: A potential crop to meet future demand scenario for alternative sustainable protein. *Journal of the Science of Food and Agriculture* 101(3): 831–842. https://doi.org/10.1002/JSFA.10716.

Sadiq, I. Z., S. A. Maiwada, D. Dauda, Y. M. Jamilu, and M. A. Madungurum. 2015. Comparative nutritional analysis of black fonio (*Digitaria iburua*) and white fonio (*Digitaria exili*). *International Research Journal of Biological Sciences* 4: 4–9.

Serna-Saldivar, S. O. 2010. *Cereal Grains. Properties, Processing and Nutritional Attributes*. CRC Press, 538–562.

Sharma, N., and K. Niranjan. 2017. Foxtail millet: Properties, processing, health benefits and uses. *Food Reviews International*. https://doi.org/10.1080/87559129.2017.1290103.

Sharma, N., S. K. Goyal, T. Alam, S. Fatma, A. Chaoruangrit, and K. Niranjan. 2017. Effect of High pressure soaking on water absorption, gelatinization, and gelatinization, and biochemical properties of germinated and non-germinated foxtail millet grains. *Journal of Cereal Science* 83: 162–170. https://doi.org/10.1080/87559129.2017.1290103.

Sharma, S., D. C. Saxena, and C. S. Riar. 2016. Analysing the effect of germination on phenolics, dietary fibres, minerals and γ-amino butyric acid contents of barnyard millet (*Echinochloa frumentaceae*). *Food Bioscience* 13: 60–68. https://doi.org/10.1016/J.FBIO.2015.12.007.

Sharma, S., N. Sharma, S. Handa, and S. Pathania. 2017a. Evaluation of health potential of nutritionally enriched Kodo millet (Eleusine coracana) grown in Himachal Pradesh, India. *Food Chemistry* 214: 162–168.

Shen, R., Y. Ma, L. Jiang, J. Dong, Y. Zhu, and G. Ren. 2018. Chemical composition, antioxidant, and antiproliferative activities of nine Chinese proso millet varieties. *Food and Agricultural Immunology* 29(1): 625–637. https://doi.org/10.1080/09540105.2018.1428283.

Shobana, S., and M. Malleshi. 2007. Preparation and functional properties of decorticated finger millet (*Eleusine coracana*). *Journal of Food Engineering* 79(2): 529–538. https://doi.org/10.1016/J.JFOODENG.2006.01.076.

Srilekha, K., T. Kamalaja, K. U. Maheswari, and R. N. Rani. 2019. Evaluation of physical, functional and nutritional quality parameters of kodo millet flour. *Journal of Pharmacognosy and Phytochemistry* 8(4): 192–195.

Suma, P. F., and A. Urooj. 2015. Isolation and characterization of starch from pearl millet (Pennisetum typhoidium) flours. *International Journal of Food Properties* 18(12): 2675–2687. https://doi.org/10.1080/10942912.2014.981640.

Ugare, R., B. Chimmad, R. Naik, P. Bharati, and S. Itagi. 2014. Glycemic index and significance of barnyard millet (*Echinochloa frumentacae*) in type II diabetics. *Journal of Food Science and Technology* 51(2): 392–395.

Velu, G., K. N. Rai, and K. L. Sahrawat. 2008. Variability for grain iron and zinc content in a diverse range of pearl millet populations. *Journal of Crop Improvement* 35(2): 186–191.

Verma, S., S. Srivastava, and N. Tiwari. 2015. Comparative study on nutritional and sensory quality of barnyard and foxtail millet food products with traditional rice products. *Journal of Food Science and Technology* 52(8): 5147–5155.

Wang, L., P. Gulati, D. Santra, D. Rose, and Y. Zhang. 2018. Nanoparticles prepared by proso millet protein as novel curcumin delivery system. *Food Chemistry* 240: 1039–1046. https://doi.org/10.1016/J.FOODCHEM.2017.08.036.

Wankhede, D. B., A. Shehnaj, and M. R. Raghavendra Rao. 1978. Carbohydrate composition of finger millet (Eleusine coracana) and foxtail millet (*Setaria italica*). *Plant Foods for Human Nutrition* 28(4): 293–303.

Yang, Q., P. Zhang, Y. Qu, X. Gao, J. Liang, P. Yang, and B. Feng. 2018. Comparison of physicochemical properties and cooking edibility of waxy and non-waxy proso millet (*Panicum miliaceum* L.). *Food Chemistry* 257: 271–278.

Yankah, N., F. D. Intiful, and E. M. A. Tette. 2020. Comparative study of the nutritional composition of local brown rice, maize (obaatanpa), and millet—A baseline research for varietal complementary feeding. *Food Science & Nutrition* 8(6): 2692–2698. https://doi.org/10.1002/FSN3.1556.

Zhu, F. 2020a. Fonio grains: Physicochemical properties, nutritional potential, and food applications. *Comprehensive Reviews in Food Science and Food Safety* 19(6): 3365–3389. https://doi.org/10.1111/1541-4337.12608.

Zhu, Y., J. Chu, Z. Lu, F. Lv, X. Bie, C. Zhang, and H. Zhao. 2018. Physicochemical and functional properties of dietary fiber from foxtail millet (*Setaria italica*) bran. *Journal of Cereal Science* 79: 456–461.

Chapter 7

Phytochemicals in Millets and Related Health Benefits

7.1 INTRODUCTION

Millets are not just important for their good nutritional value but also for the presence of various phytochemicals. Millets contain various phytochemicals like phenols, sterols, peptides, polycosanols, etc. These are generally secondary plant metabolites or components of plant cell walls. These phytochemicals are of significance as they are useful for human health in various ways. Certain phenolic compounds act as help in the defence mechanisms of plants which help in tackling pests and insects. Plant sterols are components of the cell membrane and mostly the wax components of bran oil. Dietary fibre has been found in good amounts in all millets.

Recent scientific evidence has linked several phytochemicals with better health indices. Phytochemicals helps in maintaining human health by modulating various mechanisms in the body. Phenolic acids protect against various types of cancers and protect heart health. Plant sterols are good for the cardiovascular system. Dietary fibre helps in weight reduction and maintaining gut health. Complexes of these phytochemicals also account for numerous benefits. For example, phenolic acid-bound arabinoxylans exhibit prebiotic and immunomodulatory effects (Srinivasan 2021). Fructo-oligosaccharides exhibit lipid peroxidation activity, improve lipid profile, lower the risk of heart diseases, and relieve constipation. The health impacts of nutritional and phytochemicals in millets are discussed in detail in the upcoming chapter.

The presence of phytochemicals in millets is an added advantage as these are climate-resilient crops which survive under stressful conditions of heat and drought, better than virtually all other cereal crops. The current chapter will focus on the type and chemistry of phytochemicals in various millets and their relation to health.

7.2 ANTIOXIDANT POTENTIAL OF MILLETS

Millets are well known for their antioxidant nature. Li et al. (2021) studied the DPPH radical scavenging activities of free and bound portions of proso millet and found them to be in the range of 19.78–26.94% and 31.15–85.09% respectively. The total antioxidant activity of proso millet varied with the colour. Results from the ABTS radical cation scavenging activity of proso millet ranged between 0.26 and 1.09 mmol Trolox/g. The black- and grey-coloured varieties had better antioxidant capacity as estimated by DPPH, ABTS, and FRAP in proso millet in comparison to the red and white varieties (Li et al. 2021). Slama et al. (2020) assessed the antioxidant activity of pearl millet oil and found it to be higher than the grain (79.2% DPPH inhibition) with an IC$_{50}$ value of 25.20. The antioxidant activity of pearl millet oil analysed in the study was greater than two Tunisian varieties of olive oil (Nakbi et al. 2010). The DPPH scavenging activity (IC$_{50}$ value) was found to be in the range of 114.37–132.68 and 113.24–118.47 µg/ml for soluble and bound fractions extracted from three cultivars of foxtail millet. The reducing power was in the range of 39.47–48.26 mg ascorbic acid equivalent/100 g and 14.25–24.26 mg ascorbic acid equivalent/100 g for soluble and bound fractions respectively. Peroxide scavenging in foxtail millet cultivars was in the range of 39.66–46.63% and 10.08–11.7% for soluble and bound fractions respectively. Metal chelating (µmol EDTA equivalents/g) was found to be in the range of 3.41–5.67 and 2.09–3.29 respectively (Pradeep and Sreerama 2018). In the same study, the antioxidant properties of little millet were also analysed. The DPPH scavenging activity (IC$_{50}$ value) was found to be in the range of 102.91–114.26 µg/ml and 176.29–191.28 µg/ml for soluble and bound fractions extracted from three cultivars of little millet. The reducing power was in the range of 69.62–75.43 mg ascorbic acid equivalent/100 g and 37.49–41.06 mg ascorbic acid equivalent/100 g for soluble and bound fractions respectively. Peroxide scavenging in little millet cultivars was in the range of 66.17–75.55% and 12.7–14.51% for soluble and bound fractions respectively. Metal chelating (µmol EDTA equivalents/g) was found to be in the range of 9.21–1.26 and 5.73–8.11 for soluble and bound fractions of little millet respectively (Pradeep and Sreerama 2018). Oluwasesan et al. (2022) assessed the antioxidant potential of fonio millet and compared it with finger and pearl millet. Fonio millet was found to have the highest antioxidant activity (% inhibition value = 47.909 ± 3.472) and the greatest activity against haemoglobin glycosylation (29.469 ± 0.399%), alpha-amylase (43.729 ± 0.410%), and glucosidase (55.835 ± 2.198%). The antioxidant potential of different millets is given in Table 7.1.

7.3 PHYTOCHEMICALS IN MILLETS

This section deals with the chemistry and health impact associated with different types of phytochemicals in millets with special reference to phenols, peptides, etc. Li et al. (2021) identified 672 metabolites in different varieties of proso millet including phenolic acids, flavonoids, alkaloids, terpenoids, steroids, etc. These compounds are responsible for various types of health benefits including anti-obesity, anti-diabetic, cardioprotective, anti-allergic, anti-hypertensive, anti-cancer effects, and many more. Table 7.2 gives the health benefits of some phytochemicals and summarizes the mechanism of action as proposed in various studies.

7.3.1 Phenolic Compounds

The phenolic compounds are found in plants and are important for various reasons like structural components, intermediaries of various metabolic pathways, natural defence molecules, signalling molecules, etc. The research literature extensively investigated the impact of polyphenols on health and found significant evidence on the beneficial effect against various diseases.

Millets, like other cereals, contain phenolic acids and tannins as the predominant phenols. The phenolic compounds in millets are mainly concentrated in the bran layer of the grain. During processing most of the bran is removed leaving behind starchy endosperm. The endosperm is devoid of the majority of nutrients. Similar to other grasses, the phenolic component

TABLE 7.1 ANTIOXIDANT ACTIVITY OF DIFFERENT TYPES OF MILLETS

Millet	Antioxidant activity (% DPPH inhibition)	ABTS (μmol Trolox equivalents/g, DW)	ORAC (μmol Trolox equivalents/g, DW)	Hydrogen peroxide assay	References
Pearl millet	19.8	16.7	60.3 (defatted meal)	15.3	Oluwasesan et al. 2022, Ahmed et al. 2021
Finger millet (μmol trolox equivalents/g, DW)	14.23–21.48	18.96–26.75	116.82–161.06	22.4	Xiang et al. 2019
Proso millet	58–297	2.82–9.75 (free fraction) 9.56–15.8 (bound fraction)	69.4		Kim et al. 2010; Yuan et al. 2022; Chandrasekara and Shahidi 2011
Foxtail millet	196–318		88.36		Kim et al. 2010; Chandrasekara and Shahidi 2011
Barnyard millet	25.4–359.6	11.7			Kim et al. 2011
Kodo millet	44.21		95.7 (defatted meal)	4.52	Sharma et al. 2021a; Ahmed et al. 2021
Little millet	90.2		91.28		Pradeep and Guha 2011; Chandrasekara and Shahidi 2011
Teff	27.52	9.3–10.3		10.3	Koubova et al. 2018; Kataria et al. 2022
Fonio	60.64	22.9		31.6	Oluwasesan et al. 2022

TABLE 7.2 PHYTOCHEMICALS IN MILLETS

Phytochemical	Health benefits	Mechanism of action	Reference
Phenolic acids	Antimicrobial, anticancer, anti-inflammatory, anti-mutagenic, *etc.*	Act as hydrogen atom donors changing free radical into a less reactive phenoxyl radical	Chandrashekara and Shahidi 2012; Kumar and Goel 2019
Flavonoids	Protect the cardiovascular system, antidiabetic, antiobesity, anticancer	Act as a potent antioxidant due to the presence of multiple hydroxyl groups, conjugation and delocalization of electrons	Suma and Urooj 2015; Bangoura et al. 2013; Ballard and Maróstica 2019
Tannins	Anti-oxidant, anti-cancerous, anti-allergic, anti-inflammatory, anti-helminthic and anti-microbial	Act as pro-oxidant through redox cycling and by the production of oligomeric compounds	Suma and Urooj 2015; Siwela et al. 2007; Sharma et al. 2021
Xylooligo-saccharides	Prebiotic, antioxidant, anti-diabetic, immune stimulation, lowering cholesterol, anticancer	Presence of ester linked phenolic acids and sugar linked acetyl groups imparts strong antioxidant activity	Devi et al. 2014; Palaniappan et al. 2017; Palaniappan et al. 2021
Fibre	Prebiotic, antioxidant, anti-diabetic, immune stimulation, lowering cholesterol, anticancer	High water absorption and unique phytochemical composition	Amadou et al. (2013); Devi et al. (2014)
Bioactive peptides	Anti-hypertensive, antimicrobial	Bioactive peptides have the ability of inactivation, scavenging, chelation, and anzymatic elimination of free radicals or transition metals	Majid and Priyadarshini (2020); Amadou et al. (2013)

Source: Modified from Liang and Liang 2019.

concentration in the bran is much greater than the seed coat of dicots like legumes (Vogel 2008). The majority of the phenolic acids are either esterified or etherified to the polysaccharides of the cell wall or lignin where these act as cross-linkers. Such phenolic forms cannot be extracted by organic solvents and need alkali for their hydrolysis. The forms which can be extracted are usually esterified to monosaccharides like glycerides and glycosides. Phenolic acids are rare in free form in most cereals but can be found in some millets. Flavonoids are also present but in comparatively smaller quantities (Dykes and Rooney 2006). These phenols act as major free radical scavengers, as metal chelators, and as oxygen radical quenchers. Hence, the phenolic compounds are known mainly for their antioxidant activities. The total phenol contents of various millets have been presented in Table 7.3. The total phenol content in raw samples of pearl millet was found in the range of 268.5–420 mg/100 g on dry weight basis. On cooking, very little decrease in the total phenolic content was observed (247.5–335 mg/100 g of dry weight) (Nambiar et al. 2012). In finger millet, total polyphenol content was reported to be 1.2 mg/g (Hithamani et

TABLE 7.3 CONTENT OF PHENOLIC COMPOUNDS IN DIFFERENT MILLETS

Phenolic compound	Pearl millet	Finger millet	Proso millet (ultra-fine pulverized sample) μg/g extract residue	Foxtail millet (ultra-fine pulverized sample) μg/g extract residue	Kodo millet	Little millet	Teff	Fonio
Gallic acid	–	10.69	5.93	22.72	1.8	1.74	25.6	6.6
Pyrogallol			34.36	71.22				
Catechin				12.92			25.4	
Protocatechiuc acid	15.36	14.6	–	1.34–108.89	6.84	ND	207.9	8.69
p-Hydroxy benzoic acid	–	3.62	–		1.41	ND	1.1	
Gentisic acid		4.54	–		ND	0.65		
Vanillic acid	–	–	0.99	2.87	25.4	ND	4	3.68
Caffeic acid		1.40	0.99	1.07–6.18		–	0.5	4.75
Syringic acid	6.68	10.17			ND	ND	4.4	ND
p-Coumaric acid	54.91	19.07	1.38	1.52		–	18.6–244.2 (μg/g) db	1.25
Rutin			7.73	7.03			1.4	
Ferulic acid	28.95	12.88	0.20	10.08	10.3	13.54	153.6–475 (μg/g) db	25.14
Sinapic acid	501	24.78					6.3	5.2
Salicylic acid	181.9	413	4.74	13.97			–	ND
t-Cinnamic acid	43.43	31.87	1.58	2.33–3128.72		ND	ND	
Naringenin			1.60	1.10			–	
Quercetin			1.61	–			30.8	
Total phenols	832.2	555.6	72.49	153.92			690.5	60.82
References	Hithamani and Srinivasan 2014	Hithamani and Srinivasan 2014	Seo et al. 2011	Seo et al. 2011	Chandrasekara and Shahidi 2011	Pradeep and Guha 2011	Ravisankar et al. 2018; Koubova et al. 2018	N'Dri et al. 2013

al. 2014). Hassan et al. (2020) detected catechin, epicatechin, protocatechuic acid, tryptophan, p-hydroxybenzoic acid, procyanidin B1 and B2, and kaempferol glycoside in two varieties of finger millet procured from Zimbabwe. The total phenolic content of methanolic extract of fonio millet was detected to be 57.96 ± 6.84 mg gallic acid equivalent per gram of dried weight (Omaji et al. 2021). Slama et al. (2020) assessed the antioxidant and phytochemical content of pearl millet oil and found that pearl millet contained higher content of polyphenols (2.20 mg gallic acid equivalent per gram) than foxtail (0.47 mg gallic acid equivalent per gram) and proso millet (0.29 mg gallic acid equivalent per gram). The total phenolic content in foxtail and little millet was detected in the range of 19.42–20.8 μmol ferulic acid equivalents/g and 23.95–24.12 μmol ferulic acid equivalents/g respectively (Pradeep and Sreerama 2018).

7.3.1.1 Phenolic Acids

The activity phenols exhibit depends upon the chemistry and position of the phenolic ring and the extent of hydroxylation in addition to other structural variations. Depending on the substitution pattern on the rings (benzene and heterocyclic) and the degree of condensation to other compounds, more than 5000 phenolic acids and flavonoids have been identified. Generally, phenolic acids are classified into benzoic acid derivatives and cinnamic acid derivatives. Millets exhibit great genetic diversity and therefore differ widely in their phenolic composition as well; the variation however, is not as diverse as sorghum. Nambiar et al. (2012) identified five phenolic acids in pearl millet – vanilic acid, syringic acid, melilotic acid, para-hydroxyl benzoic acid, and salicylic acid. Slama et al. (2020) analysed the type of phenolic acids present in pearl millet oil and detected *Trans*-cinnamic acid (633 μg/g of extract) in the highest quantities followed by protocatechuic, hydroxybenzoic, gentistic, gallic, and caffeic acids. In finger millet procured from Zimbabwe, protocatechuic acid was predominant in the range of 20.9–23.7 mg/kg. *p*-hydroxybenzoic acid was detected with a range of 13.5–16.8 mg/kg (Hassan et al. 2020). Both these compounds were detected at the retention time of 7.35 and 9.178 respectively.

More than 60% of the phenolic acids in millets are present in the bound form while the remainder are found in free form (Zhang et al. 2015). Millets including fonio were reported to contain higher amounts of free and bound phenolics in comparison to sorghum (N'Dri et al. 2013). As the insoluble bound fraction of the millet grain, hydroxycinnamic acids are predominant. A potent example of hydroxycinnamic acid is ferulic acid which is a powerful antioxidant and possesses high anti-inflammatory activity. Apart from the monomer compounds, even the dimers of ferulic acid in millets were shown to have high levels of antioxidant activities. An advantage of both monomers and dimers of ferulic acid is that these possess strong antioxidant activities even in their bound form and hence provide better function even without the aid of acid or microbial digestion for better release in the colon. In finger millet, water-soluble feraxans were reported which exhibited an antioxidant activity 5000 times greater than the sulphated polysaccharides (Rao et al. 2006). Unbound phenolic acids are easily digested by microbes in the colon and therefore directly influence the systemic health of a person. Ferulic acid is the dominant bound phenolic in millets including finger millet (Subba Rao and Muralikrishna 2002; Chandrasekara and Shahidi 2011). The effect of heat on bound phenolic has different impacts on different millets. In a study conducted by N'Dri et al. (2013), heat treatment increased the content of total bound phenolics in fonio, whereas there was no influence on millet. In fonio, all bound acids depicted increase except gallic acid. The increase in ferulic acid was observed to be more than fivefold. In millets, the major bound phenolics lost were identified to be gallic acid and protocatechuic acid with a loss of 64 and 32% respectively (N'Dri et al. 2013). Li et al. (2021) reported hydroxybenzoic acids and hydroxycinnamic acid derivatives in bound form in proso millet. The bound phenolic acids present in the highest quantities were found to be caffeic, ferulic, p-coumaric, isoferulic, and vanillic acid. Chlorogenic acid, cryptochlorogenic, and syringin were detected only in white varieties of proso millet but missing in black, grey, and red varieties. The bound phenolic compounds in foxtail millet were detected to be in the range of 147.72 to 250.13 µg/g and between 125.13 and 290.99 µg/g in little millet. The bound fractions of foxtail millet were found to have benzoic acid derivatives in the range of 6.68–37.47 µg/g and cinnamic acid derivatives in the range of 147.72–250.13 µg/g. In case of the bound fraction of little millet, the content of benzoic acid derivatives and cinnamic acid derivatives was between 9.02 and 16.73 µg/g and 114.12 and 274.26 µg/g respectively (Pradeep and Sreerama 2018).

Most of the unbound (i.e. unbound to the cell wall) phenolic acids are found in combination with carbohydrate monomers like glycerol and glucose. Free unconjugated monomeric phenolic acids are comparatively less stable and therefore are limited in millets. In millets, direct structural conjugation of the phenolic acids is not available and hence they are analysed through indirect analysis. As already mentioned, a majority of the phenols are the bound forms accounting for more than 70% of the total phenolics in grains. These are responsible for the extensive degree of cross linking of cereal cell wall polysaccharides. Fonio millet contains protocatechuic acid as the predominant soluble phenolic acid and p-coumaric and gallic acid in other millets in general (Chandrasekara and Shahidi 2011). However, variation has been observed among studies in this regard based on varietal differences. The process of cooking, in general, enhanced the total soluble phenolic content in millets with the exception of vanillic and protocatechuic acids, which showed a decreasing trend. In fonio, a contrast was observed where total phenolic acid content declined probably due to a loss of major compounds viz., protocatechuic acid (47%) and vanillic acid (42%) (N'Dri et al. 2013). Li et al. (2021) analysed four varieties of proso millet for their free phenolic acid content and found the presence of hydoxybenzoic, hydoxycinnamic acid derivatives, and chlorogenic acid isomers. The most abundant free phenolic acids identified were vanillic acid, 4-hydroxybenzoic acid, p-coumaric acid, ferulic acid, isoferulic acid, and p-hydroxycinnamic acid. The study also reported the presence of isoferulic acid and p-hydroxycinnamic acid. The dark varieties of proso millet (black, grey, and red) had 0.25–1.04 times higher ferulic acid content than white varieties. Similar was the trend reported for syringic acid, syringaldehyde, sinapinaldehyde, and methyl ferulate. The total soluble phenolic content of foxtail millet and little millet was found in the range of 117.94–261.84 µg/g and 174.69–291.09 µg/g for foxtail and little millet respectively. In the soluble fraction of foxtail millet, two phenolic acid derivatives were identified, namely benzoic acid derivatives and cinnamic acid derivatives in the range of 7.24–21.11 µg/g and 110.7–261.84 µg/g respectively. The content of these derivatives in the soluble fraction of little millet was found to be 8.78–50.29 µg/g and 174.69–291.09 µg/g respectively. Benzoic acid derivatives identified in foxtail and little millet were gallic, p-hydroxybenzoic, and vanillic acid. Whereas the cinnamic acids identified were caffeic, chlorogenic, ferulic, sinapic, and p-coumaric acid (Pradeep and Sreerama 2018).

The antioxidant activity of free phenolic acids is influenced by a number of factors including the number of phenolic hydroxyl moieties, the location of hydroxyl moiety on aromatic ring, and the type of substituent. The substitution of ortho-hydroxyl groups with electron-donating groups, for example methoxyl group, was found to enhance the total antioxidant activity of the compound (Liang and Liang 2019). Also, it has been observed that, apart from phenolic acids, the antioxidant activity is significantly enhanced by the presence of sugars with uronyl or acetyl moieties. The influence is not just due to the presence of the moieties but also due to the level or nature of polymerization.

Apart from the above discussed phenols, tannins also act as potent antioxidants, especially those with high molecular weight. Many studies have reported the presence of tannins in different millet varieties, coloured ones especially. High content of tannins has been reported in the dark-coloured varieties. Sripriya et al. (1996) reported high tannin content in brown-coloured finger millet varieties in comparison to white-coloured ones. Condensed tannins were reported in the range of 28.09–34.47 mg catechin equivalents/100 g in three varieties of foxtail millet and between 89.12 and 107.56 mg catechin equivalents/100 g (Pradeep and Sreerama 2018).

7.3.1.2 Flavonoids

Another group related to phenolic acids which potentially have high antioxidant capacity are the flavonoids. These are the most heterogeneous group of phenols and are found in abundance in fruits, vegetables, and even in some cereals. These are a class of phenolic compounds which possess a basic C6–C3–C6 structure, with two aromatic rings joined by a three-carbon heterocyclic ring. In cereals and millets, flavonoids which are generally found are unsubstituted at the C-3 position of the heterocyclic ring. For this reason, these are also referred to as 3-deoxyflavanoids or flavones. Such flavonoids like catechin derivatives and anthocyanins are found only in certain species and genetic variants. Millets contain different types of flavonoids like anthocyanidins, chalcones, aminophenolics, flavanols, flavonols, flavones, and flavanones, etc. (Chandrasekara et al. 2011). These either occur as free aglycones or in conjunction with glycosides. The predominant flavonoids in major cereals were identified as flavones-C-glycosides, whereas flavones, luteolin, and tricin were isolated from barnyard millet (Watanabe et al. 1999, Brazier-Hicks et al. 2009). Millets, including whole and pearled, are good sources of flavonoids which are responsible for several health-benefiting properties of millets. A high content of flavonoids in the skin or outer region of most fruits and seeds helps in protection against pathogens and pests. This is evidenced by the fact that most plants produce increased quality of flavonoids when damaged or under attack. Nambiar et al. (2012) identified tricin, acacetin, 3, 4 Di-OMe luteolin, and 4-OMe tricin flavonoids in pearl millet. Slama et al. (2020) assessed the total flavonoid content of oil extracted from the Medinine ecotype of pearl millet and found it to be 0.84 mg catechin per gram. Quecertin and catechin were the major flavonoids detected in pearl millet oil (Slama et al. 2020). The presence of these flavonoids indicate the chemopreventive potential of pearl millet. The total flavonoid content of finger millet was reported to be 5.54 mg/g (Hithamani and Srinivasan 2014). Li et al. (2021) identified 97 different flavonoids and their metabolites in proso millet. Tectorigenin was detected only in the red proso millet variety and liquiritigenin was found only in black-coloured varieties. Genistin and tricetin were detected in white, grey, and red varieties but not in black. The total flavonoid content of methanolic extract of fonio millet was detected to be 38.75 ± 9.76 mg quercetin equivalent per gram of dried weight (Omaji et al. 2021). Pradeep and Sreerama (2018) assessed the soluble and bound flavonoid content in foxtail and little millet. The soluble fraction of flavonoid was found to be in the range of 5.41–8.99 μmol catechin equivalents/g and 9.92–13.64 μmol catechin equivalents/g in foxtail millet and little millet respectively. The total flavonoid content in the bound fraction was found to be in the range of 0.83–1.07 μmol catechin equivalents/g and 0.79–1.45 μmol catechin equivalents/g for foxtail and little millet respectively. The flavonoids identified in foxtail and little millet were catechin, myricetin, daidzein, luteolin, quercetin, apigenin, naringenin, and kaempferol (Pradeep and Sreerama 2018).

7.3.1.3 Flavones

Flavones are the class of flavonoids which are found in all cereals including millets. Most of the flavones exist in the form of ortho- and cis-glycosides of apigenin and luteolin. However, grains like fonio and sorghum also contain some free forms. Some millet varieties were also found to contain tricin in small amounts. Cereal flavones have certain characteristic structural features which are responsible for the prevention of inflammatory diseases and cancer (Agah et al. 2017). In general, cereals possess minute quantities of flavones but in millets (except finger millet) flavones are the predominant flavonoid compounds. Ravisankar et al. (2018) identified various flavone compounds in white and brown teff varieties of Ethiopia and the USA. The total flavones reported in white and brown teff varieties were in the range of 1398 and 2049 µg/g on dry weight basis. White teff was reported to have only apigenin derivatives whereas brown teff had mainly luteolin derivatives. Diglycoside derivatives accounted for approximately half of the total flavones in teff. Monoglycosides and acylated derivatives were the next major derivatives reported. Flavones possess high levels of C-glycosides which means they have potential for thyroid function inhibition. Li et al. (2021) identified 46 flavones out of which 3 were the most abundant, namely isovitexin, butin, and 4,2′,4′,6′-tetrahydroxychalcone in four varieties of proso millet. Apigenin and luteolin were identified in grey, black, and red varieties but not in white proso millet varieties.

7.3.1.4 Flavonones and Flavonols

Flavanones lack a double bond between the second and third carbons in their basic 2,3-dihydroflavone structure, making them C2 chiral and not planar like conjugated flavonoids. Such variations in orientation make the difference in the bioactive properties of these compounds due to their modified ability to interact with different receptors (Agah et al. 2017). Li et al. (2021) identified four flavonones, namely naringenin, naringenin 7-O-glucoside, 4′,5,7-trihydroxyflavanone, and 4′-hydroxy-5,7-dimethoxyflavanone, in four varieties of proso millet.

Flavonols are flavonoids which possess 2-phenyl-3,4-dihydro-2H-chromen-3-ol backbone. Limited evidence in millets shows the presence of these in the form of condensed tannins. Condensed tannins, polymers of flavan-3-ols and derivatives, are also referred to as proanthocyanidins due to their ability to hydrolyse anthocyanidins. Limited research is available on the characterization and degree of polymerization in millets. Finger millet tannin polymerization usually occurs through C4–C8 bonds, also known as the B-type linkage similar to sorghum (Chandrasekara and Shahidi 2011). In millets, tannins are found in between the pericarp and aleurone layer of the endosperm in the form of a separate layer called pigmented testa. Li et al. (2021) identified three flavonols, namely Di-O-methyl-quercetin, kaempferol 3-O-glucoside, and kaempferol 3-O-galactoside, in four varieties of proso millet.

7.3.2 Phytosterols

Phytosterols are concentrated in the bran layer of millets and can be extracted as a component of bran oil. Phytosterols are beneficial for the human body in many ways mainly because of their cholesterol-lowering effects (Demonty et al. 2009). They exhibit ferric-reducing activity and hydroxyl radical scavenging activity. Squalene has been reported to be a potent singlet oxygen quencher and also acts as an antidote for the reduction of drug-induced toxicities. Analysis of foxtail millet seeds exhibited five isomers of vitamin E. The total phytosterol content analysed in foxtail millet was 443 to 569 mg/kg. Among campesterol, stigmasterol, and β-sitosterol, β-sitosterol was identified to be the major form found in approximately 85% of the samples. The precursor of phytosterols, squalene was also reported and was found to be in the range of 6.8 to 10.2 mg/kg (Bhandari and Lee 2013; Duodu and Awika 2019). The content of phytosterols reported in foxtail millet is higher or similar when compared to grains like maize and barley

(Ryan et al. 2007). The total phytosterol content of foxtail millet seed increases in early phases and then starts to decrease with seed development. Yuan et al. (2021) identified the mechanism and using transcriptome analysis identified 152 genes associated with fatty acid metabolism. Out of these, 62 were found to be related to phytosterol biosynthesis. The overexpression of CAS1, STM1, EGR6, and DWF1 genes leads to phytosterol accumulation in the S2 stage (Yuan et al. 2021). Foxtail millet bran contains 56.6% β-sitosterol followed by lower quantities of campesterol and stigmasterol (15.9 and 5.6% respectively) (Pang et al. 2014). The phytosterol content of Italian finger millet is between 477 and 568 mg/kg and consists of campesterol, stigmasterol, and β-sitosterol. β-sitosterol constitutes 85% (Bhandari and Lee 2013; Sharma et al. 2021). Yin et al. (2021) noticed that the refining process of millet bran oil reduced the phytosterol content by 50% approximately. Similar results were obtained by Lin et al. (2015). Esche et al. (2013) analysed four varieties of proso millet for their free sterol/stanol content which was found to be between 518.4 and 741.8 μg/100 mg oil. Sitosterol, campesterol, stigmasterol, cycloartanol, sitostanol, avenasterol, and campestanol were identified in proso millet. For millets, data on phytosterol content is extremely limited and needs to be explored. Novel compositions of steryl ferulates were identified in foxtail and barnyard millet and compared with other cereals using high-performance liquid chromatography and atmospheric pressure chemical ionization-mass spectra. The steryl ferulate content of whole foxtail millet was detected to be 80% of brown rice. Foxtail millet contained mainly stigmastanyl ferulate (73%) and some quantity of stigmastanyl coumarate (10%). The total content of steryl ferulates was 28.9 mg-oryzanol equivalent/100 g in foxtail millet, 0.2 mg-oryzanol equivalent/100 g in proso millet, and 4.8 mg-oryzanol equivalent/100 g in barnyard millet (Tsuzuki et al. 2018).

7.3.3 Bioactive Peptides

Bioactive peptides are hydrolysed fragments of proteins (2–20 amino acid sequence and molecular weight less than 6 kDa) known for their proven antioxidant potential. These compounds have a positive impact on health and are considered biologically active regulators (Gan et al. 2017; Sánchez and Vázquez 2017). Amino acids like histidine, leucine, glycine and proline, tryptophan, proline, valine, and cysteine act as antioxidants and play a crucial role in radical scavenging activity. Bioactive peptides containing valine-alanine-proline epitope have been found to have the capacity to inhibit angiotensin-1 converting enzyme which makes then potentially antihypertensive. Takumi et al. (1996) analysed the polypeptide compositions of prolamine from Italian and Japanese cultivars and identified peptides ranging from 13 to 23 kDa. Millet-derived biopeptides contain good quantities of peptidic amino acid residues that have the capability to transfer electrons to high-energy free radicals, positively contributing to the antioxidant properties (Matsui et al. 2018). Millet-derived bioactive peptides also exhibited anti-hypertensive, anti-proliferative, and antimicrobial effects (Majid and Priyadarshini 2020).

In pearl millet, the isolation of bioactive peptides was done using trypsin and the bioactive peptides were subjected to various antioxidant assays like DPPH, ABTS, ferrous chelating activity, and hydroxyl activity. Bioactive peptides have proven antioxidant potential. Bioactive peptides were isolated using protein hydrolysing enzyme trypsin from pearl millet. The antioxidant potential of these bioactive peptides was assessed using different assays like DPPH, ABTS, ferrous chelating activity, hydroxyl activity, etc. The antioxidant activity using DPPH assay was found to be 67.66%, ABTS was found to be 78.81%, and iron chelating and hydroxyl activity were reported to be 51.20% and 60.95% respectively. The reducing power was reported at a maximum of 0.375 nm respectively. The peptide with antioxidant potential was found to have a sequence of

SDRDLLGPNNQYLPK (Agrawal et al. 2016). Similar metal chelating activity was reported from protein hydrolysates derived from foxtail millet. Foxtail millet prolamin hydrolysates were prepared in a study and were subjected to ultrasonic and heat treatment to increase the degree of hydrolysis. The study reported that the fraction with molecular weight <1 kDa hydrolysed using alcalase possessed the highest antioxidant activity. Two novel peptide sequences with antioxidant potential were identified as Pro-Phe-Leu-Phe and Ile-Ala-Leu-Leu-Ile-Pro-Phe with molecular weights of 522.3 and 785.5 Da respectively. The fraction with low molecular weight (Pro-Phe-Leu-Phe) exhibited high antioxidant activity as shown by DPPH radical scavenging ability (149 μM TE/g protein) and an ORAC value of 1180 μM TE/g protein. These peptide sequences exhibited an antioxidant effect by the reduced formation of malonaldiahyde and increased content of glutathione (Ji et al. 2019). Hu et al. (2020) assessed the antioxidant and anti-inflammatory properties of foxtail millet peptides prepared using in-vitro gastrointestinal digestion. The study identified seven peptides with high antioxidant potential. Two peptides with sequences EDDQMDPMAK and QNWDFCEAWEPCF had better capacity to inhibit inflammatory markers like nitric oxide, tumour necrosis factor-α, and interleukin-6 in a RAW 264.7 cell model (Hu et al. 2020).

Agrawal et al. (2019) experimented to purify, identify, and characterize novel antioxidant peptides from finger millet. Protein isolates were prepared using hydrolysis using enzymes. Hydrolysates prepared using trypsin (17.47%) were found to have a higher degree of hydrolysis in comparison to pepsin. After fractionation using ultrafiltration, the component with molecular weight less than 3 kDa possessed higher antioxidant activity (61.79%). This component was further fractionated into five fractions among which GF_B had the highest antioxidant activity (61.79%). The sequences to two novel peptides with antioxidant potential were found to be TSSSLNMAVRGGLTR and STTVGLGISMRSASVR. Molecular docking studies revealed that the interaction of serine and threonine amino acids with free radicals is responsible for the antioxidant potential.

Yun and Park (2018) analysed the antioxidant activities of brown teff hydrolysates prepared by the action of three proteases, namely protamax, flavourzyme, and alcalase. The highest phenol, flavonoid, and total antioxidant capacity was obtained by hydrolysates obtained from treatment with 1 wt% protamax.

7.3.4 Fibre and Its Components

Millets contain indigestible food components which act as potent prebiotic factors. These include components of dietary fibre arabinoxylans, arabinogalactans, β-glucans, fructans, hemicelluloses, etc., which possess prebiotic potential (Kumar et al. 2015). The insoluble fibre also exhibited antioxidant activity. The yellow varieties of foxtail millet had higher antioxidant activity in comparison to white varieties (Bangoura et al. 2013). Their antioxidant potential is attributed to insoluble fibre bonding with the phytochemical components. Polysaccharides from foxtail millet exhibited higher scavenging potential for 1,1-diphenyl-2-picrylhydrazyl and hydroxyl radicals in basic potential (Zhu et al. 2015). Arabinoxylans can be utilized by various microorganisms like *Lactobacilli*, *Bacteroides*, *Bifidobacterium animalis*, *Faecalibacterium prausnitzii*, *Roseburia eubacterium*, and non-pathogenic *clostridia* (Sharma et al. 2018, Sharma et al. 2021b). In finger millet, arabinoxyalans were extracted with a maximum yield of 5.68% when extracted using 0.5% sodium hydroxide as a solvent with a substrate to solvent ratio of 10 ml/g and an extraction temperature and time of 60°C and 10 minutes (Bijalwan et al. 2016). Arabinoxylans on hydrolysis produce xylooligosaccharides which are known to have a prebiotic effect on various species of *Lactobacillus* like *L. brevis* and *L. plantarum* (Stojanovski 2012). Millets are also rich in fructo-oligosaccharides which support the growth of *Bifidobacterium* and *Lactobacillus* and inhibit the growth of *E. coli* and *Salmonella* in the large intestine (Manosroi et al. 2014).

Xylooligosaccharides are the low-degree polymers (degree of polymerization – 2 to 10) of D-xylose which are linked together by beta 1-4 glucosidic linkage. Examples include xylobiose, xylotriose, and xylotetrose. These are extracted by hydrolysis reactions involving arabinoxylans derived from cereal and millet bran. The antioxidant activity of xylooligosaccharides as assessed by DPPH, beta-carotene emulsion, and FRAP assays was found to be much higher in comparison to other cereals like wheat, rice, and maize (Veenashri and Muralikrishna 2011). The potent antioxidant activity of xylooligosaccharides comes from the fact that these form complexes with phenolic acids like caffeic and ferulic (Samantha et al. 2012). This is evident from finger millet where the high antioxidant activity correlates with an increase in the content of syringic and ferulic acid (Veenashri and Muralikrishna 2011; Palaniappan et al. 2017). Singh and Eligar (2021) derived bioactive feruloylated xylooligosaccharides from pearl millet bran and assessed their antioxidant potential. The study isolated ferulic acid (25.17 mg/g), arabinose (227.5 mg/g), and xylose (359.9 mg/g) as major monosaccharides and glucose (27.2 mg/g), galactose (27.2 mg/g), and glucuronic acid (14 mg/g) as minor components. The total arabinoxylan content was found to be 51.5% in pearl millet feruloylated xylooligosaccharides. The anti-glycation potential of these was also assessed by studying fructosamine inhibition and protein carbonyl formation using a bovine serum albumin/glucose model in different stages of glycation and it was found to be a potent antiglycation molecule. Palaniappan et al. (2017) tried to produce xylooligosaccharides from the finger millet seed coat using xylanase of *Thermomyces lanuginosus* and observed a yield of 72%. The xylooligosaccharide fraction obtained presented antioxidant activity of 75% and effective prebiotic potential of *Lactobacillus plantarum*. Bijalwan et al. (2016) isolated hydoxycinnamic acid-bound arabinoxylans from the bran of kodo millet, finger miller, proso millet, and foxtail millet. The study revealed that arabinoxylans derived from kodo millet were comparatively low branched consisting of monosubstituted (14.6%), disubstituted (1.2%), and unsubstituted (41.2%) Xyl$p$ residues. DPPH and FRAP assays revealed that the antioxidant potential of low substituted arabinoxylans was comparatively higher than those obtained from finger millet, proso millet, barnyard millet, and foxtail millet. The xylose:arabinose ratio was higher for kodo millet (2.38) in comparison to all other millets analysed. For instance, foxtail millet has a xylose:arabinose ratio of 1.83 followed by pearl millet (1.77), barnyard millet (1.70), and finger millet (0.91) (Bijalwan et al. 2016). The phenolic acid-bound arabinoxylans and fructans extracted from kodo millet and little millet exhibited higher antioxidant activity and exhibited the capability to modulate immune system pathways in RAW 264.7 cells (Srinivasan et al. 2021). Similar reports were earlier obtained for barnyard millet and foxtail millet as well (Srinivasan et al. 2020).

7.4 EFFECT OF PROCESSING ON BIOAVAILABILITY AND FUNCTIONALITY

With advancing research, it has been understood that possessing a certain beneficial phytochemical and delivering health benefits to the human body are not directly correlated. The health impact delivered by any phytochemical per se depends upon the bioavailability of that compound. Bioavailability is the part of the ingested compound which actually reaches systemic circulation and is utilized by the body (Carbonell-Capella et al. 2014). It depends upon the extent of release of the compound from the food matrix in gastrointestinal conditions and then absorption by the intestines. The stability of the compound also determines the bioavailability and hence final fate of the product. It can be assessed with both in-vivo and in-vitro methods. This section deals with the bioavailability of different phytochemicals as influenced by various factors.

In general, phenolic compounds like flavonoids and proanthocyanidins have the capability to combine with proline-rich salivary proteins and form polyphenol-protein complexes in the oral cavity (Karas et al. 2017). Such complexes formed can either be soluble or form precipitates depending upon their molecular size; this does not affect the final absorption in the small intestine however. The same has been studied in quercetin (Cai and Bennik 2006). The stability of phenolic compounds in the acidic environment of the stomach depends on the nature of individual components. For example, proanthocyanidins can be hydrolysed to their monomeric units on exposure to the acidic environment of the stomach. Whereas for others, these are generally stable and can be absorbed as such from the stomach itself. In the small intestine, enzymes hydrolyse flavonoid glycosides into aglycones and sugar which can further be absorbed via passive diffusion (Lewandowska et al. 2013, Farrell et al. 2012). Lewandowska et al. (2013) have also reported that passive absorption is possible as aglycones are less lipophilic in comparison to flavonoid glycosides. Most of the metabolism of the phenolic compounds occurs in the small intestine via glucuronidation, methylation, and sulphation, while the main absorption site is the colon. In the large intestine, colonic bacteria act upon these compounds and break them down into low molecular weight compounds through various types of reactions (Roowi et al. 2010).

The functionality of the phytochemicals is also influenced by the type of processing operation performed. Generally, millet products are prepared by adopting milling as the first step. In this, millets are first hulled which removes a major portion of the seed coat. As most phytochemicals are concentrated in the seed coat, dehulling leads to major losses in them. Dehulling reduces the phytochemical content of the grain by 80%. Other methods which millets are subjected to are germination and fermentation. The germination of millets for a period of 24 hours reduced the phytochemical content by 40% approximately (Chethan and Malleshi 2007). The total polyphenol content in two pearl millet cultivars (Standard and Ugandi) was found to be reduced by 59% on fermentation for a period of 14 hours (El Hag 2002). Fermentation using mixed cultures of *Saccharomyces diastaticus*, *Saccharomyces cerevisiae*, *Lactobacillus brevis*, and *Lactobacillus fermentum* showed no change in the phytochemical content. In another study, fermentation using mixed cultures of *S. diastaticus* and *L. brevis* presented a decrease in the polyphenol content (Khetarpaul and Chauhan 1989).

Chandrasekara and Shahidi (2012) tried to assess the bioavailability of phenolic compounds from cooked grains of various millets (finger, kodo, proso, foxtail, and pearl millet) by using simulated in-vitro digestion and microbial fermentation. After subjecting millets to the given environments, the authors found enhanced total phenols, flavonoids, and antioxidant activities in the digested samples. This indicates that these compounds were released on digestion and fermentation and the authors found better potential bioavailability of phenolic components in all millets. However, the study did not report the actual values for these components.

Hithamani and Srinivasan (2014) assessed the impact of domestic processing on the polyphenolic content of finger millet and pearl millet. The study showed that, on open pan boiling of finger millet, total polyphenolic content was reduced by 12–15%. Domestic processing methods like pressure cooking and boiling reduced the content of polyphenols by half. However, the study also reported an increase in phenolic acids on sprouting and roasting. The effect on bioaccessibility was also studied. Domestic operations like pressure cooking, open pan boiling, and microwave processing reduced the bioaccessibility of polyphenols by 3–35%. On sprouting, the polyphenol bioaccessibility was reported to increase by 67% for finger millet and 20% for pearl millet. Similar results were obtained for flavonoids as well. The process of germination enhances the fibre content of finger millet, kodo millet, and proso millet. In these millets, germination and roasting decreased the content of tannins and phytosterols (Prabha et al. 2018).

Kumar et al. (2021) assessed the effect of germination on the antioxidant activity and total phenolic content (% as gallic acid equivalent) of finger millet. The antioxidant activity increased by 14% approximately after 96 hours of germination. The total phenol content first decreased from 1.5 to 0.65% after 48 hours of germination, after which the levels again started increasing and reached 1.13% (gallic acid equivalent) after 96 hours of germination. A similar trend was reported for tannins (mg/100 g as tannic acid) which decreased from 173.7 mg to 64.3 mg after 48 hours followed by an increase to 94.66 mg after 96 hours of germination time. The authors attributed this trend to the solubilization and leaching of phytochemicals from the seed coat in the initial hours followed by a loss of carbohydrates in later stages of germination. Also, polyphenol oxidases mobilization during germination might be responsible for polyphenol degradation during germination. Pradeep and Guha (2011) reported enhanced total phenolic (from 429.9 to 453.3 mg/100 g), total flavonoid (from 334.9 to 350.2 mg/100 g), and tannin (from 283.4 to 332.1 mg/100 g) content in little millet on germination.

Malting and blanching have also been reported to decrease polyphenol content in millets. A three-fold decrease in polyphenolic compounds was reported (protocatechuic acid) on blanching pearl millet (Akanbi et al. 2019). Steaming enhanced the total phenolic (from 429.9 to 485.7 mg/100 g), flavonoid (from 334.9 to 410.1 mg/100 g), and tannin (from 283.4 to 308.6 mg/100 g) content in little millet (Pradeep and Guha 2011). Singh et al. (2018) studied the effect of roasting on total phenolic content of finger millet and found that the content decreased from 314.24 to 223.31 mg/100 g. In little millet, Pradeep and Guha (2011) reported an increase in the content of total phenolics, flavonoids, and tannins in little millet extracts.

7.5 CONCLUSION

Millets are rich in phytochemicals and the content of these compounds is much higher than staple cereals. They contain phenolic acids, flavonoids, tannins, phytosterols, bioactive peptides, and xylooligosaccharides as the major phytochemicals. The antioxidant activity of these grains is not just limited to these phytochemicals but also related to the presence of nutrients like beta-carotene and tocopherols. The literature studied in this chapter clearly indicates that the content of these phytochemicals is directly influenced by the processing method applied. In general, these phytochemicals were found to be reduced after processing with the exception of a few studies where different results were reported. The identification of phytochemicals is poorly studied in minor millets. Hence, future research must focus on the identification and bioavailability of these components.

REFERENCES

Agah, S., H. Kim, S. U. Mertens-Talcott, and J. M. Awika. 2017. Complementary cereals and legumes for health: Synergistic interaction of sorghum flavones and cowpea flavonols against LPS-induced inflammation in colonic myofibroblasts. *Molecular Nutrition and Food Research* 41(7). DOI: 10.1002/mnfr.201600625.

Agrawal, H., R. Joshi, and M. Gupta. 2016. Isolation, purification and characterization of antioxidative peptide of pearl millet (*Pennisetum glaucum*) protein hydrolysate. *Food Chemistry* 204: 365–372.

Agrawal, H., R. Joshi, and M. Gupta. 2019. Purification, identification and characterization of two novel antioxidant peptides from finger millet (*Eleusine coracana*) protein hydrolysate. *Food Research International* 120: 697–707.

Ahmed, I. A., N. Uslu, F. Al Juhaimi, M. M. Özcan, M. A. Osman, H. A. Alqah, E. E. Babiker, and K. Ghafoor. 2021. Effect of roasting treatments on total phenol, antioxidant activity, fatty acid compositions, and phenolic compounds of teff grains. *Cereal Chemistry* 98(5): 1027–1037.

Akanbi, T. O., Y. Timilsena, and S. Dhital. 2019. Bioactives from millet: Properties and effects of processing on bioavailability. In *Bioactive Factors and Processing Technology for Cereal Foods*, . J. Wang, B. Sun, R. Tsao (eds), 171–183. Singapore: Springer.

Amadou, I., G. W. Le, T. Amza, J. Sun, and Y. H. Shi. 2013. Purification and characterization of foxtail millet-derived peptides with antioxidant and antimicrobial activities. *Food Research International* 51(1): 422–428.

Ballard, C. R., and M. R. Maróstica. 2019. Health benefits of flavonoids. In M. R. S. Campos (ed.), *Bioactive Compounds* (pp. 185–201). Woodhead Publishing. ISBN 9780128147740. DOI: 10.1016/B978-0-12-814774-0.00010-4.

Bangoura, M. L., J. Nsor-Atindana, and Z. H. Ming. 2013. Solvent optimization extraction of antioxidants from foxtail millet species' insoluble fibers and their free radical scavenging properties. *Food Chemistry* 141(2): 736–744. DOI: 10.1016/j.foodchem.2013.03.029.

Bhandari, S. R., and Y. S. Lee. 2013. The contents of phytosterols, squalene, and vitamin E and the composition of fatty acids of Korean Landrace *Setaria italica and Sorghum Bicolar Seeds*. *Korean Journal of Plant Resources* 26(6): 663–672.

Bijalwan, V., U. Ali, A. K. Kesarwani, K. Yadav, and K. Mazumder. 2016. Hydroxycinnamic acid bound arabinoxylans from millet brans-structural features and antioxidant activity. *International Journal of Biological Macromolecules* 88: 296–305.

Brazier-Hicks, M., K. M. Evans, M. C. Gershater, H. Puschmann, P. G. Steel, and R. Edwards. 2009. The C-glycosylation of flavonoids in cereals. *Journal of Biology and Chemistry* 284(27): 17926–17934. DOI: 10.1074/jbc.M109.009258.

Cai, K., and A. Bennick. 2006. Effect of salivary proteins on the transport of tannin and quercetin across intestinal epithelial cells in culture. *Biochemical Pharmacology* 72(8): e974–e980.

Carbonell-Capella, J. M., M. Buniowska, F. J. Barba, M. J. Esteve, and A. Frigola. 2014. Analytical methods for determining bioavailability and bioaccessibility of bioactive compounds from fruits and vegetables: A review. *Comprehensive Reviews in Food Science and Food Safety* 13(2): e155–e171.

Chandrasekara, A., and F. Shahidi. 2011. Determination of antioxidant activity in free and hydrolyzed fractions of millet grains and characterization of their phenolic profiles by HPLC-DAD-ESI-MSn. *Journal of Functional Foods* 3(3): 144–158. DOI: 10.1016/j.jff.2011.03.007.

Chandrasekara, A., and F. Shahidi 2012. Bioaccessibility and antioxidant potential of millet grain phenolics as affected by simulated in vitro digestion and microbial fermentation. *Journal of Functional Foods* 4(1): 226–237.

Chethan, S., and N. G. Malleshi. 2007. Finger millet polyphenols: Characterization and their nutraceutical potential. *American Journal of Food Technology* 2: 618–629.

Demonty, I., R. T. Ras, H. C. M. van der Knaap, G. S. M. J. E. Duchateau, L. Meijer, P. L. Zock, J. M. Geleijnse, E. A., and Trautwein. 2009. Continuous dose-response relationship of the LDL-cholesterol lowering effect of phytosterol intake. *Journal of Nutrition* 139: e271–e284.

Devi, P. B., R. Vijayabharathi, S. Sathyabama, N. G. Malleshi, and V. B. Priyadarisini. 2014. Health benefits of finger millet (*Eleusine coracana* L.) polyphenols and dietary fiber: A review. *Journal of Food Science and Technology* 51(6): 1021–1040.

Duodu, K. G., and J. M. Awika. 2019. Phytochemical-related health promoting attributes of sorghum and millets. In *Sorghum and Millets*, J. R. N. Taylor and K. G. Doudu (eds), 225–258. Duxford:Elsevier.

Dykes, L., and L. W. Rooney. 2006. Sorghum and millet phenols and antioxidants. *Journal of Cereal Science* 44(3): e236–e251.

El Hag, M. E., A. H., El Tinay, and N. E. Yousif. 2002. Effect of fermentation and dehulling on starch, total polyphenols, phytic acid content and in vitro protein digestibility of pearl millet. *Food Chemistry* 77(2): 193–196.

Esche, R., B. Scholz, and K. H. Engel. 2013. Online LC–GC analysis of free sterols/stanols and intact steryl/stanyl esters in cereals. *Journal of Agricultural and Food Chemistry* 61(46): 10932–10939.

Farrell, T. L., M. Gomez-Juaristi, L. Poquet, K. Redeuil, K. Nagy, M. Renouf, and G. Williamson. 2012. Absorption of dimethoxycinnamic acid derivatives in vitro and pharmacokinetic profile in human plasma following coffee consumption. *Molecular Nutrition and Food Research* 56(9): e1413–e1423.

Gan, R. Y., H. B. Li, A. Gunaratne, Z. Q. Sui, and H. Corke. 2017. Effects of fermented edible seeds and their products on human health: Bioactive components and bioactivities. *Comprehensive Reviews in Food Science and Food Safety* 16(3): e489–e531.

Hassan, Z. M., N. A. Sebola, and M. Mabelebele. 2020. Assessment of the phenolic compounds of pearl and finger millets obtained from South Africa and Zimbabwe. *Food Science and Nutrition* 8(9): 4888–4896.

Hithamani, G., and K. Srinivasan. 2014. Effect of domestic processing on the polyphenol content and bioaccessibility in finger millet (*Eleusine coracana*) and pearl millet (*Pennisetum glaucum*). *Food Chemistry* 164: 55–62.

Hu, S., J. Yuan, J. Gao, Y. Wu, X. Meng, P. Tong, and H. Chen. 2020. Antioxidant and anti-inflammatory potential of peptides derived from in vitro gastrointestinal digestion of germinated and heat-treated foxtail millet (*Setaria italica*) proteins. *Journal of Agricultural and Food Chemistry* 68(35): 9415–9426.

Ji, Z., R. Feng, and J. Mao. 2019. Separation and identification of antioxidant peptides from foxtail millet (*Setaria italica*) prolamins enzymatic hydrolysate. *Cereal Chemistry* 96(6): 981–993.

Karas, M., A. Jakubczyk, U. Szymanowska, U. Złotek, and E. Zielinska. 2017. Digestion and bioavailability of bioactive phytochemicals. *International Journal of Food Science and Technology* 52: e291–e305.

Kataria, A., S. Sharma, and B. N. Dar. 2022. Changes in phenolic compounds, antioxidant potential and antinutritional factors of Teff (*Eragrostis* tef) during different thermal processing methods. *International Journal of Food Science and Technology* 57(11): 6893–6902.

Khetarpaul, N., and B. M. Chauhan. 1989. Effect of germination and pure culture fermentation on HCl-extractability of minerals of pearl millet (*Pennisetum typhoideum*). *International Journal of Food Science and Technology* 24(3): 327–331.

Kim, J. Y., K. C. Jang, B. R. Park, S. I. Han, K. J. Choi, S. Y. Kim, S. H. Oh, J. E. Ra, T. J. Ha, J. H. Lee, J. Hwang, H. W. Kang, and W. D. Seo. 2011. Physicochemical and antioxidative properties of selected barnyard millet (*Echinochloa utilis*) species in Korea. *Food Science and Biotechnology* 20(2): 461–469.

Kim, J. Y., T. K. Hyun, and M. Kim. 2010. Anti-oxidative activities of sorghum, foxtail millet and proso millet extracts. *African Journal of Biotechnology* 9(18): 2683–2690.

Koubová, E., M. Mrázková, D. Sumczynski, and J. Orsavová. 2018. In vitro digestibility, free and bound phenolic profiles and antioxidant activity of thermally treated *Eragrostis tef* L. *Journal of the Science of Food and Agriculture* 98(8): 3014–3021.

Kumar, A., A. Kaur, K. Gupta, Y. Gat, and V. Kumar. 2021. Assessment of germination time of finger millet for value addition in functional foods. *Current Science* 120(2): 406.

Kumar, A., V. Tomer, A. Kaur, and V. K. Joshi. 2015. Synbiotics: A culinary art to creative health foods. *International Journal of Food and Fermentation Technology* 5(1): 1–14.

Kumar, N., and N. Goel. 2019. Phenolic acids: Natural versatile molecules with promising therapeutic applications. *Biotechnology Reports* 24: e00370.

Lewandowska, U., K. Szewczyk, E. Hrabec, A. Janecka, and S. Gorlach. 2013. Overview of metabolism and bioavailability enhancement of polyphenols. *Journal of Agriculture and Food Chemistry* 61(50): e12183–e12199.

Li, W., L. Wen, Z. Chen, Z. Zhang, X. Pang, Z. Deng, T. Liu, and Y. Guo. 2021. Study on metabolic variation in whole grains of four proso millet varieties reveals metabolites important for antioxidant properties and quality traits. *Food Chemistry* 357: 129791.

Liang, S., and K. Liang. 2019. Millet grain as a candidate antioxidant food resource: A review. *International Journal of Food Properties* 22(1): 1652–1661. https://doi.org/10.1080/10942912.2019.1668406.

Lin, T. C., S. H. Huang, and L. T. Ng. 2015. Soaking conditions affect the contents of tocopherols, tocotrienols, and c-oryzanol in pigmented and non-pigmented brown rice. *Journal of the American Oil Chemists' Society* 92(11–12): 1681–1688.

Majid, A., and C. G. Priyadarshini. 2020. Millet derived bioactive peptides: A review on their functional properties and health benefits. *Critical Reviews in Food Science and Nutrition* 60(19): 3342–3351.

Manosroi, J., N. Khositsuntiwong, and A. Manosroi. 2014. Biological activities of fructooligosaccharide (FOS)-containing *Coix lachryma-jobi* Linn. extract. *Journal of Food Science and Technology* 51(2): 341–346.

Matsui, R., R. Honda, M. Kanome, A. Hagiwara, Y. Matsuda, T. Togitani, N. Ikemoto, and M. Terashima. 2018. Designing antioxidant peptides based on the antioxidant properties of the amino acid side-chains. *Food Chemistry* 245: 750–755. DOI: 10.1016/j.foodchem.2017.11.119.

Nakbi, A., M. Issaoui, S. Dabbou, N. Koubaa, A. Echbili, M. Hammami, and N. Attia. 2010. Evaluation of antioxidant activities of phenolic compounds from two extra virgin olive oils. *Journal of Food Composition and Analysis* 23(7): 711–715. DOI: 10.1016/j.jfca.2010.05.003.

Nambiar, V. S., N. Sareen, M. Daniel, and E. B. Gallego. 2012. Flavonoids and phenolic acids from pearl millet (Pennisetum glaucum) based foods and their functional implications. *Functional Foods in Health and Disease* 2(7): 251–264.

N'Dri, D., T. Mazzeo, M. Zaupa, R. Ferracane, V. Fogliano, and N. Pellegrini. 2013. Effect of cooking on the total antioxidant capacity and phenolic profile of some whole-meal African cereals. *Journal of the Science of Food and Agriculture* 93(1): 29–36.

Oluwasesan, M. B., A. B. Ogbesejana, A. Balkisu, M. Osibemhe, B. Musa, and O. O. Stephen. 2022. Polyphenolic fractions from three millet types (Fonio, finger millet, and pearl millet): Their characterization and biological importance. *Clinical Complementary Medicine and Pharmacology* 2 (1): 100020.

Omaji, G. O., M. Osibemhe, B. O. Orji, L. E. Ilouno, B. O. Abdulrahman, and A. O. Ajadi. 2021. Antioxidant potential of crude powder, methanol and aqueous extracts of Fonio millet (Digitaria exilis) grains. *Nigerian Journal of Biotechnology* 38(1): 91–97.

Palaniappan, A., U. Antony, and M. N. Emmambux. 2021. Current status of xylooligosaccharides: Production, characterization, health benefits and food application. *Trends in Food Science and Technology* 111: 506–519.

Palaniappan, A., V. G. Balasubramaniam, and U. Antony. 2017. Prebiotic potential of xylooligosaccharides derived from finger millet seed coat. *Food Biotechnology* 31(4): 264–280.

Pang, M., S. He, L. Wang, X. Cao, L. Cao, and S. Jiang. 2014. Physicochemical properties, antioxidant activities and protective effect against acute ethanol-induced hepatic injury in mice of foxtail millet (Setaria italica) bran oil. *Food & Function* 5(8): 1763–1770.

Prabha, D. C., P. Chitra, and S. Sujatha. 2018. Bioavailability of nutraceuticals and phytocomponents in the roasted and germinated form of selected millets. *Research Journal of Life Sciences, Bioinformatics, Pharmaceutical and Chemical Sciences* 4(1): 248.

Pradeep, P. M., and Y. N. Sreerama. 2018. Phenolic antioxidants of foxtail and little millet cultivars and their inhibitory effects on α-amylase and α-glucosidase activities. *Food Chemistry* 247: 46–55.

Pradeep, S. R., and M. Guha. 2011. Effect of processing methods on the nutraceutical and antioxidant properties of little millet (*Panicum sumatrense*) extracts. *Food Chemistry* 126(4): 1643–1647.

Rao, R. S. P., and G. Muralikrishna. 2006. Water Soluble feruloyl Arabinoxylans from Rice and Ragi: Changes upon Malting and Their Consequence on Antioxidant activity. *Phytochemistry* 67(1): 91–99.

Ravisankar, S., K. Abegaz, and J. M. Awika. 2018. Structural profile of soluble and bound phenolic compounds in teff (*Eragrostis tef*) reveals abundance of distinctly different flavones in white and brown varieties. *Food Chemistry* 263: 265–274.

Roowi, S., A. Stalmach, W. Mullen, M. E. Lean, C. A. Edwards, and A. Crozier. 2010. Green tea flavan-3-ols: Colonic degradation and urinary excretion of catabolites by humans. *Journal of Agriculture and Food Chemistry* 58: e1296–e1304.

Ryan, E., K. Galvin, T. P. O'Connor, A. R. Maguire, and N. M. O'Brien. 2007. Phytosterol, squalene, tocopherol content and fatty acid profile of selected seeds, grains, and legumes. *Plant Foods for Human Nutrition* 62(3): 85–91. DOI: 10.1007/s11130-007-0046-8.

Sa´nchez, A., and A. Va´zquez. 2017. Bioactive peptides: A review. *Food Quality and Safety* 1: e29–e46.

Samanta, A. K., N. Jayapal, A. P. Kolte, S. Senani, M. Sridhar, K. P. Suresh, and K. T. Sampath. 2012. Enzymatic production of xylooligosaccharides from alkali solubilized xylan of natural grass (*Nervosum sehima*). *Bioresource Technology* 112: 199–205. DOI: 10.1016/j.biortech.2012.02.036.

Seo, M. C., J. Y. Ko, S. B. Song, J. S. Lee, J. R. Kang, D. Y. Kwak, B. G. Oh, Y. N. Yoon, M. H. Nam, H. S. Jeong, and K. S. Woo. 2011. Antioxidant compounds and activities of foxtail millet, proso millet and sorghum with different pulverizing methods. *Journal of the Korean Society of Food Science and Nutrition* 40(6): 790–797.

Sharma, K., V. Kumar, J. Kaur, B. Tanwar, A. Goyal, R. Sharma, Y. Gat, and A. Kumar. 2021. Health effects, sources, utilization and safety of tannins: A critical review. *Toxin Reviews* 40(4): 432–444.

Sharma, P., C. Bhandari, S. Kumar, B. Sharma, P. Bhadwal, and N. Agnihotri. 2018. Dietary fibers: A way to a healthy microbiome. In *Diet, Microbiome and Health*, A. M. Holban and A. M. Grumezescu (eds), 299–345. Cambridge: Academic Press,.

Sharma, R., S. Sharma, B. N. Dar, and B. Singh. 2021. Millets as potential nutri-cereals: A review of nutrient composition, phytochemical profile and techno-functionality. *International Journal of Food Science & Technology* 56(8): 3703–3718.

Sharma, S., R. Jan, C. S. Riar, and V. Bansal. 2021. Analyzing the effect of germination on the pasting, rheological, morphological and in-vitro antioxidant characteristics of kodo millet flour and extracts. *Food Chemistry* 361: 130073.

Singh, A., and S. M. Eligar. 2021. Bioactive feruloylated xylooligosaccharides derived from pearl millet (Pennisetum glaucum) bran with antiglycation and antioxidant properties. *Journal of Food Measurement and Characterization* 15(6): 5695–5706.

Singh, N., J. David, D. K. Thompkinson, B. S. Seelam, H. Rajput, and S. Morya. 2018. Effect of roasting on functional and phytochemical constituents of finger millet (Eleusine coracana L.). *The Pharma Innovation Journal* 7(4): 414–418.

Siwela, M., J. R. Taylor, W. A. de Milliano, and K. G. Duodu. 2007. Occurrence and location of tannins in finger millet grain and antioxidant activity of different grain types. *Cereal Chemistry* 84(2): 169–174.

Slama, A., A. Cherif, F. Sakouhi, S. Boukhchina, and L. Radhouane. 2020. Fatty acids, phytochemical composition and antioxidant potential of pearl millet oil. *Journal of Consumer Protection and Food Safety* 15(2): 145–151.

Srinivasan, A., J. Aruldhas, S. S. Perumal, and S. P. Ekambaram. 2021. Phenolic acid bound arabinoxylans extracted from Little and Kodo millets modulate immune system mediators and pathways in RAW 264.7 cells. *Journal of Food Biochemistry* 45(1): e13563.

Srinivasan, A., S. P. Ekambaram, S. S. Perumal, J. Aruldhas, and T. Erusappan. 2020. Chemical characterization and immunostimulatory activity of phenolic acid bound arabinoxylans derived from foxtail and barnyard millets. *Journal of Food Biochemistry* 44(2): e13116.

Sripriya, G., K. Chandrasekhara, V. S. Murty, and T. S. Chandra. 1996. ESR spectroscopic studies on free radical quenching action of finger millet (*Eleusine coracana*). *Food Chemistry* 57(4): 537–540. DOI: 10.1016/S0308-8146(96)00187-2.

Stojanovski, S., M. Ananieva, T. Mandadzhieva, I. Koliandova, I. Iliev, and I. Ivanova. 2012. Utilization of xylooligosaccharides from different Lactobacillus strains. *Journal of BioScience and Biotechnology* 1: 147–150.

Subba Rao, M. V. S. S. T., and G. Muralikrishna. 2002. Evaluation of the antioxidant properties of free and bound phenolic acids from native and malted finger millet (Ragi, Eleusine coracana Indaf- 15). *Journal of Agriculture and Food Chemistry* 50(4): 889–892.

Suma, P. F., and A. Urooj. 2015. Isolation and characterization of starch from pearl millet (Pennisetum typhoidium) flours. *International Journal of Food Properties* 18(12): 2675–2687. DOI: 10.1080/10942912.2014.981640.

Takumi, K., J. Udaka, M. Kanoh, T. Koga, and H. Tsuji. 1996. Polypeptide compositions and antigenic homologies among prolamins from Italian, common and Japanese millet cultivars. *Journal of the Science of Food and Agriculture* 72(2): 141–147.

Tsuzuki, W., S. Komba, E. Kotake-Nara, M. Aoyagi, H. Mogushi, S. Kawahara, and A. Horigane. 2018. The unique compositions of steryl ferulates in foxtail millet, barnyard millet and naked barley. *Journal of Cereal Science* 81: 153–160.

Veenashri, B. R., and G. Muralikrishna. 2011. In vitro anti-oxidant activity of xylo-oligosaccharides derived from cereal and millet brans–A comparative study. *Food Chemistry* 126(3): 1475–1481.

Vogel, J. 2008. Unique aspects of the grass cell wall. *Current Opinion in Plant Biology* 11(3): e301–e307.

Watanabe, M. 1999. Antioxidative phenolic compounds from Japanese barnyard millet (Echinochloautilis) grains. *Journal of Agriculture and Food Chemistry* 47(11): 4500–4505. DOI: 10.1021/jf990498s.

Xiang, J., F. B. Apea-Bah, V. U. Ndolo, M. C. Katundu, and T. Beta. 2019. Profile of phenolic compounds and antioxidant activity of finger millet varieties. *Food Chemistry* 275: 361–368.

Yin, R., Y. Fu, L. Yousaf, Y. Xue, J. Hu, X. Hu, and Q. Shen. 2021. Crude and refined millet bran oil alleviate lipid metabolism disorders, oxidative stress and affect the gut microbiota composition in high-fat diet-induced mice. *International Journal of Food Science & Technology* 57(5): 2600–2610.

Yuan, Y., C. Liu, G. Zhao, X. Gong, K. Dang, Q. Yang, and B. Feng. 2021. Transcriptome analysis reveals the mechanism associated with dynamic changes in fatty acid and phytosterol content in foxtail millet (*Setaria italica*) during seed development. *Food Research International* 145: 110429.

Yuan, Y., J. Xiang, B. Zheng, J. Sun, D. Luo, P. Li, and J. Fan. 2022. Diversity of phenolics including hydroxycinnamic acid amide derivatives, phenolic acids contribute to antioxidant properties of proso millet. *LWT-Food science and Technology* 154: 112611.

Yun, Y. R., and S. H. Park. 2018. Antioxidant activities of brown teff hydrolysates produced by protease treatment. *Journal of Nutrition and Health* 51(6): 599–606.

Zhang, L. Z., and R. H. Liu. 2015. Phenolic and carotenoid profiles and antiproliferative activity of foxtail millet. *Food Chemistry* 174: 495–501. DOI: 10.1016/j.foodchem.2014.09.089.

Zhu, A., L. Tang, Q. Fu, M. Xu, and J. Chen. 2015. Optimization of alkali extraction of polysaccharides from foxtail millet and its antioxidant activities in vitro. *Journal of Food Biochemistry* 39(6): 708–717. DOI: 10.1111/jfbc.12183.

Chapter 8

Anti-Nutritional Factors and Methods to Improve Nutritional Value of Millets

8.1 INTRODUCTION

Food is essential for one's survival and growth. The world is constantly facing the burden of malnutrition. Food sufficiency is not the final aspect; the availability of nutrients is vital for malnutrition eradication. Certain factors in millets have been identified which limit the availability of nutrients.

Plants synthesize various secondary metabolites for protection against herbivores, insects, and pathogens. These secondary metabolites also act as anti-nutritional factors which limit the availability of nutrients and reduce the productivity in animals. In humans, these anti-nutritional factors including tannins, non-starch polysaccharides, and enzyme inhibitors influence the digestibility of nutrients (Kumar et al. 2016). The anti-nutritional factors and their mechanisms are given in Figure 8.1.

Anti-nutrients are compounds which act to reduce nutrient intake and interfere with the digestion of a specific nutrient or with the absorption or utilization of nutrients or other harmful effects. Other symptoms reported on the consumption of anti-nutrients are nausea, bloating, headaches, rashes, and nutritional deficiencies, etc. However, these compounds if eaten wisely can impart many health benefits. Like many other cereals, millets also contain a variety of phytochemicals which act as anti-nutritional factors. As discussed in previous chapters, millets contain a variety of nutrients in good amounts and various compounds which protect against diseases. Owing to these attributes, millets can be considered a potential tool for malnutrition eradication. However, the factor which challenges their potential is the presence of high amounts of anti-nutritional factors in millets. This chapter deals with the anti-nutritional factors present in various types of millets. Methods to improve the nutritional value of millets are also discussed here.

8.2 PHYTATES

Phytates are present in cereals, millets, legumes, etc. Seeds, usually in their husk, store phosphorus in the form of phytic acid (Figure 8.2) and its salts.

Phytic acid was first discovered in 1855. Seed bran is the major source of phytates. Phytic acid accounts for up to 80% of the total seed phosphorus and it contributes approximately 1.5% to

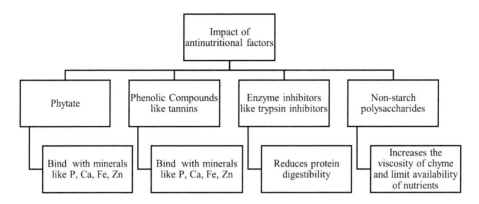

Figure 8.1 Anti-nutritional factors in millet.

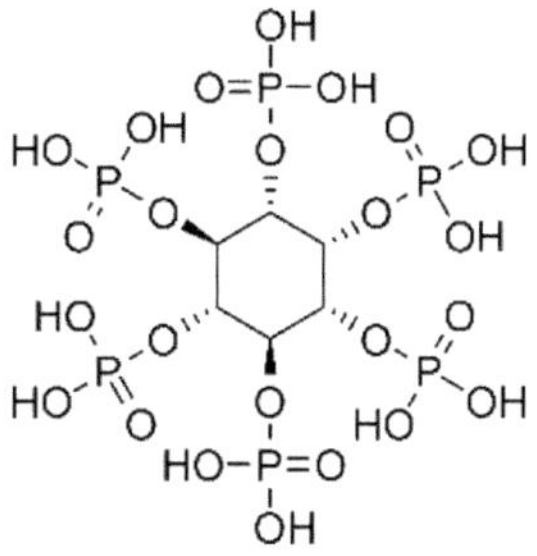

Figure 8.2 Structure of phytic acid.

the dry seed weight; however this depends upon the type and variety of seed. Ahmed et al. (2010) studied two varieties of pearl millet and identified that phytate phosphorus was approximately 27% of the total phosphorus present in the seeds. In monocots, it accumulates in the aleurone layer with little deposition in the embryo (Bohn et al. 2008). In dicots, phytic acid remains in the protein fraction. Phytates are generally hydrolysed by phytases into phosphate, inositol, and other micronutrients required for the emergence of seedling (Popova and Mihaylova 2019). Phytase activity in humans is very low (Iqbal et al. 1994).

The position of the phosphate group in phytate varies with the pH which it is in. A change in orientation of the phosphate group has been observed at pH between 0.5 and 9 and above 9.5 (Feizollahi et al. 2021). The structure contains high-density phosphate anions and 12 replaceable protons. This arrangement makes phytic acid's anionic phosphate group strongly bind to metal cations of calcium, iron, potassium, magnesium, and zinc making them unavailable (Figure 8.2). Phytic acid acts as a strong chelator making very stable chelates (Dost and Karaca 2016). The phytate phosphorus content in *Ugandi* and *Dembi* yellow varieties of pearl millet was found to be 299.85 and 287.44 mg/100 g (Ahmed et al. 2010).

As stated above, the phytate content of bran is much higher than other parts of the grain, but variations have been reported in the bran fractions as well. Such data could not be found for millets; hence an example of wheat has been taken. Pearling of wheat has revealed that the outer 0.4% of the layer contains high zinc content. The next outer 4–8% of the layer reported high contents of phytate, iron, and also high phytase activity (Liu et al. 2007; Bohn et al. 2008).

Phytic acid binding with minerals leads to the formation of salts which are soluble at acidic but not at physiological pH and, hence, have poor absorption in the intestine. At gastric pH, phytic acid remains undissociated limiting the availability of nutrients especially minerals. This becomes a problem in regions where cereals, millets, and legumes are consumed as staples, leading to various nutritional deficiencies. The availability of minerals from phytic acid depends on the relative concentration, processing methods used, the presence of other anti-nutritional factors, phytase content, and strength of binding. A 10:1 ratio of phytic acid:iron leads to 92% iron inhibition. Also, as reported by Engle-Stone et al. (2005), the type of iron also plays a major role. The ferrous form (as in $FeSO_4$) shows higher solubility in comparison to the ferric form (as in $FeCl_3$) in complexing with phytic acid. Apart from iron, phytic acid affects the bioavailability of zinc, calcium, magnesium, etc. Studies have indicated that consumption of a phytate-deficient diet ensures the absorption of 20% more zinc and 60% more magnesium from food (Coulibaly et al. 2011). The extent is such that various countries have even established a recommended daily intake for phytic acid. The recommended daily intake in the United States, the United Kingdom, Finland, Italy, and Sweden is 631–746, 370, 219, and 180 mg/day respectively (Nissar et al. 2017; Feizollahi et al. 2021). Phytic acid content in various types of millets is given in Table 8.1. The phytate content of teff was found to be 541.45 mg/100 g (Asres et al. 2018). Pal et al. (2016) assessed the phytate content of 11 germplasms (K1 to K10 and KK1) of kodo millet in kharif season and reported that K6 had the lowest phytic acid content (114.6 mg/100 g).

Dephytinization can be done using different processing treatments which will be discussed in the next section. Apart from processing treatments, phytates can be hydrolysed from phytases obtained from microbes, plant, and animal tissues. Plant phytase extraction is not economical and also exhibits very poor activity in humans. Microbial phytases have reported the best results at gastric pH (Gupta et al. 2015; Sandberg and Scheers 2016). Appropriate phytases can be added to the diet of an individual for benefits.

8.3 TANNINS

Tannins are a group of polyphenolic, non-nitrogenous compounds with a bitter/astringent taste. Tannins generally have a molecular weight in the range of 500–3000 Daltons. They act as anti-nutritional compounds and interfere with the availability of nutritional components like protein, alkaloids, etc. These reduce protein digestibility by either making them unavailable or by inhibiting digestive enzymes like trypsin and chymotrypsin. Tannins also act as a strong inhibitor of iron absorption. In-vivo studies indicate that a molar ratio of 1:10 of tannic acid:iron leads to 90% inhibition of iron absorption (Glahn et al. 2002). Based upon their structure, tannins can be divided into condensed and hydrolysable. Millets contain mostly condensed tannins. Condensed tannins after hydrolysis yield flavans while hydrolyasable tannins yield gallic acid. The binding capacity of tannins depend upon their chemical structure, pH, molecular size, etc. (Kumar et al. 2016). Tannin protein interaction involves hydrogen bonding and hydrophobic interactions. The larger the molecular size of tannins, the better the protein precipitation and its binding with tannins. Above 5000 Daltons, tannins become insoluble and lose their precipitating capability (Thakur et al. 2019). Tannins also possess anti-amylase activity and bind with B-vitamins, iron, etc. Other harmful effects of tannins include low food intake, inhibition of digestive enzymes, enhanced protein excretion, disturbances in the digestive tract, etc. Among millets, finger millet contains the highest quantity of tannin with the content of 0.04% to 3.74% catechin equivalents. A similar result was obtained by Bhuvaneshwari et al. (2020), where finger millet was found to

TABLE 8.1 CONTENT OF ANTI-NUTRITIONAL FACTORS IN DIFFERENT MILLETS

Millet type	Phytic acid	Total phenol	Tannin	Oxalates	Enzyme inhibitors	References
Pearl millet	647 mg/100 g	2.15–304 mg GAE/100 g			7.33 %	El Hag et al. 2002; Almaski et al. 2017; Osman 2011
Finger millet	0.49 g/100 g	2.3–136.4 mg FAE/100 g	17.65 mg CEQ/100 g	30 mg/100 g	207.35–234.23 TIU g^{-1}	Ravindran 1991; Nakarani 2021
Proso millet	2.2–8.3 mg/g	1.2 mg/g				Devisetti et al. 2014
Foxtail millet	1.9–11.7 mg/g	0.5–33.17 mg GAE/100 g		25–28 mg/100 g	0.19 TIU/mg	Devisetti et al. 2014; Ravindran 1991; Nazni and Shobana 2016
Barnyard millet	92.07–97.88 mg/100 g	0.6–129.5 mg FAE/100 g	59.54 CEQ/100 g		0.11 TIU/mg	Kumar et al. 2020; Nazni and Shobana 2016
Kodo millet	225 mg/g	143 mg/100 g	50.46 TAE/g			Patil et al. 2020
Little millet		1.51–429.9 mg GAE/100 g	283.4 mg CEQ/100 g			Pradeep and Guha 2011; Almaski et al. 2017
Teff	541.45 mg/100 g		0.87 mg/100 g			Asres et al. 2018
Fonio	513.7 mg/100 g	1.96 mg/g	16 mg CE/100 g dry matter	0.71 mg/g		Nassarawa and Ahmad 2019; Baye 2014

have a tannin content of 2.077 mg/g which was higher in comparison to foxtail millet (1.971 mg/g), kodo millet (1.42 mg/g), little millet (1.223 mg/g), and pearl millet (1.863 mg/g). This content may vary with the variety though. Dark-coloured varieties (brown and dark brown) possess high quantities of tannins, whereas the white varieties possess negligible amounts (Rao and Deosthale 1988, Kumar et al. 2016). Tannin content in proso millet was reported to be 73.41 mg TA/100 g (Sharma et al. 2022b). Pal et al. (2016) assessed the tannin content of 11 germplasms (K1 to K10 and KK1) of kodo millet in kharif season and reported that K6 had the lowest tannin content (102.22 mg/100 g). The tannin content of finger millet was reported to be 0.25 g/100 g. The tannin content of kodo millet, little millet, and Japanese millet was found to be 0.41, 0.5, and 0.3 g/100 g respectively (Gupta et al. 2013). Tannin content in teff was reported to be 0.87 mg/100 g (Asres et al. 2018). C-glycosylflavones, mainly glycosylvitexin, glycosylorientation, and vitexin, were identified in pearl millet and exhibited goitrogenic effects (Salunke et al. 1983).

The process of nixtamatization reduced the tannin content of crisp fired pearl millet dough wafers from 7.44 µg CAE/g to 3.05 µg CAE/g. The concentration of lime used for the process was 2%. High cooking temperatures lead to pericarp removal and saturate tannin content (Gaytán-Martínez et al. 2017; Pandey et al. 2022).

8.4 OXALATES

Oxalates are present as anti-nutritional factors in millets. Oxalic acid is synthesized in small quantities by plants and animals. These are formed by the binding of oxalic acid with various minerals. Plant-derived oxalates occur as water-soluble and -insoluble forms. Oxalic acid binds with potassium and sodium to form soluble complexes. Binding with minerals like calcium, magnesium, and iron, oxalic acid forms insoluble complexes. Plants synthesize oxalates for various functions like protection, calcium regulation, and detoxification, etc. Soluble or bound oxalates have chelating capability with dietary minerals making them less available (Petroski and Minich 2020). Insoluble salts cannot be excreted out through the urinary tract after being processed by the digestive tract. The accumulation of calcium oxalate in the kidney may lead to kidney stones. Some people even have sensitivity to oxalates, and their consumption even in small quantities can cause burning in eyes, ears, and throat. Consumption in large amounts in such people can cause abdominal pain, muscle weakness, nausea, and diarrhoea (Popova and Mihaylova 2019). Oxalates also form complexes with proteins inhibiting peptic digestion (Kakade and Hathan 2015).

Millets have high oxalate content. In *Kalukombu* and MRB varieties of pearl millet, it was observed that the bran rich fraction had a higher amount of oxalate in comparison to the semi-refined flour and whole flour (Suma and Urooj 2014). Dehulled proso millet contains 21–23 mg/100 g of oxalates of which the majority are in soluble form (60–91%). Dark-coloured varieties have high oxalate content (Kalinova 2007). Finger millet has 0.27% of anti-nutritional factors as oxalate (Rathore et al. 2019). The oxalate content of finger millet was reported to be 0.02 g/100 g. The oxalate content of kodo millet, little millet, and Japanese millet was found to be 0.052, 0.026, and 0.02 g/100 g respectively (Gupta et al. 2013).

8.5 NON-STARCH POLYSACCHARIDES

Non-starch polysaccharides are polymeric high molecular weight carbohydrates that are different from the composition of starch (amylose and amylopectin) in terms of the glycosidic bonds present. The molecular weight of these compounds ranges from 8000 to 9000 million. These polysaccharides are predominantly present in the cell wall of millets. Xylan in food inhibits nutrient utilization by enhancing chyme viscosity and reducing the movement of chyme in the GI tract. Also, it forms complexes with minerals like iron, calcium, and zinc reducing their bioavailability (Jaworski et al. 2015; Debon and Tester 2001).

8.6 ENZYME INHIBITORS

Enzyme inhibitors are natural compounds present specifically in legumes and cereals which affect the activity of the digestive enzymes. Protein inhibitors are those which disrupt protein hydrolysis making the digestive enzymes either fully or partially unavailable. Plant serpins are among the biggest family of plant inhibitors. Trypsin and chymotrypsin are the major enzymes affected. Serpins attack the most reactive sites of these enzymes via catalytic mode. Two major types of inhibitors include Bowman-Birk and Kunitz trypsin inhibitor (Budhwar et al. 2020). The type and variety of millet determine which enzyme will be affected. For instance, the inhibitor extracted from finger millet possesses more anti-tryptic activity in comparison to anti-chymotryptic activity (Kumar et al. 2016). Finger millet contains pentosane content of 6.2–7.2%. Native finger millet varieties were found to have more hexoses in comparison to pentosans (Malleshi et al. 1986). The protease-inhibiting activity of millets is influenced by variety and genetic variations. The anti-tryptic activity of proso millet was reported to be 732 trypsin inhibitory units per gram of flour which is much higher than the reported anti-chymotrypsin activity (Kalinová 2007).

8.7 METHODS TO IMPROVE NUTRITIONAL VALUE

Different processing methods are performed on millets before consumption. These include traditional methods like soaking and germination and modern methods like irradiation and the use of exogenous enzymes. This section deals with the impact that different processing treatments have on the anti-nutritional factors of millets. Table 8.2 summarizes the effect on anti-nutritional factors like phytates, tannins, and trypsin inhibitors, etc.

8.7.1 Soaking

Soaking is usually used as a pretreatment for millets to facilitate further processing or cooking operations. Also it is crucial for the germination and fermentation of millets. Soaking can also be combined with antimicrobial treatments with compounds like 0.2% formaldehyde, 1% sodium hypochlorite to prevent microbial and especially mould growth. There are a couple of factors which affect phytate reduction on soaking – endogenous phytase activity and leaching. Phytates are generally soluble in water and their content can be minimized by discarding soaked water. The leaching of phytate depends upon the pH of the medium and the nature of bonding. The phytic acid content of proso millet reduced to 7.58 to 8.77 mg/g. Soaking at 25°C for a period of 12 hours in a water:seed ratio of 15:1 reduced the tannin content of proso millet from 73.41 to 60.3 mg TA/100 g (Sharma et al. 2022b).

Bhuvaneshwari et al. (2020) assessed the effect of soaking on the anti-nutritional factors in pearl, kodo, finger, little, and foxtail millet. On soaking for a period of 48 hours, the decline in

TABLE 8.2 EFFECT OF PROCESSING ON ANTI-NUTRITIONAL FACTORS IN MILLETS

Millet type	Processing treatment and condition	Impact on anti-nutritional factor/availability of nutrient	Reference
Pearl millet	Dehulling	80–84% increase in in-vitro digestibility; 50% reduction of phytate and tannin	El Hag et al. 2002; ElShazali et al. 2011
Pearl millet	Fermentation at 30°C for 14 hours	79% increase in in-vitro protein digestibility	El Hag et al. 2002
Pearl millet	Germination at 25°C for 48 hours	Reduction in total phenols by 77%, tannins by 34%	Nithya et al. 2007
Pearl millet	Autoclaving at 110°C for 15 minutes	Tannin content decreases from 415.3 to 390.2 mg/100 g; polyphenols from 220 to 100 mg/100 g	Eltayeb et al. 2007
Pearl millet	Gamma radiation generated at a dose rate of 20 Gy/min at 25°C	No effect of radiation alone on pearl millet flour but radiation followed by cooking significantly reduced anti-nutrients and enhanced protein digestibility	ElShazali et al. 2011
Pearl millet	Gamma radiation at doses from 0.25 to 2 kGy with dose rate of 33 Gy per minute	The reduction in anti-nutritional factors was dose dependent. Tannins and phytates were reduced by 74.9 and 52.8% respectively on exposure to 2 kGy	Mahmoud et al. 2016
Pearl millet	Microwave treatment for 900 W, 2450 MHz for 40 to 100 seconds	Reduction in tannin and phytate by 70 and 45.53%	Singh et al. 2017
Pearl millet	Exogenous phytases	Enhanced absorption of zinc from 9.5 to 16%	Tharifkhan et al. 2021
Finger millet	Malting for 48 hours	Decreased the content of total phenols	Udeh et al. 2018
Finger millet	Germination at 25°C for 5 days	Trypsin inhibitor reduced to undetectable levels; phytates and tannins were reduced to less than half; three-fold decrease in trypsin inhibitors	Abioye et al. 2018; Mbithi-Mwikya 2000
Finger millet	Gamma radiation of doses 2.5 kGy to 10 kGy at a dose rate of 2.5 kGy per hour	Dose-dependent relationship Total phenols reduced from 166.05 to 126.68 mg/100 g. Reduction in tannins as well	Gowthamraj et al. 2021
Finger millet	Microwave treatment for 900 W, 2450 MHz for 40 to 100 seconds	Reduction in phytic acid and tannins by 58.14 and 45.18%	Singh et al. 2017
Finger millet	Ultrasound-assisted hydration	Phytates and tannins decreased by 66.98 and 62.83%	Yadav et al. 2021

(Continued)

TABLE 8.2 (CONTINUED) EFFECT OF PROCESSING ON ANTI-NUTRITIONAL FACTORS IN MILLETS

Millet type	Processing treatment and condition	Impact on anti-nutritional factor/ availability of nutrient	Reference
Barnyard millet	Soaking for 11.8 hours followed by germination for 36.5 hours at 33°C	Increase in total phenol content (54.23 mg GAE/100 g) on dry weight basis	Sharma et al. 2016
Barnyard millet	Boiling, germination, pressure cooking, and roasting	All methods increased the content of tannin, and trypsin inhibitor	Nazni and Shobana 2016
Little millet	Soaking for a period of 48 hours	39% reduction in tannin content	Bhuvaneshwari et al. 2020
Little millet	24-hour germination	71.6% reduction in tannin content	Bhuvaneshwari et al. 2020
Proso millet	Solid-state fermentation using fungi *Penicillium camemberti* at 25°C for 5 days	Reduction in phytic acid (reduction of 35%) and trypsin inhibitor (72% reduction); no significant change in tannin	Dwivedi et al. 2015
Foxtail millet	Soaking	17.5% reduction in tannin	Bhuvaneshwari et al. 2020
Foxtail millet	Boiling, germination, pressure cooking, and roasting	Germination (28 and 31%), boiling (8 and 31%), and roasting (11 and 36%) reduced the content of tannin, and trypsin inhibitor	Nazni and Shobana 2016
Foxtail millet	Germination for 24 hours	37.44% reduction in tannin	Bhuvaneshwari et al. 2020
Foxtail millet	Fermentation for 20 hours at 38°C	30.2% reduction in tannic acid; 25% reduction in saponins Similar reduction for phytates (30%)	Sharma and Sharma 2022a
Kodo millet	Soaking for 24 hours	Reduces the tannin content by 34.5%	Bhuvaneshwari et al. 2020
Kodo millet	Germination (38°C for 10 hours)	Reduction in phytic acid and tannin by 25 and 85% respectively	Sharma et al. 2017
Kodo millet	Solid-state fermentation using fungi *Penicillium camemberti* at 25°C for 5 days	No significant change in the content of tannin, phytic acid, and trypsin inhibitor on fermentation	Dwivedi et al. 2015
Fonio	Steam cooking for 10 minutes	Change in molar ratio of phytate:iron from 1.3 to 4.7 Change in molar ratio of phytate:zinc from 15.9 to 5.7	Koreissi-Dembélé et al. 2013
Fonio	Extruded snack with 140°C barrel temperature	Reduction in trypsin inhibitors by 82%	Olapade and Aworh 2012
Teff	Thermal processing like autoclaving (121°C for 15 min), roasting for 10 min, microwave (for 5 min)	Phytic acid and tannins were reduced in the range of 26–37%	Kataria et al. 2022

tannin content was reported to be approximately 41% in finger millet, 17% in foxtail millet, 35% in kodo millet, 35% in little millet, and 52% in pearl millet. The soaking of finger millet at 25°C for a period of 12 hours reduced the trypsin inhibitor activity and the content of oxalates, polyphenols, tannins, and phytates by 33, 20, and 39% respectively (Hotz and Gibson 2007; Rathore et al. 2019). High-pressure soaking is more effective in reducing anti-nutritional factors like phytates and tannins in foxtail millet (Sharma et al. 2017).

8.7.2 Processing/Milling

Dehulling can also help in phytate reduction of the grain. The extent of phytate reduction depends on the type of grain and the location of phytic acid deposition in the grain. Dehulling is effective if phytates are present in the bran or seed coat. The milling of two pearl millet varieties reduced the phytate content by 60–70%, and polyphenols by 17.5–21% (Ahmed et al. 2010). Suma and Urooj (2014) assessed the impact of milling on pearl millet. The Kalukombu variety of pearl millet after milling has phytate content in the range of 0.33 to 0.62 g percent and oxalate content from 36 to 64.8 mg percent. Hama et al. (2011) used manual and mechanical methods of decortication for the reduction of anti-nutritional factors. It was found that both methods resulted in similar decline in phytates. Bread prepared with debranned pearl millet flour showed very low content of phytates and phenols (Rathore et al. 2016).

8.7.3 Germination

Germination is preceded by an important step called soaking to enhance the moisture content to the desired level for various metabolic activities in the seeds. During the process of germination, the activation of phytase enzyme occurs which leads to the degradation of the phytate phosphate complex into inositol monophosphate leading to a decrease in the content of phytate (Ayet et al. 1999). For cereals, the phytase and acid phosphatase activity is highest on the fourth and fifth days of germination. The three-day germination of *Ugandi* and *Dembi* yellow varieties of pearl millet reduced the phytate phosphorus content in the range 12 and 16% respectively. The phytic acid content of Ugandi pearl millet reduced from 1050 to 939 mg/100 g on sprouting. In the Dembi yellow variety, phytic acid content reduced from 1020 to 865 mg/100 g on sprouting (Ahmed et al. 2010). The germination of proso millet for a period of 48 hours led to a 47% decrease in phytic acid and a 45% decrease in the level of tannin (Sharma et al. 2022b). Similar results were obtained by Morah and Etikudo (2017) who reported a decline in oxalate from 38.5 mg/100 g to 17.40 mg/100 g after 96 hours of germination of proso millet. The decrease in phytic acid content was observed to be 96% after 96 hours of germination. Even germination up to 48 hours showed a 94% reduction in phytic acid and a 31.4% reduction in oxalate. The process of germination reduced the oxalate content in finger millet by 29%. Reductions in tannin and phytic acid were reported to be 46 and 45% respectively (Hejazi and Orsat 2016). In kodo millet, germination for a period of 10 hours at 38°C reduced the phytates and tannins by 25 and 85% respectively (Sharma et al. 2018).

8.7.4 Fermentation

Fermentation is used for the development of many products from millets. It brings desirable biochemical changes triggered by microbes and released enzymes. The degradation of phytic acid on fermentation depends upon the initial concentration, variety of grain, conditions of fermentation, and microorganism used. Lactic acid bacteria reduces the pH to the optimal level for phytate degradation. In pearl millet dough, fermentation at 30°C for a period of 14 hours decreases the pH from 6.6 to 3.8 and also reduces the content of phytic acid by approximately 60% (El Hag et al. 2002). Dough obtained from two pearl millet varieties was fermented at 37°C for a period of 14 hours and then dehydrated in a hot air oven at 70°C. The decline in phytate phosphorus was between 17 and 25%. Phytic acid reduced from a level of 1050 to 887 mg/100 g on fermentation of pearl millet dough (Ahmed et al. 2010). Proso millet flour fermented with a culture of mixed lactic acid bacteria, for 20 hours at 38°C, obtained a pH of 3.0 and recorded a decline in the anti-nutritional content. The fermented flour reported a 36.5% decline in tannin content and a 50% reduction in the content of phytic acid (Sharma et al. 2022b). Malik et al. (2021) prepared a matrix of acid whey and kodo millet and proso millet flour. The matrix reduced the tannin content by 69.17% in kodo millet and 82.8% in proso millet. The decrease in the phytic acid content was found to be 9.21% in kodo millet indicating the effect of fermentation on the anti-nutritional factors. In a study reported by Purewal et al. (2019), the fermentation of pearl millet with *Rhizopus azygoporus* enhanced the total phenol content from 6.6 to 21.8 GAE/g on dry weight basis. Similar results were reported using *Aspergillus sojae* (MTCC-8779), *R. azygoporus*, etc. (Wang et al. 2022). The authors attributed the rise to the utilization of fibre by the microbes releasing the mechanically trapped phytochemicals from the fibre matrix.

Bhuvaneshwari et al. (2020) assessed the effect of germination on the anti-nutritional factors in pearl, kodo, finger, little, and foxtail millet. On germination for 24 hours after soaking for a period of 24 hours, the decline in tannin content was reported to be approximately 73.3% in finger millet, 37.4% in foxtail millet, 62.9% in kodo millet, 71.6% in little millet, and 71.5% in pearl millet. The process of fermentation at a temperature of 25°C reduced the phytate content by 20%, tannins by 52%, and trypsin inhibitors by 32% in finger millet (Rathore et al. 2019). A fermentation time of 72 hours reduced the phytate and tannin content in teff by 18 and 14% (Asres et al. 2018).

8.7.5 Heat Treatment

A reduction in blanching usually occurs due to leaching in the water. The blanching of pearl millet at a temperature of 98°C for a time duration of 30 seconds exhibited a decrease in polyphenol content to 544.5 mg/100 g from an initial content of 764 mg/100 g; whereas the content of phytic acid was reduced from 833.4 to 512.1 mg/100 g (Rani et al. 2018).

Cooking involves the use of high temperatures. Phytates are heat stable and hence generally do not degrade in cooking. However, at higher temperatures, the activation of phytases could lead to a reduction in phytates. The presence of acid, alkali, temperature, and pressure during cooking affects phytate reduction. For instance, pressure cooking is reported to reduce phytate more than cooking at atmospheric pressure. Phytates inhibit the development of cross-linkages between pectin and minerals like calcium and magnesium. This leads to the softening of the texture by weakening the cell membrane. High-phytate foods have been reported to have lower cooking time (Feizollahi et al. 2021). The impact of various types of heat treatment on the phenolic content was studied by Kalam Azad et al. (2019). The authors kept the temperature constant at 110°C and exposed proso millet to roasting, puffing, steaming, and extrusion. Among these processes, roasting increased the phenolic content the most in comparison to other methods at constant temperature. The cooking of fonio millet not only reduces the phytate content to 12.5 to 513.4 mg/100 g but also decreased the molar ratio of phytate:iron from 1.3 to 4.7 and improved the phytate:zinc ratio from 15.9 to 5.7 (Koreissi-Dembélé et al. 2013).

The popping of finger millet decreased the tannin content by 54% and the phytate content by 41% (Pragya et al. 2015; Rathore et al. 2019). Puffed proso millet was found to have higher polyphenol content (Piat et al. 2016).

Parboiling is generally done to enhance the milling yield and for the retention of nutrients. The parboiling of millet led to an increase in the free and bound phenolic content due to the migration of phenolics from the pericarp to the inner layers leading to lower losses on milling (Bora et al. 2019).

8.7.6 Extrusion

Extrusion is usually a high-temperature operation. The extrusion of finger millet flour led to a decrease in the total phenol and total flavonoid content by 46% and 22% respectively (Patil et al. 2016). Extruded snacks prepared from fonio millet were found to have lower trypsin inhibitor activities. Trypsin inhibitors reduced to 82% on extrusion at 140°C barrel temperature (Olapade and Awori 2012).

8.7.7 Radiation

Free radicals produced during irradiation are the major factor responsible for the reduction of anti-nutritional factors in millets. Microwave processing leads to a reduction of phytic acid by hydrolysing the inositol hexaphosphate to pentaphosphate and tetraphosphate (Suhag et al. 2021). The location of tannins, solubility, and heat lability affect the content on microwave processing. Microwave treatment breaks down the tannin-protein complexes and releases free tannins which are then leached in water (Suhag et al. 2021). The microwave treatment of pearl millet and finger millet (900 W, 2450 MHz) for a period of 40–100 seconds presented a decline in the tannin content to a level of 45.53 and 45.18% respectively. Microwaves have been effectively used to reduce the content of trypsin inhibitors and oxalates in other cereals (Irakli et al. 2020). The effect of gamma irradiation on the anti-nutritional content of finger millet was studied by Gowthamraj et al. (2021). Finger millet of CO14 and CO15 varieties was subjected to gamma radiation at doses of 2.5, 5, 7.5, and 10 kGy at a dose rate of 2.5 kGy per hour. The results revealed a reduction in total phenols and tannin content while the amount of total flavonoids increased. The total phenol content decreased from 166.05 to 126.68 mg/100 g at 10 kGy dosage. The decrease of total tannins was from 249.71 to 192.98 mg/100 g at the same dosage levels. Similar results were obtained by Reddy and Vishwanath (2019) in finger millet.

Mahmoud et al. (2016) assessed the impact of gamma radiation on anti-nutritional factors present in pearl millet flour. The study used gamma radiation from 0.25 to 2 kGy. The decrease in anti-nutritional factors was dose dependent. Radiation of 2 kGy reduced the tannin content by 74.9% and phytic acid by 52.8%. The generation of free radicals by gamma radiation might have reduced tannins due to the chemical degradation of the molecule and phytic acid due to cleavage in the phytate ring.

The impact of ultrasound-assisted hydration on finger millet was assessed by Yadav et al. (2021). The study optimized the ultrasound conditions and assessed the effect on anti-nutritional factors in finger millet. An ultrasound amplitude of 66% with a treatment time of 26 minutes and soaking in a millet to water ratio of 1:3 resulted in the reduction of phytate and tannins by 67 and 63% respectively (Yadav et al. 2021). The authors suggested that the effect may be due to the hydrolysis of ester linkage and the induction of tannin leaching. Ultrasound also helps in converting tannins into gallic acid and phytates into lower inositol phosphate (Bhangu et al. 2018).

8.7.8 Exogenous Enzymes

Exogenous enzymes can be intentionally added for target benefits in millets. Such treatment can reduce the anti-nutritional factors in millets. Enzymes that can be used for such function include phytases, tannases, cellulases, xylanases, proteases, and polyphenol oxidases. Not many studies have been conducted on the utilization of exogenous enzymes for the reduction of anti-nutritional factors and enhancing nutrient bioavailability (Tharifkhan et al. 2021). The addition of phytases to millet-based test meals enhanced the absorption of zinc by 40% (Brnić et al. 2017). The addition of even a small amount of phytases reduced phytate content and enhanced functional properties. The addition of phytases synthesized by *Saccharomyces cerevisiae* MTCC 5421 at a temperature of 60°C and pH of 5 exhibited a 94% reduction in phytates in pearl millet flour (Roopashri and Varadraj 2015). Polyphenol oxidases extracted from pearl millet were similar to that from wheat and have the ability to reduce the iron-binding ability of polyphenols. Tannases have the ability to reduce tannins. Cellulases make available digestive enzymes like proteases, amylases, etc., eliminating anti-nutritional factors. Xylases help in the degradation of xylase into xylooligosaccharides (Palaniappan et al. 2017).

8.8 CONCLUSION

The current chapter attempted to review the anti-nutritional factors present in millets and the effect of various processing methods to mitigate them. Millets contain various types of anti-nutritional factors which inhibit nutrient availability in various ways. Phytates bind with minerals, trypsin inhibitors inhibit digestive enzymes, non-starch polysaccharides act by increasing the viscosity of chyme and limiting mineral availability. These compounds inhibit the availability of proteins and minerals like calcium, iron, and zinc among others. Various household processing treatments have been shown to either mitigate or destroy the compounds from the grain. Since these factors are concentrated mainly in the bran portion, even simple processing treatments like dehulling, soaking, etc., have been found to be effective. The promotion of millets for consumption is highly recommended but not without the elimination of anti-nutritional factors, e.g. the degradation of phytate so as to increase iron absorption to 18%. Literature is available on major millets but not on minor millets and grains like teff and fonio. Future research must focus on this aspect of minor millets as well. The eradication of malnutrition through millets cannot be achieved without the elimination of anti-nutritional factors.

REFERENCES

Abioye, V. F., G. O. Ogunlakin, and G. Taiwo. 2018. Effect of germination on anti-oxidant activity, total phenols, flavonoids and anti-nutritional content of finger millet flour. *Journal of Food Processing & Technology* 9: 1–5.

Ahmed, A. I., A. A. Abdalla, K. A. Ibrahim, A. H. El-Tinay, and S. Shambat. 2010. Effect of traditional processing on phosphorus content and some anti nutritional factors of pearl millet (*Pennisetum glaucum* L.). *Research Journal of Agriculture and Biological Sciences* 6(3): 176–180.

Almaski, A., S. Thondre, H. Lightowler, and S. Coe. 2017. Determination of the polyphenol and antioxidant activity of different types and forms of millet. *Proceedings of the Nutrition Society* 76: E5.

Asres, D. T., A. Nana, and G. Nega. 2018. Complementary feeding and effect of spontaneous fermentation on anti-nutritional factors of selected cereal-based complementary foods. *BMC Pediatrics* 18(1): 1–9.

Ayet, G., C. Burbano, C. Cuadrado, M. M. Pedrosa, L. M. Robredo, M. Muzquiz et al. 1999. Effect of germination, under different environmental conditions, on saponins, phytic acid and tannins in lentils (Lens culinaris). *Journal of the Science of Food and Agriculture* 74(2): 273–279.

Baye, K. 2014. *Teff: Nutrient Composition and Health Benefits* (Vol. 67). International Food Policy Research Institute.

Bhangu, S. K., R. Singla, E. Colombo, M. Ashokkumar, and F. Cavalieri. 2018. Sono-transformation of tannic acid into biofunctional ellagic acid micro/nanocrystals with distinct morphologies. *Green Chemistry* 20(4): 816–821.

Bhuvaneshwari, G., A. Nirmalakumari, and S. Kalaiselvi. 2020. Impact of soaking, sprouting on antioxidant and anti-nutritional factors in millet grains. *Journal of Phytology* 12: 62–66. https://doi.org/10.25081/jp.2020.v12.6384.

Bohn, L., A. S. Meyer, and S. Rasmussen. 2008. Phytate: Impact on environment and human nutrition. A challenge for molecular breeding. *Journal of Zhejiang University. Science. Part B* 9(3): 165–191.

Bora, P., S. Ragaee, and M. Marcone. 2019. Effect of parboiling on decortication yield of millet grains and phenolic acids and in vitro digestibility of selected millet products. *Food Chemistry* 274: 718–725.

Brnić, M., R. F. Hurrell, L. T. Songré-Ouattara, B. Diawara, A. Kalmogho-Zan, C. Tapsoba et al. 2017. Effect of phytase on zinc absorption from a millet-based porridge fed to young Burkinabe children. *European Journal of Clinical Nutrition* 71(1): 137–141.

Budhwar, S., K. Sethi, and M. Chakraborty. 2020. Efficacy of germination and probiotic fermentation on underutilized cereal and millet grains. *Food Production, Processing and Nutrition* 2(1): 1–17.

Coulibaly, A., B. Kouakou, and J. Chen. 2011. Phytic acid in cereal grains: Structure, healthy or harmful ways to reduce phytic acid in cereal grains and their effects on nutritional quality. *American Journal of Plant Nutrition and Fertilization Technology* 1(1): 1–22.

Debon, S. J. J., and R. F. Tester. 2001. In vitro binding of calcium, iron and zinc by non-starch polysaccharides. *Food Chemistry* 73(4): 401–410.

Devisetti, R., S. N. Yadahally, and S. Bhattacharya. 2014. Nutrients and antinutrients in foxtail and proso millet milled fractions: Evaluation of their flour functionality. *LWT - Food Science and Technology* 59(2): 889–895.

Dost, K., and G. Karaca. 2016. Evaluation of phytic acid content of some tea and nut products by reverse-phase high performance liquid chromatography/visible detector. *Food Analytical Methods* 9(5): 1391–1397.

Dwivedi, M., V. K. Yajnanarayana, M. Kaur, and A. P. Sattur. 2015. Evaluation of anti nutritional factors in fungal fermented cereals. *Food Science and Biotechnology* 24(6): 2113–2116.

El Hag, M. E., A. H. El Tinay, and N. E. Yousif. 2002. Effect of fermentation and dehulling on starch, total polyphenols, phytic acid content and in vitro protein digestibility of pearl millet. *Food Chemistry* 77(2): 193–196.

ElShazali, A. M., A. A. Nahid, H. A. Salma, and E. B. Elfadil. 2011. Effect of radiation process on antinutrients, protein digestibility and sensory quality of pearl millet flour during processing and storage. *International Food Research Journal* 18(4): 1401.

Eltayeb, M. M., A. B. Hassn, E. E. Babiker, and M. A. Sulieman. 2007. Effect of processing followed by fermentation on antinutritional factors content of pearl millet (Pennisetum glaucum L.) cultivars. *Pakistan Journal of Nutrition* 6(5): 463–467.

Engle-Stone, R., A. Yeung, R. Welch, and R. Glahn. 2005. Meat and ascorbic acid can promote Fe availability from Fe-phytate but not from Fe-tannic acid complexes. *Journal of Agriculture and Food Chemistry* 53(26): 10276–10284.

Feizollahi, E., R. S. Mirmahdi, A. Zoghi, R. T. Zijlstra, M. S. Roopesh, and T. Vasanthan. 2021. Review of the beneficial and anti-nutritional qualities of phytic acid, and procedures for removing it from food products. *Food Research International* 143: 110284.

Gaytán-Martínez, M. et al. 2017. Effect of nixtamalization process on the content and composition of phenolic compounds and antioxidant activity of two sorghums varieties. *Journal of Cereal Science* 77: 1–8. https://doi.org/10.1016/j.jcs.2017.06.014.

Glahn, R. P., G. M. Wortley, P. K. South, and D. D. Miller. 2002. Inhibition of iron uptake by phytic acid, tannic acid, and ZnCl2: Studies using an in vitro digestion/Caco-2 cell model. *Journal of Agricultural and Food Chemistry* 50(2): 390–395.

Gowthamraj, G., C. Jubeena, and N. Sangeetha. 2021. The effect of γ-irradiation on the physicochemical, functional, proximate, and anti-nutrient characteristics of finger millet (CO14 & CO15) flours. *Radiation Physics and Chemistry* 183: 109403.

Gupta, R. K., S. S. Gangoliya, and N. K. Singh. 2015. Reduction of phytic acid and enhancement of bioavailable micronutrients in food grains. *Journal of Food Science and Technology* 52(2): 676–684.

Gupta, S., S. K. Shrivastava, and M. Shrivastava. 2013. Study of biological active factors in some new varieties of minor millet seeds. *International Journal of Innovative Engineering Technology* 3: 115–117.

Hama, F., C. Icard-Vernière, J. P. Guyot, C. Picq, B. Diawara, and C. Mouquet-Rivier. 2011. Changes in micro and macronutrient composition of pearl millet and white sorghum during in field versus laboratory decortication. *Journal of Cereal Science* 54(3): 425–433.

Hejazi, S. N., and V. Orsat. 2016. Malting process optimization for protein digestibility enhancement in finger millet grain. *Journal of Food Science and Technology* 53(4): 1929–1938.

Hotz, C., and R. S. Gibson. 2007. Traditional food-processing and preparation practices to enhance the bioavailability of micronutrients in plant-based diets. *Journal of Nutrition* 37(4): 1097–1100.

Iqbal, T. H., K. O. Lewis, and B. T. Cooper. 1994. Phytase activity in the human and rat small intestine. *Gut* 35(9): 1233–1236.

Irakli, M., A. Lazaridou, and C. G. Biliaderis. 2020. Comparative evaluation of the nutritional, antinutritional, functional, and bioactivity attributes of rice bran stabilized by different heat treatments. *Foods* 10(1): 57. https://doi.org/10.3390/ foods10010057.

Jaworski, N. W., H. N. Lærke, K. E. Bach Knudsen, and H. H. Stein. 2015. Carbohydrate composition and in vitro digestibility of dry matter and nonstarch polysaccharides in corn, sorghum, and wheat and coproducts from these grains. *Journal of Animal Science* 93(3): 1103–1113.

Kakade, S. B., and B. S. Hathan. 2015. Finger millet processing: Review. *International Journal of Agriculture Innovations and Research* 3(4): 1003–1008.

Kalam Azad, M. O., D. I. Jeong, M. Adnan, T. Salitxay, J. W. Heo, M. T. Naznin, et al. 2019. Effect of different processing methods on the accumulation of the phenolic compounds and antioxidant profile of broomcorn millet (*Panicum miliaceum* L.) Flour. *Foods* 8(7): 230.

Kalinová, J. 2007. Nutritionally important components of proso millet (*Panicum miliaceum* L.). *Food* 1(1): 91–100.

Kataria, A., S. Sharma, and B. N. Dar. 2022. Changes in phenolic compounds, antioxidant potential and antinutritional factors of Teff (*Eragrostis tef*) during different thermal processing methods. *International Journal of Food Science & Technology* 57(11): 6893–6902.

Koreissi-Dembélé, Y., N. Fanou-Fogny, P. J. Hulshof, and I. D. Brouwer. 2013. Fonio (*Digitaria exilis*) landraces in Mali: Nutrient and phytate content, genetic diversity and effect of processing. *Journal of Food Composition and Analysis* 29(2): 134–143.

Kumar, A., A. Mazeed, D. Kumar, R. Kumar, P. S. Verma, and N. B. Lothe. 2020. Evaluation of yield potential and nutritional quality of various cultivars of barnyard millet (*Echinochloa frumentacea* L.) grown under subtropical India. *Emerging Life Science Research* 6(2): 54–59.

Kumar, S. I., C. G. Babu, V. C. Reddy, and B. Swathi. 2016. Anti-nutritional factors in finger millet. *Journal of Nutrition and Food Science* 6(03): 5–6.

Liu, Z. H., H. Y. Wang, X. E. Wang, G. P. Zhang, P. D. Chen, and D. J. Liu. 2007. Phytase activity, phytate, iron, and zinc contents in wheat pearling fractions and their variation across production locations. *Journal of Cereal Science* 45(3): 319–326.

Mahmoud, N. S., S. H. Awad, R. M. Madani, F. A. Osman, K. Elmamoun, and A. B. Hassan. 2016. Effect of γ radiation processing on fungal growth and quality characteristics of millet grains. *Food Science & Nutrition* 4(3): 342–347.

Malik, S., K. Krishnaswamy, and A. Mustapha. 2021. Development and functional characterization of complementary food using kodo and proso millet with acid whey from Greek yogurt processing. *Journal of Food Processing and Preservation* 46(9): e16051.

Malleshi, N. G., H. S. R. Desikachar, and R. N. Tharanathan. 1986. Free sugars and non-starch polysaccharide of finger millet (*Eleusine coracana*), pearl millet (*Pannisetum typhoideum*) and foxtail millet (*Setaria italica*) and their malts. *Food Chemistry* 20(4): 253–261.

Mbithi-Mwikya, S., J. Van Camp, Y. Yiru, and A. Huyghebaert. 2000. Nutrient and antinutrient changes in finger millet (*Eleusine coracan*) during sprouting. *LWT – Food Science and Technology* 33(1): 9–14.

Morah, F., and U. P. Etukudo. 2017. Effect of sprouting on nutritional value of Panicum miliaceum (proso millet). *Edorium Journal of Nutrition and Dietetics* 4: 1–4.

Nakarani, U. M., D. Singh, K. P. Suthar, N. Karmakar, P. Faldu, and H. E. Patil. 2021. Nutritional and phytochemical profiling of nutracereal finger millet (*Eleusine coracana* L.) genotypes. *Food Chemistry* 341(2): 128271.

Nassarawa, S. S., and G. Ahmad. 2019. Comparative of phytochemical and antioxidant properties of selected millet varieties in katsina metropolis, Nigeria. *Annals: of Food Science and Technology* 20(4): 820–831.

Nazni, P., and D. R. Shobana. 2016. Effect of processing on the characteristics changes in barnyard and foxtail millet. *Journal of Food Processing and Technology* 7(3): 1–9.

Nissar, J., T. Ahad, H. Naik, and S. Hussain. 2017. A review phytic acid: As antinutrient or nutraceutical. *Journal of Pharmacognosy and Phytochemistry* 6(6): 1554–1560.

Nithya, K. S., B. Ramachandramurty, and V. V. Krishnamoorthy. 2007. Effect of processing methods on nutritional and anti-nutritional qualities of hybrid (COHCU-8) and traditional (CO7) pearl millet varieties of India. *Journal of Biological Sciences* 7(4): 643–647.

Olapade, A. A., and O. C. Aworj. 2012. Chemical and nutritional evaluation of extruded complementary foods from blends of fonio (*Digitaria exilis Stapf*) and cowpea (*Vigna unguiculata L. Walp*) Flours. *International Journal of Food and Nutrition Science* 1(3): 4–8.

Osman, M. A. 2011. Effect of traditional fermentation process on the nutrient and antinutrient contents of pearl millet during preparation of Lohoh. *Journal of the Saudi Society of Agricultural Sciences* 10(1): 1–6.

Pal, M., S. K. Z. Rizvi, R. P. Singh, and R. N. Kewat. 2016. Biochemical and antinutritional factors of kodo millet (*Paspalum scrobiculatum* L.). *Asian Journal of Biological Science* 11(1): 172–175.

Palaniappan, A., V. G. Balasubramaniam, and U. Antony. 2017. Prebiotic potential of xylooligosaccharides derived from finger millet seed coat. *Food Biotechnology* 31(4): 264–280.

Pandey, S., R. Ghosh, D. Suryawansh, and R. Waghmare. 2022. Evaluation of physio-chemical characteristics and acrylamide content in crisp fried dough wafers made from nixtamalized pearl millet. *Journal of Microbiology, Biotechnology and Food Sciences* 11(6): e3186.

Patil, R. B., K. G. Vijayalakshmi, and D. Vijayalakshmi. 2020. Physical, functional, nutritional, phytochemical and antioxidant properties of kodo millet (*Paspalum scrobiculatum*). *Journal of Pharmacognosy and Phytochemistry* 9(5): 2390–2393.

Patil, S. S., E. Varghese, S. G. Rudra, and C. Kaur. 2016. Effect of extrusion processing on phenolics, flavonoids and antioxidant activity of millets. *International Journal of Food and Fermentation Technology* 6(1): 177–184. https://doi.org/10.5958/2277-9396.2016.00040.4.

Petroski, W., and D. M. Minich. 2020. Is there such a thing as "anti-nutrients"? A narrative review of perceived problematic plant compounds. *Nutrients* 12(10): 2929.

Piat, B., D. Ogrodowska, and R. Zadernowski. 2016. Nutrient content of puffed proso millet (*Panicum miliaceum L.*) and Amaranth (*Amaranthus cruentus L.*) grains. *Czech Journal of Food Sciences* 34(4): 362–369.

Popova, A., and D. Mihaylova. 2019. Antinutrients in plant-based foods: A review. *The Open Biotechnology Journal* 13(1): 68–76.

Pradeep, S. R., and M. Guha. 2011. Effect of processing methods on the nutraceutical and antioxidant properties of little millet (*Panicum sumatrense*) extracts. *Food Chemistry* 126(4): 1643–1647.

Pragya, U. D., V. E. Pradesh, and T. K. Singh. 2015. Chemical composition of finger millet of food and nutritional security. *International Journal of Food Science and Microbiology* 3(6): 92–98.

Purewal, S. S., K. S. Sandhu, R. K. Salar, and P. Kaur. 2019. Fermented pearl millet: A product with enhanced bioactive compounds and DNA damage protection activity. *Journal of Food Measurement and Characterization* 13(2): 1479–1488. https://doi.org/10.1007/s11694-019-00063-1.

Rani, S., R. Singh, R. Sehrawat, B. P. Kaur, and A. Upadhyay. 2018. Pearl millet processing: A review. *Nutrition & Food Science* 48(1): 30–44.

Rao, P. V., and Y. G. Deosthale. 1988. In vitro availability of iron and zinc in white and colouredra (*Eleusine coracana*): Role of tannin and phytate. *Plant Foods for Human Nutrition* 38: 35–41.

Rathore, S. 2016. Millet grain processing, utilization and its role in health promotion: A review. *International Journal of Food and Nutrition Science* 5(5): 318.

Rathore, T., R. Singh, D. B. Kamble, A. Upadhyay, and S. Thangalakshmi. 2019. Review on finger millet: Processing and value addition. *Journal of Pharmaceutical Innovation* 8(4): 283–329.

Ravindran, G. J. F. C. 1991. Studies on millets: Proximate composition, mineral composition, and phytate and oxalate contents. *Food Chemistry* 39(1): 99–107.

Reddy, C. K., and K. K. Viswanath. 2019. Impact of γ-irradiation on physicochemical characteristics, lipoxygenase activity and antioxidant properties of finger millet. *Journal of Food Science and Technology* 56(5): 2651–2659.

Roopashri, A. N., and M. C. Varadaraj. 2015. Functionality of phytase of *Saccharomyces cerevisiae* MTCC 5421 to lower inherent phytate in selected cereal flours and wheat/pearl millet-based fermented foods with selected probiotic attribute. *Food Biotechnology* 29(2): 131–155.

Salunkhe, D. K., S. J. Jadhav, S. S. Kadam, J. K. Chavan, and B. S. Luh. 1983. Chemical, biochemical, and biological significance of polyphenols in cereals and legumes. *Critical Reviews in Food Science & Nutrition* 17(3): 277–305.

Sandberg, A. S., and N. Scheers. 2016. Phytic acid: Properties, uses, and determination. In *Encyclopedia of Food and Health*, B. Caballero, P. M. Finglas, and F. Toldra (eds), 365–368. Cambridge: Academic Press.

Sharma, R., and S. Sharma. 2022. Anti-nutrient & bioactive profile, in vitro nutrient digestibility, technofunctionality, molecular and structural interactions of foxtail millet (*Setaria italica* L.) as influenced by biological processing techniques. *Food Chemistry* 368: 130815.

Sharma, R., S. Sharma, and B. Singh. 2022. Modulation in the bio-functional & technological characteristics, in vitro digestibility, structural and molecular interactions during bioprocessing of proso millet (*Panicum miliaceum* L.). *Journal of Food Composition and Analysis* 107: 104372.

Sharma, S., D. C. Saxena, and C. S. Riar. 2016. Analysing the effect of germination on phenolics, dietary fibres, minerals and γ-amino butyric acid contents of barnyard millet (*Echinochloa frumentaceae*). *Food Bioscience* 13: 60–68.

Sharma, S., D. C. Saxena, and C. S. Riar. 2017. Using combined optimization, GC-MS and analytical technique to analyze the germination effect on phenolics, dietary fibers, minerals and GABA contents of Kodo millet (*Paspalum scrobiculatum*). *Food Chemistry* 233: 20–28.

Singh, A., S. Gupta, R. Kaur, and H. R. Gupta. 2017. Process optimization for anti-nutrient minimization of millets. *Asian Journal of Dairy and Food Research* 36(4): 322–326.

Suhag, R., A. Dhiman, G. Deswal, D. Thakur, V. S. Sharanagat, K. Kumar, and V. Kumar. 2021. Microwave processing: A way to reduce the anti-nutritional factors (ANFs) in food grains. *LWT* 150: 111960.

Suma, P., and A. Urooj. 2014. Nutrients, antinutrients & bioaccessible mineral content (invitro) of pearl millet as influenced by milling. *Journal of Food Science and Technology* 51(4): 756–761.

Thakur, A., V. Sharma, and A. Thakur. 2019. An overview of anti-nutritional factors in food. *International Journal of Chemical Studies* 7(1): 2472–2479.

Tharifkhan, S. A., A. B. Perumal, A. Elumalai, J. A. Moses, and C. Anandharamakrishnan. 2021. Improvement of nutrient bioavailability in millets: Emphasis on the application of enzymes. *Journal of the Science of Food and Agriculture* 101(12): 4869–4878.

Udeh, H. O., K. G. Duodu, and A. I. O. Jideani. 2018. Malting period effect on the phenolic composition and antioxidant activity of finger millet (*Eleusine coracana* L. Gaertn) flour. *Molecules* 23(9): 2091.

Wang, H., Y. Fu, Q. Zhao, D. Hou, X. Yang, S. Bai, et al. 2022. Effect of different processing methods on the millet polyphenols and their anti-diabetic potential. *Frontiers in Nutrition* 9: 780499.

Yadav, S., S. Mishra, and R. C. Pradhan. 2021. Ultrasound-assisted hydration of finger millet (*Eleusine Coracana*) and its effects on starch isolates and antinutrients. *Ultrasonics Sonochemistry* 73: 105542.

Chapter 9

Role of Millets in Disease Management

9.1 INTRODUCTION

Diet and nutrition are two of the most important factors for the maintenance of well-being. Diet provides us nutrients and non-nutritive components which not only contribute to the optimal functioning of the human body but also boost immunity and ensure healthy survival in a world full of diseases.

Millets are a group of crops suitable for arid and semi-arid conditions. They include many species of plants which are relatively distant from each other genetically. Some common examples of the types of millets include finger millet, pearl millet, barnyard millet, foxtail millet, proso millet, fonio, teff, little millet, etc. (Kumar et al. 2018). Although millets have been a part of regular cookeries since ancient times, their recent popularity can be correlated with the developments in the nutraceutical and functional food market (Kumar et al. 2020). The current obsession with functional foods is not arbitrary but factual and is evidenced in many studies published recently discussing the role of nutraceuticals and functional foods in the prevention, management, and treatment of diseases (Singh et al. 2018). Also, these are climate-resilient, sustainable, and easily grown crops (Kumar et al. 2018).

The nutritional and phytochemical profile of millets is among the best in cereals (Himanshu et al. 2018). They contain appreciable levels of macronutrients like carbohydrates, quality protein, and lipids. They are rich in dietary fibre which controls the absorption and allows the controlled release of glucose in the blood. Also, optimal dietary fibre helps in maintaining intestinal health and preventing colon cancers. Millet has high-quality protein and provides eight essential amino acids. It does not contain gluten and therefore can be safely consumed by people suffering from gluten intolerance/celiac disease. The lipids present in millets are mostly unsaturated. To utilize the health-promoting factors in millets, several functional products have been formulated and supplemented with additives to improve acceptability (Kumar et al. 2020, Kumar et al. 2020a). The phytochemicals present in millets boost their effectiveness against many diseases. In this chapter, the role of millets in disease prevention and management will be discussed in detail.

9.2 GLOBAL DISEASE PARADIGM

Diseases can be broadly classified into two categories: communicable and non-communicable, depending on the mode of transmission. Diet plays an important role in the prevention and management of communicable diseases by strengthening the immune system and of non-communicable diseases by controlling activities at the metabolic level. Non-communicable diseases were believed to be diseases of affluence but in contrast, data predicts that 60% of the burden of chronic disease will occur in developing countries.

It was projected that, by 2020, chronic diseases would account for almost three-quarters of all deaths worldwide, with more than 70% from heart disease, stroke, and diabetes (WHO 2002). It is estimated that, by 2025, approximately 228 million people will suffer from diabetes. Obesity is identified to be the risk factor/root cause of almost all non-communicable diseases. It is worrisome that the number of obese and overweight individuals is increasing alarmingly in developing countries, the public health consequences of which are already apparent.

Chronic diseases by and large are preventable. Current science-based evidence provides an adequately strong and credible foundation to link nutrition (food) and health. While the genetically determined factors are said to be non-modifiable, other risk factors like diet are modifiable. Recent advances in nutrigenomics have revealed that even genes can be controlled through diet and lifestyle and diseases which are thought to be hereditary in nature can be avoided.

Apparently, around the globe, dietary habits have undergone major changes in the past century. Traditional plant-based diets have been replaced largely by high-fat, energy-dense diets which has resulted in many health disorders. It is high time to think about plant-based foods and incorporate them into our daily diet as the consumption of foods with high fibre, nutrients, and phytochemicals can be beneficial in health promotion.

9.3 COMPOUNDS IN MILLETS WITH DISEASE PREVENTION POTENTIAL

9.3.1 Nutrients with Health Potential

The quantity and the type of nutrient present in any food determine the extent of health benefits exhibited. A high intake of carbohydrates, especially simple sugars, leads to weight gain and fat deposition leading to obesity, impaired glucose tolerance, altered lipid parameters, etc. Millets possess low amounts of carbohydrates. Varieties of barnyard and kodo millet have been shown to have carbohydrate content of as low as 49 and 56.1% respectively (Kumar et al. 2018). Free sugars are present in minute amounts and range between 1 and 1.4% in different types of millets (Nirmala et al. 2000, Leder 2004). Millet contains a good amount of resistant starch. High amounts of dietary fibre have also been reported (8–37%) which is much higher than other cereals (Table 9.1). Wide variations have been observed in different millet varieties: kodo millet (6.8–37.8%), finger millet (15–20%), and foxtail millet (18–20%) (Kumar et al. 2018, Dayakar et al. 2017). High amounts of dietary fibre (both soluble and insoluble) add bulk to the meal and enhance water absorption in the colon, creating a sense of early satiety and lowered appetite and maintaining colon health, and are also responsible for the low glycaemic index of millets. Low glycaemic index foods are directly associated with the low prevalence of obesity and cardiovascular diseases and the management of blood sugar levels. Dietary fibre like hemicelluloses, β-glucan, arabinoxylans, etc., functions as prebiotics and acts as a substrate for the colonic microflora and maintains gut health. Kumar et al. (2020) reported higher growth of *Lactobacillus* species in a finger millet and oat-based drink as compared to milk. Similar results have been obtained in other studies viz., pearl millet fibre and flour for *Lactobacillus rhamnosus*, *Bifidobacterium bifidum*, and *L. acidophilus*; and kodo millet bran hydrolysate for *L. delbrueckii* (Farooq et al. 2013, Mridula and Sharma 2015, Balakrishnan et al. 2020).

TABLE 9.1 NUTRIENT COMPOSITION OF MAJOR MILLETS AND OTHER STAPLE CEREALS

Millet type	Carbohydrates (%)	Protein (%)	Fat (%)	Ash (%)	Total dietary fibre (%)	Glycaemic index
Pearl millet	67–70.4	10.6–12.6	4.8–5.0	1.9–2.3	8.0–13.5	55
Finger millet	66.5–75.0	6–8.2	1.3–1.5	2.6–2.7	15.0–20.0	55–65
Proso millet	62.3–72.2	10.6–12.5	1.76–4.0	1.9–3.2	8.5–14.2	50.2–64.7
Foxtail millet	61.6–72.4	9.9–12.3	2.5–4.0	3.3–3.5	19.11	33
Barnyard millet	49–65.5	8.9–11.9	2.2–4.5	4–4.5	13.6	41.7–50
Kodo millet	56.1–74.0	8.3–11.6	1.3–4.2	2.6–3.3	6.8–37.8	49.5
Little millet	64.2–67	7.6–10	2.4–2.8	1.5–1.9	6.4–12.2	41.50–61.80
Rice (brown rice)	72.8–75.8	8.5–9.90	1.12–1.30	0.9–1.2	0.7–6.0	65–81
Wheat	61.2–66.8	10–12	1.42–152	1.20–1.64	2.8–12.1	44–60
Maize	63.2–66.42	8–9.5	3.60–4.12	1.02–1.34	3.9–13.4	78.5–86.3

Source: Kumar et al. 2020.

Millet contains high-quality protein averaging between 10 and 11% (Kumar et al. 2018). The health benefits exhibited by millets mainly rely on the amino acid composition. The essential amino acid content of millets is comparable with or higher than staple cereals. Like other cereals lysine and threonine are the limiting amino acids in millets; however the content of lysine in proso millet is higher than wheat (Ravindran 1992).

Millets are low in fat as well. The highest fat percentage has been reported in pearl millet (4.8–5%) and the lowest in finger millet (1.2–1.5%) (Singh and Raghuvanshi 2012). This is comparable to cereals like wheat and rice. The majority of lipids present in millets are in the form of unsaturated fatty acids, e.g. proso millet (86–89%), finger millet (74%). Unsaturated fatty acids are cardioprotective in nature and impart several other health benefits.

Millets are a rich source of micronutrients as well. Finger millet is especially rich in calcium (162–487 mg/100 g) and provides approximately 50% of the RDA. Pearl millet is the richest source of phosphorus among millets; the highest amounts of zinc in foxtail millet (60.6 mg) and iron in barnyard millet (15–19 mg/100 g) have been reported in studies (Singh and Raghuvanshi 2012).

9.3.2 Bioactive Compounds with Health Potential

Bioactives are those compounds which exhibit desired effect on the cellular and tissue level. Millets possess different bioactives in sufficient quantities which directly help in the prevention and management of diseases (Figure 9.1). These include dietary fibre, polyphenols, flavonoids, and tannins among others.

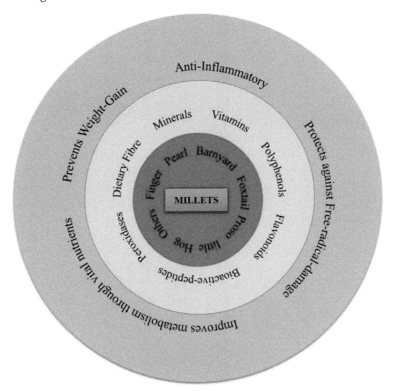

Figure 9.1 Potential health benefits of millets.

The polyphenol composition in millets is varied and is believed to be because of diversity in species. Like major cereals, maximum polyphenols are concentrated in the bran layer of millets. This explains the low phenolic content of processed millet grains. The phenolic acids in millets are tightly ester/etherified to the cell wall where they function as cross linkers. The strong linkage is broken only by acid or alkaline hydrolysis. However, unlike other cereals, millets contain high amounts of free phenolic acid comparatively. Benzoic acid derivatives are the most common phenolic acids found in millets. These include gentistic, vanillic, gallic, syringic, and protocatechuic acids (Chandrasekara and Shahidi 2011). These derivatives naturally are found in loose association with the cell wall and therefore are easily extractable. Ferulic acid derivatives are the dominant class and occur in bound form. However, the distribution and the type of phenolic acid found will depend on the species or variety of the millet grain. The bound acids usually account for more than 70% of the total phenolic acids in the grain and are mostly ferulic, diferulic, and cinnamic acids (Svensson et al. 2010). The extractable phenolic acids can be metabolized by microbes and are considered to be one of the major contributors to the effect of millet on gut microbiota and colon health.

Some millets possess flavonoids in abundance and these play an important part in processing, nutritional, and sensorial attributes. Polymeric flavonoids are known for their negative impact on micronutrient bioavailability. Proanthocyanidins are found in high quantities in finger millet and have the capability to form non-digestible complexes with proteins (Awika 2017). However, recently it has been observed that tannins also bind with starch and diminish starch digestibility significantly and exhibit strong antioxidant activity. These properties may be beneficial in the mitigation of chronic diseases. Besides tannins, millets also possess an array of flavonoids, like flavones and flavanones, which have a positive impact on health. Flavones mostly occur as O- or C-glycosides of luteolin and apigenin; free forms are significantly found in fonio millet, however; another flavone reported in some millets is tricin. The structural features of flavones in millets have been associated with improved biological activity which has been of importance in diseases linked to inflammation and cancer. Although all millet varieties accumulate flavones, pearl millet contains these in unusually high amounts leading to reports of goitre incidences, as high flavones have been associated with impaired thyroid function. The distribution of flavonones is limited in all grains including millets; however, these are found in abundance in sorghum. Finger millet contains flavanols and their derivatives. Tannins have strong affinity for proteins and form strong protein-tannin linkage complexes which are largely indigestible (Amoako and Awika 2016). These also inhibit gastric enzymes and intestinal amino acid transporters. This property can prove to be highly beneficial in weight reduction strategies and managing obesity.

Phytosterols in millets are well known for their cholesterol-lowering effects and are present in the bran layer of millet grains in easily extractable forms (Demonty et al. 2009). The common phytosterols reported in millet grains include Beta-sitosterol (56%), campesterol (16%), and stigmasterol (15%) respectively in the order of their abundance. The phytosterol content of pearl and foxtail millet has been reported to be between 0.58 and 0.44 mg/g (Ryan et al. 2007, Bhandari and Lee 2013).

Bioactive peptides are the latest identified bioactive components known for their antioxidant, antimicrobial, antihypertensive, immunomodulatory effects. Some reports have identified protein hydrolysates from foxtail millet which possess biological activity (Chen et al. 2017). Biopeptides contain amino acids like alanine, cysteine, lysine, etc., and contribute to the antioxidant effect. The ability of the biopeptides in inhibiting angiotensin-1 converting enzyme contributes to antihypertensive properties. Lactic acid fermentation and malting induce significant proteolytic properties, thereby releasing peptides. And since millet protein is specifically rich in proline, alanine, and valine, which are known to exhibit a potent acetylcholine esterase inhibitory effect, millet biopeptides may have important health-promoting properties.

9.3.3 Bioavailability of the Bioactives

The efficacy of any bioactive compound depends upon its bioavailability. Bioavailability or bioaccessibility depends on the stability of the compound under processing, storage, gastrointestinal conditions, and the extent of absorption and utilization of the compound by the body. Regarding millets, one study discussed the bioavailability of phenolic compounds. In this work of Chandrasekara and Shahidi (2012), the cooked grains of different millets like pearl, kodo, finger millet, and proso millet were cooked and subjected to in-vitro digestion for complex breakdown. The simulation of colonic fermentation was performed through microbial fermentation. These two major steps were performed to assess the bioavailability of phenolic compounds. The study revealed that the release of millet phenolics occurs on digestion and fermentation which potentially enhances their availability for absorption. The study concluded that the increase in compound identification and the change in their quantities may be due to treatments like cooking, digestion, and fermentation.

In the mouth, the complex formation of polyphenols with salivary proteins occurs; this does not affect their absorption efficiency however (Karas et al. 2017). Low pH conditions in the stomach do not affect the complex stability of the majority of the phenols. Reports of the absorption of phenolic acids in the stomach have also been identified (Farrell et al. 2012). Proanthocyanidins can be hydrolysed in the stomach. In the small intestine, upon the action of hydrolytic enzymes, flavonoid glycosides are hydrolysed into aglycones and glycosides. Both of these are absorbed either through passive absorption or facilitated transportation. The absorption of phenolics occurs mainly through the large intestine in the form of metabolites formed by the gut microbiota forming a bouquet of compounds that can be easily absorbed. No work was available to date regarding the bioavailability of polyphenols from millets.

9.4 GENERAL MECHANISM OF HEALTH PROMOTION

Millets in recent times have gained a lot of popularity attributed to their positive impact on human health. Therefore, the factors contributing to health promotion are the focus of recent research. From recent literature it is evidenced that the inclusion of millets in regular diet provides long-term health benefits against non-communicable diseases like diabetes, obesity, and cardiovascular diseases (Figure 9.2). The major mechanisms contributing to disease are highlighted in this section.

9.4.1 Protection against Free Radical Damage

Free radicals or reactive oxidant species generation is a part of the normal physiological process and there is the continuous production and removal of free radicals. When reactive oxygen species are overproduced, or the antioxidant defence systems are inefficient, the accumulation of free radicals occurs which leads to oxidative damage to cellular protein, DNA, and/or RNA (Lobo et al. 2010). The condition when the production of free radicals supersedes neutralization or removal is termed oxidative stress. Free radicals may start regulating the activities of transcription factors and may be a contributory factor in the development of many non-communicable diseases. This led to the hypothesis that the consumption of foods rich in antioxidants can help in managing the excess free radicals, thereby protecting against oxidative stress on the body. Dietary antioxidants complement the endogenous antioxidant system and minimize adversities.

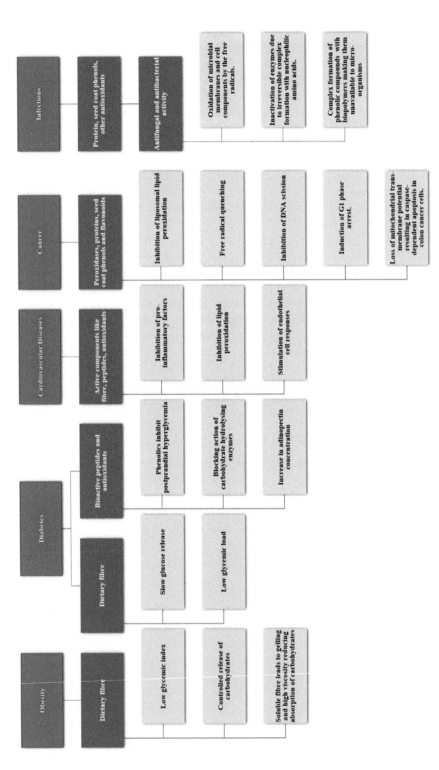

Figure 9.2 General mechanisms for the prevention and management of different diseases by millets.

Both *in-vivo* and *in-vitro* methods have been adopted to detect the antioxidant potential of millets. *In-vitro* methods include assessment of the antioxidant capacity using metal chelating assay or free radical scavenging activity. *In-vivo* studies have also been carried out to assess the antioxidant efficacy of millets. Murtaza et al. (2014) studied the effect of finger millet grain and bran using biological markers like oxidation products, enzymes, and reducing agents on Swiss albino mice fed with a high-fat diet. It was reported that finger millet grain and bran decreased the levels of oxidative stress parameters like free radicals (lipid peroxide, nitrite) and superoxide dismutase. In addition, improved levels of the reduced form of glutathione and catalase were observed in the liver, pancreas, and adipose tissue. The observations obtained were linked to the high phenolic content in finger millet grain and bran.

Millet bran possesses antioxidant properties. This was observed in a study conducted by Bijalwan et al. (2016). The authors extracted hydroxycinnamic acid bound arabinoxylans from five different types of millets viz., kodo millet, foxtail millet, finger millet, proso millet, and barnyard millet. An in-vitro model was used to study the antioxidant activity. It was reported that kodo millet bran possessed the highest amount of low-branched hydroxycinnamic acid-arabinoxylans and had the highest antioxidant activity. The authors also mentioned that the antioxidant activity of hydroxycinnamic acid-arabinoxylans depended on the structure and total phenolic acid composition.

Several studies have shown that the antioxidant potential of millets is not confined to their phytochemical content but is also due to other components like bioactive peptides. The antioxidant potential of fractionated foxtail millet protein hydrolysate was studied. The potential was estimated through tests like the linoleic acid auto-oxidation inhibition, free radical scavenging activity, and reducing power using ABTS and DPPH. The authors observed that the antioxidant potential of foxtail millet fraction (85.7%) was comparable to that of alpha-tocopherol (86.3%). On performing ABTS, DPPH, and metal-chelating activity, it was found that it possessed higher activity than other antioxidants. Amino acid profiling was also performed in the study. This revealed high antioxidant potential and the highest amount of hydrophobic amino acids (Mohamed et al. 2012).

Berwal et al. (2016) assessed the total antioxidant potential of 92 different genotypes of pearl millet using total antioxidant activity assay following DPPH and ABTS methods. The authors found an appreciable amount of total antioxidant activity irrespective of the pearl millet genotype. The authors also observed that the designated inbred lines of pearl millet had higher mean total antioxidant activity in comparison to the inbred lines. The study also compared the total antioxidant activity of pearl millet genotype with commercial samples of wheat and maize. On analysis, it was reported that the pearl millet genotypes had significantly higher average antioxidant activity (using DPPH) in comparison to wheat and maize.

In another significant study, variations in several finger millet varieties subjected to drought stress on stress markers and antioxidant potential were observed. It was reported that the plant limited reactive oxygen species and antioxidant potential even in extreme conditions like drought (Bartwal et al. 2016).

In a recent development, Salar and Purewal (2016) attempted to improve the DNA damage protection and antioxidant activity of bio-transformed pearl millet cultivar PUSA-415 by employing solid-state fermentation for 10 days using *Aspergillus oryzae* MTCC 3107. After fermentation, the grains were extracted using 50% ethanol. The samples were subjected to analysis for the assessment of DNA damage protection and the presence of bioactive compounds. DNA damage protection activity was assessed using Fenton's reagent while the bioactive compounds with antioxidant potential were assessed through total phenolic content, DPPH, ABTS, FRAP, total antioxidant capacity, and recombinase polymerase amplification assays. The authors reported that on fermentation, a significant increase was observed in the total phenolic content and condensed tannins. As revealed by high-performance liquid chromatography, fermentation resulted in the modulation of phenolic compounds like ascorbic, benzoic, cinnamic and gallic acid, catechol, and p-coumaric acid.

Xiang et al. (2019) analysed the phenolic compounds and antioxidant activity of finger millet varieties. The study first analysed ten phenolics in finger millet varieties harvested in northern Malawi. The total phenolic contents in the free fractions (114.4–179.2 mg ferulic acid equivalent/100 g) and bound fractions (58.3–123.23 mg ferulic acid equivalent/100 g) were reported. Catechin and apicatechin were reported to be the predominant flavonoids in the free fractions while ferulic acid is the major phenolic found in bound fractions. The major finding of the study also mentioned that the darker colour finger millet varieties had higher amounts of phenolic compounds and better antioxidant properties than the light colour varieties.

9.4.2 Anti-Obesogenic Effects

Obesity has widespread occurrence across the world due to mechanization, nutritional transition, and urbanization. It is said to be the seed which ensures the onset of many diseases, especially non-communicable diseases. Anecdotal evidence suggested millet and sorghum consumption were linked with early satiety in Africa (De Morais Cardodo et al. 2017). Hypoglycaemic properties are partly responsible for the antiobesogenic effect. Millets, especially pearl and finger millet, contain slow digestible starch (dietary fibre) and this may contribute to weight loss. Apart from dietary fibre another possible factor that may have profound implications for anti-obesogenic effects are phytochemicals. Recent studies have tried to analyse the link between phytochemicals in millets and their potential role in reducing weight.

Murtaza et al. (2014) assessed the impact of finger millet bran on obesity-induced oxidative stress and inflammation in mice fed on a high-fat diet for 12 weeks. It was reported that the supplementation of finger millet bran not only prevented weight gain in mice but also regulated the expression levels of different genes related to obesity. The incorporation of finger millet bran into a high-fat diet increased expression levels of PL1N1, DLK1, and ADIPOQ genes. Similar results were obtained by Sarma et al. (2017) on supplementation with kodo millet whole grain and bran.

Jun et al. (2014) studied the anti-adipogenic effects of proso millet grain on 3T3-L1 murine preadipocytes. For this, the authors used 80% ethanolic extracts of various cereal grains. A high level of toxicity was reported in proso millet. The study also reported the decline of intracellular lipid accumulation up to a level of 81%. The butanol fraction had the capacity to inhibit the differentiation of 3T3-L1 murine preadipocytes to mature adipocytes.

Arabinoxylans from finger millet are one of the compounds responsible for the antiobesity effects. Finger millet arabinoxylans were found effective in preventing high-fat diet induced adiposity and insulin resistance in mice. Sarma et al. (2018) evaluated the effect of finger millet-arabinoxylans (0.5 to 1 g/kg for 10 weeks) on metabolic derangements in male Swiss albino mice. Results indicated that finger millet prevented metabolic endotoxemia. Its supplementation prevented high-fat diet induced weight gain. Li et al. (2019) studied the effect of the supplementation of millet whole grain on the lipid profile and gut bacteria in rat fed with a high-fat diet. It was recorded that 50% millet supplementation increased the concentration of short chain fatty acids. The ethanolic extract of millet decreased the lipid accumulation in HepG2 cells and modulated lipid profile.

9.4.3 Anti-Inflammatory Properties

Persistent inflammation has been associated closely with the onset of several non-communicable diseases. Inflammation occurs as a result of the active immune response of the human body to infections. It is a complex mechanism involving the use of several molecular pathways using complex enzyme systems such as oxygenases (lipoxygenase and cyclo-oxygenase), lipases, and synthases wherein inflammatory cytokines are produced. Inflammatory cytokine levels are often used as markers to study the anti-inflammatory potential of millets. Recent literature has shown prolific evidence of the anti-inflammatory potential of millets.

Lee et al. (2014) assessed the anti-inflammatory effect of 80% ethanol extracts of proso, barnyard, foxtail millet, and sorghum. To study the anti-inflammatory effect, free radical (NO) generation was induced by lipopolysaccharide in RAW264.7 cells. Results indicate that the ethanolic extracts of barnyard millet had comparatively higher anti-inflammatory activity. Pre-treatment with methylene chloride led to the downregulation of several transcripts, proteins, and proinflammatory cytokine gene transcripts. Further analysis revealed that the phenolic compounds responsible for the effect were kaempferol and biochanin A. Formononetin was also identified in the extract but did not exhibit anti-inflammatory properties. In another study conducted on barnyard millet, Woo et al. (2015) identified and isolated novel compounds namely echinochlorins (A-C) from methylene chloride and ethyl acetate fractions of whole grains of barnyard millet. On assessing the anti-inflammatory activity on lipopolysaccharide activated murine macrophages RAW 264.7 cells, it was observed that these compounds possessed nitric oxide inhibitory activity, hence giving evidence regarding a beneficial effect against inflammation.

Similar effects were studied in hog millet by Park et al. (2011b) in murine macrophages. For the study hot water extract of hog millet was used with extract yield reported to be 5.3%. The results concluded that hog millet extract inhibited lipo-polysaccharide induced inflammatory response in RAW 264.7 macrophages. Finger millet arabinoxylans were reported to have a similar effect (Sarma et al. 2018).

Anti-inflammatory properties of millet bran were identified in several studies. Shi et al. (2017) studied the anti-inflammatory properties of polyphenolic compounds present in bound form in the inner shell of foxtail millet. The effect was studied in lipo-polysaccharide induced HT-29 cells in nude mice. The study reported that inner-shell polyphenols from foxtail millet restrained the levels of several pro-inflammatory cytokines (IL-1β, -6, -8) and enhanced the expression level of anti-inflammatory cytokines (IL-10). The effect is exhibited by the blockage of the nuclear factor-kappa B nuclear translocation. The authors concluded that bound phenols of inner shell are a potent anti-inflammatory agent which functions by signalling cascade of the ROS/miR-149/Akt/NF-κB axis. Murtaza et al. (2014) analysed the anti-inflammatory effect of finger millet bran. The addition of finger millet bran to a high-fat diet fed to mice downregulated the expression of pro-inflammatory genes like *CD 68* and *F4/80*. The expression of anti-inflammatory genes like *IL6* and *ADAM 8* was increased. Immunomodulating effects of Japanese millet bran were also reported by Hosoda et al. (2012).

9.5 MILLETS AND PROTECTION FROM DISEASES

The compounds and mechanisms behind the positive health effect of millets on the human body have been explained in the previous sections. It has been noted that the association of millet consumption and reduced incidences of lifestyle diseases can be attributed to compounds such as fibre and phytochemicals like polyphenols and flavonoids (Table 9.2). In this section, the role of millet consumption in some common lifestyle disorders will be highlighted.

9.5.1 Hypertension

Hypertension is said to be one of the major precursors of cardiovascular diseases. There is evidence that a decline of 3 mmHg in blood pressure can minimise human mortality occurring due to coronary heart disease (Wei et al. 2018).

Hou et al. (2018) studied the antihypertensive effect of foxtail millet. In the first of its kind, the study proved the anti-hypertensive effects of foxtail millet based on clinical trial. To observe the effect, 50 g of whole grain was included, partially replacing the staple in the diet of 45 hypertensive subjects. On administration for 12 weeks, it was observed that millet significantly lowered both systolic and diastolic blood pressure. It also improved other metabolic parameters.

There are several factors in millets which have been proposed to be responsible for the health of the circulatory system. As discussed in the previous section, certain biopeptides possess antihypertensive properties. Hydrolysed proteins in millets are sources of bioactive peptides which possess ACE properties. The effect of protein hydrolysates derived from foxtail millet on systolic and diastolic blood pressure was observed in hypertensive rats. For this, protein fractions were isolated from raw, fermented, and extruded foxtail millet and were found to be equally effective. A significant drop in blood pressure was observed in the rats on supplementation. The authors recorded a significant decrease in angiotensin II and angiotensin-converting enzyme activity levels (Chen et al. 2017). Wei et al. (2018) investigated the effect of nine brands of foxtail millet in rats with high-salt diet induced hypertension. Twenty-five grams of whole millet flour was supplemented for eight weeks and was shown to reduce hypertension. Millet contains 0.3 to 3% of the nutrients as phytochemicals (Hegde and Chandra 2005). The authors suggested that changes in angiotensin, the activation of the renin-angiotensin-aldosterone system, or vascular active peptides may be responsible for the effect.

9.5.2 Cardiovascular Diseases

Several studies have been conducted which provide evidence against the active role of millets in protection against cardiovascular diseases. Park et al. (2011) assess the anti-adipogenic effect in water extract of nine cereal grains on 3T3-L1 adipocytes. The authors concluded that hog millet had the highest potency against lipid accumulation. Not only this, hog millet extract was also found to alter the ratio of monounsaturated fatty acids to saturated fatty acids on adipocytes. The authors further continued their research. With these results, the authors hypothesized that hog millet extract will augment hyperlipidaemia by the augmentation of fatty acid metabolism. Park et al. (2011a) analysed the effect of a hog millet-supplemented diet on hyper-lipidaemia and lipid accumulation in the liver. Hog millet extract was fed to C57BL/6J-*ob/ob* mice ranging from 0.5 to 1% for four weeks and blood serum and lipid profiles, gene expression related to hepatic fatty acid metabolism, and the fatty acid composition of the white adipose tissue were determined. The study concluded that 1% hog millet extract significantly decreased blood triglyceride and total cholesterol levels. It also reduced the expression of genes related to lipogenesis (PPARα, FAS, etc.) and enhanced lipolysis-related gene expression (CPT1). Administration of the extract also reduced the ratio of C18:1/C18:0 fatty acids significantly.

The effect of proso millet on hyperlipidaemia was studied by Bora et al. (2018) in 48 Wister rats with high-fat diet induced hyperlipidaemia. Consumption of a diet with 40% polished and

TABLE 9.2 ROLE OF MILLETS IN DISEASE MANAGEMENT

Disease	Millet type	Study subject and method	Mechanism and inference	Reference
Hypertension	Foxtail millet	In-vivo model: rats were fed with whole seed flour at the rate of 200 mg/kg of body weight for 4 weeks	Significant reduction in blood pressure and markers like angiotensin converting enzyme activity and angiotensin II levels	Chen et al. (2017)
	Foxtail millet	Clinical trial: conducted on mildly hypertensive subjects without medication. Subjects were supplemented with 50 g whole grain for 12 weeks	Reduction in systolic and diastolic blood pressure	Hou et al. (2018)
	Foxtail millet	In-vivo model: hypertensive rats with salt-induced hypertension administered with 25 g of whole grain for 8 weeks	Reduction in blood pressure and improvement in lipid profile	Wei at al (2018)
Diabetes	Foxtail millet	Clinical trial: subjects with impaired glucose intolerance were supplemented with 50 g whole grain for 12 weeks	Significant reduction in fasting and 2-h blood glucose levels, reduction in insulin resistant, and increase in leptin levels	Ren et al. (2018)
	Kodo millet	In-vivo study: ten healthy subjects were fed with 50 g digestible carbohydrates from *idli* and *sewai* supplemented with millet grain	Significant decline in glycaemic index and load	Yadav et al. 2013
	Pearl millet	Clinical trial: diabetic subjects were supplemented with Chapati made from pearl millet and fenugreek leaves	Low glucose response	Thilakavathy and Muthuselvi (2010)
Cardiovascular disease	Hog millet	In-vitro study: 3T3-L1 cells were subjected to 1% (w/w) extract for 4 weeks	Reduction of liver lipogenic-related and increase in lipolytic gene expression	Park et al. 2011
	Proso millet	In-vivo rat model: high-fat diet induced hyperlipidaemic, 48 wiser rats with weight between 120 and 200 g, 40% proso millet flour	Reduction in low-density lipoprotein, cholesterol, and triglycerides. Increase in high-density lipoprotein	Bora et al. 2018
	Millet	Both in-vitro and in-vivo model: for in-vivo study supplementation of diet with 50% millet For in-vitro study: ethanol extract of millet	Ethanol extract reduced fat accumulation in HepG2 cells, modulation of lipid profile, increase in concentration of short chain fatty acids	Li et al. 2019

(*Continued*)

TABLE 9.2 (CONTINUED) ROLE OF MILLETS IN DISEASE MANAGEMENT

Disease	Millet type	Study subject and method	Mechanism and inference	Reference
	Pearl millet bran	In-vitro model: atherogenic diet induced hyperlipidaemic albino rats supplemented with 6 g/kg pearl millet bran	47% increase in high-density lipoprotein, decrease in blood lipids (41%), low-density lipoprotein (55%), triglyceride (48%), total cholesterol (39%)	Javed et al. 2012
	Hog millet	In-vitro model: supplementation of mice with 2% hog millet for 7 weeks	2% hog millet in diet decreased liver weight, blood triglyceride, total cholesterol. Increase in high-density lipoprotein	Park et al. 2012
Cancer	Foxtail millet	Cell line study on colon cancer cells: peroxidase extracted from foxtail millet bran strains expressed in host *E. coli* DH5α	Reversal of 5-Fu resistance in human colorectal cancer, inhibition of cell proliferation, promotion of cell apoptosis. Decrease in expression of breast cancer cell protein	Zhang et al. 2019
	Proso millet, Barnyard millet	250–1000 μg/ml extracted vanillin administered to HT-29 human colon cancer cell line	Moderate inhibition of HT-29 cell line. Cytotoxic activity on cancer cells. No toxicity in normal cells	Ramadoss and Sivalingam 2020
	Foxtail millet	Human HCT-8/Fu colorectal cancer cell line administered with bound polyphenol of inner shell with molecular weight < 200	Reversal of multidrug resistance in HCT-8 cells. Main active components – ferulic acid and p-coumaric acid. Sensitivity of chemotherapeutic drugs enhanced	Lu et al. 2018
	Foxtail millet	HCT-116 cell colon cell xenograft tumour model	ROS generation leads to cell apoptosis	Shi et al. 2015
	Proso millet	Cationic peroxidase extracted and purified from proso millet whole seeds administered to HCT116 and HT29 cell lines	The peroxidase induced receptor protein kinase 1 and 3 dependent necroptosis. It had the capacity to induce tumour necrosis factor-alpha production by means of transcriptional upregulation	Cui et al. 2018

unpolished proso millet flour improved the lipid profile significantly. Supplementation significantly improved plasma high-density lipoprotein, low-density lipoprotein, triglyceride, and cholesterol.

The lipid-lowering efficacy of pearl millet was assessed by Javed et al. (2012). To assess the hyperlipidaemic properties of pearl millet, its bran was given to hyperlipidaemic Albino rats at a maximum dose rate of 6 g/kg for a period of 15 days. This dose led to a 47% increase in high-density lipoprotein and a more than 50% reduction in low-density lipoprotein, cholesterol, and triglycerides. Similar observations were made by Narra et al. (2013) for kodo millet. Consumption up to a maximum of 600 mg/kg attenuated lipid parameters in rats with high-fat diet induced hyperlipidaemia. In these studies, the authors proposed that the lipid-lowering property is due to dietary fibre and phytochemicals.

In a recent advancement, Zhu et al. (2019) tried to modify foxtail millet dietary fibre in bran by xylanase-catalyzed hydrolysis to improve its cholesterol-binding capacity. The study optimised the conditions for hydrolysis (pH – 3.8, xylanase – 50 U/g, hydrolysis temperature and time – 50°C and 2 hour). Enzymatic treatment is an effective treatment to enhance the functional properties of dietary fibre by increasing its capacity to bind with cholesterol.

The hypolipidemic effect of a millet oil-based drink was assessed by Sun et al. (2016). A potential hypolipidemic effect in animal model was obtained for the compound drink with 1.5 mL/100 g millet oil.

Peroxidase derived from foxtail millet bran has a potential atherosclerotic effect. To assess this, Liu et al. (2020) extracted peroxidase from foxtail millet bran and studied its effect in an animal mice model. The administration of foxtail millet bran promoted the transformation of human aortic smooth muscle cells from synthetic to contractile. This led to the significant inhibition of its migration and suppression of phagocytosis in HASMCs (by 52%). In THP-1 cells, foxtail millet bran peroxidase reduced the expression of CD36 cells and STAT3. This led to the reduced secretion of inflammatory factor IL-1β and the suppression of lipid phagocytosis in THP-1 (49%).

9.5.3 Diabetes

Diabetes is characterised by high blood sugar levels. Millets have been found to be effective in controlling blood sugar levels due to components like dietary fibre and polyphenols. The anti-diabetic effect of millets has been evidenced in both pre-clinical and clinical trials. Singh et al. (2020) studied the effect of millet-based food product in blood glucose management through clinical trial. Patients with type 2 diabetes mellitus were administered food product containing 60% millets for 4 weeks and its effect on postprandial glucose levels was studied. The intervention diet significantly reduced both fasting and 2-hour postprandial blood glucose and lipid indicators. Okoyomoh et al. (2013) studied the impact of finger millet seed coat matter on blood glucose management. In the study, a dietary intervention was given to streptozotocin-induced diabetic rats. The study concluded that a diet containing 40% finger millet coat matter improved the body weight and glucose levels of the animals. Similar results were obtained by Shobana et al. (2010) on administration of 20% finger millet seed coat matter for 6 weeks.

Ren et al. (2018) established the anti-diabetic effect of foxtail millet through clinical trial. To estimate this effect, free-living subjects with impaired glucose tolerance were provided with 50 g/day of foxtail millet whole grain for a period of 12 weeks. The authors observed significant reductions in fasting (from 5.7 to 5.3 mmol/L) and 2-hour (from 10.2 to 2.3 mmol/L) blood sugar levels. An increase in serum leptin and a reduction in insulin resistance indicated anti-diabetic potential. The authors attributed these changes to low starch digestibility and moderate glycaemic index.

In a one of its kind study, Yadav et al. (2013) evaluated the hypoglycaemic properties of food products like *idli* and *upma* incorporating kodo millet. The study assessed the glycaemic index of the products in ten healthy individuals and concluded that the incorporation of kodo millet reduced the glucose response.

The anti-diabetic potential of foxtail millet was studied by Sireesha et al. (2011). The study showed that intervention with 300 mg of foxtail millet whole grain for 30 days reduced blood glucose by 70% and improved glycaemic control.

9.5.4 Cancer

Several studies have reported the anti-cancer properties of millets. Chandrasekara and Shahidi (2011) studied the inhibition potential of millet phenolic extracts on the proliferation of HT-29 colon cancer cells. Seventy percent acetone was used for extraction from pearl millet, kodo millet, little millet, foxtail millet, little millet, and finger millet. The extracts were used in a time- and dose-dependent manner on HT-29 colon cancer cells. The time-dependent anti-proliferative effect of different extracts was observed. Kodo millet was found to have the highest activity, while the lowest anti-proliferative activity was observed in pearl millet. The authors proposed that this effect was due to the phenolic acids and flavonoids.

The anti-cancerous properties of free and bound phenolics from foxtail millets were studied by Zhang and Liu (2015). This impact was studied on MDA human breast cancer cells (isolated at M D Anderson) and HepG2 human liver cancer cells. It was concluded in the study that both types of phenolics reduced cell proliferation in cancer cells. The phenolics identified in the extracts included chlorogenic, syringic, caffeic, p-coumaric, and ferulic acid. In another study conducted on foxtail millet bran, its efficacy for inhibiting the growth of human colorectal cancer cells (H-116) was studied (Shi et al. 2015). A classical method involving solvent extraction for free phenolics and alkaline hydrolysis for bound phenolics was used. In this study, it was observed that the bound phenolics significantly impacted the cancer cell proliferation in a dose- and time-dependent manner. Not only did it reduce the growth ability of HCT116 in survival colonies but it also ameliorated the levels of reactive oxygen species in HCT116 cells. The anti-carcinogenic potential of finger millet and pearl millet extract was evaluated against HepG2 liver cancer cell lines. A dose-dependent suppression of tumour cell proliferation was observed by Singh et al. (2015). Similar results were obtained for proso millet on the same cell lines (Zhang et al. 2014). The authors identified the higher content of bound ferulic acid and ρ-coumaric content as the potential agents for the anti-cancer effect.

In a recent development, Ramadoss and Sivalingam (2020) identified that vanillin extracted from proso millet and barnyard millet has anti-cancer properties. The study involved the extraction of active phenolic compounds from proso and barnyard millet and their characterization. The study revealed that the chemical structure of the compounds identified was similar to phenolic aldehyde-vanillin. The anti-proliferative potential was assessed for HT-29 cell lines for a concentration ranging from 250 to 1000 µg/ml. The cytotoxicity was observed in a dose-dependent manner by cell arrest in G0/G1 phase.

In an unprecedented study, Lu et al. (2018) observed the reversal effect of foxtail millet bran in colon cancer cells with multiple drug resistance. The study identified 12 bound phenolic compounds from the inner shell, ferulic acid and p-coumaric acid being the major ones. Bound phenolics from the inner shell of foxtail millet potentially enhanced the sensitivity of chemotherapeutic drugs by several mechanisms like the inhibition of cell proliferation, the promotion of cell apoptosis, and enhancing the accumulation of rhodamine-123 (Rh-123). The fraction of compounds with molecular weight of less than 200 were able to reverse the drug resistance in HCT-8/Fu cells. Polymerase chain reaction results indicated that the polyphenols also reduced the expression of different proteins like breast cancer protein, glycoprotein, and multidrug resistance protein.

Other bioactives in millets (apart from polyphenols) like lipids and protein fractions have also been shown to possess anti-cancerous properties. A novel protein (35 kDa) extracted from foxtail millet bran by Shan et al. (2014) exhibited anti-carcinogenic properties in colon cancer. *In-vivo* anti-tumour results indicated that this protein has the potential to induce G1 phase arrest leading to the loss of transmembrane potential in mitochondria. This results in capase-dependent apoptosis in colon cancer cells, thereby leading to its suppression. The novelty of this protein is in the fact that it is minimally toxic to normal epithelial cells. In continuation, Shan et al. (2014a) identified that a peroxidase derived from foxtail millet bran suppresses human colon cancer by inhibiting cell migration by antagonizing STAT3-mediated epithelial-mesenchymal transition. Another mechanism for the above effect was identified to be the elevation in free radicals (reactive oxygen species) in colon cancer cells on administration with foxtail millet bran-derived peroxidase (Shan et al. 2015). Cui et al. (2018) purified cationic peroxidase from proso millet seeds. The study showed that proso millet-derived cationic peroxidase was toxic to cancer cells and normal cells with more sensitivity in HT29 and HCT116 cells. These peroxidases induced necroptosis leading to cancer suppression.

Finger millet is also beneficial against leukaemia cells. Sen and Dutta (2012) extracted finger millet bifunctional inhibitor and assessed its role in the inhibition of K562 myeloid leukaemia cells. Finger millet bifunctional inhibitor is the compound extracted from *ragi* and belongs to the category of cereal α-amylase or protease inhibitors and inhibits amylase and trypsin concurrently. Results revealed that the bifunctional inhibitor presented cytogenic toxicity against K562 chronic myeloid leukaemia cells. However, this toxic effect was not observed for the healthy cells.

In-vitro and *in-vivo* studies have explicitly identified the potency of foxtail millet bran peroxidase in the suppression of colorectal cancer cells. Zhang et al. (2019) designed and cloned three sequences of foxtail millet bran peroxidase into plasmid vector and expressed in *E. coli*. Results showed that the bound peroxidase was effective in reversing 5-Fu resistance in cancer cells.

9.5.5 Infections

The anti-bacterial potential of pearl millet and maize was compared by Ndiku and Ngule (2015). The study concluded that pearl millet had 50% more types of phytochemicals compared to maize. Pearl millet was found to possess much higher anti-bacterial activity. It has the potential to inhibit *Serratia marcescens*, *Proteus vulgaris*, and *Staphylococcus epidermidis* but not *Bacillus cereus*, *Serretia liquificans*, or *E. coli*. Cysteine protease inhibitor derived from pearl millet demonstrated anti-fungal activity (Joshi et al. 1998).

The phenolics extracted from finger millet seed coat, namely benzoic acid, cinnamic acid derivatives, and flavonoid-quecertin, inhibited the proliferation of *E. coli*, *Bacillus cereus*, *Yersinia enterocolitica*, *Serretia marcescens*, and *Listeria monocytogenes*. Quercetin was especially effective against these pathogens (Banerjee et al. 2012).

The microbial inhibitory effects of peptides derived from foxtail millet were studied by Amadou et al. (2013). In this study, the peptides were obtained from foxtail millet meal by fermentation using *Lactobacillus paracasei* Fn032. These derived tyrosine/leucine-rich peptides exhibited fair inhibition of *E. coli* ATCC 8099 growth. Venkataramana et al. (2015) synthesised nanoparticles (30 nm) from foxtail millet husk. The nanoparticles were found to be effective against both Gram-positive and Gram-negative bacteria of *Staphylococcus* and *E. coli* spp., respectively.

9.6 CONCLUSIONS

Lifestyle and dietary patterns coupled with environment and other uncontrolled factors have made people more vulnerable to diseases and disorders. The majority of these diseases are preventable to a great extent by small dietary changes. The inclusion of optimal dietary components in the form of not only nutrient-rich but also phytochemical-rich foods can help in the protection and management of these diseases. Millets may be the preferable choice in such cases. Millets are rich in nutrients, but what makes them special is the presence of biologically active compounds which not only help in body function but also have therapeutic potential. The studies given in different subsections in this chapter provide an insight into the therapeutic potential of millets.

REFERENCES

Amadou, I., G. W. Le, T. Amza, J. Sun and Y. H. Shi. 2013. Purification and characterization of foxtail millet-derived peptides with antioxidant and antimicrobial activities. *Food Research International* 51(1): 422–428.

Amoako, D. B. and J. M. Awika. 2016. Polymeric tannins significantly alter properties and in vitro digestibility of partially gelatinized intact starch granule. *Food Chemistry* 208: 10.e17.

Awika, J. M. 2017. Sorghum: its unique nutritional and health promoting attributes. In: J. R. N. Tayor, and J. M. Awika (Eds.). *Gluten-Free Ancient Grains - Cereals, Pseudocereals, and Legumes: Sustainable, Nutritious, and Health-Promoting Foods for the 21st Century*. Woodhead Publishing-Elsevier, Duxford, pp. 21.e54.

Balakrishnan, R., S. R. R. Tadi, S. K. Rajaram, N. Mohan and S. Sivaprakasam. 2020. Batch and fed-batch fermentation of optically pure D (-) lactic acid from Kodo millet (*Paspalum scrobiculatum*) bran residue hydrolysate: Growth and inhibition kinetic modeling. *Preparative Biochemistry & Biotechnology* 50(4): 365–378.

Banerjee, S., K. R. Sanjay, S. Chethan and N. G. Malleshi. 2012. Finger millet (*Eleusine coracana*) polyphenols: Investigation of their antioxidant capacity and antimicrobial activity. *African Journal of Food Science* 6(13): 362–374.

Bartwal, A., A. Pande, P. Sharma and S. Arora. 2016. Intervarietal variations in various oxidative stress markers and antioxidant potential of finger millet (*Eleusine coracana*) subjected to drought stress. *Journal of Environmental Biology* 37(4): 517.

Berwal, M. K., L. K. Chugh, P. Goyal and R. Kumar. 2016. Total antioxidant potential of pearl millet genotypes: Inbreds and designated b-lines. *Indian Journal of Agricultural Biochemistry* 29(2): 201–204.

Bhandari, S. R. and Y. S. Lee. 2013. The contents of phytosterols, squalene, and vitamin E and the composition of fatty acids of Korean landrace *Setaria italica* and Sorghum bicolor seeds. *Korean Journal of Plant Research* 26: 663.e672.

Bijalwan, V., U. Ali, A. K. Kesarwani, K. Yadav and K. Mazumder. 2016. Hydroxycinnamic acid bound arabinoxylans from millet brans-structural features and antioxidant activity. *International Journal of Biological Macromolecules* 88: 296–305.

Bora, P., P. Das, P. Mohan and A. Barthakur. 2018. Evaluation of hypolipidemic property of proso millet (*Panicum miliaceum* L.) in high fat diet induced hyperlipidemia in rats. *Journal of Entomological and Zoological Studies* 6(3): 691–695.

Chandrasekara, A. and F. Shahidi. 2011a. Determination of antioxidant activity in free and hydrolyzed fractions of millet grains and characterization of their phenolic profiles by HPLC-DAD-ESI-MSn. *Journal of Functional Foods* 3(3): 144.e158.

Chandrasekara, A. and F. Shahidi. 2012. Bioaccessibility and antioxidant potential of millet grain phenolics as affected by simulated in vitro digestion and microbial fermentation. *Journal of Functional Foods* 4(1): 226.e237.

Chen, J., W. Duan, X. Ren, C. Wang, Z. Pan, X. Diao and Q. Shen. 2017. Effect of foxtail millet protein hydrolysates on lowering blood pressure in spontaneously hypertensive rats. *European Journal of Nutrition* 56(6): 2129–2138.

Cui, X., R. Wang and Z. Wang. 2018. Cationic peroxidase from proso millet induces human colon cancer cell necroptosis by regulating autocrine TNF-α and RIPK3 demethylation. *Food & Function* 9(3): 1878–1888.

Dayakar, R. B., K. Bhaskarachary, G. D. Arlene Christina, G. Sudha Devi and A. Tonapi. 2017. *Nutritional and Health Benefits of Millets*. ICAR-Indian Institute of Millets Research (IIMR), Rajendranagar, Hyderabad, p. 112.

De Morais Cardoso, L., S. S. Pinheiro, H. S. D. Martino and H. M. Pinheiro-Sant'Ana. 2017. Sorghum (Sorghum bicolor L.): Nutrients, bioactive compounds, and potential impact on human health. *Critical Reviews in Food Science and Nutrition* 57(2): 372.e390.

Demonty, I., R. T. Ras, H. C. M. van der Knaap, G. S. M. J. E. Duchateau, L. Meijer, P. L. Zock, J. M. Geleijnse and E. A. Trautwein. 2009. Continuous dose-response relationship of the LDL-cholesterolelowering effect of phytosterol intake. *Journal of Nutrition* 139(2): 271.e284.

Farooq, U., M. Mohsin, X. Liu and H. Zhang. 2013. Enhancement of short chain fatty acid production from millet fibres by pure cultures of probiotic fermentation. *Tropical Journal of Pharmaceutical Research* 12(2): 189–194.

Farrell, T. L., M. Gomez-Juaristi, L. Poquet, K. Redeuil, K. Nagy, M. Renouf and G. Williamson. 2012. Absorption of dimethoxycinnamic acid derivatives in vitro and pharmacokinetic profile in human plasma following coffee consumption. *Molecular Nutrition and Food Research* 56(9): 1413e1423.

Hegde, P. S. and T. S. Chandra. 2005. ESR spectroscopic study reveals higher free radical quenching potential in kodo millet (*Paspalum scrobiculatum*) compared to other millets. *Food Chemistry* 92(1): 177–182.

Himanshu, K., M. Chauhan, S. K. Sonawane and S. S. Arya. 2018. Nutritional and nutraceutical properties of millets: A review. *Clinical Journal of Nutrition and Dietetics* 1(1): 1–10.

Hosoda, A., Y. Okai, E. Kasahara, M. Inoue, M. Shimizu, Y. Usui, A. Sekiyama and K. Higashi-Okai. 2012. Potent immunomodulating effects of bran extracts of traditional Japanese millets on nitric oxide and cytokine production of macrophages (RAW264. 7) induced by lipopolysaccharide. *Journal of UOEH* 34(4): 285–296.

Hou, D., J. Chen, X. Ren, C. Wang, X. Diao, X. Hu, Q. Shen and Q. Shen. 2018. A whole foxtail millet diet reduces blood pressure in subjects with mild hypertension. *Journal of Cereal Science* 84: 13–19.

Javed, I., B. Aslam, M. Z. Khan, F. Muhammad and M. K. Saleemi. 2012. Lipid lowering efficacy of Pennisetum glaucum Bran in hyperlipidemic albino rats. *Pakistan Veterinary Journal* 32(2).

Joshi, B. N., M. N. Sainani, K. B. Bastawade, V. S. Gupta and P. K. Ranjekar. 1998. Cysteine protease inhibitor from pearl millet: A new class of antifungal protein. *Biochemical and Biophysical Research Communications* 246(2): 382–387.

Jun, D. Y., J. Y. Lee, C. R. Han, K. P. Kim, M. C. Seo, M. H. Nam and Y. H. Kim. 2014. Pro-apoptotic and anti-adipogenic effects of proso millet (*Panicum miliaceum*) grains on 3T3-l1 preadipocytes. *Journal of Life Science* 24(5): 505–514.

Karas, M., A. Jakubczyk, U. Szymanowska, U. Złotek and E. Zielinska. 2017. Digestion and bioavailability of bioactive phytochemicals. *International Journal of Food Science and Technology* 52: 291.e305.

Kumar, A., A. Kaur, V. Tomer, K. Gupta and K. Kaur. 2020. Effect of rose syrup and marigold powder on the physicochemical, phytochemical, sensorial and storage properties of nutricereals and milk-based functional beverage. *Journal of the American College of Nutrition* 24: 1–8.

Kumar, A., A. Kaur, V. Tomer, P. Rasane and K. Gupta. 2020a. Development of nutricereals and milk-based beverage: Process optimization and validation of improved nutritional properties. *Journal of Food Process Engineering* 43(1): e13025.

Kumar, A., V. Tomer, A. Kaur, V. Kumar and K. Gupta. 2018. Millets: A solution to agrarian and nutritional challenges. *Agriculture and Food Security* 7(1): 31.

Leder, I. 2004. Sorghum and millets. Cultivated plants, primarily as food sources. In: F. Gyargy (Ed.). *Encyclopedia of Life Support Systems, UNESCO*. Eolss Publishers, Oxford.

Lee, J. Y., D. Y. Jun, Y. H. Yoon, J. Y. Ko, K. S. Woo, M. H. Woo and Y. H. Kim. 2014. Anti-inflammatory effect of flavonoids kaempferol and biochanin A-enriched extract of barnyard millet (*Echinochloa crus-galli* var. frumentacea) grains in LPS-stimulated RAW264. 7 cells. *Journal of Life Science* 24(11): 1157–1167.

Li, S., W. Yu, X. Guan, K. Huang, J. Liu, D. Liu and R. Duan. 2019. Effects of millet whole grain supplementation on the lipid profile and gut bacteria in rats fed with high-fat diet. *Journal of Functional Foods* 59: 49–59.

Liu, F., S. Shan, H. Li and Z. Li. 2020. Treatment of peroxidase derived from foxtail millet bran attenuates atherosclerosis by inhibition of CD36 and STAT3 in vitro and in vivo. *Journal of Agricultural and Food Chemistry* 68(5): 1276–1285.

Lobo, V., A. Patil, A. Phatak and N. Chandra. 2010. Free radicals, antioxidants and functional foods: Impact on human health. *Pharmacognosy Reviews* 4(8): 118.

Lu, Y., S. Shan, H. Li, J. Shi, X. Zhang and Z. Li. 2018. Reversal effects of bound polyphenol from foxtail millet bran on multidrug resistance in human HCT-8/Fu colorectal cancer cell. *Journal of Agricultural and Food Chemistry* 66(20): 5190–5199.

Mohamed, T. K., A. Issoufou and H. Zhou. 2012. Antioxidant activity of fractionated foxtail millet protein hydrolysate. *International Food Research Journal* 19(1): 207.

Mridula, D. and M. Sharma. 2015. Development of non-dairy probiotic drink utilizing sprouted cereals, legume and soymilk. *LWT – Food Science and Technology* 62(1): 482–497.

Murtaza, N., R. K. Baboota, S. Jagtap, D. P. Singh, P. Khare, S. M. Sarma, R. K. Boparai, S. Alagesan, T. S. Chandra, K. K. Bhutani, R. K. Boparai, M. Bishnoi and K. K. Kondepudi. 2014. Finger millet bran supplementation alleviates obesity-induced oxidative stress, inflammation and gut microbial derangements in high-fat diet-fed mice. *British Journal of Nutrition* 112(9): 1447–1458.

Narra, S., B. Ramadurg and C. D. Saraswathi. 2013. Antihyperlipidemic activity of Paspalum scrobiculatum L. grains extract in albino rats. *Research Journal of Pharmacology and Pharmacodynamics* 5(6): 362–370.

Ndiku, H. M. and M. C. Ngule. 2015. Comparative study on the antibacterial and chemical constituents of Pennisetum glaucum (pearl millet) and Zea mays (maize). *International Journal of Nutrition and Metabolism* 7(4): 46–51.

Nirmala, M., M. V. S. S. T. Subbarao and G. Muralikrishna. 2000. Carbohydrates and their degrading enzymes from native and malted finger millet (Ragi, *Eleusine corcana*, Indaf-15). *Food Chemistry* 69(2): 175–180.

Okoyomoh, K., O. S. Okere, O. D. Olowoniyi and G. O. Adejo. 2013. Antioxidant and antidiabetic properties of *Eleucine coracana* (L.) Geartn. (Finger Millet) seed coat matter in streptozotocin induced diabetic rats. *ASJ International Journal of Advances in Herbal Alternative Medicine* 1(1): 1–9.

Park, M.Y., D.W. Seo, J.Y. Lee, M.K. Sung, Y.M. Lee, H.H. Jang, H.Y. Choi, J.H. Kim and D.S. Park. 2011. Effects of *Panicum miliaceum* L. extract on adipogenic transcription factors and fatty acid accumulation in 3T3-L1 adipocytes. *Nutrition Research and Practice* 5: 192–197.

Park, M. Y., H. H. Jang, J. B. Kim, H. N. Yoon, J. Y. Lee, Y. M. Lee, J. H. Kim and D. S. Park. 2011a. Hog millet (*Panicum miliaceum* L.)-supplemented diet ameliorates hyperlipidemia and hepatic lipid accumulation in C57BL/6J-ob/ob mice. *Nutrition Research and Practice* 5(6): 511–519.

Park, M. Y., H. H. Jang, J. Y. Lee, Y. M. Lee, J. H. Kim, J. H. Park and D. S. Park. 2012. Effect of hog millet supplementation on hepatic steatosis and insulin resistance in mice fed a high-fat diet. *Journal of the Korean Society of Food Science and Nutrition* 41(4): 501–509.

Park, M. Y., J. H. Kim and D. S. Park. 2011b. Anti-inflammatory activities of hog millet (Panicum miliaceum L.) in murine macrophages through IRAK-4 signaling. *The Korean Journal of Food and Nutrition* 24(2): 268–272.

Ramadoss, D. P. and N. Sivalingam. 2020. Vanillin extracted from Proso and Barnyard millets induce apoptotic cell death in HT-29 human colon cancer cell line. *Nutrition and Cancer* 72(8): 1422–1437.

Ravindran, G. 1992. Seed protein of millets: Amino acid composition, proteinase inhibitors and in-vitro protein digestibility. *Food Chemistry* 44(1): 13–17.

Ren, X., R. Yin, D. Hou, Y. Xue, M. Zhang, X. Diao, Y. Zhang, J. Wu, J. Hu, X. Hu and Q. Shen. 2018. The glucose-lowering effect of foxtail millet in subjects with impaired glucose tolerance: A self-controlled clinical trial. *Nutrients* 10(10): 1509.

Ryan, E., K. Galvin, T. P. O'Connor, A. R. Maguire and N. M. O'Brien. 2007. Phytosterol, squalene, tocopherol content and fatty acid profile of selected seeds, grains, and legumes. *Plant Foods for Human Nutrition* 62(3): 85.e91.

Salar, R. K. and S. S. Purewal. 2016. Improvement of DNA damage protection and antioxidant activity of biotransformed pearl millet (*Pennisetum glaucum*) cultivar PUSA-415 using *Aspergillus oryzae* MTCC 3107. *Biocatalysis and Agricultural Biotechnology* 8: 221–227.

Sarma, S. M., D. P. Singh, P. Singh, P. Khare, P. Mangal, S. Singh, V. Bijalwan, J. Kaur, S. Mantri, R. K. Boparai and K. Mazumder. 2018. Finger millet arabinoxylan protects mice from high-fat diet induced lipid derangements, inflammation, endotoxemia and gut bacterial dysbiosis. *International Journal of Biological Macromolecules* 106: 994–1003.

Sarma, S. M., P. Khare, S. Jagtap, D. P. Singh, R. K. Baboota, K. Podili, R. K. Boparai, J. Kaur, K. K. Bhutani, M. Bishnoi and K. K. Kondepudi. 2017. Kodo millet whole grain and bran supplementation prevents high-fat diet induced derangements in a lipid profile, inflammatory status and gut bacteria in mice. *Food and Function* 8(3): 1174–1183.

Sen, S. and S. K. Dutta. 2012. Evaluation of anti-cancer potential of ragi bifunctional inhibitor (RBI) from *Elusine coracana* on human chronic myeloid leukemia cells. *European Journal of Plant Science and Biotechnology* 6: 103–108.

Shan, S., J. Shi, Z. Li, H. Gao, T. Shi, Z. Li and Z. Li. 2015. Targeted anti-colon cancer activities of a millet bran-derived peroxidase were mediated by elevated ROS generation. *Food and Function* 6(7): 2331–2338.

Shan, S., Z. Li, I. P. Newton, C. Zhao, Z. Li and M. Guo. 2014b. A novel protein extracted from foxtail millet bran displays anti-carcinogenic effects in human colon cancer cells. *Toxicology Letters* 227(2): 129–138.

Shan, S., Z. Li, S. Guo, Z. Li, T. Shi and J. Shi. 2014a. A millet bran-derived peroxidase inhibits cell migration by antagonizing STAT3-mediated epithelial-mesenchymal transition in human colon cancer. *Journal of Functional Foods* 10: 444–455.

Shan, S., Z. Li, S. Guo, Z. Li, T. Shi and J. Shi. 2014a. A millet bran-derived peroxidase inhibits cell migration by antagonizing STAT3-mediated epithelial-mesenchymal transition in human colon cancer. *Journal of Functional Foods* 10: 444–455.

Shi, J., S. Shan, H. Li, G. Song and Z. Li. 2017. Anti-inflammatory effects of millet bran derived-bound polyphenols in LPS-induced HT-29 cell via ROS/miR-149/Akt/NF-κB signaling pathway. *Oncotarget* 8(43): 74582.

Shi, J., S. Shan, Z. Li, H. Li, X. Li and Z. Li. 2015. Bound polyphenol from foxtail millet bran induces apoptosis in HCT-116 cell through ROS generation. *Journal of Functional Foods* 17: 958–968.

Shobana, S., M. R. Harsha, K. Platel, K. Srinivasan and N. G. Malleshi. 2010. Amelioration of hyperglycaemia and its associated complications by finger millet (*Eleusine oracana* L.) seed coat matter in streptozotocin-induced diabetic rats. *British Journal of Nutrition* 104(12): 1787–1795.

Singh, P. and R. S. Raghuvanshi. 2012. Finger millet for food and nutritional security. *African Journal of Food Science* 6(4): 77–84.

Singh, R. B., J. Fedacko, V. Mojto, A. Isaza, M. Dewi, S. Watanabe, A. Chauhan, G. Fatima, K. Kartikey and A. Sulaeman. 2020. Effects of millet based functional foods rich diet on coronary risk factors among subjects with diabetes mellitus: A single arm real world observation from hospital registry. *MOJ Public Health* 9(1): 18–25.

Singh, R. B., R. R. Watson and T. Takahashi (Eds). 2018. *The Role of Functional Food Security in Global Health*. Academic Press.

Singh, N., G. Meenu, A. Sekhar and A. Jayanthi. 2015. Evaluation of antimicrobial and anticancer properties of finger millet (*Eleusine coracana*) and pearl millet (*Pennisetum glaucum*) extracts. *The Pharmaceutical Innovation, 3* 11(Part B): 82.

Sireesha, Y., R. B. Kasetti, S. A. Nabi, S. Swapna and C. Apparao. 2011. Antihyperglycemic and hypolipidemic activities of *Setaria italica* seeds in STZ diabetic rats. *Pathophysiology* 18(2): 159–164.

Sun, Q. S., C. Q. Wang, S. W. Zhang, Y. Y. Bai, T. Chen, Q. P. Ning and J. Y. Wang. 2016. Hypolipidemic effect experimental research of millet oil compound drink. *Science and Technology of Food Industry* 17: 65.

Svensson, L., B. Sekwati-Monang, D. L. Lutz, A. Schieber and M. G. Ga¨nzle. 2010. Phenolic acids and flavonoids in nonfermented and fermented red sorghum (Sorghum bicolor (L.) Moench). *Journal of Agriculture and Food Chemistry* 58(16): 9214e9220.

WHO. 2002. *The World Health Report. 2002: Reducing Risks, Promoting Healthy Life*. World Health Organization, Geneva.

Thilakavathy, S. and S. Muthuselvi. 2010. Development and evaluation of millets incorporated chappathi on glycemic response in type II diabetics. *Indian Journal of Nutrition and Dietetics* 47(2): 42–50.

Venkataramana, B., S. S. Sankar, A. S. Kumar and B. V. K. Naidu. 2015. Synthesis of silver nanoparticles using *Setaria italica* (foxtail millets) husk and its antimicrobial activity. *Research Journal of Nanoscience and Nanotechnology* 5(1): 6–15.

Wei, S., D. Cheng, H. Yu, X. Wang, S. Song and C. Wang. 2018. Millet-enriched diets attenuate high salt-induced hypertension and myocardial damage in male rats. *Journal of Functional Foods* 44: 304–312.

Woo, M. H., D. H. Nguyen, B. T. Zhao, U. M. Seo, T. T. Nguyen, D. Y. Jun and Y. H. Kim. 2015. Echinochlorins A˜ C from the grains of *Echinochloa utilis* (Barnyard Millet) and their anti-inflammatory activity. *Planta Medica* 81(11): PX66.

Xiang, J., F. B. Apea-Bah, V. U. Ndolo, M. C. Katundu and T. Beta. 2019. Profile of phenolic compounds and antioxidant activity of finger millet varieties. *Food Chemistry* 275: 361–368.

Yadav, N., K. Chaudhary, A. Singh and A. Gupta. 2013. Evaluation of hypoglycemic properties of kodo millet based food products in healthy subjects. *IOSR Journal of Pharmacy*, 3(2):14–20.

Zhang, L., R. Liu and W. Niu. 2014. Phytochemical and antiproliferative activity of proso millet. *PLoS One* 9(8): e104058.

Zhang, L. Z. and R. H. Liu. 2015. Phenolic and carotenoid profiles and antiproliferative activity of foxtail millet. *Food Chemistry* 174: 495.e501.

Zhang, X., S. Shan, H. Li, J. Shi, Y. Lu and Z. Li. 2019. Cloning, expression of the truncation of recombinant peroxidase derived from millet bran and its reversal effects on 5-Fu resistance in colorectal cancer. *International Journal of Biological Macromolecules* 132: 871–879.

Zhu, Y., C. He, H. Fan, Z. Lu, F. Lu and H. Zhao. 2019. Modification of foxtail millet (*Setaria italica*) bran dietary fiber by xylanase-catalyzed hydrolysis improves its cholesterol-binding capacity. *LWT* 101: 463–468.

Chapter 10

Extraction and Application of Millet Starch

10.1 INTRODUCTION

Starch, the primary source of energy for green plants, abundantly exists in seeds (such as cereal grains and pulses), tubers (such as potatoes), roots (such as cassava and sweet potatoes), fruits (such as bananas and squash), stems (such as sago), and leaves (such as tobacco). Millets similar to major cereals are a rich source of starch, i.e. around 60–70% (Hassan et al. 2021; Satyavathi et al. 2021). Starch is composed of amylose and amylopectin; amylose is a linear chain structure while amylopectin is a branched chain structure. Starch plays a crucial role in human diet and nutrition, and a scientific evaluation conducted by the World Health Organization (WHO) underscores its significance by highlighting that grains abundant in starch contribute 50% of the total energy (Kanehara et al. 2022; Oladele and Mbaye 2022). Millet starch can be divided into non-waxy (high amylose) and waxy (low amylose) starch (Bangar et al. 2021; Chang et al. 2023). The industrial application of millet flour is highly dependent upon the physicochemical, rheological, morphological, functional, and structural characteristics of starch which decide its diverse industrial application. Millets are gluten-free cereals and so is the starch produced from them. This increases the utilization of millet-based starch in gluten-free food products as it reduces the chances of any gluten contamination. Millet starch is also in demand for the development of low-glycaemic foods (Selladurai et al. 2022; Bangar et al. 2021; Saini et al. 2021) due to their slow digestion (especially finger millet starch) compared to other cereals (Verma et al. 2021). Furthermore, different botanically derived starches exhibit distinctive gelatinization properties that reflect different starch molecule structures and the arrangement of double-helical crystalline structures inside starch granules (Ren et al. 2021). Starch is also used in the creation of industrial adhesives, papermaking, food processing, and numerous cosmetic and oral medicinal components. It has been used as a donning/wetting agent in surgical gloves (Aggarwal et al. 2020; Gupta et al. 2019). Therefore, in this chapter, we have discussed various isolation techniques of millet starches, their functional characteristics, and the structural morphology along with suitable applications.

10.2 ISOLATION METHODS AND COMPOSITION OF MILLET STARCH

Starch is the major constituent of millets and its content varies from 51 to 80% in various millets (Kumar et al. 2022; Mahajan et al. 2021). Typically millet starches have an amylopectin content of 70–80% and an amylose content of 20–30% (Dimri and Singh 2022; Mirzababaee et al. 2022; Punia et al. 2021). In millets, the starch is closely linked to the protein matrix, and thus for extraction, the protein part is solubilized, and starch is extracted from the grains using a wide range of chemical agents and methods. The wet milling processes are typically used to extract millet starches (Kaur et al. 2023; Mahajan et al. 2021). Depending on the source of starch, steeping condition, and isolation technique, neutral, alkaline, or acidic methods can be used for starch extraction (Rhowell et al. 2021; Malik et al. 2022). The extraction process begins with the soaking of whole millet kernels (or flour) in an aqueous solution for a duration long enough to help the starch separate from other ingredients. The range of solutions for starch extraction depends on the millet's chemical components and characteristics. Generally, in the neutral method, water or 0.01% sodium azide solution is used, in the alkaline method 0.3% sodium hydroxide or sodium meta-bisulphite solution is used, and in the acidic method, 0.5% of lactic acid is used. Enzymes used for the degradation of starch are dextrinases, pullulanases, glucosidases, amylases, and glucosidases. Millet grains are washed, soaked (different solutions), and homogenized before being isolated. Afterwards, the slurry undergoes sifting and multiple rinses with water to eliminate the protein and fatty constituents. Centrifugation is then employed to recover the separated starch before the drying process (Aruna and Parimalavalli 2022; Kaur et al. 2023). In a nutshell, morphological fracture, cell rupturing, and starch purification are typically the three successive steps in the starch extraction process (Bangar et al. 2021; Li and Wei. 2020). The yield of millet starches generally ranges from 52 to 68.2% (Verma et al. 2021). The starch content of various millets is also provided in Table 10.1.

The methods of extraction have a significant impact on starch yield. It is crucial to choose a method that produces a high starch content with few protein residues. In this context, the alkaline steeping method (0.7%) results in lower protein residual (4.3%) in starch granules compared to the acidic steeping method. The extraction process of grain starch is hindered by the presence of small granules, leading to a longer settling time. Moreover, the interactions between proteins and starch in grains are more robust compared to those in legumes and tubers. The low water content of cereal starches makes their extraction more difficult. In the case of millet starches, a significant portion of the lipid composition consists of nonpolar lipids, particularly triglycerides, which make up approximately 89% of the total lipid content (Serna-Saldivar and Espinosa- Ramirez 2019; Khatun 2019). These lipids interact with the amylose portion of starch

TABLE 10.1   STARCH CONTENT OF DIFFERENT MILLETS

Type of millet	Starch yield (% dry basis)
Little millet	42–56
Finger millet	55–65
Barnyard millet	48–60
Kodo millet	47–60
Pearl millet	53–68
Proso millet	48–60

to create complexes and because of their hydrophobic connections and cohesive makeup, they can also decrease the flowability and swelling capacity of starches. Hydrophilic and hydrophobic links present in proteins influence their capacity to bind to both water and oil (Nawaz et al. 2019; Santschi et al. 2021). Thus, the chemical constitutions of starch depend on extraction from its plant sources, particularly concerning the minor components of protein, lipid, and ash. Depending on millet type, variety, and the extraction technique, the amylose and amylopectin content of the extracted starch can be different and so can be its properties. The techno-functional properties of starch are highly dependent upon its amylose content and its content varies in various millets. For example, the amylose content of finger millet is 14–39%, proso millet is 0–34%, and pearl millet is 4–38% (Bean et al. 2019). The amylose content also varies with varieties. The starch extracted from various proso millet varieties yields different starch, amylose, fat, and protein content (Li et al. 2021). Finger millet and pearl starches have long chains of amylose compared to foxtail and proso millet starches. The diverse industrial application of starch depends upon the rheological behaviour of gel, water syneresis, swelling power, solubility, refrigeration, and gelatinization. The content of amylose and amylopectin as well as their molecular structure in the starch granule determine these physicochemical characteristics. Low-amylose starches are characterized by a more viscous nature, high clarity, breakdown value, and negligible retrogradation properties (Liu et al. 2019; Tarahi et al. 2022). While, high-amylose starches exhibit the opposite tendency. Lipid-starch aggregates in food matrices play a role in impeding the hydration of amylopectin chains and retrogradation. As a result, they contribute to decreased solubility and swelling power of starches, as well as a slowdown in their gelatinization process. Additionally, these aggregates also reduce the rate of enzymatic hydrolysis of starches. The high-purity millet starches should have low values for non-starch components. Despite being a minor component, lipids are crucial in the functionality of starches. High-amylose starches release more energy (exothermic) and may form more permanent lipid-amylose aggregates. The functional and physicochemical characteristics of starch also affect cooking time and taste.

10.3 MORPHOLOGICAL CHARACTERISTICS OF MILLET STARCH

The morphological characteristics of millet starch can be effectively studied using polarized light microscopy (PLM), normal light microscopy (NLM), scanning electron microscopy (SEM), and transmission electron microscopy (TEM). Among these techniques, SEM is generally used to study the size and shape of isolated starch granules. The size of millet starch granules varies depending on the species. However, millet starch granules are primarily spherical and polygonal in shape. The four Canadian millets (foxtail, pearl, finger, and proso millets) were discovered to have starch granules that ranged in size from 2.5 to 24 μm (Shi et al. 2023; Bora et al. 2019). Most of them had polygonal shapes, with a few having spherical granule shapes. The size and shape of various millets is also provided in Table 10.2.

Yet, the type of modification or treatment that the starch undergoes has a significant impact on its shape. Environmental factors also affect the morphology of starch; for example, species grown at high altitudes, which have lower average temperatures, have larger granule sizes. Random indentations can frequently appear on the granule surface; these indentations might be caused by the existence of protein structures and surface pores (Song et al. 2020; Ashogbon 2021). These pores allow some molecules to enter the granule matrix by connecting the central cavity of the granules to the external environment. This phenomenon aids in the alteration of

TABLE 10.2 SIZE AND SHAPE OF DIFFERENT MILLET STARCHES

Type of millet	Size of granules (μm)	Findings	References
Chinese foxtail millet (*Setaria italica* Beauv.)	3–10	Study revealed that granules of millet starch were oval, spherical, and polygonal	Li et al. 2019
Proso, little, kodo, and foxtail millet	0.8–10	Most starch granules were polygonal and some were spherical	Kumari and Thayumanavan 1998
Pearl, foxtail, proso, and finger millet	0.5–24	All investigated millet starches were spherical and polygonal in shape	Annor et al. 2014
Proso, finger, and barnyard millet	7.6	The average starch granule size of all the millets was 7.6 μm	Bangar et al. 2021
Foxtail (waxy and non-waxy) millets	2.5–12	Both waxy and non-waxy starches were round and polygonal in shape	Yang et al. 2019

starch. With the access of water and other hydroxyl ions, the amorphous area containing amylose is disrupted, reducing amylose's restrictive characteristic and improving the hydration and swelling properties of starches (Xie et al. 2023). It is thought that the external portion of amylopectin unit chains mostly crystallizes as a double helix that is organized in two different ways to produce the A- and B-type polymorphs of granular starch. Wide-angle X-ray diffraction investigation frequently reveals the polymorph type. A- and B-types are combined to form the C-type. All of the millet samples examined so far displayed the A-type polymorph, similar to the majority of other cereal starches (Krishnan et al. 2021; Zhu 2020; Solaesa et al. 2019).

The physical and chemical properties of starch, such as pasting, solubility, crystallinity, and enzyme susceptibility, are influenced by the granule size. Smaller starch granules exhibit higher water-holding capacity and better solubility (Wang et al. 2023; Punia et al. 2021). The average chain length (ACL) of finger millet amylose is 180–240 glucosyl residues, with approximately 6 chains per molecule. However, for the pearl millet amylose, a chain length of 260–270 glucosyl residues with 4 chains per molecule and an average degree of polymerization (DPn) of approximately 1060–1250 are reported. Several millet starches' amylopectin content has also been thoroughly investigated. Pearl millet starch amylopectin is composed of chains of 8–21 glucosyl units. Pearl millet amylopectin's internal chain length and exterior chain length were discovered to be 5.0–6.2 and 12.0–13.8, respectively. The amylose content, ACL, and DPn of millet starches influence their different properties (Yashini et al. 2022; Iuga and Mironeasa 2020). In comparison to the non-waxy proso and foxtail millet, the waxy proso millet contains higher crystallinity and heat enthalpy because of its high amylopectin content. It suggests that waxy starches are more stable than non-waxy starches. With an increase in branch ACL, the temperature at which starches gelatinize increases. The number of short branch chains and the amount of amylopectin present, however, reduce the time required for starches to gelatinize. The functionality and potential usage of amylose and amylopectin units as ingredients in particular food product development ultimately depend on the length of their chains (Ren et. 2021; Bangar et al. 2022). For instance, waxy millet starches can be used as thickeners and in frozen food products due to their slow setback viscosity (SBV) and good stability, but non-waxy millet starches can be used in medicines due to their low breakdown viscosity (BDV) and anti-shear properties.

10.4 CRYSTALLINE BEHAVIOUR OF MILLET STARCH

Granules of starch are naturally semicrystalline, with both amorphous and crystalline regions. Various starches show multiple X-ray diffraction (XRD) spectra depending on the structure of the starch. Moreover, the characteristics and the presence of the crystalline structure of the starch granules were studied using XRD (Castillo et al. 2019; Liu et al. 2020). Sharp peaks were observed in the crystalline portions of starch, whereas dispersive peaks were visible in the amorphous portions. In this context, A, B, and C types of starch are classified depending on these peaks (Lacerda et al. 2019; Hassan et al. 2022). Since it is characteristic of cereal starches, the diffraction pattern for proso millet starch showed A-type diffraction patterns with peaks at 15°, 17°, and 23.1° (2 θ). The following aspects can be used to illustrate how different starches differ in their relative crystallinity, quantities of crystalline domains (determined by amylopectin concentration and amylopectin chain length), crystal size, and direction of the double helices inside the crystalline domains (Ma and Boye 2018). A typical A-type diffraction pattern for pearl millet starch was revealed by X-ray diffraction analyses. It includes four primary characteristic peaks, including a sharp, strong peak at 2 θ values of 15°, doublet peaks at 2 θ values of 17° and 18°, and a single peak at 2 θ value of 23°. The variances between the different types of starch granules have an impact on the degree of crystallinity. Amylose deprivation increases crystallinity without changing granular size (Shi et al. 2018; Romano and Kumar 2019).

10.5 SOLUBILITY AND SWELLING POWER OF MILLET STARCH

The interaction of the starch chains within the crystalline and amorphous regions of starch granules is characterized by solubility and swelling power (SP). The characteristics mentioned can exhibit variations based on factors such as climate, geographical location, soil conditions, and starch extraction methods employed. When heated in an excessive amount of water, the crystalline part of starch molecules becomes disrupted. This disruption leads to increased solubility and SP of starch by facilitating the formation of hydrogen bonds between water molecules and the exposed hydroxyl groups present in starch (Goud et al. 2022). Starch granules absorb water and swell when heated in the presence of water. The SP of millet starches lies between 50 and 90°C. However, the solubility and swelling power capacities of different millet starches are varied at various temperatures (Uzizerimana et al. 2021; Sharma et al. 2021). Millet starch has shown better swelling power and solubility than wheat, rye, and potato starches due to their comparatively strong granular bonding forces. Millet starches appear to have more resistance to swelling behaviour. At lower temperatures (65–70°C), the waxy starches exhibit simple breakdown, which causes the creation of thick pastes. Hydrogen bonds between the short amylopectin chains found in starches make it easy for them to bind water molecules, while amylose, its lipid complex, and long amylopectin chains create helical boundaries that prevent water molecules from entering (Ge et al. 2021; Chang et al. 2021). Because of this, starches with low amylose contents can readily be used in foods where viscosity is needed, such as soup thickeners. Moreover, the inclusion of amylose benefits materials that need retrogradation (Himashree et al. 2022; Zhang et al. 2021; Donmez et al. 2021). A crucial structural attribute of starches in food is their involvement in a range of processes, including water binding, swelling, and eventual granule breakdown. The extent of interaction between starch chains and both crystalline and amorphous components is determined by their solubility and swelling power, as indicated by studies conducted by Chen et al. (2021) and Ali et al. (2020). It has been observed that the water absorption rate and swelling power of millets are influenced by temperature. Evaluations of the solubility and swelling power of millets are often conducted across a temperature range of 60–90°C (Makroo et al. 2021; Moorthy et al. 2020). At higher temperature ranges (such as 90°C), millet starch generally showed lower swelling and solubilization values than potatoes (Thakur et al. 2022; Siroha et al. 2021). Due to the breakdown of the H-bond during heating, starch loses its crystalline structure and new bonds are then generated between the molecules. The swelling power of millet starch is directly related to the amylopectin concentration, whereas amylose dilutes and inhibits swelling (Bangar et al. 2021; Punia et al. 2021; Bakouri and Guemra 2019). As a result of the fragility of swollen starch granules caused by millet starch with low amylose concentration, which disintegrates at temperatures above 65°C, millet starch exhibits shear thinning behaviour (Dhull et al. 2022; Xiao et al. 2021). According to reports, the swelling power and solubility of starch are influenced by starch chains' molecular weight, lipid-amylose complexes, amylopectin ratios, and the relative distribution of amylose and amylopectin (Yang et al. 2022; Silva et al. 2017). Granular solubilization is defined by solubility (%) or amylose leaching (%), whereas the swelling of granules in water during heating is typically evaluated by the SP or swelling factor (SF) (Ashogbon et al. 2021). SF and SP refer to the water's inter- and intragranular components, respectively.

Genetic variation in swelling traits has been observed within the same species of millet. Proso millet varieties differ significantly in terms of starch SP and solubility (Bangar et al. 2021; Xiao et al. 2021). However, there is no clear-cut distinction in the swelling power and swelling capacity that can be used to differentiate between different millet species or to distinguish millets from other cereal crops. To reach a reliable statistical conclusion, it is essential to conduct further studies involving a broader range of genotypes from various millet species. This would facilitate a more comprehensive understanding of the variations and enable more accurate comparisons and conclusions (Kaur et al. 2023; Gowda et al. 2022; Serna-Saldivar and Espinosa-Ramirez 2019).

10.6 RHEOLOGICAL AND GELATINIZATION PROPERTIES OF MILLET STARCH

For rheological analysis of millet starch, the Brabender visco-amylograph (BVA) and rapid visco-analyzer (RVA) are frequently employed by various researchers. In both situations, starch is heated while being continuously sheared in the presence of an excessive amount of water, and a change in viscosity with a controlled temperature cycle is observed (Hussain et al. 2022; Dong et al. 2021). Temperature, shear rate, water content, and starch structure all have an impact on the pasting process. The amount of starch utilized in research varied from 6 to 10%. Across different species and genotypes of the same species, there is a significant variation in pasting characteristics (Miao et al. 2021; Obadi et al. 2021). For instance, peak viscosity for ten genotypes of foxtail millet ranged from 345 to 425 RVU, breakdown from 182 to 237 RVU, setback from 142 to 188 RVU, and pasting temperature from 72.0 to 78.5°C. In addition to dynamic rheological and pasting analysis, studies can be carried out to clarify certain additional elements of the millet starch's physicochemical properties (Zhang and Shen 2021; Zhu et al. 2018). It is yet unknown how the chemical structure of the starch components affects their rheological characteristics. Moreover, gelatinization is a major key factor for the starch composition, which decides its applications in various fields (Barbhuiya et al. 2021; De Pilli and Alessandrino, 2020). Several reports have revealed that the amylopectin and amylose content of starch is highly influenced by its functional properties. From this perspective, it has been discovered that the starch gelatinization capabilities of millet species and genotypes differ from one another (Zou et al. 2023; Tyl et al. 2018). The gelatinization properties of several millet starches using differential scanning calorimetry (DSC) demonstrated the onset gelatinization temperature (To) in the order of proso millet (68.4°C) > foxtail millet (66.7°C) > finger millet (63.9°C) > pearl millet (62.8°C). The finger millet starch has the highest enthalpy (H) of all the millet starches, followed by proso millet (13.1 J/g), pearl millet (12.3 J/g), and foxtail millet (11.8 J/g). Gelatinization is an irreversible phase transition phenomenon that happens when starch is subjected to high temperatures and water (Dimri and Singh 2022; Bean et al. 2019). Amylopectin's external chains are clustered together and uncoil and melt as a result of the gelatinization process. Low gelatinization temperature (typically 70°C) starches provide superior cooking quality (Korompokis et al. 2021). DSC, hot-stage microscopes with polarizing filters, thermal analytical techniques, X-ray scattering, and nuclear magnetic resonance (NMR) spectroscopy are only a few of the techniques used to measure the gelatinization qualities (Chakraborty et al. 2022; Nilsson et al. 2022).

10.7 APPLICATIONS OF MILLET STARCH

Starch is a low-cost material with easy availability and eco-friendly features. It has found applications in various industries like food, textiles, and pharmaceuticals (Kumar et al. 2022). The physicochemical and functional characteristics of starch determine its desired functionality or involvement in a certain field. The native starch has limited functionality and therefore it has limited industrial applications. The properties of starch can be modified by heating, oxidation, acid treatment, alkali treatment, enzymatic treatment, acetylation, etherification, acetification, etc. (Kumar et al. 2023). These starches can be utilized as a binder, as thickeners in baked food items, meat products, and snack seasonings, as a gelling agent in gels and gums, as a foam stabilizer in marshmallows, and as a fat replacer in ice creams, as flavour encapsulating agents, as emulsion stabilizers in beverages and juices, and as a crisping agent for fried snack products (Nasrollahzadeh et al. 2021; Keller 2020). Acid-modified, acetylated, esterified, pre-gelatinized, and native starches are used widely for tablet formulation as a drug delivery system. There haven't been any reports yet about millet starches' limits in food formulations. However, the availability and abundance of alternative sources of starches (such as maize, wheat, and potato) that meet the current need for starch for a variety of uses could be the main cause of its underutilization. Millets are challenging to process because of their small grain size (Saggi and Dey 2019). In comparison to the often used hydroxypropylated starch, pre-gelatinized starch was developed to make a biodegradable polymer that may be more appropriate for drilling operations in environmentally sensitive places. Moreover, a lot of millet starches contain a lot of amyloses, which reduces some of their useful qualities. For instance, it may be limited in its wider applicability because it has less swelling power than other conventional starches like rice and wheat (Kringel et al. 2020; Li et al. 2019). The starch of finger millet has a low glycaemic index (GI) and it can be used for the preparation of low-GI foods. The low GI of finger millet starch is due to the absence of pinholes on its surface that retard the entrance of starch hydrolyzing enzymes and slowing down its digestion (Verma et al. 2021). Overall, the field of millet starch requires advanced research as it is one of the non-conventional sources of starch. In the future, millet starch could be used for the development of biodegradable packaging materials, edible coatings, and various other food applications.

10.8 CONCLUSIONS

Wet milling methods are the most widely employed methods for the extraction of starch from millets. The major steps of extraction using wet milling are steeping, milling, sieving, and centrifugation. The quality of the extracted starch depends on the source of starch, steeping method, and extraction method. A good quality starch should be free from protein residues and other impurities. Alkaline steeping produces fewer protein residues compared to acidic methods. The properties of starch can be studied by techniques like PLM, NLM, SEM, and TEM. Among these, SEM is the most widely used technique to study the size and shape of millet starches. Millet starches are semicrystalline and are mostly spherical or polygonal. The size of millet starch granules ranges between 2.5 and 24 μm. The functional properties and industrial application of millet starches depend on their amylose and amylopectin content. The amylose-rich starches are good where low viscosity is required while amylopectin is added in various foods to impart viscosity and prevent retrogradation. The modified starches have also found various applications in food and other industries. However, the studies on millet starches are limited compared to rice, wheat, and potato starches, and hence more studies are needed on millet starches.

REFERENCES

Aggarwal, J., S. Sharma, H. Kamyab, and A. Kumar. 2020. The realm of biopolymers and their usage: An overview. *Journal of Environmental Treatment Techniques* 8(2): 1005–1016.

Ali, N. A., K. K. Dash, and W. Routray. 2020. Physicochemical characterization of modified lotus seed starch obtained through acid and heat moisture treatment. *Food Chemistry* 319: 126513.

Annor, G. A., M. Marcone, E. Bertoft, and K. Seetharaman. 2014. Physical and molecular characterization of millet starches. *Cereal Chemistry* 91(3): 286–292.

Aruna, M., and R. Parimalavalli. 2022. Influence of acid on the isolation of starch from pearl millet (underutilized crop) and its characteristics. *Materials Today: Proceedings* 66: 920–927.

Ashogbon, A. O. 2021. The recent development in the syntheses, properties, and applications of triple modification of various starches. *Starch-Stärke* 73(3–4): 2000125.

Ashogbon, A. O., E. T. Akintayo, A. O. Oladebeye, A. D. Oluwafemi, A. F. Akinsola, and O. E. Imanah. 2021. Developments in the isolation, composition, and physicochemical properties of legume starches. *Critical Reviews in Food Science and Nutrition* 61(17): 2938–2959.

Bakouri, H., and K. Guemra. 2019. Etherification and cross-linking effect on physicochemical properties of Zea mays starch executed at different sequences in 1-butyl-3-methylimidazolium chloride [BMIM] Cl ionic liquid media. *International Journal of Biological Macromolecules* 125: 1118–1127.

Bangar, S. P., A. K. Siroha, M. Nehra, M. Trif, V. Ganwal, and S. Kumar. 2021. Structural and film-forming properties of millet starches: A comparative study. *Coatings* 11(8): 954.

Bangar, S. P., A. O. Ashogbon, A. Singh, V. Chaudhary, and W. S. Whiteside. 2022. Enzymatic modification of starch: A green approach for starch applications. *Carbohydrate Polymers* 287: 119265.

Bangar, S. P., A. O. Ashogbon, S. B. Dhull, R. Thirumdas, M. Kumar, M. Hasan, V. Chaudhary, and S. Pathem. 2021. Proso-millet starch: Properties, functionality, and applications. *International Journal of Biological Macromolecules* 190: 960–968.

Barbhuiya, R. I., P. Singha, and S. K. Singh. 2021. A comprehensive review on impact of non-thermal processing on the structural changes of food components. *Food Research International* 149: 110647.

Bean, S. R., L. Zhu, B. M. Smith, J. D. Wilson, B. P. Ioerger, and M. Tilley. 2019. Starch and protein chemistry and functional properties. In *Sorghum and Millets (Secondnd Edition)* (Taylor, J. R. N., and K. G. Duodu, Eds.). Woodhead Publishing and AACC International Press, Washington, pp. 131–170.

Bora, P., S. Ragaee, and M. Marcone. 2019. Characterisation of several types of millets as functional food ingredients. *International Journal of Food Sciences and Nutrition* 70(6): 714–724.

Castillo, L. A., O. V. López, M. A. García, S. E. Barbosa, and M. A. Villar. 2019. Crystalline morphology of thermoplastic starch/talc nanocomposites induced by thermal processing. *Heliyon* 5(6): e01877.

Chakraborty, I., S. Rongpipi, S. I. Govindaraju, S. S. Mal, E. W. Gomez, E. D. Gomez, R. D. Kalita, Y. Nath, and N. Mazumder. 2022. An insight into microscopy and analytical techniques for morphological, structural, chemical, and thermal characterization of cellulose. *Microscopy Research and Technique* 85(5): 1990–2015.

Chang, L., N. Zhao, F. Jiang, X. Ji, B. Feng, J. Liang, X. Yu, and S. K. Du. 2023. Structure, physicochemical, functional and in vitro digestibility properties of non-waxy and waxy proso millet starches. *International Journal of Biological Macromolecules* 224: 594–603.

Chang, Q., B. Zheng, Y. Zhang, and Y. Zeng. 2021. A comprehensive review of the factors influencing the formation of retrograded starch. *International Journal of Biological Macromolecules* 186: 163–173.

Chen, L., D. J. McClements, T. Yang, Y. Ma, F. Ren, Y. Tian, and Z. Jin. 2021. Effect of annealing and heat-moisture pretreatments on the oil absorption of normal maize starch during frying. *Food Chemistry* 353: 129468.

De Pilli, T., and O. Alessandrin. 2020. Effects of different cooking technologies on biopolymers modifications of cereal-based foods: Impact on nutritional and quality characteristics review. *Critical Reviews in Food Science and Nutrition* 60(4): 556–565.

Dhull, S. B., A. Chandak, A. M. N. Collins, S. P. Bangar, P. Chawla, and A. Singh. 2022. Lotus seed starch: A novel functional ingredient with promising properties and applications in food—A review. *Starch-Stärke* 74(9–10): 2200064.

Dimri, S., and S. Singh. 2022. A brief review on millet starch. *Bhartiya Krishi Anusandhan Patrika* 37(2): 126–132.

Dong, H., Q. Zhang, J. Gao, L. Chen, and T. Vasanthan. 2021. Comparison of morphology and rheology of starch nanoparticles prepared from pulse and cereal starches by rapid antisolvent nanoprecipitation. *Food Hydrocolloids* 119: 106828.

Donmez, D., L. Pinho, B. Patel, P. Desam, and O. H. Campanella. 2021. Characterization of starch–water interactions and their effects on two key functional properties: Starch gelatinization and retrogradation. *Current Opinion in Food Science* 39: 103–109.

Ge, X., H. Shen, C. Su, B. Zhang, Q. Zhang, H. Jiang, and W. Li. 2021. The improving effects of cold plasma on multi-scale structure, physicochemical and digestive properties of dry heated red adzuki bean starch. *Food Chemistry* 349: 129159.

Goud, E. L., J. Singh, and P. Kumar. 2022. Climate change and their impact on global food production. In *Microbiome Under Changing Climate: Implications and Solutions* (Kumar, A., J. Singh, and L. Ferreira, Eds.). Woodhead Publishing, Cambridge, pp. 415–436.

Gowda, N. N., K. Siliveru, P. V. Prasad, Y. Bhatt, B. P. Netravati, and C. Gurikar. 2022. Modern processing of Indian millets: A perspective on changes in nutritional properties. *Foods* 11(4): 499.

Gulati, P., L. Sabillón, and D. J. Rose. 2018. Effects of processing method and solute interactions on pepsin digestibility of cooked proso millet flour. *Food Research International* 109: 583–588.

Gupta, P. K., S. S. Raghunath, D. V. Prasanna, P. Venkat, V. Shree, C. Chithananthan, … K. Geetha. 2019. An update on overview of cellulose, its structure and applications. *Cellulose* 201(9): 84727.

Hassan, N. A., O. M. Darwesh, S. S. Smuda, A. B. Altemimi, A. Hu, F. Cacciola, I. Haoujar, and T. G. Abedelmaksoud. 2022. Recent trends in the preparation of nano-starch particles. *Molecules* 27(17): 5497.

Hassan, Z. M., N. A. Sebola, and M. Mabelebele. 2021. The nutritional use of millet grain for food and feed: A review. *Agriculture and Food Security* 10(1): 1–14.

Himashree, P., A. S. Sengar, and C. K. Sunil. 2022. Food thickening agents: Sources, chemistry, properties and applications-A review. *International Journal of Gastronomy and Food Science* 27: 100468.

Hussain, S., A. A. Mohamed, M. S. Alamri, M. A. Ibraheem, A. A. A. Qasem, T. Alsulami, and I. A. Ababtain. 2022. Effect of cactus (Opuntia ficus-indica) and Acacia (Acacia seyal) gums on the pasting, thermal, textural, and rheological properties of corn, sweet potato, and Turkish bean starches. *Molecules* 27(3): 701.

Iuga, M., and S. Mironeasa. 2020. A review of the hydrothermal treatments impact on starch based systems properties. *Critical Reviews in Food Science and Nutrition* 60(22): 3890–3915.

Kanehara, R., A. Goto, N. Sawada, T. Mizoue, M. Noda, A. Hida, S. Tsugane, and S. Tsugane. 2022. Association between sugar and starch intakes and type-2 diabetes risk in middle-aged adults in a prospective cohort study. *European Journal of Clinical Nutrition* 76(5): 746–755.

Kaur, B., A. Singh, S. Suri, M. Usman, and D. Dutta. 2023. Minor millets: A review on nutritional composition, starch extraction/modification, product formulation, and health benefits. *Journal of the Science of Food and Agriculture* 103(10): 4742–4754.

Keller, J. D. 2020. Sodium carboxymethylcellulose (CMC). In *Food Hydrocolloids* (Keller, J. D., Ed.). CRC Press, Florida, pp. 43–109.

Khatun, A. 2019. *The Impact of Rice Protein and Lipid on In Vitro Rice Starch Digestibility* (Doctoral dissertation, Southern Cross University). Accessed July 25, 2021.

Korompokis, K., K. Verbeke, and J. A. Delcour. 2021. Structural factors governing starch digestion and glycemic responses and how they can be modified by enzymatic approaches: A review and a guide. *Comprehensive Reviews in Food Science and Food Safety* 20(6): 5965–5991.

Kringel, D. H., A. R. G. Dias, E. D. R. Zavareze, and E. A. Gandra. 2020. Fruit wastes as promising sources of starch: Extraction, properties, and applications. *Starch-Stärke* 72(3–4): 1900200.

Krishnan, V. M., A. Singh Awana, S. Goswami, T. Vinutha, R. R. Kumar, S. P. Singh, T. Sathyavathi, A. Sachdev, and S. Praveen. 2021. Starch molecular configuration and starch-sugar homeostasis: Key determinants of sweet sensory perception and starch hydrolysis in pearl millet (*Pennisetum glaucum*). *International Journal of Biological Macromolecules* 183: 1087–1095.

Krishna Kumari, S., and B. Thayumanavan. 1998. Characterization of starches of proso, foxtail, barnyard, kodo, and little millets. *Plant Foods for Human Nutrition* 53: 47–56.

Kumar, A., P. Kumari, K. Gupta, M. Singh, and V. Tomer. 2023. Recent advances in extraction, techno-functional properties, food and therapeutic applications as well as safety aspects of natural and modified stabilizers. *Food Reviews International* 39(4): 2233–2276.

Kumar, A., P. Kumari, and M. Kumar. 2022. Role of millets in disease prevention and health promotion. In *Functional Foods and Nutraceuticals in Metabolic and Non-Communicable Diseases* (R. B. Singh, S. Watanabe, and A. A. Isaza, Eds.). Academic Press, Cambridge, USA, pp. 341–357.

Lacerda, L. D., D. C. Leite, and N. P. da Silveira. 2019. Relationships between enzymatic hydrolysis conditions and properties of rice porous starches. *Journal of Cereal Science* 89: 102819.

Li, H., M. J. Gidley, and S. Dhital. 2019. High-amylose starches to bridge the "Fiber Gap": Development, structure, and nutritional functionality. *Comprehensive Reviews in Food Science and Food Safety* 18(2): 362–379.

Li, W., L. Wen, Z. Chen, Z. Zhang, X. Pang, Z. Deng, T. Liu, and Y. Guo. 2021. Study on metabolic variation in whole grains of four proso millet varieties reveals metabolites important for antioxidant properties and quality traits. *Food Chemistry* 357: 129791.

Li, Z., and C. Wei. 2020. Morphology, structure, properties and applications of starch ghost: A review. *International Journal of Biological Macromolecules* 163: 2084–2096.

Liu, D., Z. Li, Z. Fan, Z. X. Zhang, and G. Zhong. 2019. Effect of soybean soluble polysaccharide on the pasting, gels, and rheological properties of kudzu and lotus starches. *Food Hydrocolloids* 89: 443–452.

Liu, Z., C. Wang, X. Liao, and Q. Shen. 2020. Measurement and comparison of multi-scale structure in heat and pressure treated corn starch granule under the same degree of gelatinization. *Food Hydrocolloids* 108: 106081.

Ma, Z., and J. I. Boye. 2018. Research advances on structural characterization of resistant starch and its structure-physiological function relationship: A review. *Critical Reviews in Food Science and Nutrition* 58(7): 1059–1083.

Mahajan, P., M. B. Bera, P. S. Panesar, and A. Chauhan. 2021. Millet starch: A review. *International Journal of Biological Macromolecules* 180: 61–79.

Makroo, H. A., S. Naqash, J. Saxena, S. Sharma, D. Majid, and B. N. Dar. 2021. Recovery and characteristics of starches from unconventional sources and their potential applications: A review. *Applied Food Research* 1(1): 100001.

Malik, M. K., V. Kumar, P. P. Sharma, J. Singh, S. Fuloria, V. Subrimanyan, N. K. Fuloria, and P. Kumar. 2022. Improvement in digestion resistibility of Mandua starch (Eleusine coracana) after cross-linking with epichlorohydrin. *ACS Omega* 7(31): 27334–27346.

Miao, W. B., S. Y. Ma, X. G. Peng, Z. Qin, H. M. Liu, X. S. Cai, and X. D. Wang. 2021. Effects of various roasting temperatures on the structural and functional properties of starches isolated from tigernut tuber. *LWT- Food Science and Technology* 151: 112149.

Mirzababaee, S. M., D. Ozmen, M. A. Hesarinejad, O. S. Toker, and S. Yeganehzad. 2022. A study on the structural, physicochemical, rheological and thermal properties of high hydrostatic pressurized pearl millet starch. *International Journal of Biological Macromolecules* 223(A): 511–523.

Moorthy, S. N., M. S. Sajeev, R. P. K. Ambrose, and R. J. Anish. 2020. *Tropical Tuber Starches: Structural and Functional Characteristics. CABI- Digital Library*, p. 270. CABI Publishers, Wallingford, Oxfordshire.

Nasrollahzadeh, M., Nezafat, Z., N. Shafiei, and F. Soleimani. 2021. Polysaccharides in food industry. In *Biopolymer-Based Metal Nanoparticle Chemistry for Sustainable Applications* (Nasrollahzadeh, M., Ed.). Elsevier Science, Iran, pp. 47–96.

Nawaz, A., Z. Xiong, H. Xiong, S. Irshad, L. Chen, P. K. Wang, H. M. Ahsan, N. Walayat, and S. H. Qamar. 2019. The impact of hydrophilic emulsifiers on the physico-chemical properties, microstructure, water distribution and in vitro digestibility of proteins in fried snacks based on fish meat. *Food and Function* 10(10): 6927–6935.

Nazari, B., M. A. Mohammadifar, S. Shojaee-Aliabadi, E. Feizollahi, and L. Mirmoghtadaie. 2018. Effect of ultrasound treatments on functional properties and structure of millet protein concentrate. *Ultrasonics Sonochemistry* 41: 382–388.

Nilsson, K., C. Sandström, H. D. Özeren, F. Vilaplana, M. Hedenqvist, and M. Langton. 2022. Physiochemical and thermal characterisation of faba bean starch. *Journal of Food Measurement and Characterization* 16(6): 4470–4485.

Obadi, M., Y. Qi, and B. Xu. 2021. Highland barley starch (Qingke): Structures, properties, modifications, and applications. *International Journal of Biological Macromolecules* 185: 725–738.

Oladele, E. O. P., and A. Mbaye. 2022. The potential of resistant starch Type 1 for nutritional food security. In *Food Security and Safety Volume 2: African Perspectives* (O. O. Babalola, A. S. Ayangbenro, and O. B. Ojuderie, Eds.). Springer International Publishing, Gewerbestrasse, Cham, pp. 3–17

Punia, S., M. Kumar, A. K. Siroha, J. F. Kennedy, S. B. Dhull, and W. S. Whiteside. 2021. Pearl millet grain as an emerging source of starch: A review on its structure, physicochemical properties, functionalization, and industrial applications. *Carbohydrate Polymers* 260: 117776.

Rajeswari, J. R., M. Guha, A. Jayadeep, and B. S. Rao. 2015. Effect of alkaline cooking on proximate, phenolics and antioxidant activity of foxtail millet (Setaria italica). *World Applied Sciences Journal* 33(1): 146–152.

Ren, Y., T. Z. Yuan, C. M. Chigwedere, and Y. Ai. 2021. A current review of structure, functional properties, and industrial applications of pulse starches for value-added utilization. *Comprehensive Reviews in Food Science and Food Safety* 20(3): 3061–3092.

Rhowell Jr, N. T., A. P. Bonto, and N. Sreenivasulu. 2021. Enhancing the functional properties of rice starch through biopolymer blending for industrial applications: A review. *International Journal of Biological Macromolecules* 192: 100–117.

Romano, N., and V. Kumar. 2019. Starch gelatinization on the physical characteristics of aquafeeds and subsequent implications to the productivity in farmed aquatic animals. *Reviews in Aquaculture* 11(4): 1271–1284.

Saggi, S. K., and P. Dey. 2019. An overview of simultaneous saccharification and fermentation of starchy and lignocellulosic biomass for bio-ethanol production. *Biofuels* 10(3): 287–299.

Saini, S., S. Saxena, M. Samtiya, M. Puniya, and T. Dhewa. 2021. Potential of underutilized millets as Nutri-cereal: An overview. *Journal of Food Science and Technology* 58(12): 4465–4477.

Santschi, P. H., W. C. Chin, A. Quigg, C. Xu, M. Kamalanathan, P. Lin, and R. F. Shiu. 2021. Marine gel interactions with hydrophilic and hydrophobic pollutants. *Gels* 7(3): 83.

Satyavathi, C. T., S. Ambawat, V. Khandelwal, and R. K. Srivastava. 2021. Pearl millet: A climate-resilient nutricereal for mitigating hidden hunger and provide nutritional security. *Frontiers in Plant Science* 12: 659938.

Selladurai, M., M. K. Pulivarthi, A. S. Raj, M. Iftikhar, P. V. Prasad, and K. Siliveru. 2022. Considerations for gluten free foods-pearl and finger millet processing and market demand. *Grain and Oil Science and Technology* 6(2): 59–70.

Serna-Saldivar, S. O., and J. Espinosa-Ramírez. 2019. Grain structure and grain chemical composition. In *Sorghum and Millets (Secondnd Edition)* (Taylor, J. R. N., and K. G. Duodu, Eds.). Woodhead Publishing and AACC International Press, Washington, pp. 85–129.

Sharma, S., D. C. Saxena, and C. S. Riar. 2018. Characteristics of β-glucan extracted from raw and germinated foxtail (Setaria italica) and kodo (Paspalum scrobiculatum) millets. *International Journal of Biological Macromolecules* 118: 141–148.

Sharma, V., M. Kaur, K. S. Sandhu, S. Kaur, and M. Nehra. 2021. Barnyard millet starch cross-linked at varying levels by sodium trimetaphosphate (STMP): Film forming, physico-chemical, pasting and thermal properties. *Carbohydrate Polymer Technologies and Applications* 2: 100161.

Shi, K., X. Gu, W. Lu, and D. Lu. 2018. Effects of weak-light stress during grain filling on the physicochemical properties of normal maize starch. *Carbohydrate Polymers* 202: 47–55.

Shi, P., Y. Zhao, F. Qin, K. Liu, and H. Wang. 2023. Understanding the multi-scale structure and physicochemical properties of millet starch with varied amylose content. *Food Chemistry* 410: 135422.

Silva, W. M. F., B. Biduski, K. O. Lima, V. Z. Pinto, J. F. Hoffmann, N. L. Vanier, and A. R. G. Dias. 2017. Starch digestibility and molecular weight distribution of proteins in rice grains subjected to heat-moisture treatment. *Food Chemistry* 219: 260–267.

Siroha, A. K., S. Punia, S. S. Purewal, and K. S. Sandhu. 2021. *Millets: Properties, Processing, and Health Benefits*. CRC Press, Florida, p. 272.

Solaesa, Á. G., M. Villanueva, S. Beltrán, and F. Ronda. 2019. Characterization of quinoa defatted by supercritical carbon dioxide. Starch enzymatic susceptibility and structural, pasting and thermal properties. *Food and Bioprocess Technology* 12(9): 1593–1602.

Song, Z., Y. Zhong, W. Tian, C. Zhang, A. R. Hansen, A. Blennow, W. Liang, and D. Guo. 2020. Structural and functional characterizations of α-amylase-treated porous popcorn starch. *Food Hydrocolloids* 108: 105606.

Sudha, K. V., S. J. Karakannavar, N. B. Yenagi, and B. Inamdar. 2021. Effect of roasting on the physicochemical and nutritional properties of foxtail millet (Setaria italica) and Bengal gram dhal flours. *The Pharma Innovation Journal* 10(5): 1543–1547.

Tarahi, M., F. Shahidi, and S. Hedayati. 2022. Physicochemical, pasting, and thermal properties of native corn starch–mung bean protein isolate composites. *Gels* 8(11): 693.

Thakur, K., S. Sharma, and R. Sharma. 2022. Morphological and functional properties of millet starches as influenced by different modification techniques: A review. *Starch-Stärke* 75(3–4): 2200184.

Tyl, C., A. Marti, J. Hayek, J. Anderson, and B. P. Ismail. 2018. Effect of growing location and variety on nutritional and functional properties of proso millet (*Panicum miliaceum*) grown as a double crop. *Cereal Chemistry* 95(2): 288–301.

Uzizerimana, F., K. Dang, Q. Yang, M. S. Hossain, S. Gao, P. Bahati, N. G. Mugiraneza, P. Yang, and B. Feng. 2021. Physicochemical properties and in vitro digestibility of Tartary buckwheat starch modified by heat moisture treatment: A comparative study. *NFS Journal* 25: 12–20.

Verma, V. C., S. Agrawal, M. K. Tripathi, and A. Kumar. 2021. Millet starch: Current knowledge and emerging insights of structure, physiology, glycaemic attributes and uses. In *Millets and Millet Technology* (Kumar, A., M. K. Tripathi, D. Joshi, and V. Kumar, Eds.). Springer Nature, Singapore, pp. 121–142.

Wang, D., M. Zhao, Y. Wang, H. Mu, C. Sun, H. Chen, and Q. Sun. 2023. Research progress on debranched starch: Preparation, characterization, and application. *Food Reviews International* 39(9): 6887–6907..

Xiao, Y., M. Zheng, S. Yang, Z. Li, M. Liu, X. Yang, N. Lin, and J. Liu. 2021. Physicochemical properties and in vitro digestibility of proso millet starch after addition of proanthocyanidins. *International Journal of Biological Macromolecules* 168: 784–791.

Xiao, Y., X. Wu, B. Zhang, F. Luo, Q. Lin, and Y. Ding. 2021. Understanding the aggregation structure, digestive and rheological properties of corn, potato, and pea starches modified by ultrasonic frequency. *International Journal of Biological Macromolecules* 189: 1008–1019.

Xie, Q., X. Liu, H. Liu, Y. Zhang, S. Xiao, W. Ding, Q. Lyu, Y. Fu, and X. Wang. 2023. Insight into the effect of garlic peptides on the physicochemical and anti-staling properties of wheat starch. *International Journal of Biological Macromolecules* 229: 363–371.

Yang, T., P. Wang, F. Wang, Q. Zhou, X. Wang, J. Cai, M. Huang, and D. Jiang. 2022. Influence of starch physicochemical properties on biscuit-making quality of wheat lines with high-molecular-weight glutenin subunit (HMW-GS) absence. *LWT- Food Science and Technology* 158: 113166.

Yang, Q., Zhang, W., J. Li, X. Gong, and B. Feng. 2019. Physicochemical properties of starches in proso (non-waxy and waxy) and foxtail millets (non-waxy and waxy). *Molecules* 24(9): 1743. https://doi.org/10.3390/molecules24091743.

Yashini, M., S. Khushbu, N. Madhurima, C. K. Sunil, R. Mahendran, and N. Venkatachalapathy. 2022. Thermal properties of different types of starch: A review. *Critical Reviews in Food Science and Nutrition*: 1–24. doi: 10.1080/10408398.2022.2141680.

Zhang, F., and Q. Shen. 2021. The impact of endogenous proteins on hydration, pasting, thermal and rheology attributes of foxtail millet. *Journal of Cereal Science* 100: 103255.

Zhang, Y., J. Gao, Q. Qie, Y. Yang, S. Hou, X. Wang, X. Li, and Y. Han. 2021. Comparative analysis of flavonoid metabolites in foxtail millet (Setaria italica) with different eating quality. *Life* 11(6): 578.

Zhu, F. 2020. Fonio grains: Physicochemical properties, nutritional potential, and food applications. *Comprehensive Reviews in Food Science and Food Safety* 19(6): 3365–3389.

Zhu, F., R. Mojel, and G. Li. 2018. Physicochemical properties of black pepper (Piper nigrum) starch. *Carbohydrate Polymers* 181: 986–993.

Zou, J., Y. Li, F. Wang, X. Su, and Q. Li. 2023. Relationship between structure and functional properties of starch from different cassava (Manihot esculenta Crantz) and yam (Dioscorea opposita Thunb) cultivars used for food and industrial processing. *LWT- Food Science and Technology* 173: 114261.

Chapter 11

Extraction and Application of Millet Proteins

11.1 INTRODUCTION

Proteins are of preponderant importance as they play multiple roles in the human body and are important for human metabolism. Flour and dough properties are affected by the protein content and the type present in cereals. The content and type of protein in synergy with other components establish the texture and other functional properties of the final food products. Hence, the study of proteins has remained of utmost importance since prehistoric times. Recently, the role of proteins and their components have emerged in disease management and prevention as well. Research has emphasized the role of specific amino acids, bioactive peptides, and proteins in the prevention and cure of various diseases. Hence, proteins are important not only for processing but also play a key role in disease management and health improvement. Because of its importance, there is a continuous search for new protein sources. Proteins from plant sources are more sustainable and environment-friendly in comparison to animal-based proteins.

Millets, traditional cereal grains that serve as a potential source of nutrition for millions of people worldwide, are naturally rich in proteins (Kumar et al. 2022). Protein is the second major nutrient in millets with an average content of 7–12% (Kaur et al. 2023; Ramashia et al. 2021; Selokar et al. 2022; Ujong et al. 2022). The essential amino acid profile of millet protein is high compared to other cereals (Huchchannanavar et al. 2019; Anitha et al. 2020). The essential amino acids, mainly leucine, isoleucine, and methionine, isolated from proso millet were shown to have significantly higher activity in comparison to wheat (Li et al. 2021). These also have high digestibility due to the presence of fewer crosslinked prolamins (Hassan et al. 2021; Allai et al. 2022). The composition of millet proteins and their amino acid content have been discussed in detail in previous chapters. This chapter will emphasize the different types of proteins present in millets, their chemistry, method of extraction, and food applications.

11.2 PROTEIN COMPOSITION OF MILLETS

The average protein content of millet varies from 7 to 12%; however, it may be as high as 21% in some varieties (Krishnan and Meera 2018; Uwagbale et al. 2016). Morphological analysis of millet grains shows that proteins are concentrated in the embryonic tissue and endosperm (Evers and

Millar 2002). Due to its enormous rich germ and comparatively small endosperm, pearl millet has the most significant protein content of all grains ranging from 9 to 21% (Krishnan and Meera 2018; Taylor 2017; Gwamba 2016). In contrast, finger millet due to its small germ has relatively low protein content of 4.9–11.3% (Sachdev et al. 2021; Himanshu et al. 2018). The protein content and the amino acid composition of various millets have already been discussed in a previous chapter (Chapter 6). The composition of millet protein is similar to other cereal grains and can be classified into major categories based on solubility. These categories are water-soluble albumins, alcohol-soluble prolamins, dilute salt-solution soluble globulins, and dilute alkali- or acid-soluble glutelins. Glutelin (45–55%) is the major protein in millets followed by prolamin (15–30%). In pearl millet, the percentage of prolamin and prolamin-like compounds ranges from 22 to 35%, albumin and globulin range from 22 to 28%, and glutelin and glutelin-like compounds range from 28 to 32% of total nitrogen (Krishnan and Meera 2018). Sharma et al. (2023) studied the protein fractions of foxtail millet and reported glutelin (34.69%) as the major protein fraction followed by prolamin (31.10%). The content of albumin + globulin fraction was 32.41%. Akharume et al. (2020) reported prolamin (25.58–30.93%) and glutenin (20.97–21.13%) as the major protein fractions in two proso millet varieties of the USA, i.e. *Dawn* and *Plateau*. The content of albumins and globulins in the proso millet was in the range of 5.1–5.2% and 2.1–3.5%, respectively. The major storage proteins in Japanese barnyard millet are true glutelins (39.3–54.4%). The second major fraction is of albumin + globulins, i.e. 11.3–17.2%. The content of prolamins varies from 6.8 to 9.3%, and prolamins-like fractions are in the range of 7.5–11.6% of total proteins. The glutelin-like fraction is 5.9–9.1% (Monteiro et al. 1988). The major protein fraction in finger millet is prolamin which accounts for 24.6–36.2% of total protein content (Abioye et al. 2022). The second major protein fraction in finger millet is glutenin (13–25%) (Virupaksha et al. 1975). The major protein fraction in kodo millet is true glutelin, i.e. 40.7–54.4% of total protein, followed by the albumin + globulin fraction (5.6–11.7%), true prolamin (6.5–11.1%), glutelin-like (8.2–10.3%), and prolamin like (5.2–9.5%) fractions. The major protein fraction in teff is glutelin (46.6%), followed by albumin (39.1%), prolamin (29.4%), and globulin (21.1%) (Gebru et al. 2020). The major protein fraction of fonio is glutelin (14%), followed by prolamin (5.5%), albumin (3.5%), and globulin (1.8%) Ballogou et al. 2013).

11.3 EXTRACTION OF MILLET PROTEINS

The extraction and purification of millet proteins are important to understand their nature, functionality, and utilization in food, pharma, and biomaterial products. Various conventional and novel techniques are used for the efficient extraction of proteins. In general, these methods are termed "wet isolation techniques". The isolation of millet protein can broadly be classified into two phases: the sequential fractionation process and the isolate/concentrate preparative process.

Sequential millet protein fractionation works on the principle of the segregation of proteins into different fractions based on their solubility in the solvents. This method is used to isolate, identify, and characterize the proteins. This process is useful in analyzing the differences between cultivars, varieties, species, and different millet types. Two major sequential fractionation processes have been identified. The classical Osborne approach is the first method. In this approach, proteins are categorized into albumin, globulin, prolamin, and glutelin based on their solubility in water, salt solution, alcohol, acid, and alkali. This method has the disadvantage of low protein recovery. Another method is Landry and Moureaux's fractionation method. To achieve better efficiency for heterogenous millet proteins, this method uses reducing agents like 2-mercaptothanol. Complexes of detergents and alcohol/alkali are added for the disruption of the polymeric protein structure. This scheme classifies protein into five fractions (fractions I to V) (Gowthamraj et al. 2020). Foxtail millet concentrates were prepared in a study by Sharma et al. (2023). For this, millet dispersion was prepared in water (1:4) and adjusted to pH 9.5 with 1 N NaOH. After this, the dispersion was centrifuged at 5000 rpm for 30 minutes. Then, the supernatant was collected and its pH was adjusted to 4 using 1 N HCl. The concentrate was extracted by centrifuging at 5000 rpm for 15 minutes and the pellet obtained was dried. In the same study, foxtail millet protein fractions were also extracted. For this, defatted and depigmented foxtail millet flour was used. Extraction was done using 0.5M NaCl and centrifuged. The supernatant was purified into albumin and globulin. The precipitate obtained was again extracted using 0.05 M NaOH and centrifuged. The supernatant represents the alkaline soluble fraction mainly glutelin. The precipitate was again extracted using 70% isopropanol. This is the alcohol-soluble fraction and is purified into prolamin (Akharume et al. 2020; Sharma et al. 2023). In another study, the extraction of proso millet protein fractions was done using the Osborne method (Akharume et al. 2020). Pennisetin was extracted from pearl millet flour with 90% purity using 70% ethanol mixed with 1% sodium metabisulphite and 0.2% sodium hydroxide (Bibi et al. 2021).

After fractionation, the next stage is the preparatory stage wherein the isolation of protein is done for a better understanding of protein behaviour. At the laboratory scale, wet milling and/or chemical solvation is done followed by precipitation and separation. The product so obtained is further dried for use. Proso millet protein isolated using trichloroacetic acid in cold acetone or pure acetone resulted in protein fractions with poor solubility (Akharume et al. 2020). For albumin extraction from teff seed, deionized water was used in a ratio of 1:10 along with shaking for one hour. The extract was centrifuged at $6000 \times g$ for 10 min at 4°C. The supernatant was used to obtain the albumin fraction. The residue was further extracted by 1.25 M NaCl for the globulin fraction. Prolamines were extracted using 70% ethanol (Gebru et al. 2019).

Extraction and isolation using acidic conditions are not recommended for millets due to their insufficiency in cell wall degradation. This leads to poor protein solubilization and diffusion into the solvent. An alkali-based or combination extraction system favours solubilization through the rupture of cell walls and breakage of disulphide-crosslinking bond distribution within the protein matrix. A pH of 8–13.5 and an extraction temperature of 30–90°C are recommended for millets (Contreras et al. 2019).

11.4 CHARACTERIZATION OF MILLET PROTEINS/PEPTIDES

Millets are not a part of a single taxonomic group and the characteristics of proteins from millets can differ considerably with their species. Protein characteristics are generally determined by sodium dodecyl sulphate polyacrylamide gel electrophoresis and lab-on-a-chip (LOC) techniques. Studies have confirmed that millets generally do not contain high molecular weight subunits. Until now, the majority of the millets have been found to have polypeptides with a weight of only 100 kDa or less (Mohamed et al. 2009, Bagdi et al. 2011). Also, some researchers have reported specific and varietal differences in millet protein composition. Jhan et al. (2021) assessed the major polypeptide in pearl millet proteins in the range of 12–43 kDa. Polypeptides with a molecular weight of 20 kDa correspond to prolamin fractions. The major protein reported in foxtail was prolamine with proline, alanine, valine, aspartic acid, glutamic acid, phenylalanine, and leucine as the major amino acids contributing to approximately 55% of the total storage proteins (Ji et al. 2019, Hu et al. 2020). In other studies, proso millet was found to have 11 to 150 kDa prolamines and glutelin and glutelin Type-B 5 like protein having a molecular weight of 15–70 kDa with a monodispersed size distribution (6.50–43.82 nm) (Tyl et al. 2020; Sharma et al. 2023). Bagdi et al. (2011) characterized the protein of Hungarian proso millet varieties using gel electrophoresis and LOC techniques. The results indicated negligible qualitative differences among the varieties but significant quantitative differences were observed. Eight protein subunits were detected based on the mobility during electrophoretic analysis. The peptides determined had a molecular mass of 58 kDa, 50 kDa, 40 kDa, 18 kDa, 17 kDa, 12 kDa, 7 kDa, and 5 kDa in all the samples predominated by the polypeptide with 18 kDa mass. Apart from these, 18 other polypeptides were found in small concentrations. Some studies also identified proportion as the percentage of protein under different solubility classes. Table 11.1 presents the details of different fractions based on solubility in millets.

Agrawal et al. (2016) used trypsin enzyme for the isolation of peptides from pearl millet protein. The fractionation of tryptic hydrolysates into 25 fractions was done using gel filtration chromatography. The first fraction (F1) represented the shortest retention time corresponding to smaller-sized peptide fragments at 280 nm. F7 fraction was the one generated in the highest quantity (2.95 mg/mL) (Agrawal et al. 2016). The peptide sequence obtained exhibited antioxidative properties. Pearl millet protein isolate exhibited an endothermic peak at 64.79°C with enthalpies ΔH 4.34 J/g. In finger millet, Agrawal et al. (2019) isolated two novel peptides STTVGLGISMRSASVR and TSSSLNMAVRGGLTR which exhibited strong antioxidant activity. The property was attributed to low molecular weight aromatic and hydrophobic amino acids and amino acid residue interaction (serine and threonine specifically). Also, the prolamin fraction exhibited four bands α, β, γ, and δ with molecular weights of 22, 21, 20, and 19 kDa. Apart from this, 59.4% α-helix structure was observed (Kumar et al. 2017). In another study, the particle size of hydrolysates derived from pearl, finger millet, and sorghum was found to be in the range of 100–300 nm (Agrawal et al. 2020). Also, the hydrolysates exhibited higher solubility in pH 2–10. Wang et al. (2021) reported the molecular weight of proso millet proteins in the range of 10–70 kDa with distinct bands corresponding to 11S and 7S globulin fractions.

TABLE 11.1 PROTEIN COMPOSITION OF MILLETS LANDRY AND MOUREAUX SOLUBILITY PROTEIN FRACTIONATION SCHEME

	Percent contribution of protein in different solvent classes to total protein in whole seed flour			
Millet type	Protein content/soluble fraction class (defatted flour) (g/100g)	Water-salt fraction I	Alcohol fraction II	Alcohol + mercaptoethanol fraction III
Pearl millet	12.7	26	39.7	4
Proso millet	12.3	16.3	6.1	9.8
Barnyard millet	12.3	12.1	8.7	13.7
Finger millet	8.5	12.1	10.3	32.4
Kodo millet	10.5	8.9	9.5	11.1
Little millet	12.9	9.3	15.1	14.3

Source: Sachdev et al. 2021; Sachdev et al. 2023; Longvah et al. 2017; Monteiro et al. 1988; Sudharshana et al. 1988.

The secondary structure analysis of barnyard millet was done by Sharma et al. (2021). The results revealed that for barnyard millet proteins, the spectral regions dedicated to β-sheets, random coils, α-helix, and β-turn were at 1640–1600 /cm, 1640–1650 /cm, and 1680–1670 /cm, 1660–1650 /cm, and 1700–1680 /cm and 1670–1660 /cm, respectively. For proso millet, the proteins were found to be β-turn, α-helix, β-sheets, and random coils and their spectral regions were 1681–1700 /cm, 1651–1660, and 1661–1670 /cm, 1601–1640 and 1671–1680 /cm, and 1641–1650 /cm respectively (Sharma et al. 2022). β-sheets were the major secondary structures in proso millet protein.

11.5 EFFECT OF PROCESSING ON MILLET PROTEIN

Thermal processing methods are the conventional methods of food processing where food is exposed to high temperatures for extended periods. Various thermal processing methods employed for millet processing are boiling, roasting, drying, par-boiling, baking, popping, extrusion, etc. These processing methods can be applied individually or in combination (Budhwar et al. 2020). The non-thermal processing methods involve the use of methods where the processing temperature remains below room temperature. Common non-thermal processing methods are high-pressure processing (HPP), pulsed electric field (PEF), pulsed light, irradiation, oscillating magnetic fields, etc. (Jadhav et al. 2021). Both the thermal and non-thermal methods are reported to affect the protein properties. Heating operations like cooking or boiling cause the protein molecules to vibrate, which breaks the weak bonds in their structure, unravelling the protein strands, sticking together, and forming an aggregate. This improves protein digestibility (Anitha et al. 2020). The cooking of pearl millet is reported to decrease the globulin and true-prolamin fractions, and increase the prolamin-like, true-glutelin, and insoluble protein fractions (Ali et al. 2009). Poor protein extractability was observed in finger millet on thermal processing. Dharamraj and Malleshi (2011) reported a reduction in protein digestibility from 94 to 56% on heat processing. The study also reported an increase in the extraction of globulins and a decline in albumin, prolamin, and glutelin. Roasting finger millet decreased the protein content by 19% (Singh et al. 2018). Similar results were obtained for foxtail and proso millet as well in different studies (Gulati et al. 2018). It has been reported to have both positive and negative effects on protein digestibility. Bangar et al. (2022) reported increased *in-vitro* protein digestibility (45.5–65.8% and 49.3–75.4%) in two pearl millet cultivars on roasting. In contrast to this, the *in-vitro* digestibility of proso millet proteins was shown to be lower after dry and wet heating in comparison to raw grains. This might be explained by the creation of hydrophobic protein aggregates during heat processing and a significant rise in the hydrophobicity of their surfaces (Gulati et al. 2017). Parboiling of millets alters their secondary structure due to the formation of disulphide cross-linkage and reduces their protein digestibility. The digestibility of pearl and proso millet was decreased by 14–17% on parboiling (Bora et al. 2019). Pan-frying, puffing, and popping of millet have been reported to increase protein concentration (Nanje Gowda et al. 2022). Explosion puffing has been also reported to increase protein digestibility (Joye 2019). The extrusion process did not significantly affect the protein content quantitatively but qualitative changes were observed in protein structure (Yang et al. 2022). Structural changes like an increase in β-fold content and a decrease in the α-helix and β-turn content were observed. Fu et al. (2023) assessed the impact of processing on the in-vitro protein digestibility and digestible indispensable amino acid score using an in-vitro digestion protocol, INFOGEST (an international network of excellence on the fate of food in the gastrointestinal tract) protocol. Millet grains were soaked for 30 min at 25°C and cooked in water (1:1.5 w/v) for 25 minutes. The results showed a decrease in the availability of cysteine and isoleucine while increasing the availability of lysine. Traditional processes like soaking, germination, malting, fermentation, etc., also affect protein quality, digestibility, and functionality. The effect of various processing methods on protein quality is given in Table 11.2.

The non-thermal methods also change the protein quality. HPP results in protein denaturation. This leads to aggregation and dissociation of polypeptides, and modifies their surface hydrophobicity, solubility, and other physiochemical properties (He et al. 2016). The pulsed

TABLE 11.2 EFFECT OF PROCESSING METHODS ON THE MILLET PROTEIN QUALITY

Type of millet	Method of processing	Experimental conditions	Inference	References
Pearl millet	Germination	Room temperature for 72 hours	Varietal variation in protein changes. Reduction in MRB variety and no impact on K variety	Suma and Urooj 2014
	Malting	Steeping at 25°C for 24 hours	Increase in protein from 7.5 to 9.19%	Obadina et al. 2017
	Cold plasma treatment	40–45 kV for 5–15 minutes	Reduction in protein content; alteration in protein structure	Lokeswari et al. 2021
Foxtail millet	Roasting	10 minutes	Reduction in total protein	Sudha et al. 2021
	Cooking in alkaline conditions	95°C for 30 minutes in varying concentration on calcium chloride solution	Increase in protein content by 27% approx.	Rajeswari et al. 2015
	Germination	46.5 hours	Increase in protein content from 10.6 to 13.75 g/100g	Gowda et al. 2022
Proso millet	Cooking	1.83–2.73 minutes	Lower protein solubility and high content of β-sheet structure	Tyl et al. 2020
	Ultrasound	–	Increase in protein solubility	Nazari et al. 2018
Finger millet	Roasting	At temperature of 120°C for 5 minutes	Reduction in essential amino acids	Navyashree et al. 2022
	γ radiation	Dose: 15 kGy	Increase in protein content	Zhang et al. 2022
Kodo millet	Microwave heating	720, 540, and 360 W power for 150, 210, and 270 s	Alterations in protein functionality	Gopal and Bhuvana 2021
	Germination	Ambient temperature	Increase in total protein by 29.7%	Gowda et al. 2022, Sharma et al. 2018, Pradeep et al. 2015
Little millet	Pan and microwave cooking	–	Pan cooking reduced protein; increase in protein content was observed in microwave cooking	Kumar et al. 2020

electric field has been reported to improve *in-vitro* protein digestibility by 18–31% in bovine longissimus thoracis (Chian et al. 2019). Pulsed light treatments favour protein aggregation in bovine milk due to modification in the β-lactoglobulin structure (Orcajo et al. 2019). The exposure of whole and dehulled seeds of pearl millet to gamma rays generated from cobalt-60 at a dosage of 2 Gy/ min at 25°C and normal relative humidity had no significant effect on protein digestibility. However, a reduction in protein digestibility was reported in the same study when radiation was followed by cooking (El Shazali et al. 2011). Electromagnetic waves can damage the structure of proteins and cause conformational changes via the creation of free radicals or other molecules (Han et al. 2018).

Processing techniques also exhibit a significant impact on the functional properties of millet protein concentrate. Nazari et al. (2023) studied the impact of ultrasound on the functional properties of proso millet concentrates. Proso millet concentrates were exposed to high-power ultrasound (18.4, 29.58, and 73.95 W/cm^2) for 5–20 minutes. The treatment positively impacted the solubility of the concentrates. The low intensity of ultrasound reduced the foaming capacity but increased with the application at high intensity. Ultrasound treatment also increased the negative surface charge on proso millet protein concentrate as depicted by zeta potential. This leads to aggregate expansion and prevention of aggregation, hence modifying the dispersion stability. An increase in foam capacity and foam stability was reported at high ultrasound intensity. Significant improvement was also reported for emulsion activity index and emulsion stability.

11.6 APPLICATION OF MILLET PROTEINS

11.6.1 Food-Related Applications

Plant-derived proteins are sustainable sources of protein supply. Protein concentrates and isolates are used as nutritional supplements and used for the fortification of food products to enhance their nutritional value, to obtain desired functional properties, and as biopolymers in manufacturing materials for food. Also, millets can be used as an alternative to wheat in formulations for people suffering from celiac disease or gluten intolerance (Kumar et al. 2018). Millet proteins naturally have low solubility and can be used in appropriate products like baked products, protein bars, breakfast cereals, etc. Altered functional properties can be achieved through applications of techniques like high-intensity ultrasound and enzymatic hydrolysis. The modified proteins exhibit improved solubility (Kamara et al. 2010). Such products can be successfully used in soups, salads, etc.

11.6.2 Therapeutic Applications

The therapeutic potential of millet protein has been explored in recent studies. Several studies have exhibited the positive impact of millet protein on health. Millet proteins in the form of isolates, concentrates, hydrolysates, and bioactive peptides are beneficial in the treatment and management of metabolic disorders, cardiovascular diseases, inflammation, cancer, etc. (Figure 11.1).

Proso millet is reported to contain bioactive peptides with anti-lipoxygenase properties (Jakubczyk et al. 2019). Lipoxygenases are enzymes that catalyze the oxidation of polyunsaturated fatty acids, and generate hydroperoxides (free radicals), indicating the antioxidant nature of millets. He et al. (2022) assessed the anti-inflammatory properties of peptides derived from

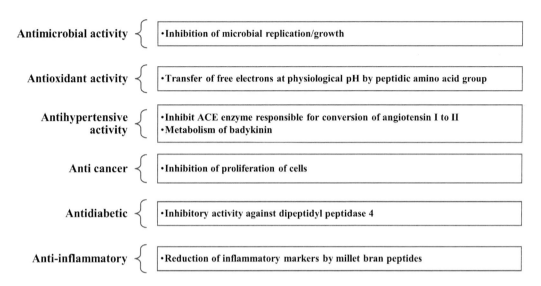

Figure 11.1 Health benefits of millet proteins.

millet bran. Millet bran peptide significantly reduced tumour necrosis factor-α, interleukin 1β, and prostaglandin in lipo-polysaccharide-induced RAW264.7 cells (He et al. 2022). Similar results were obtained in the rat model as well. In another study, foxtail millet protein hydrolysates were tested for their efficacy in treating experimental colitis in mice. The study showed that the consumption of foxtail millet protein hydrolysates orally decreased the disease activity index score and improved the symptoms of dextran sulphate sodium-induced colitis mice. The study identified 22 peptides as potential biopeptides in foxtail millet bran (Zhang et al. 2022). Bisht et al. (2016) successfully identified anti-bacterial peptides in finger, barnyard, and proso millet. Finger millet seed extract exhibited a zone of inhibition of 22.6 mm and anti-bacterial activity against *P. aeruginosa*. Protein hydrolysates from foxtail millet showed antihypertensive activity. The activity was achieved by an oral dosage of 200 mg peptides/kg body weight. The effect was better for raw and extruded samples in comparison to fermented samples (Chen et al. 2017). The mechanism of action of bioactive peptides for therapeutics is represented in Figure 11.1.

11.6.3 Other Applications

The gel-forming property of millet proteins is one of the potential features which allows the formation of gels with diverse microstructural and mechanical properties contributing to its broad spectrum of industrial applications. This development has paved the way for the creation of biocompatible carriers, which provide a promising platform for encapsulating and protecting sensitive bioactive compounds during digestion and facilitating their controlled release in the body (Feng et al. 2022; Agrawal et al. 2021).

Protein-based microparticles have found widespread and quickly expanding applications in the food industry because they can be accurately produced for usage in a wide range of food formulations and may encapsulate practically any ingredient, whether hydrophobic, hydrophilic, or even microbial. Food proteins have a lot of potential for creating novel GRAS (generally recognized as safe) matrices with the ability to combine nutraceutical ingredients and deliver controlled release orally (Kumar et al. 2022, Gwamba 2016; Taylor 2017). Several food components need to be encapsulated due to their undesirable odour, taste, instability, and the need for possible controlled release.

Pearl millet and foxtail millet nanoparticles with improved structural, functional, and thermal properties were obtained using ultrasonication treatment. An increase in nanoparticle stability was observed as indicated by increased zeta potential (from −11.25 to −15.87). Nano-protein samples were shown to have altered functional properties like solubility, emulsification, foaming, etc. (Jhan et al. 2021).

Millet protein has been used as a carrier for the delivery of lipophilic bioactive compounds. In a study conducted by Wang et al. (2018), proso millet protein was used successfully as an encapsulating material for the delivery of curcumin. The entrapment efficiency varied from 11.2 to 78.9% with millet protein extracted via ethanol giving better performance than wet milling. The encapsulation of curcumin using proso millet protein did not exhibit a negative impact on the antioxidant activity. The encapsulation of curcumin was also attempted by Chen et al. (2022) using fabricating caseinate stabilized composite nanoparticles from foxtail millet prolamin. Curcumin-loaded caseinate-stabilized foxtail millet nanodispersions were synthesized by the anti-solvent/evaporation method. The encapsulation efficiency obtained for the particles was more than 71.3%.

11.7 CONCLUSIONS

Glutelins are the major proteins in millets followed by prolamins. Extraction from the matrix and isolation steps are critical to the functionality of the isolated protein. The characterization of millet protein revealed that the majority of the millets have been found to have polypeptides with a weight of only 100 kDa or less. Thermal and non-thermal processing methods affect the digestibility and functionality of proteins. The functional properties of millet protein can be enhanced with the use of techniques like ultrasound treatment and enzymatic hydrolysis. Millet proteins find application in food products as supplements and can also be used to enhance the nutritional content of products. Millet protein peptides exhibit therapeutic potential against metabolic disorders, cardiovascular diseases, inflammation, etc. Millet protein is also used as a delivery vehicle and encapsulating material for the delivery of bioactive compounds without affecting their bioactivity. In a nutshell, millet proteins hold immense potential for food and non-food applications; however, only limited research is available in this dimension. Future studies should focus on exploring the potential of millet proteins for better utilization in food, pharmaceutical, and other industries.

REFERENCES

Abioye, V. F., G. O. Babarinde, G. O. Ogunlakin, J. A. Adejuyitan, S. J. Olatunde, and A. O. Abioye. 2022. Varietal and processing influence on nutritional and phytochemical properties of finger millet: A review. *Heliyon* 8(12). Doi: 10.1016/j.heliyon.2022.e12310.

Agrawal, H., R. Joshi, and M. Gupta. 2019. Purification, identification and characterization of two novel antioxidant peptides from finger millet (*Eleusine coracana*) protein hydrolysate. *Food Research International* 120: 697–707.

Agrawal, H., R. Joshi, and M. Gupta. 2020. Functional and nutritional characterization of *in vitro* enzymatic hydrolyzed millets proteins. *Cereal Chemistry* 97(6): 1313–1323.

Agrawal, H., R. Joshi, and M. Gupta. 2021. Optimization of pearl millet-derived bioactive peptide microspheres with double emulsion solvent evaporation technique and its release characterization. *Food Structure* 29: 100200.

Agrawal, H., R. Joshi, and M. Gupta. 2016. Isolation, purification and characterisation of antioxidative peptide of pearl millet (*Pennisetum glaucum*) protein hydrolysate. *Food Chemistry* 204: 365–372. Doi: 10.1016/j.foodchem.2016.02.127.

Akharume, F., D. Santra, and A. Adedeji. 2020. Physicochemical and functional properties of proso millet storage protein fractions. *Food Hydrocolloids* 108: 105497.

Ali, M. A., A. H. E. Tinay, I. A. Mohamed, and E. E. Babiker. 2009. Supplementation and cooking of pearl millet: Changes in protein fractions and sensory quality. *World Journal of Dairy and Food Sciences* 4(1): 41–45.

Allai, F. M., Z. Azad, G. Khalid, and B. N. Dar. 2022. Wholegrains: A review on the amino acid profile, mineral content, physicochemical, bioactive composition and health benefits. *International Journal of Food Science and Technology* 57(4): 1849–1865.

Anitha, S., M. Govindaraj, and J. Kane-Potaka. 2020. Balanced amino acid and higher micronutrients in millets complements legumes for improved human dietary nutrition. *Cereal Chemistry* 97(1): 74–84.

Bagdi, A., G. Balázs, J. Schmidt, M. Szatmári, R. Schoenlechner, E. Berghofer, and S. J. A. A. Tömösközia. 2011. Protein characterization and nutrient composition of Hungarian proso millet varieties and the effect of decortication. *Acta Alimentaria* 40(1): 128–141.

Ballogou, V. Y., M. M. Soumanou, F. Toukourou, and J. D. Hounhouigan. 2013. Structure and nutritional composition of fonio (*Digitaria exilis*) grains: A review. *International Research Journal of Biological Sciences* 2(1): 73–79.

Bangar, S. P., S. Suri, S. Malakar, N. Sharma, and W. S. Whiteside. 2022. Influence of processing techniques on the protein quality of major and minor millet crops: A review. *Journal of Food Processing and Preservation* 46(12): e17042. Doi: 10.1111/jfpp.17042.

Bibi, R., H. Mokrane, K. Khaladi, and H. Amoura. 2021. Improvement of pearl millet (ennisetum glaucum (l.) R. Br) prolamin extractability: chromatographic separation, characterization and functional properties. *Journal of Microbiology, Biotechnology and Food Sciences* 11(2): e3674–e3674.

Bisht, A., M. Thapliyal, and A. Singh. 2016. Screening and isolation of antibacterial proteins/peptides from seeds of millets. *International Journal of Current Pharmaceutical Research* 8(3): 96–99.

Bora, P., S. Ragaee, and M. Marcone. 2019. Effect of parboiling on decortication yield of millet grains and phenolic acids and in vitro digestibility of selected millet products. *Food Chemistry* 274: 718–725.

Budhwar, S., K. Sethi, and M. Chakraborty. 2020. Efficacy of germination and probiotic fermentation on underutilized cereal and millet grains. Food production. *Processing and Nutrition* 2: 1–17.

Chen, J., W. Duan, X. Ren, C. Wang, Z. Pan, X. Diao, and Q. Shen. 2017. Effect of foxtail millet protein hydrolysates on lowering blood pressure in spontaneously hypertensive rats. *European Journal of Nutrition* 56(6): 2129–2138. Doi: 10.1007/s00394-016-1252-7.

Chen, X., T. Y. Zhang, Y. C. Wu, P. X. Gong, and H. J. Li. 2022. Foxtail millet prolamin as an effective encapsulant deliver curcumin by fabricating caseinate stabilized composite nanoparticles. *Food Chemistry* 367: 130764.

Chian, F. M., L. Kaur, I. Oey, T. Astruc, S. Hodgkinson, and M. Boland. 2019. Effect of Pulsed Electric Fields (PEF) on the ultrastructure and in vitro protein digestibility of bovine longissimus thoracis. *LWT-Food Science and Technology* 103: 253–259. Doi: 10.1016/j.lwt.2019.01.005.

del Mar Contreras, M., A. Lama-Muñoz, J. M. Gutiérrez Pérez, F. Espínola, M. Moya, and E. Castro. 2019. Castro Protein extraction from agri-food residues for integration in biorefinery: Potential techniques and current status. *Bioresource Technology* 280: 459–477.

Dharmaraj, U., and N. G. Malleshi. 2011. Changes in carbohydrates, proteins and lipids of finger millet after hydrothermal processing. *LWT – Food Science and Technology* 44(7): 1636–1642.

ElShazali, A. M., A. A. Nahid, H. A. Salma, and E. B. Elfadil. 2011. Effect of radiation process on antinutrients, protein digestibility and sensory quality of pearl millet flour during processing and storage. *International Food Research Journal* 18(4): 1401.

Evers, T., and S. Millar. 2002. Cereal grain structure and development: Some implications for quality. *Journal of Cereal Science* 36(3): 261–284.

Feng, R., Y. Fan, L. Chen, Q. Ge, J. Xu, M. Yang, and K. Chen. 2022. Based on 16 S RRNA sequencing and metabonomics to reveal the new mechanism of aluminum potassium sulfate induced inflammation and abnormal lipid metabolism in mice. *Ecotoxicology and Environmental Safety* 247: 114214.

Fu, L., S. Gao, and B. Li. 2023. Impact of processing methods on the in vitro protein digestibility and DIAAS of various foods produced by millet, highland barley and buckwheat. *Foods* 12(8): 1714.

Gebru, Y. A., D. B. Sbhatu, and K. Kim. 2020. Nutritional composition and health benefits of Teff (*Eragrostis tef* (Zucc.) Trotter). *Journal of Food Quality*: Article ID 9595086. Doi: 10.1155/2020/9595086.

Gebru, Y. A., J. Hyun-Ii, K. Young-Soo, K. Myung-Kon, and K. Kwang-Pyo. 2019. Variations in amino acid and protein profiles in white versus brown teff (*Eragrostis tef*) seeds, and effect of extraction methods on protein yields. *Foods* 8(6): 202.

Gopal, D. S., and S. Bhuvana. 2021. Effect of microwave treatment on phytochemical, functional and rheological property of Kodo millet (Paspalum scrobiculatum). *The Pharma Innovation Journal* 10 (11): 397–401.

Gowda, N. N., K. Siliveru, P. V. Prasad, Y. Bhatt, B. P. Netravati, and C. Gurikar. 2022. Modern processing of Indian millets: A perspective on changes in nutritional properties. *Foods* 11(4): 499.

Gowthamraj, G., M. Raasmika, and S. Narayanasamy. 2020. Efficacy of fermentation parameters on protein quality and microstructural properties of processed finger millet flour. *Journal of Food Science and Technology* 58 (8): 3223–3234.

Gulati, P., A. Li, D. Holding, D. Santra, Y. Zhang, and D. J. Rose. 2017. Heating reduces proso millet protein digestibility via formation of hydrophobic aggregates. *Journal of Agricultural and Food Chemistry* 65(9): 1952–1959.

Gulati, P., Sabillón, L., & Rose, D. J. (2018). Effects of processing method and solute interactions on pepsin digestibility of cooked proso millet flour. *Food Research International*, 109, 583–588.

Gwamba, J. 2016. Pearl Millet: Influence of mineral biofortification and simple processing technologies on minerals and antinutrients. MSc. Diss. University of Pretoria. https://repository.up.ac.za/bitstream/handle/2263/60808/Gwamba_Pearl_2016.pdf?sequence=1&isAllowed=y (accessed June 5, 2023).

Han, Z., M. Cai, J. Cheng, and D. Sun. 2018. Effects of electric fields and electromagnetic wave on food protein structure and functionality: A review. *Trends in Food Science and Technology* 75: 1–9. Doi: 10.1016/j.tifs.2018.02.017.

Hassan, Z. M., N. A. Sebola, and M. Mabelebele. 2021. The nutritional use of millet grain for food and feed: A review. *Agriculture and Food Security* 10(1): 16.

He, R., M. Liu, Z. Zou, M. Wang, Z. Wang, X. Ju, and G. Hao. 2022. Anti-inflammatory activity of peptides derived from millet bran in vitro and in vivo. *Food and Function* 13(4): 1881–1889.

He, X., L. Mao, Y. Gao, and F. Yuan. 2016. Effects of high pressure processing on the structural and functional properties of bovine lactoferrin. *Innovative Food Science and Emerging Technologies* 38: 221–230. Doi: 10.1016/j.ifset.2016.10.014.

Himanshu, K., S. K. Sonawane, and S. S. Arya. 2018. Nutritional and nutraceutical properties of millets: A review. *Clinical Journal of Nutrition and Dietetics* 1(1): 1–10.

Hu, S., J. Yuan, J. Gao, Y. Wu, X. Meng, P. Tong, and H. Chen. 2020. Antioxidant and anti-inflammatory potential of peptides derived from *in vitro* gastrointestinal digestion of germinated and heat-treated foxtail millet (*Setaria italica*) proteins. *Journal of Agriculture and Food Chemistry* 68(35): 9415–9426.

Huchchannanavar, S., L. N. Yogesh, and S. M. Prashant. 2019. Nutritional and physicochemical characteristics of foxtail millet genotypes. *International Journal of Current Microbiology and Applied Sciences* 8(01): 1773–1778.

Jadhav, H. B., U. S. Annapure, and R. R. Deshmukh. 2021. Non-thermal technologies for food processing. *Frontiers in Nutrition* 8: 657090. Doi: 10.3389/fnut.2021.657090.

Jakubczyk, A., U. Szymanowska, M. Karaś, U. Złotek, and D. Kowalczyk. 2019. Potential anti-inflammatory and lipase inhibitory peptides generated by in vitro gastrointestinal hydrolysis of heat treated millet grains. *CyTA: Journal of Food* 17(1): 324–333.

Jhan, F., A. Gani, N. Noor, and A. Shah. 2021. Nanoreduction of millet proteins: Effect on structural and functional properties. *ACS Food Science and Technology* 1(8): 1418–1427.

Ji, Z., R. Feng, and J. Mao. 2019. Separation and identification of antioxidant peptides from foxtail millet (*Setaria italica*) prolamins enzymatic hydrolysate. *Cereal Chemistry* 96 (6): 981–993.

Joye, I. 2019. Protein digestibility of cereal products. *Foods* 8(6). Doi: 10.3390/foods8060199.

Kamara, M. T., I. Amadou, F. Tarawalie, and Z. Huiming. 2010. Effect of enzymatic hydrolysis on the functional properties of foxtail millet (*Setaria italica* L.) proteins. *International Journal of Food Science and Technology* 45 (6): 1175–1183. Doi: 10.1111/j.1365-2621.2010.02260.x.

Kaur, B., A. Singh, S. Suri, M. Usman, and D. Dutta. 2023. Minor millets: A review on nutritional composition, starch extraction/modification, product formulation, and health benefits. *Journal of the Science of Food and Agriculture* 103(10): 4742–4754. Doi: 10.1002/jsfa.12493.

Krishnan, R., and M. S. Meera. 2018. Pearl millet minerals: Effect of processing on bioaccessibility. *Journal of Food Science and Technology* 55(9): 3362–3372. Doi: 10.1007/s13197-018-3305-9.

Kumar, L., D. Pandey, and A. Kumar. 2017. Isolation, characterization and expression analysis of a nutritionally enhanced α-prolamin gene and protein during developing spikes of finger millet (*Eleusine coracana*). *Seed Science Research* 27 (4): 262–272.

Kumar, A., P. Kumari, and M. Kumar. 2022. Role of millets in disease prevention and health promotion. In R. B. Singh, S. Watanabe, and A. A. Isaza (Eds.), *Functional Foods and Nutraceuticals in Metabolic and Non-Communicable Diseases*, 341–357. Academic Press, Cambridge.

Kumar, A., V. Tomer, A. Kaur, V. Kumar, and K. Gupta. 2018. Millets: A solution to agrarian and nutritional challenges. *Agriculture and Food Security* 7(1): 1–15.

Kumar, S. R., M. B. Sadiq, and A. K. Anal. 2020. Comparative study of physicochemical and functional properties of pan and microwave cooked underutilized millets (Proso and Little). *LWT* 128: 109465.

Li, W., L. Wen, Z. Chen, Z. Zhang, X. Pang, Z. Deng, T. Liu, and Y. Guo. 2021. Study on metabolic variation in whole grains of four proso millet varieties reveals metabolites important for antioxidant properties and quality traits. *Food Chemistry* 357: 129791.

Lokeswari, R., P. S. Sharanyakanth, S. Jaspin, and R. Mahendran. 2021. Cold plasma effects on changes in physical, nutritional, hydration, and pasting properties of pearl millet (*Pennisetum glaucum*). *IEEE Transactions on Plasma Science* 49 (5): 1745–1751.

Longvah, T., R. Ananthan, K. Bhaskarachary, and K. Venkaiah. 2017. *Indian Food Composition Tables*. Hyderabad: National Institute of Nutrition.

Mohamed, T. K., K. X. Zhu, A. Issoufou, T. Fatmata, and H. M. Zhou. 2009. Functionality, in vitro digestibility and physicochemical properties of two varieties of defatted foxtail millet protein concentrates. *International Journal of Molecular Sciences* 10(12): 5224–5238.

Monteiro, P. V., L. Sudharshana, and G. Ramachandra. 1988. Japanese barnyard millet (*Echinochloa frumentacea*): Protein content, quality and SDSPAGE of protein fractions. *Journal of the Science of Food and Agriculture* 43(1): 17–25. doi: 10.1002/jsfa.2740430104.

Nanje Gowda, N. A., K. Siliveru, P. V. Vara Prasad, Y. Bhatt, B. P. Netravati, and C. Gurikar. 2022. Modern processing of Indian millets: A perspective on changes in nutritional properties. *Foods* 11(4). doi: 10.3390/foods11040499.

Navyashree, N., A. S. Sengar, C. K. Sunil, and N. Venkatachalapathy. 2022. White Finger Millet (KMR-340): A comparative study to determine the effect of processing and their characterisation. *Food Chemistry* 374: 131665.

Nazari, B., M. A. Mohammadifar, S. Shojaee-Aliabadi, E. Feizollahi, and L. L. Mirmoghtadaie. 2018. Effect of ultrasound treatments on functional properties and structure of millet protein concentrate. *Ultrasonics Sonochemistry* 41: 382–388.

Obadina, A. O., C. A. Arogbokun, A. O. Soares, C. W. P. de Carvalho, H. T. Barboza, and I. O. Adekoya. 2017. Changes in nutritional and physico-chemical properties of pearl millet (*Pennisetum glaucum*) ex-borno variety flour as a result of malting. *Journal of Food Science and Technology* 54: 4442–4451.

Orcajo, J., M. Lavilla, and I. Martínez-de-Marañón. 2019. Effect of pulsed light treatment on β-lactoglobulin immunoreactivity. *LWT- Food Science and Technology* 112: 108231. doi: 10.1016/j.lwt.2019.05.129.

Pradeep, P. M., and Y. N. Sreerama. 2015. Impact of processing on the phenolic profiles of small millets: Evaluation of their antioxidant and enzyme inhibitory properties associated with hyperglycemia. *Food Chemistry* 169: 455–463.

Rajeswari, J. R., Guha, M., Jayadeep, A., & Rao, B. S. (2015). Effect of alkaline cooking on proximate, phenolics and antioxidant activity of foxtail millet (Setaria italica). *World Appl. Sci. J, 33*(1): 146–152.

Ramashia, S. E., M. E. Mashau, and O. O. Onipe. 2021. Millets cereal grains: Nutritional composition and utilisation in sub-Saharan Africa. In A. K. Goyal (Ed.), *Cereal Grains*, 115–128. London: IntechOpen.

Sachdev, N., S. Goomer, and L. R. Singh. 2021. Foxtail millet: A potential crop to meet future demand scenario for alternative sustainable protein. *Journal of the Science of Food and Agriculture* 101(3): 831–842.

Sachdev, N., S. Goomer, L. R. Singh, V. M. Pathak, D. Aggarwal, and R. K. Chowhan. 2023. Current status of millet seed proteins and its applications: A comprehensive review. *Applied Food Research* 3(1): 100288.

Selokar, N., R. Vidyalakshmi, P. Thiviya, V. R. N. Sinija, and V. Hema. 2022. Assessment of nutritional quality of non-conventional millet malt enriched bar. *Journal of Food Processing and Predervation* 46(12): e17271.

Sharma, R., S. Sharma, and B. Singh. 2022. Modulation in the bio-functional & technological characteristics, *in vitro* digestibility, structural and molecular interactions during bioprocessing of proso millet (*Panicum miliaceum* L.). *Journal of Food Composition and Analysis* 107: 104372.

Sharma, S., R. Sharma, and B. Singh. 2021. Influence of bioprocessing treatments on phytochemical and functional properties, *in vitro* digestibility, protein secondary structure and morphological characteristics of Indian barnyard millet flour. *International Journal of Food Science and Technology* 57(8): 4744–4753.

Sharma, S., Saxena, D. C., & Riar, C. S. (2018). Characteristics of β-glucan extracted from raw and germinated foxtail (Setaria italica) and kodo (Paspalum scrobiculatum) millets. *International journal of biological macromolecules*, 118: 141–148.

Sharma, N., J. K. Sahu, A. Choudhary, M. Meenu, and V. Bansal. 2023. High intensity ultrasound (HIU)-induced functionalization of foxtail millet protein and its fractions. *Food Hydrocolloids* 134: 108083. doi: 10.1016/j.foodhyd.2022.108083.

Singh, N., J. David, D. K. Thompkinson, B. S. Seelam, H. Rajput, and S. Morya. 2018. Effect of roasting on functional and phytochemical constituents of finger millet (*Eleusine coracana* L.). *The Pharma Innovation Journal* 7 (4): 414–418.

Sudha, K. V., Karakannavar, S. J., Yenagi, N. B., & Inamdar, B. 2021. Effect of roasting on the physicochemical and nutritional properties of foxtail millet (Setaria italica) and bengal gram dhal flours. *The Pharma Innovation Journal*, 10(5): 1543–1547.

Sudharshana, L., P. V. Monteiro, and G. Ramachandra. 1988. Studies on the proteins of kodo millet (*Paspalum scrobiculatum*). *Journal of the Science of Food and Agriculture* 42(4): 315–323. doi: 10.1002/jsfa.2740420405.

Suma, P. F., and A. Urooj. 2014. Nutrients, antinutrients & bioaccessible mineral content (in vitro) of pearl millet as influenced by milling. *Journal of Food Science and Technology* 51(4): 756–761.

Taylor, J. R. N. 2017. Millets: Their unique nutritional and health-promoting attributes. In J. R. N. Taylor, and J. M. Awika (Eds.), *Gluten-Free Ancient Grains-Cereals, Pseudocereals, and Legumes: Sustainable, Nutritious, and Health-Promoting Foods for the 21st Century*, 55–103. Duxford: Woodhead Publishing-Elsevier.

Tyl, C., A. Marti, and B. P. Ismail. 2020. Changes in protein structural characteristics upon processing of gluten-free millet pasta. *Food Chemistry* 327: 127052.

Ujong, A. E., O. O. Fashakin, and R. D. Davour. 2022. Effect of a millet-based fortified complementary food on the anthropometric and biochemical indices of anemic infants (6–24 Months). *North African Journal of Food and Nutrition Research* 6(14): 198–206.

Uwagbale, E. E. D., A. S. Saratu, O. V. Akagwu, O. O. Stephen, and A. M. Lilian. 2016. African cereals and non-African cereals: A comparative review of their nutritional composition. *World* 1(2): 30–37.

Virupaksha, T. K., G. Ramachandra, and D. Nagaraju. 1975. Seed proteins of finger millet and their amino acid composition. *Journal of the Science of Food and Agriculture* 26(8): 1237–1246.

Wang, H., D. Li, C. Wan, Y. Luo, Q. Yang, and X. B. Gao. 2021. Improving the functionality of proso millet protein and its potential as a functional food ingredient by applying nitrogen fertiliser. *Foods* 10(6): 1332.

Wang, L., P. Gulati, D. Santra, D. Rose, and Y. Zhang. 2018. Nanoparticles prepared by proso millet protein as novel curcumin delivery system. *Food Chemistry* 240: 1039–1046.

Yang, T., S. Ma, J. Liu, and B. X. Sun. 2022. Wang Influences of four processing methods on main nutritional components of foxtail millet: A review. *Grain & Oil Science and Technology* 5(3): 156–165.

Zhang, B., Y. Xu, C. Zhao, Y. Zhang, H. Lv, X. Ji, and S. Wang. 2022. Protective effects of bioactive peptides in foxtail millet protein hydrolysates against experimental colitis in mice. *Food and Function* 13(5): 2594–2605.

Chapter 12

Mineral Bioavailability and Impact of Processing

12.1 INTRODUCTION

Minerals are essential nutrients that play a vital role in numerous physiological processes within the human body, including enzyme function, bone health, nerve transmission, and energy production. However, the bioavailability of minerals, which refers to the extent of their absorption and utilization by the body (Falah and Mohssan 2017), is a critical factor in ensuring their effective contribution to these functions. In the context of millets, understanding the factors that influence mineral bioavailability and the impact of processing techniques is crucial for optimizing the nutritional value of these versatile and widely used grains in the food industry (Puranik et al. 2017).

The bioavailability of minerals in millets is influenced by various factors. One such factor is the chemical form of minerals, where different forms can exhibit varying solubility and reactivity in the digestive system, consequently affecting their bioavailability (Cabrera et al. 2006). Another factor is the presence of anti-nutrients, such as phytates and oxalates, which are naturally occurring compounds in certain foods (Tharifkhan et al. 2021). These anti-nutrients can bind to minerals, forming insoluble complexes that hinder mineral absorption. In millets, the presence of anti-nutrients can significantly impact mineral bioavailability by reducing the number of minerals available for absorption (Hassan et al. 2021). Nutrient interactions also play a role in mineral bioavailability. Some minerals exhibit synergistic or antagonistic relationships, where they can enhance or inhibit the absorption of other minerals (Hassan et al. 2021). Achieving an appropriate balance of minerals is essential for optimal bioavailability and overall health (Yousaf et al. 2021). The pH of different parts of the digestive system can affect mineral solubility and subsequent absorption (Samtiya et al. 2021). The pH levels and digestive enzymes in the gastrointestinal tract also influence mineral bioavailability. Digestive enzymes, such as proteases and carbohydrate hydrolyzing (amylases, sucrases, maltases, oligosaccharidases, etc.) enzymes, aid in the breakdown of food and the release of minerals from complex structures, facilitating their absorption (Jagati et al. 2021). Individual variations in mineral bioavailability are influenced by factors such as genetic variation, age, sex, and underlying health conditions. Genetic variations can affect the expression and function of proteins involved in mineral absorption. Age-related changes in the digestive system and nutrient transporters can also impact mineral bioavailability. Additionally, certain health conditions, such as gastrointestinal disorders or mineral deficiencies, can affect the body's ability to absorb and utilize minerals effectively (Nadziakiewicza et al. 2019).

Millets are widely used in the food industry due to their versatility, ease of handling, and potential for targeted nutrient delivery. They can be used as ingredients in various food products or consumed directly as snacks or cereals (Durairaj et al. 2019). The abilities of millets like controlled release and targeted nutrient delivery make them suitable for developing functional foods. Therefore, understanding mineral bioavailability in millets and the impact of processing techniques is essential for optimizing their nutritional value. Various millet processing techniques, such as extrusion, coating, roasting, toasting, milling, grinding, etc., can have a significant impact on mineral bioavailability. Extrusion, for example, involves the application of heat, pressure, and shear forces to raw materials and can cause changes in mineral structure and composition, affecting their solubility and subsequent absorption (Rahate et al. 2021). Coating can enhance mineral stability and solubility by protecting minerals from interactions with other components during processing and digestion. Roasting and toasting involve the application of heat, which can induce changes in mineral form and solubility (Galanakis 2021). Milling and grinding processes can also affect mineral bioavailability by altering particle size, which can influence the rate and extent of mineral release during digestion (Cubadda et al. 2009).

In addition to understanding the effects of processing techniques on mineral bioavailability, it is important to explore techniques that can enhance bioavailability in millets. These techniques include the addition of chelating agents, fermentation and enzyme treatment, fortification, and micronutrient premixes, as well as the application of novel technologies like nanoencapsulation and inclusion complexes. Chelation is a process where minerals form complexes which can enhance their absorption and bioavailability compared to free forms. Chelating agents can enhance mineral absorption by forming complexes that improve solubility and stability. Fermentation and enzyme treatment can modify millet components, including anti-nutrients, to improve mineral bioavailability (Nkhata et al. 2018). The mineral content of millet-based products can also be enhanced by fortification. Fortification and the development of micronutrient premixes involve the addition of minerals to millet-based products to enhance their nutritional value. These approaches ensure a controlled and standardized mineral content in the final product, promoting bioavailability. Novel technologies like nanoencapsulation and inclusion complexes offer innovative strategies to optimize mineral stability, solubility, and absorption which can be explored in millets (Saffarionpour and Diosady 2023; Sneha and Kumar 2022). These technologies protect minerals from degradation or undesirable interactions, enhancing their bioavailability.

To evaluate mineral bioavailability in millets, various methods can be employed, including *in-vitro* digestion models, animal studies, human intervention trials, and analytical techniques. *In-vitro* digestion models simulate the digestive process and can provide insights into mineral release, absorption, and utilization (Silva et al. 2020). Animal studies and human intervention trials can further assess mineral bioavailability by measuring biomarkers or physiological responses (Picó et al. 2019). Analytical techniques, such as atomic absorption spectroscopy or inductively coupled plasma mass spectrometry, allow for the accurate quantification of mineral content in millets and their bioavailability assessment.

The understanding of mineral bioavailability in millets and the impact of processing techniques is crucial for developing nutrient-rich and bioavailable food products. Hence, this chapter discusses the various factors affecting mineral bioavailability, the methods to enhance their bioavailability, and the effect of processing techniques on mineral bioavailability.

12.2 FACTORS AFFECTING MINERALS BIOAVAILABILITY

12.2.1 Chemical Form of Minerals

Millet is a group of small-seeded grasses that are widely consumed as staple food crops in many parts of the world. When it comes to the bioavailability of minerals in millets, the chemical form of minerals also plays a crucial role, similar to other foods. The presence of oxalates and phytic acid makes insoluble complexes like calcium oxalate, ferrous oxalate, zinc phytate, etc., which are largely indigestible. The phytic acid also binds with the divalent ions of calcium, iron, and magnesium and decreases their absorption (Weaver and Haney 2010). In contrast to this, the use of chelated minerals in millet-based products can enhance their bioavailability. Chelated minerals, as mentioned earlier, are minerals bound to organic ligands such as amino acids or peptides (Adetola et al. 2023). By forming stable complexes, chelated minerals are protected from interactions with anti-nutrients and other substances in the digestive system. This protection can promote their absorption and utilization in the body (Onyegbulam et al. 2023). Manufacturers of millet-based products can incorporate chelated minerals to optimize their bioavailability, similar to how natural ligands found in foods optimize mineral absorption (Kumar et al. 2021). It is important to note that while the chemical form of minerals and strategies like chelation can enhance the bioavailability, the overall mineral content in millets should also be taken into consideration (Buturi et al. 2021). Millets are generally considered good sources of minerals such as calcium, iron, magnesium, phosphorus, and zinc (discussed in detail in Chapter 6). However, the specific mineral content can vary depending on the variety of millet and growing conditions (Gowda et al. 2022).

12.2.2 Presence of Anti-Nutrients

Anti-nutrients are naturally occurring compounds found in some foods that can interfere with mineral absorption. Common anti-nutrients include phytates, oxalates, tannins, and polyphenols (Petroski and Minich 2020). These substances have a high affinity for certain minerals, such as calcium, iron, and zinc, and can form insoluble complexes with them. These complexes are resistant to digestion and can reduce the bioavailability of minerals by preventing their release and absorption in the gut (Zhang et al. 2021). The presence of anti-nutrients in millets can significantly impact mineral bioavailability. Food processing techniques can help mitigate the effects of anti-nutrients by reducing their levels through methods such as soaking, fermenting, or thermal processing (Mrinal and Aluko 2020). Additionally, formulating millet recipes with ingredients that naturally contain lower levels of anti-nutrients or using mineral fortificants that are less prone to complex formation can further enhance mineral bioavailability (Ondiba et al. 2022).

12.2.3 Nutrient Interactions

In the context of millets, nutrient interactions play an important role in the absorption and utilization of minerals. The presence of certain minerals and other compounds can influence the bioavailability of minerals in millets, either through synergistic or antagonistic interactions. Synergistic interactions can occur when the presence of one mineral enhances the absorption or bioavailability of another mineral. For example, millets, being a plant-based food, contain non-haem iron, which is not as readily absorbed by the body as haem iron from animal sources. However, consuming millets alongside a source of vitamin C, such as citrus fruits or bell peppers, can enhance the absorption of non-haem iron in millets (Anitha et al. 2021). Vitamin C acts as a synergistic nutrient, promoting the uptake of non-haem iron and increasing its bioavailability. Conversely, antagonistic interactions can occur when the presence of one mineral hinders the absorption or utilization of another mineral. In the case of millets, a high intake of calcium, either from millets themselves or other dietary sources, can potentially inhibit the absorption of non-haem iron (Hurrel and Egli 2010). Calcium and iron compete for absorption in the gastrointestinal tract, and an excessive amount of calcium can interfere with the uptake of iron, particularly non-haem iron present in millets. Therefore, it is important to consider the balance of calcium and iron in millet formulations to optimize the bioavailability of both minerals. Achieving an appropriate balance of minerals in millet formulations is crucial for optimizing bioavailability and ensuring adequate intake of all essential minerals. This involves considering the presence of synergistic nutrients like vitamin C that can enhance mineral absorption and being mindful of potential antagonistic interactions, such as high calcium levels inhibiting iron absorption. A high dietary fibre content can also reduce the mineral bioavailability. Diets rich in non-soluble dietary fibres are reported to reduce the absorption of minerals in the gastrointestinal tract by their entrapment or binding (Baye et al. 2017). The presence of soluble fibres like pectin and guar gum has no significant effect on mineral bioavailability (Adams et al. 2018). By formulating millet-based meals that take these nutrient interactions into account, it is possible to maximize the bioavailability and utilization of minerals present in millets, thus supporting overall nutritional well-being.

12.2.4 pH and Digestive Enzymes

The bioavailability of minerals in millets is influenced by the pH levels across various segments of the digestive system, such as the stomach, small intestine, and large intestine. The solubility and subsequent absorption of minerals are determined by the specific pH environment in these digestive compartments (Goff 2018). Each mineral may have distinct pH requirements for optimal absorption, with some minerals benefiting from an acidic environment, while others require alkaline conditions. Millet consumption is subject to the influence of multiple factors that can alter the pH levels in the gastrointestinal tract. These factors encompass the composition of millet-based meals, medication usage, and the presence of digestive disorders, all of which can modify the environment in which mineral absorption occurs. Digestive enzymes also play a pivotal role in the bioavailability of minerals within millets. Proteases, for instance, facilitate the breakdown of proteins into amino acids, which can subsequently bind to minerals and enhance the absorption of trace minerals (Goff et al. 2021). Similarly, the enzymes responsible for the digestion of complex carbohydrates indirectly affect mineral bioavailability by influencing overall digestion and absorption processes (Adams et al. 2018). By considering the diverse pH conditions of different digestive segments and acknowledging the impact of digestive enzymes, we can better comprehend the bioavailability of minerals in millets. By optimizing these factors, we can promote the solubility, absorption, and effective utilization of minerals contained within millets, thus contributing to overall nutritional well-being.

12.2.5 Individual Variation in Bioavailability

Individual variations in mineral bioavailability can impact the absorption and utilization of minerals from millets. Factors such as genetic variation, age, sex, and underlying health conditions play a role in these variations (Anitha et al. 2021). Genetic differences can affect the expression and function of proteins involved in mineral transport and absorption, influencing individual mineral requirements and bioavailability. Aging can lead to changes in the digestive system, including reduced stomach acid production and alterations in gut microbiota composition, which can affect mineral absorption (Rusu et al. 2020). Sex differences, driven by hormonal factors, can also influence mineral metabolism and absorption. Additionally, underlying health conditions like gastrointestinal disorders, malabsorption syndromes, or mineral deficiencies can impair mineral absorption from millets (Madla et al. 2021). For example, individuals with celiac disease may experience reduced mineral absorption due to intestinal damage. Understanding these variations is essential for tailoring dietary recommendations and personalized approaches to ensure adequate mineral intake from millets, taking into account individual needs and optimizing mineral bioavailability (Caeiro et al. 2022).

12.2.6 Physical Properties of Grain

Physical properties of millets such as size, shape, density, and porosity can influence their interaction with the digestive system. The impact of physical properties on the availability has been studied in various animal models. Particle size of the grain affects transit time in the gastrointestinal tract. Smaller sized grains have shorter transit time limiting the opportunity for complete mineral absorption in animal model and vice versa (Henze et al. 2021). The shape and surface characteristics of the millets can also play a role in mineral bioavailability (Ubeyitogullari et al. 2022). Particles with irregular surfaces or rough textures may provide more surface area for enzymatic digestion, leading to increased mineral release and absorption. Additionally, the porosity of the food material can impact the diffusion of digestive enzymes and the subsequent access to minerals (Xiong et al. 2022). Hence, when taking bioavailability of nutrients into consideration, these parameters should also be considered for effective mineral delivery.

12.3 EFFECT OF PROCESSING TECHNIQUES ON MINERAL BIOAVAILABILITY

12.3.1 Extrusion

Extrusion is a widely used processing technique in the food industry that involves the application of heat, pressure, and shear forces to raw materials. This process can have implications for the bioavailability of minerals in millets (He et al. 2020). During extrusion, the heat and pressure can cause changes in the structure and composition of minerals, affecting their solubility and subsequent absorption in the digestive system (Rahate et al. 2021). The heat generated during extrusion can lead to mineral transformations, such as the conversion of insoluble forms into more soluble forms (Kamau et al. 2020). This increased solubility can enhance the bioavailability of minerals in the finished millets. Additionally, the shear forces exerted during extrusion can disrupt the crystal structure of minerals, making them more accessible for enzymatic digestion and absorption. However, it is important to note that excessive heat and prolonged processing times during extrusion can lead to mineral degradation and loss. Therefore, optimizing the extrusion conditions is crucial to maximizing mineral bioavailability while preserving their nutritional integrity (Lai et al. 2022; Rahmani et al. 2022).

12.3.2 Coating

Coating is a processing technique used to enhance the functionality and stability of millets. Coating materials can provide a protective barrier around the millet, protecting minerals from environmental factors, such as oxidation or moisture (Peng et al. 2020). Coatings can also control the release of minerals, allowing for sustained or targeted nutrient delivery. In terms of mineral bioavailability, the coating process can have several effects (Kalia et al. 2020). Firstly, the coating can alter the dissolution characteristics of minerals, affecting their release in the digestive system. Coatings with controlled-release properties can regulate the rate at which minerals are released, ensuring a sustained and prolonged availability for absorption (Yuan et al. 2019). Furthermore, the coating can protect minerals from interactions with other substances in the gastrointestinal tract, such as anti-nutrients or substances that may form insoluble complexes with minerals (Siddiqui et al. 2022). By preventing these interactions, the coating can improve mineral bioavailability. Coatings can also enhance the stability of minerals during processing and storage, preventing degradation or loss of bioactivity. This preservation of mineral integrity can contribute to their improved bioavailability in the final millet product (Karakaş et al. 2015).

12.3.3 Roasting and Toasting

Roasting and toasting are heat treatment methods commonly used in food processing. These techniques involve subjecting the millets to dry heat, resulting in chemical and physical changes in the raw materials. In terms of mineral bioavailability, roasting and toasting can impact several factors (Pateiro et al. 2021). Firstly, the heat applied during these processes can increase the solubility of minerals by breaking down complex structures or converting insoluble forms into more soluble forms (Kamau et al. 2020). This increased solubility can enhance mineral bioavailability. Roasting and toasting can also reduce the levels of anti-nutrients present in the millet (Muhammad et al. 2022). Heat treatment can degrade or inactivate certain anti-nutrients, such as phytase or oxalates, which can interfere with mineral absorption (Mrinal and Aluko 2020). By reducing the anti-nutrient content, roasting and toasting can enhance mineral bioavailability. Additionally, these heat treatment methods can modify the matrix structure of the millets (Olaifa et al. 2022). The changes in texture and composition can influence the release and accessibility of minerals during digestion, potentially improving their bioavailability (Caetano-Silva et al. 2021).

12.3.4 Milling and Grinding

Milling and grinding are particle size reduction techniques commonly used in food processing. These processes involve reducing the size of the millets to achieve the desired particle size distribution (Rousseau et al. 2020). Milling and grinding can impact mineral bioavailability through several mechanisms. Firstly, these processes increase the surface area of the millet, exposing more mineral particles to the digestive enzymes (Goff 2018). The increased surface area facilitates the enzymatic digestion and subsequent release of minerals, enhancing their bioavailability (Sugiarto et al. 2021). Furthermore, particle size reduction can improve the dispersibility and solubility of minerals in the digestive system. Finely ground millets have a higher likelihood of dissolving quickly, allowing for better mineral absorption (Xu et al. 2021). However, it is important to note that excessive milling or grinding can lead to mineral losses or alterations in their chemical composition. Therefore, the duration and intensity of milling or grinding should be carefully controlled to ensure optimal mineral bioavailability while preserving the nutritional quality of the millet (Tan et al. 2022).

Overall, processing techniques such as extrusion, coating, roasting, toasting, milling, and grinding can all have significant impacts on mineral bioavailability in millets. Understanding the effects of these techniques and optimizing processing conditions are essential for developing millet-based food products with enhanced mineral bioavailability and nutritional quality.

12.4 TECHNIQUES TO ENHANCE MINERAL BIOAVAILABILITY

12.4.1 Addition of Chelating Agents

Chelating agents are compounds that can be added to millets to enhance mineral bioavailability. These agents can bind with minerals, forming stable complexes that are more easily absorbed by the body (Yilmaz and Ağagündüz 2020). Chelating agents act by protecting minerals from interactions with other substances in the gastrointestinal tract that may hinder their absorption, such as anti-nutrients or other minerals (Zhang et al. 2021). Commonly used chelating agents include organic compounds such as amino acids (e.g. lysine, methionine) and organic acids (e.g. citric acid, malic acid) (Wang et al. 2023). These agents can bind to minerals, particularly trace minerals like iron, zinc, and copper, forming chelates that are more resistant to interference from anti-nutrients. Chelated minerals are often more soluble and have improved bioavailability compared to their inorganic forms (Fasae et al. 2021). The addition of chelating agents during millet formulation can enhance mineral bioavailability by improving the stability and solubility of minerals, facilitating their absorption and utilization by the body.

12.4.2 Fermentation and Enzyme Treatment

Fermentation and enzyme treatments are techniques that can be employed to improve mineral bioavailability in millets. Fermentation involves the controlled microbial growth and metabolism of raw materials, while enzyme treatments utilize specific enzymes to modify the composition and structure of the millets (Yu et al. 2023). During fermentation, microorganisms produce enzymes that can degrade anti-nutrients, such as phytates and oxalates, which can inhibit mineral absorption (Mrinal and Aluko 2020). The microbial activity can break down these anti-nutrients into simpler compounds, reducing their binding capacity and enhancing mineral bioavailability. Enzyme treatments involve the addition of specific enzymes to the millet formulation to facilitate the breakdown of complex compounds and enhance mineral solubility (Petroski and Minich 2020). For example, phytase is an enzyme that can degrade phytates, improving the availability of minerals like phosphorus, iron, and zinc. Proteases and carbohydrates can also be used to break down proteins and complex carbohydrates, respectively, thereby releasing minerals bound to these compounds and improving their bioavailability (Kumar et al. 2021). Both fermentation and enzyme treatments offer effective strategies to enhance mineral bioavailability by reducing the levels of anti-nutrients and modifying mineral forms, leading to improved solubility and absorption (Sarkhel and Roy 2022).

12.4.3 Fortification and Micronutrient Premixes

Fortification involves the addition of specific minerals to millets to increase their nutrient content and improve bioavailability. Fortification can be done by incorporating mineral salts or mineral-rich ingredients into the millet formulation (Drewnowski et al. 2021). Micronutrient premixes are pre-formulated blends of essential minerals and vitamins that are added to millets to ensure a balanced nutrient profile (Sawyer 2019). These premixes contain minerals in bioavailable forms, optimized ratios, and appropriate doses for specific nutritional requirements. Fortification and micronutrient premixes can enhance mineral bioavailability by ensuring an adequate and bioavailable mineral content in the millets (Keyata et al. 2021). By incorporating minerals in forms that are readily absorbed and utilized by the body, fortification can address micronutrient deficiencies and improve overall nutritional status (Jaiswal et al. 2022).

12.4.4 Novel Technologies for Mineral Delivery

Emerging technologies offer innovative approaches to enhance mineral bioavailability in millets. Two such technologies are nanoencapsulation and inclusion complexes. Nanoencapsulation involves the encapsulation of minerals within nano-sized particles or matrices (Galanakis 2021). This technology protects minerals during processing and digestion, preventing their degradation or interaction with other substances. Nanoencapsulation can also improve mineral solubility, stability, and targeted delivery to specific sites in the gastrointestinal tract, thereby enhancing their bioavailability (Shahidi and Pan 2022). Inclusion complexes involve the formation of stable complexes between minerals and other molecules, such as cyclodextrins or proteins. These complexes protect minerals from interactions with anti-nutrients and enhance their solubility and absorption (Farcas et al. 2023). Inclusion complexes can also control the release of minerals, allowing for sustained and controlled delivery in the digestive system. These novel technologies offer promising strategies to overcome challenges related to mineral bioavailability in millets (Ahmed et al. 2023). By utilizing advanced delivery systems, they can optimize mineral stability, solubility, and absorption, ultimately improving their bioavailability and nutritional impact.

In summary, techniques such as the addition of chelating agents, fermentation, enzyme treatment, fortification, micronutrient premixes, and novel technologies like nanoencapsulation and inclusion complexes can significantly enhance mineral bioavailability in millets (Galanakis 2021). These approaches offer innovative ways to optimize mineral stability, solubility, and absorption, ensuring that millets provide a reliable source of essential minerals for human nutrition.

12.5 EVALUATING MINERAL BIOAVAILABILITY IN MILLETS

12.5.1 *In-Vitro* Digestibility Methods

In-vitro digestion methods are commonly used to evaluate the bioavailability of minerals in millets. These methods simulate the conditions of the gastrointestinal tract, including the enzymatic digestion and physiological processes that occur during digestion (Ammerman 1995). *In-vitro* digestion models involve a series of sequential steps that mimic the oral, gastric, and intestinal phases of digestion. The millets are subjected to simulated saliva, gastric fluids, and intestinal enzymes to assess the release and solubility of minerals (Mulet-Cabero et al. 2020). The extent of mineral release from the millets can be measured using various techniques, such as atomic absorption spectroscopy or inductively coupled plasma mass spectrometry. By assessing mineral release and bioaccessibility, *in-vitro* digestion methods provide valuable information on the potential availability of minerals from millets (Balaram 2021). These methods allow researchers to compare different millet formulations, processing techniques, and ingredients to optimize mineral bioavailability.

12.5.2 *In-Vivo* Studies and Animal Models

Animal-based studies play a crucial role in evaluating mineral bioavailability in millets. These studies involve feeding millet to animals and assessing the absorption, tissue distribution, and utilization of minerals (Kurtz and Feeney 2020). Animal models provide insights into the physiological processes involved in mineral absorption and metabolism. They can help researchers understand factors that influence mineral bioavailability, such as interactions with other nutrients, the presence of anti-nutrients, or the impact of gut microbiota (Cao et al. 2021). *In-vivo* studies also allow for the assessment of the long-term effects of millet consumption on mineral status and overall health. Animals can be used to study the effects of millet formulations on growth, development, and nutrient deficiencies (van der Poel et al. 2020). While animal models provide valuable information, it is important to acknowledge that there may be species differences in mineral metabolism and absorption compared to humans (Costa et al. 2021). Therefore, findings from animal studies should be interpreted with caution and confirmed through human studies.

12.5.3 Human Intervention Trials

Human intervention trials are considered the gold standard for evaluating mineral bioavailability in millets. These trials involve feeding millet to human subjects and assessing the impact on mineral status, absorption, and utilization (Swann et al. 2020). Human intervention trials can be conducted using various study designs, such as randomized controlled trials. Participants consume the millet under controlled conditions, and their mineral status is monitored through blood and urine samples (Adaptive Platform Trials Coalition 2019). Researchers can measure changes in mineral levels and assess markers of mineral utilization, such as absorption efficiency or biomarkers of mineral function. These trials provide direct evidence of the bioavailability of minerals from millets in humans (Zheng et al. 2023). They also allow researchers to evaluate the impact of individual factors, such as age, sex, genetics, and health conditions, on mineral absorption and utilization. Human intervention trials provide valuable data for establishing dietary recommendations and guidelines for mineral intake from millets (Armstrong et al. 2020). They contribute to our understanding of the nutritional impact of millets and help optimize their formulations for enhanced mineral bioavailability (Shao and Ortas 2023).

12.5.4 Analytical Techniques for Bioavailability Assessment

Analytical techniques are essential for quantifying and tracking mineral absorption and utilization in bioavailability studies. Various methods can be used to measure mineral concentrations in biological samples, assess isotopic enrichment, and determine mineral balance (Mattar et al. 2022). Atomic absorption spectroscopy, inductively coupled plasma mass spectrometry, and related techniques enable the quantification of mineral concentrations in samples such as blood, urine, or tissues (de Souza et al. 2022). These methods provide accurate measurements of mineral levels, allowing researchers to assess changes in mineral status before and after millet consumption.

Stable isotope tracers are employed to study mineral absorption and utilization. Isotopically labelled minerals are administered to subjects, and their appearance in biological samples can be tracked using mass spectrometry techniques (Moser-Veillon et al. 2020). Isotope ratio mass spectrometry allows researchers to determine the fractional absorption of minerals and assess their utilization by different organs or tissues. These analytical techniques provide valuable data on mineral absorption, utilization, and metabolism in bioavailability studies (Patrick et al. 2022). They contribute to our understanding of the fate of minerals from millets within the human body and help evaluate the efficacy of millet formulations in delivering bioavailable minerals (Kurakula and Rao 2020).

Evaluating mineral bioavailability in millets requires a multidimensional approach. *In-vitro* digestion methods, animal studies, human intervention trials, and analytical techniques all play critical roles in assessing mineral release, absorption, and utilization. Collectively, these approaches provide a comprehensive understanding of mineral bioavailability in millets and contribute to the development of nutritionally optimized millet formulations.

12.6 CONCLUSION AND FUTURE PERSPECTIVES

In conclusion, this chapter has provided a comprehensive overview of the factors affecting the bioavailability of minerals in millets. These factors include the chemical form of minerals, the presence of anti-nutrients, nutrient interactions, pH and digestive enzymes, and individual variations in bioavailability. Each of these factors can significantly impact the solubility, absorption, and utilization of minerals in the body. By understanding these factors, researchers and food manufacturers can make informed decisions regarding millet formulations and processing techniques to optimize mineral bioavailability. The chapter has also discussed commonly used millet processing techniques, such as extrusion, coating, roasting, toasting, milling, and grinding. These techniques can influence mineral bioavailability through changes in solubility, matrix structure, anti-nutrient content, and particle size. It is important to consider the effects of these processing techniques on mineral bioavailability during the development of millet-based food products. Moving forward, future research should focus on several key areas to further enhance mineral bioavailability in millets. Firstly, innovative processing technologies should be explored to improve the solubility, stability, and targeted delivery of minerals. Technologies such as nanoencapsulation and inclusion complexes offer promising approaches to enhance mineral bioavailability and should be further investigated. Additionally, personalized nutrition approaches should be considered to account for individual variations in mineral bioavailability. Factors such as genetics, age, sex, and health conditions can influence how individuals absorb and utilize minerals. Understanding these individual variations and tailoring millet formulations accordingly can optimize mineral bioavailability and promote personalized nutrition. Furthermore, the impact of processing techniques on other nutritional components, such as vitamins, antioxidants, and bioactive compounds, should be investigated in conjunction with mineral bioavailability. Lastly, more comprehensive studies, including human intervention trials, should be conducted to evaluate the long-term effects of mineral bioavailability in millets on human health. These studies can provide valuable insights into the impact of millets on mineral status, disease prevention, and overall nutritional well-being.

REFERENCES

Adams, S., C. T. Sello, G. Qin, D. Che, and R. Han. 2018. Does dietary fiber affect the levels of nutritional components after feed formulation? *Fibers* 6(2): 29. doi: 10.3390/fib6020029.

Adaptive Platform Trials Coalition. 2019. Author correction: Adaptive platform trials: definition, design, conduct, and reporting considerations. *Nature Reviews. Drug Discovery* 18(10): 808.

Adetola, O. Y., R. N. John, and K. G. Taylor. 2023. Can consumption of local micronutrient-and absorption enhancer-rich plant foods together with starchy staples improve bioavailable iron and zinc in diets of at-risk African populations? *International Journal of Food Sciences and Nutrition* 74(2): 188–208.

Ammerman, C. B. 1995. Methods for estimation of mineral bioavailability. In *Bioavailability of Nutrients for Animals* (Ammerman, C. B., D. H. Baker, and A. J. Lewis, Eds.), Academic Press, Cambridge, pp. 83–94.

Anitha, S., J. Kane-Potaka, R. Botha, D. I. Givens, N. L. B. Sulaiman, S. Upadhyay, M. Vetriventhan, T. W. Tsusaka, D. J. Parasannanavar, T. Longvah, A. Rajendran, K. Subramaniam, and R. K. Bhandari. 2021. Millets can have a major impact on improving iron status, hemoglobin level, and in reducing iron deficiency anemia-A systematic review and meta-analysis. *Frontiers in Nutrition* 8: 725529. doi: 10.3389/fnut.2021.725529.

Armstrong, J., J. K. Rudkin, N. Allen, D. W. Crook, D. J. Wilson, D. H. Wyllie, and A. M. O'Connell. 2020. Dynamic linkage of COVID-19 test results between Public Health England's Second Generation Surveillance System and UK Biobank. *Microbial Genomics* 6(7): mgen000397. doi: 10.1099/mgen.0.000397.

Baye, K., J. P. Guyot, and C. Mouquet-Rivier. 2017. The unresolved role of dietary fibers on mineral absorption. *Critical Reviews in Food Science and Nutrition* 57(5): 949–957.

Balaram, V. 2021. Current and emerging analytical techniques for geochemical and geochronological studies. *Geological Journal* 56(5): 2300–2359.

Buturi, C. V., R. P. Mauro, V. Fogliano, C. Leonardi, and F. Giuffrida. 2021. Mineral biofortification of vegetables as a tool to improve human diet. *Foods (Basel, Switzerland)* 10(2): 223.

Cabrera, C., R. Artacho, and R. Giménez. 2006. Beneficial effects of green tea-a review. *Journal of the American College of Nutrition* 25(2): 79–99.

Caeiro, C., C. Pragosa, M. C. Cruz, C. D. Pereira, and S. G. Pereira. 2022. The role of pseudocereals in celiac disease: Reducing nutritional deficiencies to improve well-being and health. *Journal of Nutrition and Metabolism*: 8502169. doi: 10.1155/2022/8502169.

Caetano-Silva, M. E., F. M. Netto, M. T. Bertoldo-Pacheco, A. Alegría, and A. Cilla. 2021. Peptide-metal complexes: Obtention and role in increasing bioavailability and decreasing the pro-oxidant effect of minerals. *Critical Reviews in Food Science and Nutrition* 61(9): 1470–1489.

Cao, H., C. Zuo, Y. Huang, L. Zhu, J. Zhao, Y. Yang, Y. Jiang, and F. Wang. 2021. Hippocampal proteomic analysis reveals activation of necroptosis and ferroptosis in a mouse model of chronic unpredictable mild stress-induced depression. *Behavioural Brain Research* 407(113261): 113261.

Costa, G. T., Q. D. J. S. Vasconcelos, G. C. Abreu, A. O. Albuquerque, J. L. Vilar, and G. F. Aragão. 2021. Systematic review of the ingestion of fructooligosaccharides on the absorption of minerals and trace elements versus control groups. *Clinical Nutrition ESPEN* 41: 68–76.

Cubadda, F., F. Aureli, A. Raggi, A., and M. Carcea. 2009. Effect of milling, pasta making and cooking on minerals in durum wheat. *Journal of Cereal Science* 49(1): 92–97.

de Souza, R. M., C. A. T. Toloza, and R. Q. Aucélio. 2022. Fast determination of trace metals in edible oils and fats by inductively coupled plasma mass spectrometry and ultrasonic acidic extraction. *Journal of Trace Elements and Minerals* 1: 100003.

Drewnowski, A., G. S. Garrett, R. Kansagra, N. Khan, R. Kupka, A. V. Kurpad, V. Mannar, R. Martorell, M. B. Zimmermann, and Double Fortified Salt Consultation (DFS) Steering Group. 2021. Key considerations for policymakers-iodized salt as a vehicle for iron fortification: Current evidence, challenges, and knowledge gaps. *The Journal of Nutrition* 151(Suppl 1): 64S–73S. doi: 10.1093/jn/nxaa377.

Durairaj, M., G. Gurumurthy, V. Nachimuthu, K. Muniappan, and S. Balasubramanian. 2019. Dehulled small millets: The promising nutricereals for improving the nutrition of children. *Maternal and Child Nutrition* 15(Suppl 3): e12791. doi: 10.1111/mcn.12791.

Falah, S., and S. N. Mohssan. 2017. Essential trace elements and their vital roles in human body. *Indian Journal of Advance in Chemistry Science* 5(3): 127–136.

Farcas, A., A. M. Resmerita, M. Balan-Porcarasu, C. Cojocaru, C. Peptu, and I. Sava. 2023. Inclusion complexes of 3, 4-ethylenedioxythiophene with Per-Modified β-and γ-Cyclodextrins. *Molecules* 28(8): 3404. doi: 10.3390/molecules28083404.

Fasae, K. D., A. O. Abolaji, T. R. Faloye, A. Y. Odunsi, B. O. Oyetayo, J. I. Enya, J. A. Rotimi, R. O. Akinyemi, A. J. Whitworth, and M. Aschner. 2021. Metallobiology and therapeutic chelation of biometals (copper, zinc and iron) in Alzheimer's disease: Limitations, and current and future perspectives. *Journal of Trace Elements in Medicine and Biology: Organ of the Society for Minerals and Trace Elements (GMS)* 67: 126779. doi: 10.1016/j.jtemb.2021.126779.

Galanakis, C. M. 2021. Functionality of food components and emerging technologies. *Foods (Basel, Switzerland)* 10(1): 128. doi: 10.3390/foods10010128.

Gowda, N. A. N., K. Siliveru, P. V. V. Prasad, Y. Bhatt, B. P. Netravati, and C. Gurikar. 2022. Modern processing of Indian millets: A perspective on changes in nutritional properties. *Foods (Basel, Switzerland)* 11(4): 499.

Hassan, Z. M., N. A. Sebola, and M. Mabelebele. 2021. The nutritional use of millet grain for food and feed: A review. *Agriculture and Food Security* 10(1): 16.

He, J., N. M. Evans, H. Liu, and S. Shao. 2020. A review of research on plant-based meat alternatives: Driving forces, history, manufacturing, and consumer attitudes. *Comprehensive Reviews in Food Science and Food Safety* 19(5): 2639–2656.

Henze, L. J., N. J. Koehl, H. Bennett-Lenane, R. Holm, M. Grimm, F. Schneider, W. Weitschies, M. Koziolek, and B. T. Griffin. 2021. Characterization of gastrointestinal transit and luminal conditions in pigs using a telemetric motility capsule. *European Journal of Pharmaceutical Sciences: Official Journal of the European Federation for Pharmaceutical Sciences* 156: 105627.

Hurrell, R., and I. Egli. 2010. Iron bioavailability and dietary reference values. *The American Journal of Clinical Nutrition* 91(5): 1461S–1467S. doi: 10.3945/ajcn.2010.28674F.

Goff, J. 2021. 10 mechanisms by which amino acids may enhance mineral absorption in animals. *Journal of Animal Science* 99(Suppl 1): 13–14. doi: 10.1093/jas/skab054.023.

Goff, J. P. 2018. Invited review: Mineral absorption mechanisms, mineral interactions that affect acid–base and antioxidant status, and diet considerations to improve mineral status. *Journal of Dairy Science* 101(4): 2763–2813.

Jagati, P., I. Mahapatra, and D. Dash. 2021. Finger millet (Ragi) as an essential dietary supplement with key health benefits: A review. *International Journal of Home Science* 7(2): 94–100.

Jaiswal, D. K., R. Krishna, G. K. Chouhan, A. P. de Araujo Pereira, A. B. Ade, S. Prakash, S. K. Verma, R. Prasad, J. Yadav, and J. P. Verma. 2022. Bio-fortification of minerals in crops: Current scenario and future prospects for sustainable agriculture and human health. *Plant Growth Regulation* 98(1): 5–22.

Kalia, A., S. P. Sharma, H. Kaur, and H. Kaur. 2020. Novel nanocomposite-based controlled-release fertilizer and pesticide formulations: Prospects and challenges. In *Multifunctional Hybrid Nanomaterials for Sustainable Agri-Food and Ecosystems* (Abd-Elsalam, K. A., Ed.), Elsevier, Amsterdam, Netherlands, pp. 99–134.

Kamau, E. H., S. G. Nkhata, and E. O. Ayua. 2020. Extrusion and nixtamalization conditions influence the magnitude of change in the nutrients and bioactive components of cereals and legumes. *Food Science and Nutrition* 8(4): 1753–1765.

Karakaş, F., B. Vaziri Hassas, and M. S. Çelik. 2015. Effect of precipitated calcium carbonate additions on waterborne paints at different pigment volume concentrations. *Progress in Organic Coatings* 83: 64–70.

Keyata, E., Y. B. Olika, G. Tola, and S. F. Bultosa. 2021. Optimization of nutritional and sensory qualities of complementary foods prepared from sorghum, soybean, karkade and premix in Benishangul-Gumuz region, Ethiopia. *Heliyon* 7(9): e07955. doi: 10.1016/j.heliyon.2021.e07955.

Kumar, A., M. Rani, S. Mani, P. Shah, D. B. Singh, H. Kudapa, and R. K. Varshney. 2021. Nutritional significance and antioxidant-mediated antiaging effects of finger millet: Molecular insights and prospects. *Frontiers in Sustainable Food Systems* 5: 684318. doi: 10.3389/fsufs.2021.684318.

Kurakula, M., and G. S. N. K. Rao. 2020. Pharmaceutical assessment of polyvinylpyrrolidone (PVP): As excipient from conventional to controlled delivery systems with a spotlight on COVID-19 inhibition. *Journal of Drug Delivery Science and Technology* 60: 102046.

Kurtz, D. M., and W. P. Feeney. 2020. The influence of feed and drinking water on terrestrial animal research and study replicability. *ILAR Journal* 60(2): 175–196.

Lai, S., T. Zhang, Y. Wang, K. Ouyang, H. Hu, X. Hu, H. Xiong, and Q. Zhao. 2022. Effects of different extrusion temperatures on physicochemical, rheological and digestion properties of rice flour produced in a pilot-scale extruder. *International Journal of Food Science and Technology* 57(10): 6773–6784.

Madla, C. M., F. K. H. Gavins, H. A. Merchant, M. Orlu, S. Murdan, and A. W. Basit. 2021. Let's talk about sex: Differences in drug therapy in males and females. *Advanced Drug Delivery Reviews* 175: 113804.

Mattar, G., A. Haddarah, J. Haddad, M. Pujola, and F. Sepulcre. 2022. New approaches, bioavailability and the use of chelates as a promising method for food fortification. *Food Chemistry* 373(A): 131394. doi: 10.1016/j.foodchem.2021.131394.

Moser-Veillon, P. B., K. Y. Patterson, and C. Veillon. 2020. Use of mineral stable isotopes in the study of nutrient homeostasis during human pregnancy and lactation. In *Kinetic Models of Trace Element and Mineral Metabolism During Development* (Subramanian, K. N., and M. E. Wasteney, Eds.), CRC Press, Taylor and Francis Group, Oxfordshire, pp. 1–9.

Mrinal, R. E., and T. Aluko. 2020. Plant food anti-nutritional factors and their reduction strategies: An overview. *Food Production, Processing and Nutrition* 2: 1–14.

Muhammad, A., D. Ahmedu Ameh, H. Chuks Nzelibe, S. M. Hassan, and A. Mansir. 2022. Impact of processing (toasting and fermentation) on the proximate, anti-nutrient and mineral composition of sickle pod (*Senna obtusifolia*) seeds. *SLU Journal of Science and Technology* 3(1 and 2): 1–15.

Mulet-Cabero, A. I., L. Egger, R. Portmann, O. Ménard, S. Marze, M. Minekus, and S. L. Feunteun. 2020. A standardised semi-dynamic in vitro digestion method suitable for food-an international consensus. *Food and Function* 11(2): 1702–1720.

Nadziakiewicza, M., S. Kehoe, and P. Micek. 2019. Physico-chemical properties of clay minerals and their use as a health promoting feed additive. *Animals (Basel)* 9(10): 714. doi: 10.3390/ani9100714.

Nkhata, S. G., E. Ayua, E. H. Kamau, and B. Shingiro. 2018. Fermentation and germination improve nutritional value of cereals and legumes through activation of endogenous enzymes. *Food Science and Nutrition* 6(8): 2446–2458. doi: 10.1002/fsn3.846.

Olaifa, O. P., A. J. Adeola, O. A. Adeniji, O. S. Ogundana, and Y. M. Yuneid. 2022. Effects of different processing methods on nutrient and anti-nutrient compositions of *Entada aricana* seed. *Nigeria Agricultural Journal* 53(2): 274–277.

Ondiba, R., E. O. Nyakwama, E. Ogello, Z. Kembenya, and K. Gichana. 2022. Future demand and supply of aquafeed ingredients: Outlines to commercialize non-conventional protein ingredients to enhance aquaculture production for food security in sub-Saharan Africa. *Aquatic Ecosystem Health and Management* 25(4): 75–84.

Onyegbulam, C., T. Mba, O. U. Chidinma, C. O. Estella, C. Ezugwu, and C. Vincent Obisike. 2023. Evaluation of proximate, minerals, vitamins and antinutrient composition of Combretum platypterum (Welw.) Hutch. and Dalziel. (Combretaceae) leaves. *Asian Journal of Research in Medical and Pharmaceutical Sciences* 12(2): 1–10.

Pateiro, M., B. Gómez, P. E. S. Munekata, F. J. Barba, P. Putnik, D. B. Kovačević, and J. M. Lorenzo. 2021. Nanoencapsulation of promising bioactive compounds to improve their absorption, stability, functionality and the appearance of the final food products. *Molecules (Basel, Switzerland)* 26(6): 1547.

Patrick, B. M., A. M. Gillanders, C. Sturrock, D. S. Izzo, J. A. Oxman, and K. Lueders-Dumont. 2022. Reading the biomineralized book of life: Expanding otolith biogeochemical research and applications for fisheries and ecosystem-based management. *Reviews in Fish Biology and Fisheries* 33: 411–449.

Peng, T., R. Xiao, Z. Rong, H. Liu, Q. Hu, S. Wang, X. Li, and J. Zhang. 2020. Polymer nanocomposite-based coatings for corrosion protection. *Chemistry, An Asian Journal* 15(23): 3915–3941.

Petroski, W., and D. M. Minich. 2020. Is there such a thing as "anti-nutrients"? A narrative review of perceived problematic plant compounds. *Nutrients* 12(10): 2929.

Picó, C., F. Serra, A. M. Rodríguez, J. Keijer, and A. Palou. 2019. Biomarkers of nutrition and health: New tools for new approaches. *Nutrients* 11(5): 1092. doi: 10.3390/nu11051092.

Puranik, S., J. Kam, P. P. Sahu, R. Yadav, R. K. Srivastava, H. Ojulong, and R. Yadav. 2017. Harnessing finger millet to combat calcium deficiency in humans: Challenges and prospects. *Frontiers in Plant Science* 8: 1311.

Rahate, K. A., M. Madhumita, and P. K. Prabhakar. 2021. Nutritional composition, anti-nutritional factors, pretreatments-cum-processing impact and food formulation potential of faba bean (*Vicia faba* L.): A comprehensive review. *LWT-Food Science and Technology* 138: 110796.

Rahmani, A., P. Mohammad, K. Gahlot, A. A. Moustakas, C. S. P. Kazmi, and V. K. Ojha. 2022. Pretreatment methods to enhance solubilization and anaerobic biodegradability of lignocellulosic biomass (wheat straw): Progress and challenges. *Fuel* 319: 123726.

Rousseau, S., C. Kyomugasho, M. Celus, M. E. G. Hendrickx, and T. Grauwet. 2020. Barriers impairing mineral bioaccessibility and bioavailability in plant-based foods and the perspectives for food processing. *Critical Reviews in Food Science and Nutrition* 60(5): 826–843.

Rusu, I. G., R. Suharoschi, D. C. Vodnar, C. R. Pop, S. A. Socaci, R. Vulturar, M. Istrati, I. Moroșan, A. C. Fărcaș, A. D. Kerezsi, C. I. Mureșan, and O. L. Pop. 2020. Iron supplementation influence on the gut microbiota and probiotic intake effect in iron deficiency-A literature-based review. *Nutrients* 12(7): 1993.

Saffarionpour, S., and L. L. Diosady. 2023. Preparation and characterization of an iron–β-cyclodextrin inclusion complex: Factors influencing the host–guest interaction. *Food and Function* 14(11): 5062–5077.

Samtiya, M., K. Soni, S. Chawla, A. Poonia, S. Sehgal, and T . Dhewa. 2021. Key anti-nutrients of millet and their reduction strategies: An overview. *Acta Scientifci Nutritional Health* 5(12): 68–80.

Sarkhel, S., and A. Roy. 2022. Phytic acid and its reduction in pulse matrix: Structure–function relationship owing to bioavailability enhancement of micronutrients. *Journal of Food Process Engineering* 45(5): e14030. doi: 10.1111/jfpe.14030.

Sawyer, A. 2019. Impacts of a nutritional water supplement and threonine to lysine ratios on growth performance of nursery pigs. Ph.D. Thesis, Oklahoma State University, Oklahoma.

Shahidi, F., and Y. Pan. 2022. Influence of food matrix and food processing on the chemical interaction and bioaccessibility of dietary phytochemicals: A review. *Critical Reviews in Food Science and Nutrition* 62(23): 6421–6445.

Shao, I., and I. Ortas. 2023. Impact of mycorrhiza on plant nutrition and food security. *Journal of Plant Nutrition* 46(13): 3247–3272. doi: 10.1080/01904167.2023.2192780.

Siddiqui, S. A., T. Alvi, A. Biswas, S. Shityakov, T. Gusinskaia, F. Lavrentev, K. Dutta, M. K. I. Khan, J. Stephen, and M. Radhakrishnan. 2022. Food gels: Principles, interaction mechanisms and its microstructure. *Critical Reviews in Food Science and Nutrition*: 1–22. doi: 10.1080/10408398.2022.2103087.

Silva, J. G. S., A. P. Rebellato, E. T. D. S. Caramês, R. Greiner, and J. A. L. Pallone. 2020. In vitro digestion effect on mineral bioaccessibility and antioxidant bioactive compounds of plant-based beverages. *Food Research International* 130: 108993. doi: 10.1016/j.foodres.2020.108993.

Sneha, K., and A. Kumar. 2022. Nanoemulsions: Techniques for the preparation and the recent advances in their food applications. *Innovative Food Science and Emerging Technologies* 76: 102914. doi: 10.1016/j.ifset.2021.102914.

Sugiarto, Y., N. M. S. Sunyoto, M. Zhu, I. Jones, and D. Zhang. 2021. Effect of biochar in enhancing hydrogen production by mesophilic anaerobic digestion of food wastes: The role of minerals. *International Journal of Hydrogen Energy* 46(5): 3695–3703.

Swann, J. R., M. Rajilic-Stojanovic, A. Salonen, O. Sakwinska, C. Gill, A. Meynier, P. Fança-Berthon, B. Schelkle, N. Segata, C. Shortt, K. Tuohy, and O. Hasselwander. 2020. Considerations for the design and conduct of human gut microbiota intervention studies relating to foods. *European Journal of Nutrition* 59(8): 3347–3368.

Tan, J., Ö. Cizer, J. De Vlieger, H. Dan, and J. Li. 2022. Impacts of milling duration on construction and demolition waste (CDW) based precursor and resulting geopolymer: Reactivity, geopolymerization and sustainability. *Resources, Conservation and Recycling* 184: 106433.

Tharifkhan, S. A., A. B. Perumal, A. Elumalai, J. A. Moses, and C. Anandharamakrishnan. 2021. Improvement of nutrient bioavailability in millets: Emphasis on the application of enzymes. *Journal of the Science of Food and Agriculture* 101(12): 4869–4878.

van der Poel, A. F. B., M. R. Abdollahi, H. Cheng, R. Colovic, L. A. den Hartog, D. Miladinovic, G. Page, K. Sijssens, J. F. Smillie, M. Thomas, W. Wang, P. Yu, and W. H. Hendriks. 2020. Future directions of animal feed technology research to meet the challenges of a changing world. *Animal Feed Science and Technology* 270: 114692.

Wang, Y., P. Zhao, Y. Zhou, X. Hu, and H. Xiong. 2023. From bitter to delicious: Properties and uses of microbial aminopeptidases. *World Journal of Microbiology and Biotechnology* 39(3): 72. doi: 10.1007/s11274-022-03501-3.

Weaver, C. M., and E. M. Haney. 2010. Nutritional basis of skeletal growth. In *Osteoporosis in Men*, Second Edition (Orwoll, E. S., J. P. Bilezikian, and D. Vanderschueren, Eds.), Academic Press, Cambridge, pp. 119–129.

Xu, Q., F. Zheng, X. Cao, P. Yang, Y. Xing, P. Zhang, H. Liu, G. Zhou, X. Liu, and X. Bi. 2021. Effects of airflow ultrafine-grinding on the physicochemical characteristics of Tartary buckwheat powder. *Molecules (Basel, Switzerland)* 26(19): 5841.

Yilmaz, B., and D. Ağagündüz. 2020. Bioactivities of hen's egg yolk phosvitin and its functional phosphopeptides in food industry and health. *Journal of Food Science* 85(10): 2969–2976.

Yousaf, L., D. Hou, H. Liaqat, and Q. Shen. 2021. Millet: A review of its nutritional and functional changes during processing. *Food Research International (Ottawa, Ont.)* 142: 110197.

Yu, Q., J. Qian, Y. Guo, H. Qian, W. Yao, and Y. Cheng. 2023. Applicable strains, processing techniques and health benefits of fermented oat beverages: A review. *Foods (Basel, Switzerland)* 12(8): 1708. doi: 10.3390/foods12081708.

Yuan, G., Y. Cao, H.-M. Schulz, F. Hao, J. Gluyas, K. Liu, T. Yang, Y. Wang, K. Xi, and F. Li. 2019. A review of feldspar alteration and its geological significance in sedimentary basins: From shallow aquifers to deep hydrocarbon reservoirs. *Earth-Science Reviews* 191: 114–140.

Zhang, Y. Y., R. Stockmann, K. Ng, and S. Ajlouni. 2021. Opportunities for plant-derived enhancers for iron, zinc, and calcium bioavailability: A review. *Comprehensive Reviews in Food Science and Food Safety* 20(1): 652–685.

Zheng, J., F. Wu, F. Wang, J. Cheng, H. Zou, Y. Li, J. Du, and J. Kan. 2023. Biomarkers of micronutrients and phytonutrients and their application in epidemiological studies. *Nutrients* 15(4): 970.

Chapter 13

Millet-Based Traditional Foods

13.1 INTRODUCTION

Millets are one of the oldest cereal crops domesticated by mankind. They are also among the most ancient foods known to humans (Ganapathy et al. 2021). Their consumption might be new to the Western world but they have remained a staple food for millions of Asians and Africans since time immemorial. There is evidence which indicates millets were adopted in Central Sudan between 7500 and 6500 years ago (Madella et al. 2014). Millets were one of the sacred grains of ancient China and they were the staple food in North-Central China by around 5000 BCE (Krishi Jagran 2020). A mention of millet production and consumption is also found in the Hindu religious book *Yajurveda* which is thought to have been written about 4500 BCE. Evidence of millet cultivation in the Korean Peninsula around 3500–2000 BCE is also reported (Koirala 2020). Millets were also cultivated and consumed in many parts of Russia and Eurasia before the popularity of rice and wheat (Karnataka State Department of Agriculture, Bengaluru, India and ICAR-Indian Institute of Millets Research, Hyderabad, India 2018). A variety of millet-based traditional foods (both fermented and non-fermented) are produced and consumed in these regions. The fermented products are either lactic acid or alcoholic products. The major millet-based traditional fermented non-alcoholic food preparations are *ben-saagla*, *bushera*, *dosa*, *dhokla*, *fura*, *idli*, *injera*, *koko*, *mangisi*, *ogi*, *oshikundu*, *paddu*, *rabadi*, *uji*, *uthapam*, etc. (Kumar et al. 2020; Amadou et al. 2011). The popular millet-based alcoholic beverages are *boza*, *burukutu*, *jaandh*, *kodo ko jaanr*, *merissa*, *okatokele*, *pito*, *sur*, and *xia mi jiao*, etc. (Joshi et al. 2015; Amadou et al. 2011). Major non-fermented traditional products prepared from millets are *dambu*, *halwa*, *kheer*, *kurrakan kanda*, *laddu*, *masavusu*, *pongal*, *porridge*, and *kheer*, etc. This chapter provides comprehensive information on millet-based major traditional products prepared and consumed all over the world.

13.2 FERMENTATION

Fermentation is the process of the breakdown of complex organic substrates into simpler ones by the action of microbial enzymes. It is one of the most ancient and inexpensive methods for food preservation. It improves the nutritional value and digestibility of raw products by reducing the content of antinutritional factors like phytic acid, oxalic acid, and tannins, etc. (Kumar et al. 2021; Tou et al. 2007). It also produces many flavouring compounds that improve the sensory characteristics of the foods. The fermentation of cereals, fruits, and vegetables is a cultural and traditional practice within indigenous communities of India, Africa, and most developing countries. Fermentation is generally carried out spontaneously by the microflora present in the food or surrounding atmosphere; however, in some cases, some traditional starter cultures like *dhaeli*, *marcha*, etc., are also added. The major types of millet-based fermented foods are either lactic acid fermented or alcoholic fermented.

13.2.1 Lactic Acid Fermented Foods

This section covers the foods which are primarily fermented by lactic acid bacteria and have an acidic taste. Details on the various traditionally produced millet-based lactic acid fermented products, their country of origin, raw materials used for their production, consumption patterns, and associated microorganisms are also provided in Table 13.1. A pictorial representation of the major lactic acid fermented millet-based products is also provided in Figure 13.1.

13.2.1.1 Ben-Saagla

Ben-saagla is a pearl millet-based fermented gruel that is prepared mainly in the West African country Burkina Faso. It is generally used as a complementary food for infants and children. The various preparatory steps are washing of the grains (optional), soaking, the addition of spices like ginger, pepper, black pepper, pepper, aniseed (*Pimpinella anisum*), and mint (*Mentha*), etc., wet milling, kneading, and sieving of moistened flour, and fermenting the settled, but diluted slurry. The soaking and settling are long processes and take 16 and 11 hours, respectively. The various microorganisms reported in fermented *ben-saagla* samples are *L. fermentum*, *L. plantarum*, and *Pediococcus pentosaceus*. The supernatant collected from the settling step is heated to a boil and the paste of the sediment is added to this. This is further boiled for 7 minutes. It has a solid content of 8–10 g/100 ml and contains lactic and acetic acid with small quantities of ethanol. *Ben-saagla* with improved nutritional and sensorial properties can be prepared by replacing pearl millet with groundnut and carrying out the fermentation under controlled conditions (Tou et al. 2007).

13.2.1.2 Bushera

Bushera is a beverage of the western highlands of Uganda that is prepared from flour of the germinated grains of sorghum or millet or a mixture of both. The preparatory steps can be divided into three major parts, i.e. germination of grains, preparation of paste or slurry, and fermentation. For the germination, the grains are soaked overnight in water, the excess water is drained, the wet grains are mixed with wood ash (one-tenth of the grain quantity), and a heap of grains is made on banana leaves or papyrus mats and left to germinate (2–4 days). The germinated grains are sun-dried, the radicle and plumule are removed, and they are ground to flour. If *bushera* is to be prepared only from millets then the mixing of wood ash is not done and only those grains which are to be used for malt are germinated. In the second step, hot boiled water is added to the flour of the germinated grains with continuous stirring to make a paste or slurry of the required viscosity and the mixture is boiled for 2–5 minutes. After boiling, the mixture is cooled and more flour of germinated grains is added to initiate fermentation. The fermentation is carried out at room temperature (27–30°C) and the fermentation time depends on the desired quality attributes of the end product. A fermentation period of 12–24 hours results in a sweet *bushera*, while fermentation of about 2–4 days is required to produce a sour *bushera* (Muyanja et al. 2003). The major microorganisms reported in *bushera* are from the genera *Lactobacillus*, *Lactococcus*, *Leuconostoc*, *Enterococcus*, and *Streprococcus*. The product is consumed by both children and

TABLE 13.1 MILLET-BASED LACTIC ACID FERMENTED FOODS, THEIR COUNTRY OF ORIGIN, RAW MATERIALS USED, CONSUMPTION PATTERN, AND MICROBIOLOGY

Product name	Country of origin	Raw materials used	Consumption pattern	Microbiology	References
Ben-saagla	Burkina Faso	Pearl millet ginger, pepper, black pepper, pepper, aniseed (*Pimpinella anisum*), and mint (*Mentha*)	Complementary food for infants and children	*Lactobacillus fermentum, L. plantarum,* and *Pediococcus pentosaceus*	Tou et al. 2007
Bushera	Uganda	Sorghum or millet and wood ash	Consumed by both children and adults	*Lactobacillus* spp., *Lactococcus* spp., *Leuconostoc* spp., *Enterococcus* spp., and *Streptococcus* spp.	Muyanja et al. 2003, Amadou et al. 2011
Dhokla	India	Bengal gram flour, pearl millet flour, finger millet flour, drumstick leaves, and curd	Snack, side dish, or main course	*Leuconostoc mesenteroides, Lactobacillus fermentum, Lb. Lactis, Lb. delbrueckii,* and *Hansenula silvicola*	Gupta et al. 2017, Lohekar and Arya 2014, Joshi et al. 1989
Fura/Fura da nono	Nigeria, Ghana and Burkina Faso	Pearl millet, spices like powdered ginger or black pepper, sugar, and fermented whole milk (*Kindrimo*) or fermented skim milk (*nono*)	Mid-day meal, refreshing drink, or a weaning food for infants	*Lactobacillus* spp., *Pediococcus* spp., *Streptococcus* spp., *Leuconostoc* spp., *Enterococcus* spp., *Issatchenkia orientalis, Saccharomyces cerevisiae, Pichia anomala, Candida tropicalis, S. pastorians, Yarrowia lipolytica, Galactomyces geotricum*	Yusuf et al. 2020, Amadou et al. 2011, Jideani et al. 2001

(Continued)

TABLE 13.1 (CONTINUED) MILLET-BASED LACTIC ACID FERMENTED FOODS, THEIR COUNTRY OF ORIGIN, RAW MATERIALS USED, CONSUMPTION PATTERN, AND MICROBIOLOGY

Product name	Country of origin	Raw materials used	Consumption pattern	Microbiology	References
Ibyer: ibyer-i-angen (sour type), ibyer-i-nyohon (sweet type)	Nigeria	Maize (Zea mays), sorghum (Sorghum bicolor), or millet	Consumed by both children and adults	–	Kure and Wyasu 2013
Idli	India	Barnyard millet, black gram dhal, and fenugreek seeds	Snack, side dish, or main course	Leuconostoc mesenteroides and Streptococcus thermophilus	Vanithasri and Kanchana 2013
Injera	Ethiopia	Teff and "ersho", a seed culture from the previous batch	Major food at lunch and dinner. Also served at family gatherings like marriages, birthday parties, and death anniversaries	Candida milleri, Rhodotorula mucilaginosa, Kluyveromyces marxianus, Pichia naganishii, and Debaromyces hansenii	Neela and Fanta 2020, Ashenafi 1994
Koko	Ghana	Pearl millet or sorghum, ginger (Zingiber officinale), black pepper (Pepper nigrum), pepper (Capsicum annuum), cloves (Syzygium aromaticum), and xylopia (Xylopia aethiopica)	Lunch or an in-between meal	Weisella confusa, Lactobacillus plantarum, L. fermentcum, L. brevis, L. salivarius, Enterobacter colcae, Saccharomyces cerevisiae, Candida mycoderma, and Acinetobacter	Danson et al. 2019, Haleegoah et al. 2016, Amadou et al. 2011

(Continued)

TABLE 13.1 (CONTINUED) MILLET-BASED LACTIC ACID FERMENTED FOODS, THEIR COUNTRY OF ORIGIN, RAW MATERIALS USED, CONSUMPTION PATTERN, AND MICROBIOLOGY

Product name	Country of origin	Raw materials used	Consumption pattern	Microbiology	References
Koozh	India	Finger millet or pearl millet	Breakfast or lunch	Weissella paramesenteroides, Lactobacillus fermentum, Vibrio parahaemolyticus, and Listeria monocytogenes	Antony et al. 2020, Subastri et al. 2015, Thirumangaimannan and Gurumurthy 2013
Mangisi	Zimbabwe	Finger millet	Consumed as an energy drink at any time of the day	LAB, yeasts, and moulds	Zvauya et al. 1997
Ogi	Nigeria	Maize, millet, or sorghum	Weaning food for babies and breakfast food for children and elders	Aspergillus sp., Penicillium sp. Cephalosporium, Fusarium, Rhizophus stolonifer, Mucor mucedo, Saccharomyces cerevisiae, Candida mycoderma, Corynebacterium sp., Lactobacillus plantarum, L. fermentum, and Leuconostoc fermentum	Bolaji et al. 2017, Bolaji et al. 2015, Okara and Lokoyi 2012, Ijabadeniyi 2007
Oshikundu	Namibia	Pearl millet or sorghum	Produced as a traditional practice on the initiation of young girls into womanhood and is served at weddings and social interactions to welcome guests	Lactobacillus plantarum, L. lactis spp. lactis, L. delbreuckii spp. delbrueckii, L. fermentum, L. pentosans, and L. curvatus spp. Cuvatus	Mishairabgwi and Cheikhyoussef 2017

(Continued)

TABLE 13.1 (CONTINUED) MILLET-BASED LACTIC ACID FERMENTED FOODS, THEIR COUNTRY OF ORIGIN, RAW MATERIALS USED, CONSUMPTION PATTERN, AND MICROBIOLOGY

Product name	Country of origin	Raw materials used	Consumption pattern	Microbiology	References
Paddu	India	Little millet, black gram *dhal*, Bengal gram *dhal*, red gram *dhal*, flaked rice, and fenugreek	Breakfast food	Yeast	Madalageri et al. 2016
Rabadi	India	Germinated pearl millet flour, skim milk, mixed *dahi* culture	–	–	Modha and Pal 2011
Togwa	East Africa	Maize flour and finger millet malt	Consumed as a refreshing and weaning food by both adults and children	*Lactobacillus fermentum, L. brevis, L. fermentum, L. cellobiosas, Pediococcus entosaceus, Weissela confusa, Saccharomyces cerevisiae, Candida pelliculosa,* and *C. tropicalis*	Kitabatake et al. 2003, Mugula et al. 2003
Uji	Kenya	Finger millet, sorghum, maize, or cassava (*Manihot esculenta* Crantz. L.)	Weaning food	Lactobacilli and yeast	Wanjala et al. 2016,
Uthappam	India	Semolina and finger millet flour	Breakfast food	–	Rastogi and Joshi 2015
Zoom-koom	Burkino Faso	Sorghum, millet, mint, and ginger	Street food: non-alcoholic beverage	*Lactobacillus, Leuconostoc, Lactococcus, Pediococcus,* and *Weisella* species	Amadou 2019, Tapsoba et al. 2017

Figure 13.1 Millet-based major fermented foods.

adults (Amadou et al. 2011). This product is also made commercially available in Uganda by a small company named Badam Foods Limited, located in Kwanda village of Central Uganda (CGIAR 2020).

13.2.1.3 Dhokla

Dhokla is a popular breakfast food in Gujarat, India. It can also be eaten as a snack, side dish, or main course. Traditionally, it is prepared by the fermentation of the batter of rice and Bengal gram (*Cicer arietinum*) dhal. The rice can also be substituted with semolina or other cereals and Bengal gram can be replaced with chickpea (Lohekar and Arya 2014). The preparation starts with the soaking of rice and Bengal gram dhal in water and grinding them coarsely to a batter. The batter is kept for overnight fermentation at a low temperature and on the next day, curd and spices are mixed into the fermented batter and it is cooked in steam for 15–20 minutes (Chandra et al. 2018). The major microorganisms reported in the *dhokla* batter are *Leuconostoc mesenteroides*, *Lactobacillus fermentum*, *Lb. Lactis*, *Lb. delbrueckii*, and *Hansenula silvicola* (Joshi et al. 1989). The fermented batter is steamed in a pie dish, cut into diamond shapes, and seasoned. It has a tangy flavour with a slightly sweet taste. The recipes have been also standardized for the preparation of millet-based *dhokla*. Roopa et al. (2017) optimized the recipe for the production of kodo millet-based *dhokla* and observed that a good quality *dhokla* can be prepared using 50% *besan* (Bengal gram dhal flour), 40% kodo millet, and 10% lentil (*Lens culinaris*). The millet-based *dhokla* had overall acceptability of 8.4 on a 9-point hedonic scale. Pathak et al. (2000) optimized the recipe for the development of *dhokla* mix powder and reported that a good quality *dhokla* mix powder can be prepared by blending a mix of foxtail and barnyard millet, legumes, and fenugreek seeds in the proportions of 55%, 35%, and 10%, respectively. Gupta et al. (2017) optimized the recipe for *dhokla* preparation using *besan*, pearl millet flour, finger millet flour, drumstick leaves, and curd and observed that a good quality *dhokla* with high sensory scores can be prepared using these ingredients in ratios of 40:20:20:10:10, respectively. The millet-fortified *dhokla* had high crude fibre and ash content and hence possessed a hypoglycaemic effect.

13.2.1.4 Dosa

Dosa is a popular food of South India that is traditionally prepared from a fermented batter of rice (*Oryza sativa*) and black gram (*Phaseolus mungo*) in the ratios of 6:1 to 10:1. In some places, a part of the rice is replaced with millets or a mixture of rice, wheat, maize, or millets. Black gram dhal could also be substituted with other pulses like green gram (*Vigna radiata*) or Bengal gram. The ingredients are finely ground and mixed with an adequate amount of water for the batter preparation. The batter is fermented overnight and the microorganisms reported in the *dosa* batter are *Leuconostoc mesenteroides*, *Streptococcus faecalis*, *Lactobacillus fermentum*, *Bacillus amyloliquefaciens*, *Saccharomyces cerevisiae*, *Debaromyces hansenii*, and *Trichosporon beigelli* (Soni et al. 1986). On the next day, the fermented batter is cooked in the form of a pancake on a hot flattened metal surface (a specially designed *tva* for *dosa* preparation). It is served with *sambhar* (a South Indian dish consisting of lentils and vegetables cooked with tamarind and other spices) and coconut *chutney*. Different variants of *dosa*, i.e. stuffed with potatoes (*Solanum tuberosm*), *paneer* (a cheese-like product of India prepared by the acid and heat coagulation of milk), or other vegetables, are also prepared and they are known as *masala dosa*, *paneer dosa*, and *vegetable dosa*, respectively. Several studies have been also optimized by researchers to prepare millet-based *dosa* and *dosa* mixes. Narayanan et al. (2016) prepared millet *dosa* using foxtail millet (90 g) and black gram dhal (20 g) and observed that millet-based *dosa* had a low glycaemic index (59.25) in comparison to rice-based *dosa* (77.86). An increase in the protein and crude fibre content was also found in millet-based *dosa*. A *dosa* dry mix was prepared by mixing equal proportions (33:33:33) of the flours of rice, deculticuled black gram dhal, and a mix of flours of 48-hour germinated millets (finger millet, foxtail millet, proso millet, kodo millet, and little millet). The millet mix fortified *dosa* was nutritionally superior to the rice *dosa* and had overall acceptability of 7.7 on a 9-point hedonic scale (Krishnamoorthy et al. 2013). Roopa et al. (2017) substituted the rice and Bengal gram with little millet (20–70%) and lentils (*Lens culinaris*) (10–30%) in an instant *dosa* mix and found that it had sensory acceptability of 8.5 on a 9-point hedonic scale. This product is also commercially available.

13.2.1.5 Fura/Fura da Nono

Fura is millet- or maize-based semi-solid dumpling that is produced and consumed in the Sahel region of West Africa. The major *fura*-producing countries are Nigeria, Ghana, and Burkina Faso. To prepare pearl millet-based *fura* the grains are slightly moistened with water and ground using the traditional method of mortar and pestle or a local disc attrition mill. The ground grains are dried in the sun and the hull is removed. The dried grains are again ground to a fine flour by using the pestle or mortar or a hammer mill. The flour is blended with spices like powdered ginger or black pepper, water is added, and it is kneaded to smoothen the dough. The dough is made into small balls and is cooked for 30 minutes in boiling water. While still hot, water is added to the balls and they are kneaded again until a smooth, slightly elastic, and cohesive mass is obtained. This mass is again moulded into 25–30 g *fura* balls and *fura* porridge is prepared by crumbling the *fura* balls into fermented whole milk (Kindrimo) or fermented skim milk (nono). Sugar may be added to this mixture and it is known as "*fura da nono*". This is served as a mid-day meal, refreshing drink, or weaning food for infants. These days instant *fura* is also available in the African market (Yusuf et al. 2020; Amadou et al. 2011, Jideani et al. 2001).

13.2.1.6 Ibyer

Ibyer is maize (*Zea maize*), sorghum (*Sorghum bicolor*), or millet-based gruel that is indigenous to the Tiv people of Nigeria (Adebiyi et al. 2016; Kure and Wyasu 2013). It is prepared by the cooking of fermented cereal flour or wet-milled paste in water. There are two types of *ibyer*, i.e. sweet type and sour type. The sweet type is known as *ibyer-i-nyohon* and is prepared by mixing the fermented maize or millet flour with water (1:1.50) to prepare a slurry and cooking it with additional water (400 ml/100 g) for 10–15 minutes with continuous stirring. This is allowed to cool for about 5 minutes and the sweet *ibyer* is ready for serving. The sour type is known as *ibyer-i-angen* and its preparatory steps are similar to that of sweet *ibyer*; the only difference is that the slurry is allowed to ferment for 12 hours before cooking. This product is consumed by both children and adults alike but it is a poor source of energy and nutrients. The nutrient density of *ibyer* can be enhanced by the addition of malt or protein-rich legumes like soybean (Kure and Wyasu 2013). To enhance the nutritional value of *ibyer* fortification with ginger powder is also practiced. Adakole et al. (2021) optimized the process for the development of ginger powder-fortified *ibyer* and found that a good quality *ibyer* with highly acceptable sensory characteristics like appearance, aroma, mouthfeel, consistency, taste, and overall acceptability can be obtained with a fermented millet flour and ginger powder ratio of 95:5. An increase in protein, fat, fibre, and ash and a decrease in carbohydrate content were also found on the addition of ginger powder. The microbiological quality of the product was also improved as the total bacterial and fungal load was reduced.

13.2.1.7 Idli

Idli is a popular steamed pudding of South India that is prepared from a thick fermented batter of polished rice and split dehusked black gram dhal (2:1–4:1). The preparation starts with the separate overnight soaking of both ingredients in water and grinding the next morning separately on a stone pestle and mortar. The rice is ground coarsely and a fine paste of black gram dhal is prepared. The ground pastes are then mixed, salted to taste, and fermented for 12–18 hours. The fermentation is dominated by *Leuconostoc mesenteroides* and *Streptococcus thermophilus* as these microorganisms outnumber the initial contaminants and produce lactic acid ($\geq 1.0\%$) and carbon dioxide. This causes the leavening of the batter which is then steamed in flat plates or perforated cups to obtain soft, spongy, tasty, and easily digestible *idlis*. The effect of the substitution of rice with millets has been studied by many researchers. Vanithasri and Kanchana (2013) studied the effect of the substitution of rice with barnyard millet on the nutritional and sensory properties of *idli*. The *idli* was prepared using two recipes. The standard *idli* was prepared using 100 g rice, 50 g black gram dhal, 5 g fenugreek seeds, and water while the rice was replaced with 50 g of parboiled barnyard millet in the fortified *idli*. The fortified *idli* had an increased content of protein, fat, fibre, calcium, phosphorus, and iron. The sensory scores of the fortified *idli* were also on par with the control *idli*. Gupta et al. (2017) optimized the recipe for semolina *idli* with the incorporation of finger millet, pearl millet, and drumstick leaves. The study concluded that good quality *idli* can be prepared using semolina and the flours of finger millet, pearl millet, drumstick leaves, and curd in the ratios 40:20:20:10:10, respectively. The optimized *idli* recipe also scored better on sensory parameters like colour, taste, and flavour, in comparison to the control prepared using semolina and curd (90:10). The fortification also increased the carbohydrates, protein, fat, polyphenols, and minerals like calcium and iron in comparison to the control. Chelliah et al. (2017) studied the effect of the addition of finger millet and pearl millet (10% w/w) on the fermentation time, nutritional content, and microbial quality of *idli*. The fortification of the basic ingredients with finger millet and pearl millet flours reduced the fermentation time to 6 hours and 8 hours, respectively. Finger millet-fortified batter had a pH of 4.32 and acidity of 0.45% after 6 hours of fermentation, while the pearl millet-fortified batter had a pH of 4.53 and titratable acidity of 0.45% after 8 hours of fermentation. However, the control batter could produce only a pH of 5.32 and acidity of 0.27% after 6 hours of fermentation. The millets-based batter also had a higher total bacterial count, LAB count, and viable yeast count as compared to control. The finger millet and pearl millet fortification also increased dietary fibre (28 and 23%), calcium (113 and 56%), and iron (51 and 258%) compared to the control. Krishnamoorthy et al. (2013) formulated an *idli* dry mix by blending rice flour, millet mix flour, and decuticuled black gram powder in the ratio 33:33:33. For the preparation of millet mix powdered finger millet, foxtail millet, proso millet, little millet, and kodo millet were germinated for 48 hours, dried, and milled to flour. A higher content of carbohydrates (72–74%), protein (15–18%), fat (5.0–6.2%), crude fibre (3.0–4.9%), and ash (1–2%) were found in millet-fortified *idli* in comparison to standard *idli*. The millet mix *idli* had a score of 7.7 for overall acceptability on the hedonic scale. Ohariya et al. (2017) standardized the recipe for the kodo millet-based *idli* mix. The study concluded that good quality *idlies* can be prepared from *idli* mix with 70% kodo rice, 15% black gram, and 10% full-fat soy flour. Similar research on the fortification of *idli* with millets has also been conducted by other researchers.

13.2.1.8 Injera

Injera is a teff-based Ethiopian bread. The part of teff can also be replaced with other cereals (barley, wheat, sorghum, maize) and millets in different proportions. The levels of teff and other cereals in the household-level preparation of *injera* depend on previous experience, traditional process, and the financial status of the family. The preparation starts with the grinding of the cereals to the flour, mixing with water (two times the amount of flour), and the addition of "*ersho*" (a seed culture from the previous batch) at the rate of 16% of the weight of flour. The major microorganisms reported in *ersho* are *Candida milleri*, *Rhodotorula mucilaginosa*, *Kluyveromyces marxianus*, *Pichia naganishii*, and *Debaromyces hansenii* (Ashenafi 1994). This mixture is hand stirred properly to form a thin watery paste and is left for fermentation (30–72 hours) in a wooden, clay, or metal container known as "*bohaka*". On the completion of fermentation, a part of the batter (about one-fifth) is mixed and boiled to produce "*absit*". The *absit* is cooled to about 46°C and is mixed back into the primary fermented batter and is allowed to ferment again for 2 hours. This step results in gas production and the rise of the paste. After this, the paste is spread in a circular movement from the periphery to the centre on a clay-made hot plate (90–95°C) greased with rapeseed oil and is griddled for 2–3 minutes after covering with a lid known as "*kidan*". The device used for pouring is known as "*mazoria*" and the hot plate is known as "*metad*". It is consumed as a major food at lunch and dinner and is served at family gatherings like marriages, birthday parties, and death anniversaries. It is usually served with a traditional Ethiopian dish "*wot*" that is prepared from a mix of vegetables, meat, spices, and sauces. *Injera* can be stored for 2–3 days in a cool, dry, and ventilated place (Neela and Fanta 2020).

13.2.1.9 Koko

Koko is a pearl millet- or sorghum-based porridge prepared and consumed mainly in the western parts of Africa. It originated in the northern parts of Ghana, where millets, sorghum, rice, and cowpeas are mostly grown (Haleegoah et al. 2016). The various steps in *koko* preparation are the overnight soaking of millet grains, discarding the excess water in the morning, the wet milling of soaked grains, the mixing of spices like ginger (*Zingiber officinale*), black pepper (*Pepper nigrum*), pepper (*Capsicum annuum*), cloves (*Syzygium aromaticum*), and xylopia (*Xylopia aethiopica*), and the addition of water to the ground material to make a slurry. The slurry is fermented and is allowed to sediment for 2–3 hours. The top layer of the liquid obtained after sedimentation is collected and boiled for 1–2 hours. To this, the sedimented bottom layer is added until the desired consistency is obtained. The product is ready for consumption on the next day of soaking (Amadou et al. 2011). It is mainly consumed at lunch or an in-between meal. It is served hot with "*koose*" fried cowpea (*Vigha unguiculata*) paste or "*maasa*", a paste of fried millet and sorghum or corn paste (Haleegoah et al. 2016). Danson et al. (2019) studied the nutritional composition of *koko* and found that it is a good source of protein and tryptophan, i.e. 2.2% and 0.35%, respectively. It is also a source of probiotics and the major microorganisms reported in the product are *Weisella confusa*, *Lactobacillus plantarum*, *L. fermentcum*, *L. brevis*, *L. salivarius*, *Enterobacter colcae*, *Saccharomyces cerevisiae*, *Candida mycoderma*, and *Acinetobacter* (Amadou et al. 2011).

13.2.1.10 Koozh

Koozh is a millet-based traditional porridge of Tamil Nadu, India. It is mainly prepared by the fermentation of finger millet (Thirumangaimannan and Gurumurthy 2013); however, other cereals like pearl millet and rice can also be used for its production. Finger millet *koozh* is consumed in Chennai and Salem and pearl millet *koozh* is consumed in Madurai. For its preparation, two parts of millet flour are mixed with one part of water, and fermentation is carried out at room temperature for about 12 to 15 hours. On the next day, broken parboiled rice is boiled in an earthen pot, and the fermented millet slurry is added to it and mixed. The mixture is cooled, hand crushed, and fermented overnight at room temperature. The microorganisms reported in *koozh* are *Weissela paramesenteroides*, *Lb. fermentum*, *Vibrio parahaemolyticus*, and *Listeria monocytogenes* (Antony et al. 2020). The fermented mixture is then shaped into balls called *kali* and stored at room temperature. The balls are mixed with tap water in the ratio of 1:6 w/v and salted to prepare *koozh*. It has a tangy flavour. This is then consumed with vegetables and accompaniments like chili pepper (*Capsicum annuum*), turkey berry (*Solanum torvum*), sliced onion (*Allium cepa*), cluster beans (*Cyamopsis tetragonoloba*), green coriander (*Coriandrum sativum*), chutney, and pickles. A single ball can be used to prepare about 2–3 L of *koozh* by diluting with water. It is rich in energy and is considered a good source of food for working-class people. It gives stamina over an extended period and is also helpful in the management of diabetes mellitus. It is also available commercially as street food (Antony et al. 2020; Ilango and Antony 2014).

13.2.1.11 Mangisi

Mangisi is a finger millet-based traditional fermented food of Zimbabwe. To prepare *mangisi*, the *masvusvu* [non-fermented beverage prepared by the cooking of finger millet and water (1:4)] is cooled at room temperature for 30 minutes, 2 litres of fresh clean water is added, the mixture strained through a sieve, and the filtrate is transferred to an earthenware pot (unwashed pot from the previous lot) and fermented at room temperature for 8 hours. The major microorganisms reported in *mangisi* belong to LAB, yeasts, and moulds. The final product has a titratable acidity of 0.67%, pH of 3.98, and lactic acid concentration of 4.10 g/L. It has a sweet-sour taste (Zvauya et al. 1997).

13.2.1.12 Ogi

Ogi is a cereal-based (maize, millet, sorghum) fermented gruel that is prepared and consumed in Nigeria mainly as a weaning food for babies. It is also enjoyed as a breakfast food by children and elders. Although the term *ogi* is used to refer to all gruels made with this technology, it specifically refers to the product made from maize. The product prepared from millets is known as *ogi gero* and that prepared from sorghum is known as *ogi baba*. The traditional process of *ogi* preparation involves the washing of grains (to remove adhered dust/dirt and impurities), steeping in water (1–3 days at room temperature in earthenware, plastic, or enamel-coated pots), wet milling of the softened grains (using a traditional pestle or mortar or an electrically powered grinder), the addition of water to obtain a slurry, sieving to separate the hull and slurry (the material is passed through a muslin cloth), and sedimentation of the slurry (24–48 hours at room temperature to obtain a starch paste). The sedimented starch paste can be transformed into *ogi* porridge/gruel by introducing small quantities of hot water (Bolaji et al. 2017; Bolaji et al. 2015; Okara and Lokoyi 2012). *Ogi* prepared from maize has a cream or yellow colour, *ogi* prepared from millets has a dirty grey colour, and the sorghum-based product has a reddish-brown colour. The product is smooth in texture and has a sour taste that resembles yogurt. The sour taste develops due to fermentation during the steeping and sedimentation step. The microorganisms involved in the fermentation during soaking are *Aspergillus* sp., *Penicillium* sp., *Cephalosporium*, *Fusarium*, *Rhizophus stolonifer*, *Mucor mucedo*, *Saccharomyces cerevisiae*, *Candida mycoderma*, *Corynebacterium* sp., *Lactobacillus plantarum*, *Lb. fermentum*, and *Leuconostoc fermentum*. The dominant microorganisms in the secondary fermentation, i.e. in the sedimentation step, are *Lb. plantarum*, *Lb. fermentum*, and *S. cerevisiae* (Okara and Lokoyi 2012; Ijabadeniyi 2007). The product is also known as "*akamu*" in South-Eastern Nigeria and *ogi* prepared from millet or sorghum is known as "*furali*" in some regions of Northern Nigeria. Millet-based *ogi* has a better nutritional profile, but millets are considered unsuitable for traditional *ogi* production because they result in lower yields due to high soluble solids loss (Banigo and Muller 1972). Akingbala et al. (2002) optimized the process for the development of pearl millet-based *ogi* and reported that a good quality (enhanced nutritional value and yields) *ogi* can be prepared by milling the millet grains to a size of 1 mm, adding distilled water to flour (500 g millet flour dry basis + 500 ml distilled water), making a slurry/paste, and fermenting the paste for 48 hours at room temperature. The fermented paste is dried in a forced oven at 60°C for 24 hours and reground to a size of 0.8 mm. The dried powder can be made into gruel or porridge by the addition of small quantities of water. This is still a household product and is not available commercially.

13.2.1.13 Oshikundu

Oshikundu is a pearl millet- or sorghum-based popular non-alcoholic fermented beverage in Namibia. In the traditional process of its preparation, the pearl millet or sorghum grains are cleaned, ground to flour, and boiled water is added to the flour with continuous stirring. A mixture of pearl millet and sorghum flour can also be used. The mixture is left to cool at room temperature after which sorghum flour is added to the mixture and thoroughly stirred. Additionally, pearl millet bran can also be added at this stage. The consistency and volume of the mixture are adjusted by the addition of water and back-slopping is done by the addition of previously fermented *oshikundu* and fermented for 4–6 hours. The major microorganisms involved in the fermentation are *Lb. plantarum*, *Lb. lactis* spp. *lactis*, *Lb. delbrueckii* spp. *delbrueckii*, *Lb. fermentum*, *Lb. pentosans*, *Lb. curvatus* spp. *curvatus*. The final product has a titratable acidity of 1.2–1.7% and pH of 3.3–3.7, total solids are in the range of 2–4.2%, and the alcohol content is 1–1.6%. This beverage is produced as a traditional practice on the initiation of young girls into womanhood and is served at weddings and social interactions to welcome guests (Mishairabgwi and Cheikhyoussef 2017).

13.2.1.14 Paddu

Paddu is rice- and pulse-based popular South Indian breakfast food that is prepared by the shallow frying of a mixture of fermented rice and pulses, usually *dosa* batter. The batter is cooked in a special mould to a round shape and the final product has a golden brown appearance. Rice can be substituted with millets and other cereals to increase the nutritional value of *paddu* (Kumar et al. 2020). Madalageri et al. (2016) prepared rice (rice – 74.58%, black gram *dhal* – 18.64%, Bengal gram *dhal* – 1.86%, red gram *dhal* – 1.86%, flaked rice – 1.94%, and fenugreek seeds) and little millet (little millet – 81.48%, black gram *dhal* – 13.58%, Bengal gram *dhal* – 1.36%, red gram *dhal* – 1.36%, flaked rice – 1.4%, and fenugreek – 0.82%) based *paddu* and compared both for nutritional, microbiological, and sensorial attributes. It was observed that little millet-based *paddu* had more protein (14.50%), fat (5.01%), ash (2.87%), and crude fibre (1.53%) in comparison to rice-based *paddu*. The protein, fat, ash, and crude fibre content of rice *paddu* was 12.17%, 4.89%, 1.67%, and 1.06%, respectively. Yeast was the dominating fermenting microorganism in the little millet-based *paddu*, while the fermentation of rice *paddu* batter was dominated by bacteria. The sensorial acceptability of little millet-based *paddu* was comparable to rice *paddu*.

13.2.1.15 Rabadi
Rabadi is a traditional lactic acid fermented beverage of the north-western semiarid regions of India. It is prepared by mixing varying proportions of cereal flour with sour buttermilk and fermenting the mixture by placing it under sunlight for 3–4 hours. When enough lactic acid is produced, the fermentation is ceased by boiling the fermented mix. The product may be consumed directly on cooling or it may be diluted with water or buttermilk. The process for pearl millet-based *rabadi* was standardized by Modha and Pal (2011). In this study, germinated pearl millet flour ranging from 3 to 8% was added to skim milk, mixed and inoculated with 3% mixed *dahi* culture, and fermented at 37°C for 12 hours. It was found that the *rabadi* prepared using 5.3% germinated millet flour and 72% water had the highest overall acceptability. The optimized product had a total solids content of 8.7% which included protein, fat, and ash content at the rate of 2.2%, 0.65%, and 1.3%, respectively.

13.2.1.16 Togwa
Togwa is a traditional non-alcoholic beverage of East Africa that is produced from maize flour and finger millet malt. To prepare *togwa* water is heated to about 45°C and at this temperature maize flour (11–16%) is added with continuous stirring. This mixture is heated to 80°C, cooled to 50–60°C, finger millet malt (3–5%) is added, and the temperature is maintained at 50°C for 15 minutes. The fresh *togwa* is transferred to a container, capped, and kept at room temperature overnight (Kitabatake et al. 2003). During the incubation spontaneous fermentation takes place and lactic acid bacteria and yeasts are the dominating fermenting microorganisms. The major microorganisms reported in *togwa* are *Lactobacillus fermentum*, *L. brevis*, *L. fermentum*, *L. cellobiosas*, *Pediococcus entosaceus*, *Weissela confusa*, *Saccharomyces cerevisiae*, *Candida pelliculosa*, and *C. tropicalis*. The pH of *togwa* is reduced to 3.34–3.10 after fermentation (Mugula et al. 2003). The maize can also be replaced with other starchy cereals and in some regions, cassava is also used. The product is consumed as a refreshing and weaning food by both adults and children (Amadou et al. 2011).

13.2.1.17 Uji
Uji is a cereal-based [finger millet, sorghum, maize, or cassava (*Manihot esculenta* Crantz. L.)] beverage/thin porridge that is mainly used as a weaning food in Kenya. Unroasted finger millet and sorghum are the most widely preferred grain choices for *uji* preparation, while maize is the least preferred. *Uji* prepared from unblended cassava possess a bland taste and flavour, and are generally not preferred. In general, the blend of these cereals and cassava is used for *uji* preparation where the largest share is of finger millet and the other crops constitute a small part.

To prepare the *uji*, a slurry is prepared by mixing equal quantities of flour and water and this slurry is added slowly with continuous stirring to a container containing boiling water. The mix is heated until a thick but free-flowing paste is obtained and there is no foaming on the surface. During this process the gelatinization of starch takes place and the starch-rich slurry transforms into porridge. The final product has a flour to water concentration of about 8–10% w/v (Wanjala et al. 2016). It can also be supplemented with fruits, vegetables, or milk and is a principal weaning food for children of the age group of 6–12 months (Huffman et al. 2000). Fermented *uji* can also be prepared by carrying out the spontaneous fermentation of a slurry (30–40 g flour/100 ml water) of unblended or composite flours at room temperature (25–35°C) for 24–48 hours. The initial stages of fermentation are dominated by coliforms and fungi; however, these are replaced by lactobacilli after some time and the major fermentation is carried out by lactobacilli. Some yeasts have been also reported in the *uji* batter. The fermented slurry can be dried and reconstituted to *uji* whenever required (Wanjala et al. 2016).

13.2.1.18 Uthappam

Uthappam also known as *uttapam*, *uttappa*, or *oothapam* is a popular breakfast food of South Asia that is prepared from fermented batter of common rice and Bengal gram dhal. Although the ingredients used in *uthappam* preparation are similar to *idli* and *dosa*, the batter is a little thicker and it is topped with vegetables like onion, tomato, capsicum, carrot, green chili, sweet corn, etc. The rice and Bengal gram *dhal* can also be substituted with semolina or other cereals for *uthappam* preparation. Rastogi and Joshi (2015) prepared *uthappam* using semolina (75%) and finger millet flour (25%) and found that the *uthappam* prepared using this blend had high overall acceptability of 9.2 on a 10-point rating scale. A good quality *uthappam* with high sensory characteristics can be prepared by fortifying semolina with the flours of finger millet (20%), pearl millet (20%), and drumstick leaves (10%). An increase in carbohydrates, protein, fat, polyphenols, vitamin C, and minerals (calcium and iron) was also observed in the fortified *uthappam* (Gupta et al. 2017).

13.2.1.19 Zoom-Koom

Zoom-koom is a sorghum/millet-based fermented, non-alcoholic beverage of Burkina Faso. It is mainly consumed as street food. For its production, sorghum/millet grains are cleaned, soaked (16 h) in twice the amount of water; mint (3g/100 g) and ginger (6 g/100 g) are added for flavour, and the mixture is ground to paste and made into a dough. The dough is fermented at room temperature for 10–12 hours and on completion of fermentation, water equivalent to three times the volume of fermented dough is added to it.. The suspension is filtered by passing through a muslin cloth and *zoom-koom* is collected. The major fermenting microorganisms belong to *Lactobacillus*, *Leuconostoc*, *Lactococcus*, *Pediococcus*, and *Weisella* species. It can also be sugared to obtain sweet *zoom-koom* (Amadou 2019, Tapsoba et al. 2017).

13.2.2 Alcoholic Beverages

Traditionally, alcoholic fermentation is usually carried out by back-slopping or by the addition of a traditional starter such as *dhaeli*, *marcha*, etc. A list of the major alcoholic beverages produced from millets, their country of origin, raw material used, alcohol content, consumption pattern, and the associated microorganism is also provided in Table 13.2. A diagrammatic representation of the traditional utensils used for the fermentation, storage, and distillation of alcoholic beverages is also provided in Figure 13.2.

230 Millets

TABLE 13.2 MILLET-BASED ALCOHOLIC FERMENTED FOODS, THEIR COUNTRY OF ORIGIN, RAW MATERIALS, CONSUMPTION PATTERN, AND MICROBIOLOGY

Product name	Country of origin	Raw materials used	Alcohol content	Consumption pattern	Microbiology	References
Boza	Turkey, Kazakhstan, Kyrgyzstan, Albania, Bulgaria, Macedonia, Montenegro, Bosnia and Herzegovina, Romania, and Serbia	Millet, maize, wheat	General: 0.02–0.79%, Egypt: 7%	As an energy drink or mood-altering device	Saccharomyces carlsbergensis, S. cerevisiae, Lactobacillus spp., Leuconostoc spp., Streptococcus spp., Micrococcus spp.	Bayat and Yildz 2019, Petrova and Petrov 2017, Arici and Daglioglu 2002
Omalodu	Namibia	Sorghum and pearl millet	–	Served to welcome guests and at traditional get-togethers such as weddings and on the birth of babies	–	Mishiarabgwi and Cheikhyoussef et al. 2017
Jaandh	Nepal	Finger millet and marcha (traditional starter culture)	2.1–18.9%	Considered to be a good tonic for postnatal women	Saccharomyces spp., Lactobacillus spp., Mucor spp., Pichia spp.	Amadou 2019, Dahal et al. 2005, Thapa et al. 2015, Amadou 2019
Kodo ko jaanr	Nepal, Bhutan, and India	Finger millet and marcha (traditional starter culture) (2%)	4–7%	Ailing and postnatal women consume the extract of kodo ko jaanr to regain their strength	Pichia anomala, S. cerevisiae, Candida globrata, S. fibuligera, Pediococcus pentosaceus, and Lactobacillus bifermentans	Thapa and Tamang 2004, Thapa and Tamang 2006

(Continued)

TABLE 13.2 (CONTINUED) MILLET-BASED ALCOHOLIC FERMENTED FOODS, THEIR COUNTRY OF ORIGIN, RAW MATERIALS, CONSUMPTION PATTERN, AND MICROBIOLOGY

Product name	Country of origin	Raw materials used	Alcohol content	Consumption pattern	Microbiology	References
Merissa	Sudan	Sorghum or millet	6%	–	LAB and *Saccharomyces*	WHO 2004, Dirar 1978
Okatokele	Namibia	Pearl millet and sugar	–	–	–	Mishihairabgwi and Cheikhyoussef 2017
Pito and burukutu	Nigeria, Republic of Benin, and Ghana	Sorghum or mixture of sorghum and millet and sediment of previous brew as starter	3–6%	–	*Saccharomyces cerevisiae, Lactobacillus fermentum, L. plantarum, L. acidophilus, L. lactis* subsp. *lactis, L. brevis*	Ajiboye et al. 2014, Atter et al. 2014, Alo et al. 2012
Sur	India	Finger millet, jaggery, and *dhaeli* (traditional inoculum)	5–16%	It is considered as a tonic by local people	*Saccharomyces cerevisiae* and *Zygosaccharomyces bisporus*	Joshi et al. 2015, Kumar 2013, Thakur et al. 2004
Xiao mi jiao	Taiwan	Foxtail millet and starter culture made up of rice and plant material [leaf sap of *Citrus grandis* Osbeck, *Diospyros discolor* Willd., *Limnophila rugosa* (Roth) Merr.]	–	–	*Lactobacillus brevis, Enterococcus faecium, Pediococcus pentosaceus, Lactococcus garvieae, Lactococcus carnosum, Lactococcus lactis, L. mesenteroides, P. stilerii, Weisella soli,* and *Weisella cibaria*	Chao et al. 2013

Figure 13.2 Traditional equipment used for fermentation. (a) A traditional earthen pot used for the fermentation of *sur* in Himachal Pradesh, India. (b) A traditional pot used for the storage of *sur*. (c) A *juggad* technology developed by the villagers of Sirmour, Himachal Pradesh, India, for the distillation of alcoholic beverages.

13.2.2.1 Boza

Boza is a millet-based alcoholic beverage that is produced and consumed primarily in Turkey, Kazakhstan, Kyrgyzstan, Albania, Bulgaria, Macedonia, Montenegro, Bosnia and Herzegovina, Romania, and Serbia. The name "*boza*" is derived from the Persian word "*buze*" which means millet. Other names like *bozan, bozas, buza, bouza,* and *busa* are also used in Romania, Greece, Russia, France, and Germany, respectively (Bayat and Yildz 2019). The Turks of middle Asia call this product *bassoi* (Petrova and Petrov 2017). This is produced both at the household and commercial levels. Millet flour is the most preferred cereal as it produces the best quality *boza* with the most desirable taste. Other cereals like maize, rye, wheat, rice, or a mix of millets and all these cereals can also be used. For the preparation, grains are cleaned, washed, and placed into a large and deep pot, potable grade water is added to this (ten times the volume of grains), boiled (until mushy), and strained through a fine strainer along with pressing, and the filtrate is collected, cooled, and fermented. To carry out the fermentation, yeast is mixed with the same quantity of sugar, and this mixture is dissolved in hot water and added to the fermentation pot containing the filtrate. Additional sugar and water can also be added on the completion of fermentation to obtain the desired consistency. The commercial process of *boza* preparation is more or less similar to the traditional process but it provides better control over boiling time and fermentation conditions and results in a better quality product. The product has a creamy-white to light yellow colour, a sweet-sour taste, and an acidic-alcoholic odour. The alcohol content is generally low (0.02–0.79%), except in Egypt where it can contain alcohol content as high as 7% by volume. It is generally served in its pure form but in Turkey, it is served with cinnamon and roasted chickpeas (Bayat and Yildz 2019; Petrova and Petrov 2017).

13.2.2.2 Omalodu

Omalodu is sorghum- and pearl millet-based traditional beer that is produced by the Oshiwambo- and Rukwangali-speaking people of Namibia. For its production sorghum malt is mixed with cold water, boiled for about 2 hours, and filtered through a sieve with fine pores. The filtrate is cooled, transferred to a traditional pot called "*oshitoo*", and fermented for 6–24 hours. On the completion of fermentation, a small quantity of pearl millet flour is added and the mixture is again fermented for 1 hour. Following the fermentation, the product is ready. This drink is served to welcome guests and at traditional get-togethers such as weddings and on the birth of babies (Mishairabgwi and Cheikhyoussef et al. 2017).

13.2.2.3 Jaandh

Jaandh is a finger millet-based traditional non-distilled alcoholic beverage of Nepal. It is mainly produced by the *Mangarantis* ethnic groups (Dahal et al. 2005). For its preparation, finger millet seeds are steamed to soften, spread on the banana leaves, cooled, a traditional starter known as *"marcha"* is sprinkled in powdered form, mixed well, and the mix is then piled into a heap and is left for fermentation for 24 hours at room temperature. The mass is then transferred to an earthen pot, the pot is made airtight, covered with leaves and straw, and is left for fermentation. The finger millet grains can also be supplemented with small quantities of corn or wheat to aid the fermentation. On the completion of fermentation, the seeds are kneaded to remove seed coats. To prepare *jaandh*, the grits are placed into bamboo vessels traditionally called *"toongba"* and hot or cold water is added (Amadou 2019, Dahal et al. 2005). The beverage is ready to drink after 10 minutes and has a slightly acidic and sweet alcoholic taste. This product's alcohol content ranges from 2.1 to 18.9% (Thapa et al. 2015). It is considered to be a good tonic for postnatal women (Amadou 2019).

13.2.2.4 Kodo Ko Jaanr

"Jaanr" is a common Nepalese name for all kinds of alcoholic beverages. *Kodo ko jaanr* is a finger millet-based alcoholic beverage that is commonly produced in Nepal, Bhutan, and the Eastern Himalayan regions (Darjeeling hills of West Bengal, Sikkim) of India. In the traditional method of preparation, the finger millet grains are cleaned, washed, and cooked in excess water for about 30 minutes. On completion of cooking, the excess water is removed and the grains are spread over a bamboo mat traditionally known as *"mandro"* for cooling (20–25°C). The cooled grains are mixed uniformly with the dry powder (2%) of the traditional starter culture known as *"marcha"*, packed in a bamboo basket lined with fresh fern or banana leaves, covered with sack clothes, and saccharified at room temperature for 2–4 days. The saccharified mass is transferred into a specially made bamboo basket called *"septu"* or into an earthen pot. The container is made air-tight and fermented at room temperature for 3–4 days during summer and 5–7 days in winter. The major microorganisms reported are *Pichia anomala, S. cerevisiae, Candida globrata, S. fibuligera, Pediococcus pentosaceus,* and *Lactobacillus bifermentans*. To prepare a drink, 200–500 g of fermented finger millet is put into a traditional vessel called *"toongbaa"*, lukewarm water is added and a milky white extract of *kodo ko jaanr* is sipped after 10–15 minutes from the hole near the bottom using a narrow bamboo straw called *"pipsing"*. It has a sweet alcoholic flavour (Thapa and Tamang 2004).

13.2.2.5 Merissa

Merissa is a traditional alcoholic beverage of Sudan that is prepared from sorghum or millet (Sawadogo-Lingani 2021; WHO 2004). The traditional process of *merissa* preparation has three major steps, i.e. the preparation of *"ajeen"*, the fermentation of *"deoba"*, and the fermentation of *"merissa"*. Sorghum or millet grains are soaked overnight, germinated for about 2 days, and sun-dried, and a malt is prepared and milled to a fine flour. About one-third of the flour is mixed with water, just enough to moisten it, and it is fermented at room temperature for 36 hours. The major fermenting microorganisms in this flour are LAB. This naturally fermented sourdough is called *"ajeen"*. It is then cooked in a hollow steel container with continuous stirring to a dark brown colour without the addition of water. This product is slightly overcooked, extremely sour, and has a pleasant caramelized flavour. This is known as "*soorij*". About 5% of malt flour is added to it, an equal amount of water is added, and about 5% of good *merissa* is added to it, and the mixture is left for fermentation at room temperature for 4–5 hours. This fermented product is known as "*deoba*". It is a very sour, dark, thick suspension. To prepare *merissa*, the remaining two-thirds of the malted flour is cooked into two equal lots, i.e. one part is half cooked to a greyish brown paste, and the second half is cooked to a brown paste. Both of these are mixed to obtain "*futtara*", i.e. a gelatinized solid material. "*Futtara*" is then mixed with 5% malt flour and successively added to "*deboba*". It is fermented for 8–10 hours and the alcohol-producing yeast found in this belongs to the genera *Saccharomyces*. After fermentation is complete, it is strained through cloth bags. The filtrate so obtained is known as "*merissa*" and the coarse material is known as "*mushuk*". Merissa is used as a drink and *mushuk* as an animal feed. The alcohol content of *merissa* is about 6%. It has a short storage life of one day only and develops a sharp sour flavour on the same day (Dirar 1978).

13.2.2.6 Okatokele
Okatokele is a pearl millet-based beer produced by the Oshiwambo-speaking people of Namibia. For its preparation water is added to pearl millet flour, sugar is added, mixed properly, and the mixture is fermented in plastic buckets for 8–24 hours at room temperature. The drink is ready for consumption after fermentation.

13.2.2.7 Pito and Burukutu
Pito and *burukutu* are popular traditional alcoholic beverages of West Africa especially Nigeria, the Republic of Benin, and Ghana. These are mainly prepared from sorghum but a mixture of sorghum and millet can also be used (Ajiboye et al. 2014; Alo et al. 2012). To prepare these beers, the grains are soaked in water (24–48 hours), malted (germinated for 4–5 days, sundried, and ground to flour), mashed (malt flour is mixed with water and boiled for 3–4 hours), cooled, filtered, naturally fermented (10–12 hours), more water is added, and the mixture is cooked (approx. for 3 hours) and cooled (20–29°C); sediment from the previous brew is added as a starter and it is left again for fermentation at room temperature for 12–24 hours. Adjuncts like the flour of cassava can also be added at the mashing stage. On the completion of fermentation two products are obtained, i.e. *pito* (a clear supernatant) and *burukutu* (a thick brown suspension) (Lyumugabe et al. 2012). The alcohol content of these beverages ranges between 3 and 6% (Alo et al. 2012).

13.2.2.8 Sur
Sur is a finger millet-based traditional alcoholic beverage of Himachal Pradesh, India. The finger millet grains are milled to a fine flour, mixed with water, and kneaded to a dough. The dough is allowed to ferment spontaneously at room temperature for 7–8 days or until a visual growth of fungus/moulds is observed. The fermented flour is then baked into "*rotis*" (unleavened bread), cooled, broken into small pieces, transferred to a previously sterilized earthen pot, and water is added. The pots are left undisturbed for 10–12 hours and jaggery is added in the form of syrup. To this, "*dhaeli*", a traditional inoculum, is added (2–3%), mixed properly, sealed airtight, and left for fermentation. A uniform temperature is maintained in the pots by covering them with woollen cloths. The fermentation is considered complete when the hissing sound of bubbles ceases to come out (Joshi et al. 2015). The major fermenting microorganisms in *sur* are *S. cerevisiae* and *Zygosaccharomyces bisporus* (Thakur et al. 2004). The alcohol content in *sur* varies from 5 to 16% v/v (Kumar 2013). This beverage is also offered to the local god *Hurang Narayan*.

13.2.2.9 Xiao Mi Jiao

Xiao mi jiao is a foxtail millet-based indigenous alcoholic beverage of Taiwan. This drink is mainly produced by the Amis ethnic group of Hualien County, Taiwan. The process of its production varies among different tribes; however, the general process involves the steaming of millet and mixing with a traditional starter culture (0.4%) made up of rice and plant material [leaf sap of *Citrus grandis* Osbeck, *Diospyros discolor* Willd., *Limnophila rugosa* (Roth) Merr.]. The major microorganisms reported in the starter are *Lactobacillus brevis*, *Enterococcus faecium*, *Pediococcus pentosaceus*, *Lactococcus garvieae*, *L. carnosum*, *L. lactis*, *L. mesenteroides*, *P. stilerii*, *Weisella soli*, and *Weissela cibaria* (Chao et al. 2013). This mixture is incubated for 3 days and on the fourth day 2.5 times of its volume of water is added to this and the mixture is further fermented for 5 days at room temperature with at least one thorough stirring per day. On the completion of fermentation, the liquid is separated by filtration and is used as an alcoholic beverage. Good quality aged beverages require fermentation times as long as 1 year (Amadou 2019).

In addition to these beverages, millets can also be used as a substitute for sorghum in sorghum-based traditional beers like *dolo* and *togo*.

13.3 NON-FERMENTED TRADITIONAL FOODS

Millet-based non-fermented traditional products are prepared mostly by baking, cooking, or roasting millets. Some of the millet-based non-fermented foods are discussed in the following sections. Details on millet-based non-fermented foods are also provided in Table 13.3. A pictorial representation of some of the millet-based products is also provided in the Figure 13.3.

13.3.1 Dambu

Dambu is a granulated dumpling prepared by the steaming of millet, maize, or sorghum. The millets are dehulled, milled, and moistened, spices are added, and the mixture is steamed for 30 minutes. The coarse particles are sprinkled into fermented milk and sugar may be added to taste. This product is produced at both household and commercial levels and is consumed mostly in Northern Nigeria (Ho et al. 2009). It is very prone to spoilage and has a storage life of only 2 days at room temperature (Amadou et al. 2011).

13.3.2 Halwa

Halwa is an Indian sweet dish that is prepared mainly from roasted wheat semolina and sugar. Yadav et al. (2011) optimized the recipe for the preparation of pearl millet *halwa* and reported that a good quality pearl millet *halwa* can be prepared by using 100 g of pearl millet semolina, 38.6 g of vanaspati ghee, 88.7 g of sugar, and 151 ml of water.

13.3.3 Kheer/Basundi/Pyasam

Kheer is milk and rice-based traditional Indian dessert that is prepared by the partial dehydration of whole milk in a "*karahi*" (a cooking utensil with a broad open mouth). The milk is boiled in a *karahi*, and cleaned and washed rice is added to it. Once the rice is cooked, sugar is added and it is further concentrated (Bhosale et al. 2021). It may also contain raisins, saffron, coconut powder, cardamom, and dry fruits for flavouring and garnishing. The process for the development of

TABLE 13.3 MILLET-BASED NON-FERMENTED FOODS, THEIR COUNTRY OF ORIGIN, AND RAW MATERIALS USED

Product name	Country of origin	Raw materials used	References
Roti	India	Pearl millet, sorghum, or finger millet	Kumar et al. 2020
Pongal	India	Foxtail millet or little millet and green gram dhal	Ambati and Sucharitha 2019, Fatima and Rao 2019
Halwa	India	Pearl millet semolina (100 g), vanaspati ghee (38.6 g), and sugar (88.7 g)	Yadav et al. 2011
Kheer/Basundi/Pyasam	India	Sugar (15 g), dairy whitener (30 g), and pearl millet flour (20g)	Bunkar et al. 2014
Laddu	India	Roasted Bengal gram and pearl millet flour (75:25), sugar, and ghee	Singh and Mehra 2017, Puri et al. 2020
Kurrakan kanda	Sri Lanka	*Ragi* flour, jaggery, milk, grated coconut, and cardamom powder	Goel 2017
Masavusvu	Zimbabwe	Finger millet	Amadou et al. 2011, Zvauya et al. 1997
Dambu		Millet, maize or sorghum, spices, fermented milk, and sugar	Amadou et al. 2011

Figure 13.3 Millet-based non-fermented foods.

pearl millet-based *kheer* was optimized by Jha et al. (2013) using response surface methodology. In this study, it was found that a good quality pearl millet *kheer* with high overall acceptability (7.29) can be produced by using 18.49% of dairy whitener and 6.0% pearl millet. The process for the development of pearl millet-based ready-to-constitute kheer mix powder was optimized by Bunkar et al. (2014). It was found that a formulation containing 15 g sugar, 30 g dairy whitener, and 20 g pearl millet flour results in a good quality ready-to-constitute *kheer* mix powder. Bhosale et al. (2021) optimized three concentrations of roasted finger millet flour for the preparation of *kheer* with the blending of paneer. It was found in the study that a good quality *kheer* with high sensorial properties can be prepared with 1.5% finger millet flour, 8% paneer shreds, and 8% sugar. The finger millet *kheer* was also nutritionally superior to rice-based *kheer*.

13.3.4 Kurrakan Kanda

Kurrakan kanda is a millet and milk-based traditional porridge of Sri Lanka. To make this porridge, dry roast ragi flour is slowly added to boiling water, cooked until softened, and then jaggery is added and thoroughly mixed. This is then mixed with milk, grated coconut, and cardamom powder before being removed from the heat. The product is served hot (Goel 2017). The process for the development of malted finger millet porridge was also optimized by Jain et al. (2016). It was found in the study that 5% malted *ragi* flour in milk was the optimum level to develop porridge.

13.3.5 Laddu/Ladoo

Laddu/ladoo is a ball-shaped traditional Indian sweet that is prepared by mixing flour, sugar, and ghee. These are generally prepared from Bengal flour but millet-based *laddus* such as *ragi ladoo* (finger millet-based *ladoo*) are also popular in many parts of India. The other millets can also be used in varying concentrations for the preparation of *ladoo*. A mixture of cereals, millets, and pulses can also be used. Puri et al. (2020) prepared *ragi* balls enriched with peanuts and dates and it was found that good quality *ragi* balls with high overall acceptability (7.5–8.7/9) can be prepared with different concentrations of finger millet (20–35 g/100 g), jaggery (10–15 g/100 g), dates (15–25 g/100 g), and coconut powder (10–15 g/100 g). Singh and Mehra (2017) prepared *laddu* with roasted Bengal gram and pearl millet flour in the ratios 100:0, 75:25, 50:50, and 25:75. It was observed in the study that the incorporation of pearl millet flour at 25% level produced *laddus* with an overall acceptability at par with control (100% Bengal gram flour).

13.3.6 Masavusvu

Masavusvu is a sweet non-fermented beverage of Zimbabwe prepared from finger millet malt. To prepare *masavusu*, one part of finger millet malt flour is mixed with four parts of water and cooked with continuous stirring for about 80 minutes. After cooking *masavusu* is ready for consumption. It is also known as unfermented *mangisi* (Amadou et al. 2011; Zvauya et al. 1997).

13.3.7 Pongal

Pongal is a traditional breakfast food of India that is prepared from rice and green gram. The rice can be replaced with foxtail millet or little millet for the preparation of *pongal* (Ambati and Sucharitha 2019; Fatima and Rao 2019).

13.3.8 Roti

Millets have remained the staple diet of many ethnic groups of India since ancient times. Millets like pearl millet and finger millet are used for the preparation of an Indian flatbread known as *"roti"*. Pearl millet *roti* (*bajre ki roti*) is very popular in Rajsthan and its bordering states. Finger millet *roti* (*kode ki roti/mandal ki roti/nchani ki roti*) is mainly consumed in Himachal Pradesh and Uttrakhand (Kumar et al. 2020).

In addition to these, millets can also be used for the preparation of other traditional Indian foods like *upama* and *papad*. Millets can also be consumed after roasting and popping (Kora 2019).

13.4 CONCLUSIONS

The history of millet cultivation is as old as the history of humankind. They were the first grains domesticated by humans. Millet-based food and beverages are indispensable parts of many cultures/tribes of Asia and Africa and many of these products are still produced, consumed, and relished. These products are not only a good source of energy but they also ensure nutritional security. Many millet-based porridges like *fura, koko, koozh, kurrakan kanda, ogi, uji*, and *ugali* serve as weaning foods. Millet-based alcoholic beverages like *boza, kodo ko jannr, jaandh, sur*, etc., are generally produced and consumed in the hilly regions where they are supposed to help combat the harsh winter season. These beverages also play an important role in socio-cultural life as some of these beverages are used to welcome guests to marriages and birthday parties. Some of these are also served at death anniversaries. *Sur*, a finger millet-based alcoholic beverage of Himachal Pradesh, India, is also offered to local gods. In a nutshell, we can say that millet-based food and beverages are deep-rooted in Asian and African cultures.

REFERENCES

Adakole, M. I., A. F. Ogori, J. K. H. Ikya, V. Upev, G. Sardo, J. Naibaho, M. Korus, G. Bono, C. O. R. Okpala, and A. T. Girgih. 2021. Fermented millet "Ibyer" beverage enhanced with ginger powder: An assessment of microbiological, pasting, proximate, and sensorial properties. *Applied Sciences* 11(7):3151.

Adebiyi, J. A., A. O. Obadina, O. A. Adebo, and E. Kayitesi. 2016. Fermented and malted millet products in Africa: Expedition from traditional/ethnic foods to industrial value added products. *Critical Reviews in Food Science and Nutrition* 58(3):463–74.

Ajiboye, T. O., G. A. Iliasu, A. O. Adeleye, F. A. Abdussalam, S. A. Akinpelu, S. M. Ogunbode, S. O. Jimoh, and O. B. Oloyede. 2014. Nutritional and antioxidant dispositions of sorghum/millet-based beverages indigenous to Nigeria. *Food Science and Nutrition* 2(5):597–604.

Akingbala, J. O., P. I. Uzo-Peters, C. N. Jaiyeoba, and G. S. H., Baccus-Taylor. 2002. Changes in the physical and biochemical properties of pearl millet (*Pennisetum americanum*) on conversion to *ogi*. *Journal of the Science of Food and Agriculture* 82(13):1458–64.

Alo, M. N., U. A. Eze, and N. E. Eda. 2012. Microbiological qualities of *burukutu* produced from a mixture of sorghum and millet. *American Journal of Food and Nutrition* 2:96–102.

Amadou, I. 2019. Millet based fermented beverages processing. In *Fermented Beverages*, eds. A. M. Grumezescu, and A. M. Holban, 433–72. Cambridge: Woodhead Publishing.

Amadou, I., O. S. Gbadamosi, and G. W. Le. 2011. Millet-based traditional processed foods and beverages—A review. *Cereal Foods World* 56(3):115–21.

Ambati, K., and K. V. Sucharitha. 2019. Millets-review on nutritional profiles and health benefits. *International Journal of Recent Scientific Research* 10(7):33943–8.

Antony, U., S. Ilango, R. Chelliah, S. R. Ramakrishnan, and K. Ravichandran. 2020. Ethnic fermented foods and beverages of Tamil Nadu. In *Ethnic Fermented Foods and Beverages of India: Science History and Culture*, ed. J. P. Tamang, 539–60. Singapore: Springer.

Arici, M., and O. Daglioglu. 2002. Boza: A lactic acid fermented cereal beverage as a traditional Turkish food. *Food Reviews International* 18(1):39–48.

Ashenafi, M. 1994. Microbial flora and some chemical properties of ersho, a starter for teff (*Eragrostis tef*) fermentation. *World Journal of Microbiology and Biotechnology* 10(1):69–73.

Atter, A., K. Obiri-Danso, and W. K. Amoa-Awua. 2014. Microbiological and chemical processes associated with the production of *burukutu* a traditional beer in Ghana. *International Food Research Journal* 21(5):1769–76.

Banigo, E. O. I., and H. G. Muller. 1972. Manufacture of ogi (a Nigerian fermented cereal porridge). Comparative evaluation of corn, sorghum and millet. Canadian Institute of Food Science and Technology Journal 5(4):217–21.

Bayat, G., and G. Yıldız. 2019. The special fermented Turkish drink: Boza. *Journal of Tourism and Gastronomy Studies* 7(4):2438–46.

Bhosale, S., R. J. Desale, and A. Mukhekar. 2021. Physico-chemical composition of millet-based kheer blended with paneer. *The Pharma Innovation Journal* 10(4):320–4.

Bolaji, O. T., L. Adenuga-Ogunji, and T. A. Abegunde. 2017. Optimization of processing conditions of ogi produced from maize using response surface methodology (RSM). *Cogent Food and Agriculture* 3(1):1407279.

Bolaji, O. T., P. A. Adepoju, and A. P. Olalusi. 2015. Economic implication of industrialization of a popular weaning food ogi production in Nigeria: A review. *African Journal of Food Science* 9(10):495–503.

Bunkar, D. S., A. Jha, and A. Mahajan. 2014. Optimization of the formulation and technology of pearl millet based 'ready-to-reconstitute' kheer mix powder. *Journal of Food Science and Technology* 51(10):2404–14.

CGIAR. 2020. How Uganda's millet drink 'bushera' saw a revival and India's peanut 'chikki' went to Zambia. CGIAR research program on grain legumes and dryland cereals (GLDC). https://www.cgiar.org/news-events/news/how-ugandas-millet-drink-bushera-saw-a-revival-and-indias-peanut-chikki-went-to-zambia-2/ (accessed September 19, 2021).

Chandra, A., A. K. Singh, and B. Mahto. 2018. Processing and value addition of finger millet to achieve nutritional and financial security. *International Journal of Current Microbiology and Applied Sciences* 7:2901–10.

Chao, S. H., H. Y. Huang, Y. H. Kang, K. Watanabe, and Y. C. Tsai. 2013. The diversity of lactic acid bacteria in a traditional Taiwanese millet alcoholic beverage during fermentation. *LWT - Food Science and Technology* 51(1):135–42.

Chelliah, R., S. R. Ramakrishnan, D. Premkumar, and U. Antony. 2017. Accelerated fermentation of Idli batter using Eleusine coracana and Pennisetum glaucum. *Journal of Food Science and Technology* 54(9):2626–37.

Dahal, N. R., T. B. Karki, B. Swamylingappa, Q. I. Li, and G. Gu. 2005. Traditional foods and beverages of Nepal—A review. *Food Reviews International* 21(1):1–25.

Danson, D., O. Ehoneah, I. Ayensu, A. Brobbey, J. K. Adu, and S. O. Bekoe. 2019. Determination of tryptophan content in Hausa Koko (spicy millet porridge): A Ghanaian beverage. *International Journal of Phytopharmacy* 9(4):e5287.

Dirar, H. A. 1978. A microbiological study of Sudanese merissa brewing. *Journal of Food Science* 43(6):1683–6.

Fatima, Z., and A. Rao. 2019. Development, organoleptic evaluation and acceptability of products developed by incorporating foxtail millet. *Journal of Food Science and Nutrition Research* 2(2):128–35.

Ganapathy, K. N., K. Hariprasanna, and V. A. Tonapi. 2021. Breeding for enhanced productivity in millets. In *Millets and Pseudo Cereals: Genetic Resources and Breeding Advancements*, eds. M. Singh, and S. Sood, 39–63. Cambridge: Woodhead Publishing.

Goel, B. R. 2017. Ragi porridge-kurakkan kanda (Sri Lanka's finger millet porridge). http://deviprasadams.blogspot.com/2017/10/ragi-porridge-kurakkan-kenda-sri-lankas.html (accessed November 12, 2021).

Gupta, A., R. Singh, R. Prasad, J. Tripathi, and S. Verma. 2017. Nutritional composition and polyphenol content of food products enriched with millets and drumstick leaves. *International Journal of Food and Fermentation Technology* 7(2):337–42.

Haleegoah, J. A. S., G. Ruivenkamp, G. Essegbey, G. Frempong, and J. Jongerden. 2016. Street-vended local foods transformation: Case of Hausa Koko, waakye and Ga kenkey in urban Ghana. *Advances in Applied Sociology* 6(3):90–100.

Ho, A., I. A. Jideani, and J. U. Humphrey. 2009. Quality of Damhu prepared with different cereals and groundnut. *Journal of Food Science and Technology* 46(2):166–8.

Huffman, S. L., R. Oniang'o, and V. Quinn. 2000. Improving young child feeding with processed complementary cereals and behavioural change in urban Kenya. *Food and Nutrition Bulletin* 21(1):75–81.

Ijabadeniyi, A. O. 2007. Microorganisms associated with ogi traditionally produced from three varieties of maize. *Research Journal of Microbiology* 2(3):247–53.

Ilango, S., and U. Antony. 2014. Assessment of the microbiological quality of koozh, a fermented millet beverage. *African Journal of Microbiology Research* 8(3):308–12.

Jagran, K. 2020. Millets: The miracle grain. https://krishijagran.com/health-lifestyle/millets-the-miracle-grains/. (accessed January 10, 2022).

Jain, S., R. S. Dabur, and S. Bishnoi. 2016. Development of milk based malted finger millet (*Ragi*) porridge: Effects of malting of finger millet on compositional attributes. *Haryana Veterinarian* 55(2):155–9.

Jha, A., A. D. Tripathi, T. Alam, and R. Yadav. 2013. Process optimization for manufacture of pearl millet-based dairy dessert by using response surface methodology (RSM). *Journal of Food Science and Technology* 50(2):367–73.

Jideani, V. A., I. Nkama, E. B. Agbo, and I. A. Jideani. 2001. Survey of fura production in some northern states of Nigeria. *Plant Foods for Human Nutrition* 56(1):23–36.

Joshi, N., S. H. Godbole, and P. Kanekar. 1989. Microbial and biochemical changes during dhokla fermentation with special reference to flavour compounds. *Journal of Food Science and Technology* 26:113–5.

Joshi, V. K., A. Kumar, and N. S. Thakur. 2015. Technology of preparation and consumption pattern of traditional alcoholic beverage 'Sur' of Himachal Pradesh. *International Journal of Food and Fermentation Technology* 5(1):75–82.

Karnataka State Department of Agriculture, Bengaluru, India and ICAR-Indian Institute of Millets Research, Hyderabad, India. 2018. The story of millets. https://www.millets.res.in/pub/2018/The_Story_of_Millets.pdf. (accessed January 10, 2022).

Kitabatake, N., D. M. Gimbi, and Y. Oi. 2003. Traditional non-alcoholic beverage, Togwa, in East Africa, produced from maize flour and germinated finger millet. *International Journal of Food Sciences and Nutrition* 54(6):447–55.

Koirala, U. 2020. Health and nutritional aspect of underutilized high-value food grain of high hills and mountains of Nepal. In *Nutritional and Health Aspects of Food in South Asian Countries*, eds. J. Prakash, V. Waisundra, and V. Prakash, 195–209. London: Academic Press. Elsevier Traditional and Ethnic Food Series. Series eds. H. Lelived, V. Anderson, V. Prakash, J. Prakash and B. Meulen.

Kora, A. J. 2019. Applications of sand roasting and baking in the preparation of traditional Indian snacks: Nutritional and antioxidant status. *Bulletin of the National Research Centre* 43(1):1–11.

Krishnamoorthy, S., S. Kunjithapatham, and L. Manickam. 2013. Traditional Indian breakfast (Idli and Dosa) with enhanced nutritional content using millets. *Nutrition and Dietetics* 70(3):241–6.

Kumar, A. 2013. Refinement of the technology of the traditional Sur production in Himachal Pradesh (Doctoral dissertation, M.Sc. Thesis UHF, Nauni).

Kumar, A., A. Kaur, K. Gupta, Y. Gat, and V. Kumar. 2021. Assessment of germination time of finger millet for the value addition in functional foods. *Current Science* 120(2):406–13.

Kumar, A., V. Tomer, A. Kaur, K. Sharma, and S. Dimple. 2020. Fermented foods based on millets: Recent trends and opportunities. In *Advances in Fermented Foods and Beverages*, eds. G. K. Sharma, A. D. Semwal, and J. R. Xavier, 191–228. New Delhi: New India Publishing Agency.

Kure, O. A., and G. Wyasu. 2013. Influence of natural fermentation, malt addition and soya fortification on the sensory and physicochemical characteristics of Ibyer-Sorghum gruel. *Advances in Applied Science Research* 4(1):345–9.

Lohekar, A. S., and A. B. Arya. 2014. Development of value added instant dhokla mix. *International Journal of Food and Nutritional Scicences* 3(4):78–83.

Lyumugabe, F., J. Gros, J. Nzungize, E. Bajyana, and P. Thonart. 2012. Characteristics of African traditional beers brewed with sorghum malt: A review. *Biotechnologie, Agronomie, Société et Environnement* 16(4):509–30.

Madalageri, D. M., N. B. Yenagi, and G. Shirnalli. 2016. Evaluation of little millet Paddu for physico-chemical nutritional, microbiological and sensory attributes. *Asian Journal of Dairy and Food Research* 35(1):58–64.

Madella, M., J. J. García-Granero, W. A. Out, P. Ryan, and D. Usai. 2014. Microbotanical evidence of domestic cereals in Africa 7000 years ago. *PLOS ONE* 9(10):e110177.

Misihairabgwi, J., and A. Cheikhyoussef. 2017. Traditional fermented foods and beverages of Namibia. *Journal of Ethnic Foods* 4(3):145–53.

Modha, H., and D. Pal. 2011. Optimization of Rabadi-like fermented milk beverage using pearl millet. *Journal of Food Science and Technology* 48(2):190–6.

Mugula, J. K., S. A. M. Nnko, J. A. Narvhus, and T. Sørhaug. 2003. Microbiological and fermentation characteristics of togwa, a Tanzanian fermented food. *International Journal of Food Microbiology* 80(3):187–99.

Muyanja, C. M. B. K., J. K. Kikafunda, J. A. Narvhus, K. Helgetun, and T. Langsrud. 2003. Production methods and composition of Bushera: A Ugandan traditional fermented cereal beverage. *African Journal of Food, Agriculture, Nutrition and Development* 3(1):10–9.

Narayanan, J., V. Sanjeevi, U. Rohini, P. Trueman, and V. Viswanathan. 2016. Postprandial glycaemic response of foxtail millet dosa in comparison to a rice dosa in patients with type 2 diabetes. *Indian Journal of Medical Research* 144(5):712–7.

Neela, S., S. W. Fanta, and S. W. 2020. Injera (An ethnic, traditional staple food of Ethiopia): A review on traditional practice to scientific developments. *Journal of Ethnic Foods* 7(1):1–15.

Ohariya, P., A. Singh, and L. P. Rajput. 2017. Quality attributes of instant kodo-soy idli mix as affected by fermentation period. *International Journal of Chemical Studies* 5(4):1611–15.

Okara, J. O., and O. O. Lokoyi. 2012. Developing an efficient method for ogi production: Towards educating the rural women. *The Nigerian Journal of Research and Production* 20(1):1–7.

Pathak, P., S. Srivastava, and S. P. Grover. 2000. Development of food products based on millets, legumes and fenugreek seeds and their suitability in the diabetic diet. *International Journal of Food Sciences and Nutrition* 51(5):409–14.

Petrova, P., and K. Petrov. 2017. Traditional cereal beverage Boza fermentation technology, microbial content and healthy effects. In *Fermented Foods*, eds. R. C. Ray, and D. Montet, 284–305. Florida: CRC Press.

Puri, A. P., S. V. Maske, D. Paresh, and A. Lal. 2020. Development and formulation of ragi balls enriched with peanuts and dates. *Journal of Pharmacogonosy and Phytochemistry* 9(2):2411–15.

Rastogi, M., and M. Joshi. 2015. Effect of Ragi (*Eleusione coracana*) for the development of value added products and their nutritional implication. *Asian Journal of Home Science* 10(1): 1–5.

Roopa, S. S., H. Dwivedi, and G. K. Rana. 2017. Development and physical, nutritional and sensory evaluation of instant mix (Dosa). *Technofame- a Journal of Multidisciplinary Advance Research* 6(1):109–13.

Sawadogo-Lingani, H., J. Owusu-Kwarteng, R. Glover, B. Diawara, M. Jakobsen, and L. Jespersen. 2021. Sustainable production of African traditional beers with focus on dolo, a West African sorghum-based alcoholic beverage. *Frontiers in Sustainable Food Systems* 5:143.

Singh, U., and A. Mehra. 2017. Sensory evaluation of Ladoo prepared with pearl millet. *International Journal of Home Science* 2:610–2.

Soni, S. K., D. K. Sandhu, K. S. Vilkhu, and N. Kamra. 1986. Microbiological studies on dosa fermentation. *Food Microbiology* 3(1):45–53.

Subastri, A., C. Ramamurthy, A. Suyavaran, R. Mareeswaran, P. Mandal, S. Rellegadla, and C. Thirunavukkarasu. 2015. Nutrient profile of porridge made from *Eleusine coracana* (L.) grains: Effect of germination and fermentation. *Journal of Food Science and Technology* 52(9):6024–30.

Tapsoba, F. W. B., H. Sawadogo-Lingani, D. Kabore, D. Compaore-Sereme, and M. H. Dicko. 2017. Effect of the fermentation on the microbial population occurring during the processing of zoom-koom, a traditional beverage in Burkina Faso. *African Journal of Microbiology Research* 11(26):1075–85.

Thakur, N., Savitri, and T. C. Bhalla. 2004. Characterization of some traditional fermented foods and beverages of Himachal Pradesh. *Indian Journal of Traditional Knowledge* 3(3):325–35.

Thapa, N., K. K. Aryal, M. Paudel, R. Puri, P. Thapa, S. Shrestha, and B. Stray-Pedersen. 2015. Nepalese home-brewed alcoholic beverages: Types, ingredients, and ethanol concentration from a nation wide survey. *Journal of Nepal Health Research Council* 13(29):59–65.

Thapa, S., and J. P. Tamang. 2004. Product characterization of kodo ko jaanr: Fermented finger millet beverage of the Himalayas. *Food Microbiology* 21(5):617–22.

Thapa, S., and J. P. Tamang. 2006. Microbiological and physio-chemical changes during fermentation of kodo ko jaanr, a traditional alcoholic beverage of the Darjeeling hills and Sikkim. *Indian Journal of Microbiology* 46(4):333–41.

Thirumangaimannan, G., and K. Gurumurthy. 2013. A study on the fermentation pattern of common millets in Koozh preparation–a traditional south Indian food. *Indian Journal of Traditional Knowledge* 12(3):512–7.

Tou, E. H., C. Mouquet-Rivier, C. Picq, A. S. Traoré, S. Trèche, and J. P. Guyot. 2007. Improving the nutritional quality of ben-saalga, a traditional fermented millet-based gruel, by co-fermenting millet with groundnut and modifying the processing method. *LWT – Food Science and Technology* 40(9):1561–69.

Vanithasri, J., and S. Kanchana. 2013. Studies on the quality evaluation of idli prepared from barnyard millet (*Echinochloa frumentacaea*). *Asian Journal of Home Science* 8(2):373–8.

Wanjala, W. G., A. Onyango, M. Makayoto, and C. Onyango. 2016. Indigenous technical knowledge and formulations of thick (ugali) and thin (uji) porridges consumed in Kenya. *African Journal of Food Science* 10(12):385–96.

WHO. 2004. WHO global status report on alcohol. https://www.who.int/substance_abuse/publications/en/sudan.pdf (accessed November 6, 2021).

Yadav, D. N., S. Balasubramanian, J. Kaur, T. Anand, and A. K. Singh. 2011. Optimization and shelf-life evaluation of pearl millet based halwa dry mix. *Journal of Food Science and Engineering* 1(4):313.

Yusuf, A. B., Z. M. Kalgo, B. H. Gulumbe, B. M. Danlami, B. Aliyu, C. Obi, and A. A. Salihu. 2020. Assessment of microbiological quality of *fura da nono* produced in Kebbi State, Nigeria. *Equity Journal of Science and Technology* 7(2):45–8.

Zvauya, R., T. Mygochi, and W. Parawira. 1997. Microbial and biochemical changes occurring during production of masvusvu and mangisi, traditional Zimbabwean beverages. *Plant Foods for Human Nutrition* 51(1):43–51.

Chapter 14

Millet-Based Processed Foods

14.1 INTRODUCTION

Millets can grow on low-fertility soils with limited requirements for water, pesticides, and fertilizers. This makes them one of the most suitable crops for drylands and semi-arid areas (Kumar et al. 2018). They also have the smallest carbon footprint among all the cereals and are environmentally sustainable. The nutritional value of millets is also superior to staple cereals like polished rice, refined wheat flour, and maize (Kane-Potaka et al. 2021). The dietary fibre and mineral content of millets are several fold higher than staple cereals (Kumar et al. 2020). High dietary fibre is associated with a low glycaemic index (GI) and can help to control diabetes. Dietary fibre also plays an important role in the management of obesity and cardiovascular diseases. The high mineral content of millets can help to combat the long-prevailing deficiency of minerals like calcium, iron, and zinc. All these attributes of millets make them the climate-resilient, nutrient-dense smart foods of the future (ICRISAT 2021). Efforts to popularize the cultivation and consumption of millets have been observed throughout the world. India, the largest producer of millets in the world, has dispensed of the use of the nomenclature "coarse cereals" for millets and now millets are renamed as "nutricereals" (Financial Express 2018) and *Shree Annas* (Business Standard 2023). India also celebrated the year 2018 as the "National Year of Millets" for promoting the cultivation and consumption of nutricereals (Government of India 2018). The United Nations is alsocelebrating the year 2023 as the "International Year of Millets" (United Nations Digital Library 2021). These worldwide efforts have helped millets to regain their lost identity and increased cultivation and consumption of millets have been observed in the past few years.

The major problem in the food application of millets is the limited availability of millet-based processed foods in the market. Food technologists from all over the world are trying either to develop new millet-based processed foods or fortify well-known processed foods like bread, biscuits, cookies, cereal bars, etc., with millets. The use of millets in the development of instant infant foods or weaning foods has been also exploited. Millets have a low GI, are gluten-free, and hence can also be used for the production of multigrain and gluten-free flours. Millet malt can also be used for the preparation of millet-based beverages (Kumar et al. 2021; Kumar et al. 2020a). Being a rich source of prebiotics like arabinoxylans, arabinogalactans, β-glucans, fructans, fructo-oligosaccharides, inulin, and xylo-oligosaccharides, millets are also exploited for the development of synbiotic foods (Kumar et al. 2020b). In addition to these, millets are also used for the production of alcoholic beverages. In this chapter, the potential of millets for the production of processed foods is discussed. Information on the millet-based products available in the market is also compiled.

14.2 MILLET-BASED PROCESSED PRODUCTS

Millets can be processed into a variety of products like biscuits, bread, cakes, cereal bars, cookies, extruded products, muffins, multigrain flours, noodles, weaning foods, synbiotic beverages, alcoholic beverages, etc.

14.2.1 Non-Fermented Products

14.2.1.1 Biscuits

Biscuits are flat, crisp baked products, usually prepared from a mixture of wheat flour, sugar, shortening, sodium chloride, sodium bicarbonate, ammonium bicarbonate, and water. They represent the largest category of snack foods and are widely consumed. The wheat flour can be either completely replaced with millet flour to produce gluten-free biscuits or millet can be blended with wheat flour to produce biscuits with improved nutritional value. This section discusses the studies on the development of millet-based biscuits and the effect of millets on dough rheology and the nutritional and sensorial properties of biscuits. Singh and Nain (2020) prepared biscuits using composite flours of pearl millet and wheat in the ratios of 50:50 and compared it with control (biscuit prepared using 100% wheat flour). The other ingredients used for the biscuit preparation were sugar (40 g), fat (35 g), milk, and 1.5 g of ammonium bicarbonate per 100 g of flour. The ingredients were mixed, kneaded to the dough, sheeted, and baked (170°C for 15–20 minutes). The nutritional analysis of biscuits revealed that the pearl millet-based biscuits had low fat (43.88 g/100 g), high protein (10.92 g/100 g), and high iron (5.26 mg/100 g) content in comparison to 100% wheat flour-based biscuits (51.7 g/100 g fat, 10.73 g/100 g protein content, and 4.1 mg/100 g iron). The sensory scores of the texture (7.8/9) and overall acceptability (7.7/9) of the pearl millet-based biscuits were also better in comparison to wheat flour-based biscuits. Inyang et al. (2018) prepared the biscuits using three different ratios, i.e. 75:25:00, 75:00:25, and 50:25:25 of the flours of whole wheat, fonio, and kidney bean, respectively, and compared their physical properties, proximate composition, and mineral and sensorial characteristics with biscuits made from 100% wheat flour. It was found in the study that the physical properties (weight, diameter, thickness, and spread ratio), nutritional composition (crude protein, crude fat, ash, crude fibre, carbohydrate, and calorific value), and sensory characteristics (appearance, taste, texture, crispiness, and overall acceptability) of the fonio-based biscuits were similar to the biscuits prepared from 100% wheat flour (control). An increase in calcium and magnesium and a decrease in phosphorus, iron, and zinc were also observed on the addition of fonio. However, an increase in thickness and a decrease in the spread ratio of the biscuits were found with the addition of kidney bean flour. Increases in protein, ash, calcium, magnesium, and iron were also found with the addition of kidney bean flour. The scores for the sensory characteristics were also high in the kidney bean-based biscuits. The overall acceptability of biscuits prepared from composite flours of wheat, fonio, and kidney bean was also higher than the control and fonio-based biscuits. The conditions for the development of teff-based gluten-free biscuits were optimized by Teshome et al. (2017). The various ingredients of the recipe were teff flour (250 g), mayonnaise (45 g), sugar (60 g), salt (2.5 g), baking powder (5 g), milk powder (30 g), whole egg (50 g), and water (30 ml). The various processing conditions, i.e. baking temperature, baking time, and biscuit thickness, were also optimized using response surface methodology (RSM). It was found in the study that the best conditions for the development of biscuits were a baking temperature of 174°C, a baking time of 9 minutes, and a thickness of 4.5 mm. Saha et al. (2011) prepared biscuits using composite flours of finger millet and wheat in two ratios, i.e. 60:40 and 70:30 w/w. The other ingredients used in the recipe were sugar (35 g), shortening (30 g), sodium chloride (0.6 g), sodium bicarbonate (0.3 g), ammonium bicarbonate (0.6 g), and water (20 ml) per 100 g of composite flour. An increase in the adhesiveness of the dough was observed with an increase in the wheat ratio from 30 to 40. However, the effect on the resistance of extension of dough was very low. The water absorption capacity was highest at the finger millet to wheat flour ratio of 60:40. The expansion and breaking strength of biscuits were higher in the biscuits prepared with a 70:30 ratio than with a 60:40 ratio. Among the various finger millet varieties (VL-146, VL-149, VL-204, VL-315, VL-324, PES-400) studied by the researchers, the variety VL-324 produced the biscuits with the highest overall acceptability. Krishnan et al. (2011) prepared composite flours (90:10 and 80:20) using wheat flour and finger millet seed coat (particle size <105 μm) of native, malted, and hydrothermally treated finger millet. The composite flour (300 g) was further mixed with sugar (90 g), shortening (60 g), water containing 1.5 g sodium bicarbonate, ammonium carbonate (1.5 g), and sodium chloride (3 g), kneaded to the dough, sheeted to a thickness of 3.5 mm, shaped into circles of 51 mm diameter, and baked at 205°C for 10 minutes. On the sensory evaluation of the biscuits, it was found that the biscuits prepared using 10% seed coat matter from native and hydrothermally processed millets and 20% from malted millet were acceptable. The biscuits prepared from the composite flour had a mild grey colour, crisp texture, high protein (9.5–12%), high dietary fibre (40–48%), and calcium (700–860 mg/100 g) in comparison to the control, i.e. biscuits prepared from 100% wheat flour. Anju and Sarita (2010) prepared millet-based biscuits by replacing 45% of the refined wheat flour with either foxtail millet flour or barnyard millet flour. The other ingredients were hydrogenated fat (23%), sugar (14%), eggs (5.5%), curd (11.5%), and water (30%). On sensory, nutritional, glycaemic, and storage studies of the biscuits, it was revealed that the overall acceptability of millet-based biscuits (6.7–7.80/9) was at par with the biscuits prepared from refined wheat flour (6.84–7.9). The content of crude fibre (2.01–2.03%), total ash (1.1–1.31%), and total dietary fibre (9.12–10.24%) was higher than refined wheat flour-based biscuits having 0.23% crude fibre, 0.66% total ash, and 5.22% total dietary fibre. The glycaemic index of foxtail millet flour-based biscuits (50.8) was lower than barnyard millet flour- (68) and refined wheat flour- (68) based biscuits. Sehgal and Kawatra (2007) prepared biscuits using refined wheat flour, blanched pearl millet flour, and green gram in two ratios, i.e. 50:40:10 and 30:60:10, respectively. The biscuits prepared using both of these ratios were liked very much by panellists. Pearl millet flour and green gram-based biscuits were also nutritionally rich in comparison to the biscuits prepared from 100% refined wheat flour. The biscuits prepared using high levels of pearl millet flour (60%) were also reported to be high in antinutritional factors.

14.2.1.2 Bread
Bread is one of the major staple foods of the world. Two major types of bread, i.e. leavened and non-leavened, are popular throughout the world. The process for non-leavened bread is not standardized and it varies from place to place; however, the process of leavened bread is more or less similar in the various parts of the world. The major steps in the production of leavened bread are mixing, kneading, fermenting, proofing, and baking (Devani et al. 2016). Refined wheat flour is the most widely preferred flour for the production of bread due to its high gluten content. However, the demand for white bread is decreasing continuously as it lacks dietary fibre, minerals, and phytochemicals. The market for bread fortified with malted grains, cereal bran, millets,

and pseudocereals is continuously increasing (Devani et al. 2016, Rozylo 2014, Singh and Mishra 2014). Zięć et al. (2021) prepared bread with blends of wheat and teff flours. Wheat flour was substituted with teff flour at concentrations of 5, 10, and 15%. The other ingredients used were 650 g water, 30 g yeast, and 20 g salt per kg of blended flour. A decrease in bread volume, bread yield, and weight of the cold bread and an increase in the baking loss and crumb moisture were observed with an increase in the concentration of teff flour from 5 to 15%. Teff flour-based blends also had less hardness and chewiness in comparison to the bread prepared from 100% wheat flour. The addition of teff flour also increased the dietary fibre as well as mineral content. The bread prepared using all three concentrations of teff flour was sensorially acceptable. Li et al. (2020) prepared bread using steamed millet flour (steamed at 100°C for 10 minutes under atmospheric pressure) and millet bran dietary fibre and studied its effect on dough development, steamed bread quality, and digestion (*in vitro*). The content of steamed millet flour was kept constant at 25%, millet bran was added at the rate of 0, 2, 4, 6, 8, and 10%, and then made to 100% with the addition of wheat flour. The addition of millet flour and dietary fibre affected the farinographical properties of the dough adversely. The rate of water absorption was decreased by 4.8% on the addition of millet flour. On the addition of dietary fibre, the water absorption rate first increased (2–6%) and then decreased (8–10%). Dough development time was decreased on the addition of millet flour and dietary fibre but the change was non-significant ($P \leq 5\%$). The dough stability time was reduced by 2.2–3.68 minutes. The lowest stability value of 5.87 minutes was for the mixed dough containing 10% dietary fibre. The increase in dietary fibre also increased the softening degree and decreased the farinograph quality number. The strength and kneading resistance of dough were weakened. The toughness and the gas retention capacity of the dough also decreased. The extension ratio increased with the addition of millet flour and increasing content of dietary fibre (2–6%) and then decreased at dietary fibre content of 8% and above. The millet fortified bread had a yellow and darker colour, increased hardness, gumminess, and chewiness, and decreased springiness, cohesiveness, and resilience. The addition of dietary fibre increased the content of resistant starch and decreased the content of rapidly digested starch and slowly digested starch. A significant decrease in the starch hydrolysis rate, the hydrolysis index, and the glycaemic index was also found. However, the sensorially acceptable concentration, i.e. wheat flour + steamed millet flour (25%) + 2% dietary fibre, had a high GI (74.79) value. Drábková et al. (2017) prepared bread using three composite blends of wheat and fonio flour (97.5:2.5, 95:5, and 90:10). There was an increase in dietary fibre, from 3.40% in wheat flour to 3.58% in wheat flour fortified with fonio. The addition of fonio flour at the rate of 2.5% improved the rheological behaviour and viscoelastic properties of dough and the bread prepared using this blend had a 20% higher specific volume in comparison to control (100% wheat flour). The volume of the bread prepared from 5% fonio flour was comparable to the control. The increase in fonio flour also decreased starch retrogradation in comparison to control. Based on various parameters studied, the researchers concluded that the composite flour of wheat and fonio (95:5) had higher baking potential and machinability and could be recommended for bread preparation. Devani et al. (2016) prepared bread by blending 10, 20, 30, 40, and 50% of finger millet flour with wheat flour. The bread prepared using 80% wheat flour and 20% finger millet flour had the highest scores for appearance, crust colour, crumb colour, taste, texture, and overall acceptability. The blending of finger millet flour at this level improved the crude fibre and calcium content of the bread. Man et al. (2016) investigated the effect of the incorporation of millet flour (10, 20, and 30%) on the quality of the bread. The physicochemical and sensory attributes of the bread were improved on fortification with millet flour. The addition of 30% millet flour produced bread with acceptable volume, crumb structure, and sensorial attributes. Różyło (2014) reported that good-quality bread can be prepared by substituting wheat flour with millet flour up to 30%. The addition of millet flour at concentrations of 5 and 15% improved the bread loaf volume significantly (P ≤ 5%). However, the higher concentrations of millet flour, i.e. 20, 25, and 30%, reduced the loaf volume significantly (P ≤ 5%). The use of millet flour at these concentrations also decreased the crumb quality and produced dustier, crumbly bread with a lower elasticity. The addition of gluten at the rate of 3% or the use of high-gluten wheat varieties was reported to produce bread with good quality attributes. Singh et al. (2012) studied the suitability of millet-wheat composite flours for the preparation of bread. Two composite flours, i.e. CF1 and CF2, were standardized using RSM for the bread preparation. The composite flour CF1 was developed by mixing the flours of barnyard millet, wheat, and gluten in the ratio 61.8:31.4:6.8 while CF2 was developed using flours of barnyard millet, finger millet, proso millet, and wheat in the ratio of 9.1:10.1:10.2:69.6, respectively. The wheat flour was substituted with cooked fermented millet flour in three concentrations, i.e. 10, 15, and 20% (Ranasalva and Vishvanathan 2014). The bread prepared with all three concentrations was reported to have textural and physical properties similar to the market bread. Karuppasamy et al. (2013) prepared bread by incorporating wheat flour and the flours of kodo millet, little millet, and foxtail millet at the concentrations of 10, 20, 30, 40, 50, 60, and 70% and optimized the best concentration for bread preparation based on sensorial attributes. It was found in the study that the substitution of wheat flour with millet at the rate of 20% produced the bread with the most acceptable sensory properties. The bread prepared by the substitution of wheat flour with 20% kodo millet, little millet, and foxtail millet was further analyzed for its physicochemical, sensorial, and storage properties. The addition of millet flour decreased the extensibility and the hardness of dough. An increase in the height, weight, specific volume, and bulk density, and a decrease in the springiness, cohesiveness, and resilience of the bread were observed with the addition of millet flour. The fibre content of bread was also increased on the addition of millet flour (i.e. 1.31 g/100 g for kodo millet-based bread, 1.46 g/100 g for little millet-based bread, and 1.53 g/100 g for foxtail millet-based bread) in comparison to 100% wheat flour bread (0.63 g/100 g bread). A small increase in the calcium content was found in the kodo (4.56%) and foxtail (9.75%) millet-based bread. A 50% increase in the iron content was also observed in the little millet-based bread. The addition of millets also increased the staleness of bread on storage. Mohammed et al. (2009) prepared bread with blends of wheat and teff flours in the ratios of 100:0, 95:5, 90:10, 85:15, and 80:20. A decrease in protein content and an increase in ash content and the falling number were observed as the percentage of teff flour was increased from 5 to 20%. The increase in teff flour also decreased the gluten content significantly (P ≤ 0.05). There was no significant change in the water absorption (%) of dough with an increased concentration of teff flour; however, a significant decrease (P ≤ 0.05) in the dough development time (minutes) was found as the concentration of teff flour was increased above 10%, i.e. 15 and 20%. It was concluded from the study that using a blend of wheat and teff flours in the ratio of 95:5 produced bread with quality as good as control.

14.2.1.3 Cakes

Cakes are generally baked products with a tender crumb and sweet taste. The moisture content of cakes typically ranges between 18 and 28%, which is lower than bread but higher than cookies (Xu et al. 2020). The major ingredients used to prepare cakes are refined wheat flour, sweetening agent (sugar), binding agent, egg, oil, liquid flavour, and leavening agent (yeast or baking powder) (Desai et al. 2010). Fathi et al. (2016) utilized different concentrations (25, 50, 75, and 100%) of proso millet flour treated at two different moisture contents (20 and 30%) and two temperatures (100 and 120°C) for the preparation of gluten-free pound cake. The effect of the addition of treated millet flour on the physicochemical and sensorial properties of batter and cake was also studied. The use of heat-moisture-treated flour decreased the specific gravity (1.056 for untreated flour and 0.908 for 100% flour treated at 30% moisture and 100°C) and consistency (21.48 Pa s for untreated flour and 16.87 Pa s for the flour treated at 20% moisture and 100°C) of the batter. An increase in volume and a decrease in the hardness of cake were also found on the addition of treated flour. The highest volume (165.73 cm^3) and least hardness (417.66 g) were obtained for cake prepared with a blend containing 75% flour treated with 30% moisture at 100°C. The highest total acceptability (7.07/9) was obtained for cake prepared using a blend containing 50% flour treated with 30% moisture at 100°C. Cakes prepared by substituting wheat flour with 30, 40, 50, 60, and 80% malted finger millet flour were studied for sensory acceptability and nutritional composition. The cakes with up to 50% flour substituted with malted finger millet flour had high overall acceptability of 4.32 on a 5-point scale, which was at par with the overall acceptability of the control (4.38). The cakes with 50% flour substituted with malted finger millet flour had high fibre (1.95%), calcium (140.60 mg/100 g), phosphorus (432 mg/100 g), and iron (7.58 mg/100 g) contents compared to the control with 0.30%, 117.2, 400, and 1.84 mg/100 g of crude fibre, calcium, phosphorus, and iron, respectively (Desai et al. 2010).

14.2.1.4 Cereal Bars

Cereal bars are a compressed mixture of cereals and dried fruits that generally contains glucose syrup as a binding agent. Glucose also provides instant energy while the cereals aid in a slower release of energy (Silva et al. 2013). Millet bar prepared from roasted finger millet (20%), roasted pearl millet (20%), puffed sorghum (30%), roasted groundnut (10%), and jaggery (40g) was studied for sensory acceptability. The roasted millet bar had an overall acceptability of 8.9 on a 9-point hedonic scale, which dropped to 7.9/9 after 60 days of room temperature storage (Karuppasamy and Latha 2020). Three pearl millet protein bars containing 25, 27.5, and 30% of steamed and oven-dried pearl millet were prepared by Samuel and Peerkhan (2020). The other ingredients used were whey protein, roasted peanuts, raisins, honey, and sugar powder (amounts not revealed). All the ingredients were mixed, poured into rectangular silicone moulds, refrigerated for 2 hours, and then baked at 180°C for 30 minutes. The prepared bars had an energy content of 332–379 kcal, protein content of 15.74–18.32 g, calcium content of 74.53–83.87 mg, and phosphorus content of 555.93–603.80 mg/100 g of the bar. The organoleptic analysis of the bars revealed that the proportion of ingredients does not affect the sensory properties of bars. However, a marginal increase in the hardness, cohesiveness, and chewiness of the bar was found with an increase in the percentage of pearl millet. The highest overall acceptability (8.6/9) was obtained for the bar prepared using 25% pearl millet. This was more than the overall acceptability (7.6/9) of the commercial bar. Kavitha et al. (2018), formulated a probiotic millet fruit bar with puffed *bajra* (pearl millet) (25 g), puffed *cholum* (white sorghum) (25 g), wheat flakes (15 g), *varagu* (kodo millet) flakes (10 g), *ragi* (finger millet) flakes (15 g), roasted groundnuts (8 g), guava or strawberry pulp with intermediate moisture (15 g), skimmed milk powder (2 g), jaggery (85 g), liquid glucose (10 g), probiotic culture (*Lactobacillus acidophillus*, *Lactobacillus delbrukii*, and *Streptococcus thermophiles*) (1g), and water (25 g). The cereals, i.e. puffed *bajra*, puffed *cholam*, wheat flakes, *ragi* flakes, and *varagu* flakes, were ground into grits and fried slightly in an open pan. To this mix of grits, a jaggery solution, prepared by boiling 85 g of jaggery, 25 ml of water, and 8 g of glucose to a softball stage, was added and then fruit pulp was added to the mix and mixed well with the jaggery solution. After this, three different probiotic cultures (1 ml) mixed with 5 ml skim milk were added to the mix separately and the contents were transferred to a greased tray and cooled at room temperature. The prepared bars had moisture, carbohydrates, protein, fat, energy, calcium, and iron content of 10.45–10.62 g, 67.70–68.15 g, 10.00–10.25 g, 3.35–3.40 g, 340–352 kcal, 80.00–80.32 mg/100 g, and 4.00–4.15 mg/100 g, respectively. The overall acceptability of probiotic millet bars with *L. acidophillus*, *L. delbrukii*, and *S. thermophiles* was 8.60, 8.10, and 8.0, respectively on a 9-point hedonic scale. Among the three studied strains, *L. acidophilus* had the highest viability of 3.5×10^8 CFU/ml followed by *L. delbrukii* (1.5×10^8 CFU/ml) and *S. thermophiles* (1.0×10^8 CFU/ml), respectively. Sobana (2017), studied the formulation of cereal bars for athletes using different concentrations of millets (35–50 g), cereals (20–30 g), and pulses (15–25 g). The millets used were a mixture of little millet, sorghum, pearl millet, and foxtail millet in equal quantities. The cereals used were a mixture of wheat flakes, oat flakes, and red rice flakes in equal quantities. The pulses used were a mixture of Bengal gram (*Cicer arietinum* L.), green gram (*Vigna radiata* L.), and cowpea (*Vigna unguiculata* L.) in equal quantities. Other ingredients such as a mixture of nuts, seeds, and chocolate chips (5 g), palm jaggery (40 g), peanut butter (10 g), and skimmed milk powder (15 g) were used in constant concentrations. It was found that 40 g of millets, 20 g of cereals, and 25 g pulses had the highest overall acceptability score of 41/45. The carbohydrate, protein, fat, calcium, and iron contents were 72.5 g/100 g, 13.7 g/100 g, 6.1 g/100 g, 159.5 mg/100 g, and 2.93 mg/100 g, respectively. The prepared bar had a shelf-life of 45 days at room temperature.

14.2.1.5 Cookies

Cookies are generally biscuits with a softer texture. They are usually prepared with refined wheat flour, fat, and sugar. They have low moisture and a longer shelf-life (Florence Suma et al. 2014). The cookie dough is typically high in fat, sugar, and water, making the dough soft and pourable (Manley 1998). Kulkarni et al. (2021) developed low-gluten cookies by substituting wheat flour with various proportions of pearl millet flour (20, 40, 60, 80, and 100%). The cookies prepared from the composite flour with a pearl millet flour composition of 60% had the highest overall acceptability (7.7/9), which was at par with the control (8.7/9). The cookies prepared from 60% pearl millet flour also had high calcium (21.0 mg/100 g), phosphorus (153.3 mg/100 g), and iron (3.8 mg/100 g) content in comparison to the control with 18.3, 86.8, and 2.5 mg/100 g of calcium, phosphorus, and iron, respectively. Millet cookies prepared using foxtail millet flour (30 g) with refined wheat flour (70 g), sugar (50 g), butter (29.4 g), baking powder (1.47 g), and milk (9.82 ml) were found to have higher calcium (1470.25 mg/100 g) and total phenolic content [9.66 mgGAE/g dry weight (dw)] compared to control (100% refined wheat flour) with 1070.33 mg/100 g calcium and 0.98 mg GAE/g dw total phenols (Marak et al. 2019). Sinha and Sharma (2017) studied the sensorial and nutritional aspects of cookies prepared with 30 and 50% substitutions of wheat flour with finger millet flour. It was found that 50% of wheat flour substituted with finger millet flour had overall acceptability of 7.8 on a 9-point hedonic scale which was close to the control (8.2/9). The 50% finger millet flour cookies had higher contents of fibre (1.09%), calcium (94 mg/100 g), iron (3.80 mg/100 g), phosphorus (100.34 mg/100 g), and β-carotene (16 μg/100 g) compared to control with 0.43%, 35 mg/100 g, 3.5 mg/100 g, 56.33 mg/100 g, and 9 μg/100 g of fibre, calcium, iron, phosphorus, and β-carotene, respectively. Cookies formulated with semi-refined pearl millet (*Pennisetum typhoideum*) flour (49.6%), fat (24.7%), sugar (24.7%), and vanilla powder (1%) had an improved nutritional quality compared to the control [prepared with refined wheat flour (49.6%), fat (24.7%), sugar (24.7%), and vanilla flavour (1%)]. The protein (8.50 mg/100 g), iron (6.71 mg/100 g), calcium (29.36 mg/100 g), and phosphorus (190 mg/100 g) content of millet cookies was higher compared to the control with 6.80, 2.48, 18.26, and 86.70 mg/100 g of protein, iron, calcium, and phosphorus, respectively (Florence Suma et al. 2014). Rai et al. (2014) prepared cookies using blends of pearl millet:rice, pearl millet:maize, and pearl millet:sorghum in the ratio of 50:50

and compared their quality characteristics against control (cookies prepared with 100% wheat flour). The protein content in the pearl millet-based cookies was higher (7.3–7.4%) compared to the control (6.91%). The cookies prepared from the blends of pearl millet with rice, maize, and sorghum had overall acceptability of 7.7, 7.9, and 8.6, respectively, on a 9-point hedonic scale. The sensory score of all three blends was higher than control (7.6/9). Swami et al. (2013) optimized the recipe for the preparation of finger millet cookies using non-linear regression analysis, Surface version 7.0 (Golden Software Inc., USA). The constant variables of the study were finger millet flour (100 g), sugar (30 g), vanilla flavour (2 ml), and baking powder (0.7 g). The independent variables of the study were vegetable oil (30, 35, and 40%), baking temperature (180, 200, and 220°C), and baking time (15, 20, 25, 30, 35, 40, 45, and 50 minutes). The highest value for hardness (1770.67 g) was observed for cookies prepared with 30% oil concentration that were baked at 180°C for 40 minutes. The lowest hardness of 652.14 g was observed for cookies with 40% oil concentration, baked at 180°C for 50 minutes. The yellowness index was highest (57.31) in cookies containing 40% oil, baked at 180°C for 50 minutes. The lowest yellowness index (50.58) was in cookies with 35% oil, baked at 200°C for 30 minutes. At low temperatures and with increased oil concentrations (30–40%), baking time increased. Cookies with 30% oil concentration and baked at 200°C for 20 minutes had the highest overall acceptability (22.1/30). The software-generated optimized conditions for the preparation of cookies were 30–32% oil concentration, 192–205°C baking temperature, and 20–34 minutes of baking time. The cookies prepared under these conditions had a hardness of 1300–1520 g and a yellowness index of 53.2–54.0. McWatters et al. (2003) prepared cookies using different compositions of the flours of wheat, fonio, and cowpea, i.e. wheat-fonio (50:50), wheat-cowpea (50:50), wheat-fonio-cowpea (33:33:33), wheat-fonio (25:75), wheat-cowpea (75:25), wheat-fonio-cowpea (25:50:25), fonio-cowpea (75:25), and fonio-cowpea (50:50), and studied their physical and sensorial characteristics. The cookies with 100% wheat flour had the highest spread factor of 5.86 and the lowest (4.39) was in the cookies prepared from wheat-cowpea in the ratio of 75:25. The lightest colour was observed in 100% wheat and 50:50 wheat-cowpea cookies. The cookies with a 25:75 ratio of wheat and fonio were the darkest among all. The highest sensory score for flavour (6.7/9) and overall acceptability (6.5/9) was obtained for cookies prepared with a wheat-fonio ratio of 50:50. These scores were comparable to 100% wheat flour cookies with scores of 7.1/9 and 6.9/9 for flavour and overall acceptability. The cookies prepared from other blends were found sensorially unacceptable.

14.2.1.6 Extruded Millet Products

Extrusion cooking is a high-temperature, short-time process that uses a combination of moisture, pressure, temperature, and mechanical shear to plasticize and cook moistened, expansive, starchy, and/or protein-rich food materials in a tube. This results in a change in chemical characteristics like viscosity, a reduction in cereal gruels, and increased nutrient density (Balasubramanian et al. 2014). Extruded products also have a porous and crunchy texture, high solubility, and low swelling power (Ushakumari et al. 2004). This technology is generally used to produce snacks, pasta, breakfast cereals, and other textured foods (textured vegetable protein and textured soy protein) (Choton et al. 2020). Millets can be extruded to add variety to the ready-to-eat products. Ganesan et al. (2021) prepared extruded snacks using finger millet (10%), pearl millet (10%), cornflour (73–77%), and squid powder (3–7%). The extrusion conditions, i.e. heater 1 temperature (55°C, 60°C, and 65°C), heater 2 temperature (120°C, 125°C, and 130°C), and screw speed (250 rpm, 275 rpm, and 300 rpm), were also optimized using RSM, Design-Expert Software, Version 11.1.2.0. Based on the software-generated results, the optimum product parameters were found to be heater 1 temperature of 64.10°C, heater 2 temperature of 129.98°C, screw speed of 317.87 rpm, and squid powder of 6.56%. The optimized product had a bulk density and expansion ratio of 1.33 g/cm^3 and 5.19, respectively. The optimized product had overall acceptability of 4.35 on a 5-point scale. Functional extruded snacks with different ratios, i.e. 100:0, 90:10, 85:15, 80:20, 75:25, and 70:30, of pearl millet flour and defatted almond cake were formulated by Naseer et al. (2022). The barrel temperature and screw speed for extrusion were 120°C and 420 rounds per minute, respectively. The pearl millet to defatted almond cake ratio of 80:20 had the highest overall acceptability (8.2) on a 9-point hedonic scale. The protein content, total phenolic content, total flavonoid content, and antioxidant activity (DPPH radical scavenging activity) of the 80:20 extruded snack were 15.45%, 56.91 mg/100 g, 18.29 mg/100 g, and 89.74%, respectively. Extrusion technology was used to prepare instant low-glycaemic rice using blends of broken rice, finger millet, barnyard millet, and quinoa. The extruder process parameters such as screw speed (30–50 rpm), die head temperature (110–140°C), and feed moisture content (26–34% wb) were studied based on the water absorption index, water solubility index, total colour change (ΔE), and cooking characteristics such as cooking time, cooking loss, and water absorption ratio. A feed moisture content of 34% (wb), screw speed of 30 rpm, and die head temperature of 110°C were optimized to produce low-GI rice. The prepared low-GI rice had a cooking time of 3.65 minutes, cooking loss of 12.05%, water absorption ratio of 2.79, water absorption index of 2.96 g/g, water solubility index of 4.16%, and ΔE of 18.20 (Yadav et al. 2021). Adebanjo et al. (2020) prepared flakes by extruding different blends of pearl millet and carrot flour (100:0, 95:5, 90:10, 85:15, and 80:20). The composite flour had high moisture content compared to control. The flakes prepared from 80:20 millet:carrot flour had the highest moisture (4.59%) content in comparison to the flakes prepared from 100% millet flour (3.54%). With an increase in carrot powder content, the ash and crude fibre content of the flakes increased while the fat, protein, and carbohydrate content of the flakes decreased. The flakes prepared from an 80:20 blend of millet and carrot flour also had the highest total carotenoids (4.36 μg/g) content. Increases in the L* (50.9–54.62), a* (0.76–3.75), and b* (10.39–24.31) values of the flakes were also observed with an increase in the content of carrot flour. The highest overall acceptability (7.25 on a 9-point hedonic scale) was obtained for the flakes prepared with an 85:15 blend of millet and carrot flour. The study concluded that good-quality flakes can be prepared from a blend of pearl millet and carrot flour. The effect of the substitution of three concentrations (10%, 20%, and 30%) of extruded pearl millet flour into unextruded pearl millet flour was studied on the quality of flatbread. It was found that the flatbreads made with 20% extruded flour had higher sensory scores for taste (8), texture (8), and overall acceptability (6.5) compared to the control with scores of 5, 4, and 4.8 for taste, texture and overall acceptability, respectively, on a 9-point hedonic scale. The iron content increased to 10.23 mg/100 g in 20% extruded flour composite compared to unextruded flour (4.1 mg/100 g). The calcium and fibre content decreased significantly ($P \leq 0.05$) from 345 mg/100 and 20.12% in unextruded flour to 94.3 mg/100 g calcium and 10.32% fibre in the composite flour containing 20% extruded flour (Kumar et al. 2020c). Suri et al. (2020) optimized ready-to-eat iron-rich extruded snacks using a central composite design of RSM, Design-Expert Software, Version 10.0.8. The effects of the blend ratio (2:1–6:1), barrel temperature (100–115°C), and amla (*Emblica officinalis*) (10–14 g/100 g of the total flour) on bulk density, crispiness, colour difference, iron, and ascorbic acid and protein content of the extruded snacks were studied. The optimized formulation was a blend of barnyard millet and defatted soy flour in the ratio 6:1 with amla flour (12.25 g/100 g of total flour) and rice flour (20 g/100 g of total flour) at a barrel temperature of 115°C. The iron, protein, and ascorbic acid contents of the optimized product were 15.71 mg/199g, 18.91 g/100 g, and 30.61 mg/100 g, respectively. The bulk density, crispiness, and colour difference of the optimized extruded product were 0.081 g/cm^3, 47.97, and 12.97, respectively. Sukumar and

Athmaselvi (2019) prepared extruded snacks using finger millet, rice, and corn (49:30:30) as base material and the effect of the fortification of this blend with banana flour (1, 2, 3, and 4 g) was also studied. To all of the blends 3% cheese was added and extruded. The extrusion process parameters like barrel temperature and extruder rpm were also optimized using a central composite design, Design-Expert Software, Version 10. The independent variables of the study were barrel temperature (118–122°C), extruder rpm (345–355), and banana flour (1–4 g). The dependent variables of the study were hardness, facturability, colour, expansion ratio, bulk density, water solubility index, water absorption index, and water holding capacity. The barrel temperature was found to have a non-significant ($P \leq 0.05$) effect on hardness and fracturability, while screw speed and banana flour concentration had a significant effect on these parameters. The L^* value of extrudate decreased significantly ($P \leq 0.05$) with an increase in the concentration of banana flour and barrel temperature. The bulk density decreased with an increase in barrel temperature and increases in the water solubility index, water absorption index, water holding capacity, and expansion ratio were observed with an increase in barrel temperature, screw speed, and banana flour concentration. Based on the dependent variables, the optimized process conditions were barrel temperature of 119°C, screw speed of 346 rpm, and banana flour concentration of 3.67 g. The product prepared using the optimized recipe and process conditions had an overall acceptability of 9/9 on a hedonic scale. This was better compared to the control (8) prepared without banana powder. Patil et al. (2016) studied the effect of the addition of unextruded and extruded finger millet flour (10%, 20%, and 30%) on the quality of bread. Bread prepared with the fortification of 20% extruded millet flour was found to have higher loaf height (6.41 cm), specific volume (3.51 cm^3/g), and moisture content (37.56 g/100 g wet basis) compared to bread prepared from unextruded finger millet flour with a loaf height of 4.36 cm, a specific volume of 2.74 cm^3/g, and moisture content of 34.56 g/100 g. Seth and Rajamanickam (2012) optimized the recipe for the preparation of soy-millet-rice-based extruded snacks using Design-Expert Software, Version 6.0.11. The rice was kept constant, i.e. 1:2.33 or 30%, to the rest of the other ingredients. The variables in the study were *ragi* (finger millet) (40–50%), sorghum (10–20%), and soy (5–15%) The responses of the study were bulk density, expansion ratio, water absorption index, and water solubility index. The optimized values for *ragi*, sorghum, and soy were 42.03, 14.95, and 12.97%, respectively, with desirability of 0.98. At these concentrations, the values of bulk density, expansion ratio, water absorption index, and water solubility index were 116.77 g/cm^3, 2.69, 6.14 g/g DM, and 30.96%, respectively.

14.2.1.7 Muffins

Muffins are spongy, soft-textured, sweet baked products with porous structures. These are highly appreciated by consumers for their good taste (Rahman et al. 2015). They are usually prepared from refined wheat flour but other cereals like millets can also be used for their development. Ashwini et al. (2016) prepared muffins using whole wheat flour (WWF), modified (protease treated) whole wheat flour (MWWF), pearl millet flour (PMF), and a combination of blends (CB) (modified WWF:PMF 50:50). The effect of emulsifiers (0.5 g/100 g flour), i.e. polysorbate-60 (PS-60), glycerol monostearate (GMS), sodium stearoyl-2-lactylate (SSL), and hydrocolloids (0.5 g/100 g flour) like carboxymethylcellulose (CMC), guar gum (GG), gum Arabic (GA), and a combination of additives (CAD), i.e. 50:50 of CMC and PS-60, was also studied on the muffins prepared using CB. The bulk density, volume, and lightness value of batter were 0.95 g/cc, 75 ml, and 56.2 for WWF and 0.98 g/cc, 70 ml, and 46.0 for PMF, respectively. The enzyme modification had a non-significant ($P \leq 0.05$) effect on batter density (0.94 g/cc) and a significant ($P \leq 0.05$) effect on batter volume (80 ml) and lightness value (49.8) in comparison to WWF. The blend of modified

WWF and PMF, i.e. CB, had higher batter density (0.97 g/cc) and the change in volume and lightness was non-significant (P ≤ 0.05) in comparison to PMF. The addition of emulsifiers had a non-significant effect on batter density; however, the addition of stabilizers improved the batter density (0.97–0.99 g/cc). The addition of emulsifiers and hydrocolloids also improved the volume and lightness value, i.e. 90–100 ml and 47.5–48.2 for emulsifiers and 90–95 ml and 45.3–51.3 in the case of hydrocolloids, respectively. The highest batter volume (120 ml) and lightness value (52.3) were obtained for CB+CAD. The use of emulsifiers and hydrocolloids also improved the sensory scores and the highest scores (90.5/100) for overall acceptability were obtained for muffins prepared using CB+CAD. The ash content of CB+CAD muffins was also higher (1.60%) in comparison to WWF muffins (1.20%). Jyotsana et al. (2016) prepared muffins using blends of finger millet flour and whey protein in the ratios 100:0, 95:5, 90:10, and 85:15 and compared the batter rheology, muffin texture, and other quality characteristics against wheat flour muffins. The other ingredients were sugar (84 g), calcium propionate (0.5 g), hydrocolloids and emulsifiers (0.5 g), glacial acetic acid (0.4 ml), pineapple essence (0.4 ml), and water (58 ml). The batter prepared from 100% finger millet flour was thin and entrapped less air, and the specific density (0.60 g/cc), viscosity (18,000 cP), and volume (55 ml) of the batter prepared from finger millet flour were lower than the wheat flour-based batter (0.85 g/cc specific density, 27,000 cP viscosity, and 150 ml volume, respectively). An increase in the specific density (0.70–1.25 g/cc) and volume (18,300–20,000 cP) of batter was found with an increase in whey protein content. The maximum improvement in batter volume (90 ml) was observed in the finger millet flour and whey protein blend of 90:10. Further, the effect of emulsifiers [sodium stearoyl-2-lactylate (SSL), distilled glycerol monostearate (DGMS), polysorbate 60 (PS-60)], stabilizers [guar gum (GG), xanthan gum (XG), hydroxypropylmethylcellulose (HPMC)], and a combination of additives (DGMS + HPMC) at a rate of 0.5% on the blends of 90% finger millet flour and 10% whey protein was also studied. The addition of emulsifiers improved the batter viscosity (19,800–20,500 cP) and the volume (90–115 ml) of muffins. Among the emulsifiers, the highest volume was obtained for DGMS. The stabilizers also improved the viscosity up to 20,000–22,200 cP. The highest increase in the muffin volume (135 ml) was found where the combination of HPMC and DGMC was used. The addition of whey protein, as well as emulsifiers and stabilizers, decreased the hardness and gumminess of muffins. The least hardness (6 N) was found with the combination of additives (DGMS and HPMC). Millet-based muffins were lacking the chewiness property which was also improved on the addition of emulsifiers and stabilizers. An improvement in cohesiveness and springiness was also observed on the addition of whey protein, emulsifiers, and stabilizers. The sensory score of 100% millet-based muffins was 64.5/100 which was improved to 86.5/100 on the addition of optimized concentrations of whey protein, emulsifiers, and stabilizers. The muffins prepared using the optimized concentrations of finger millet flour, whey protein, emulsifiers and stabilizers also had high ash (2.86%), protein (14%), dietary fibre (8.59%), calcium (340 mg/100 g), and phosphorus (260 mg/100 g) content in comparison to the wheat flour-based muffins. Goswami et al. (2015) prepared muffins by replacing refined wheat flour with barnyard millet flour. This study had 11 treatments containing refined wheat flour and barnyard millet flour in the ratios 0:100, 10:90, 20:80, 30:70, 40:60, 50:50, 60:40, 70:30, 80:20, 90:10, and 100:0. An increase in the specific gravity of batter and decreases in the weight, hardness, springiness, resilience, cohesiveness, chewiness, and baking height of muffins were found with an increase in the content of barnyard millet flour. The crumb was darker in the barnyard millet flour-based muffins in comparison to the 100% refined wheat flour-based muffins. The score for the overall acceptability was reduced for barnyard millet flour muffins (7.09 for 100% barnyard millet flour) in comparison to the muffins prepared from whole wheat (8.24). An increase in minerals (1.75 g/100 g) and dietary fibre

(2.1 g/100 g) was also found in barnyard millet flour muffins in comparison to refined wheat flour-based muffins with 0.69 g/100 g minerals and 0.14 g/100 g dietary fibre, respectively. The muffins were stable at refrigerated and ambient temperature for 15 days.

14.2.1.8 Multigrain Flour

Multigrain flour is obtained by milling different grains at required proportions according to the desired final product or nutritional needs (Blaga et al. 2018). Multigrain flours formulated with wheat (20–35%), oats (10–25%), finger millet (15–20%), pearl millet (15–20%), and buckwheat (15–20%) flour were studied for their sensory and nutritional aspects in the form of biscuits and chapattis. The formulation with wheat (30%), oats (20%), finger millet (15%), pearl millet (15%), and buckwheat (20%) flour was found to have higher overall acceptability for biscuits (8.3) and chapatti (7.76) on a 9-point hedonic scale compared to control biscuits (7.5) and chapatti (7.38) prepared from 100% wheat flour. The protein (15.31 g/100 g) and fibre (12.35 g/100 g) were highest in chapatti prepared with this blend (Agrawal et al. 2016). The stability of multigrain flour with sorghum (50 g), wheat (45 g), *ragi* (1 g), black gram (2 g), and fenugreek (1 g) was studied with and without tertiary butyl hydroxy quinine (TBHQ). The multigrain flour with TBHQ (0.02%) did not have any significant difference in quality compared to the control on storage for 120 days. The *roti* prepared with the multigrain flour containing TBHQ was able to retain its quality up to 120 days and a reduction of quality was observed in the control (without TBHQ) after 90 days (Dayakar Rao et al. 2015).

14.2.1.9 Noodles

Noodles are a popular food worldwide and the first noodles were prepared from millets (Diao and Jia 2017); however, nowadays, noodles are mostly prepared from refined wheat flour. Such noodles are low in dietary fibre, minerals, and vitamins. This can be overcome by fortification with millets (Hymavathi et al. 2019). The fortification of rice noodles with various concentrations of finger millet flour (10%, 20%, and 30%) was studied by Chen et al. (2021). Rice noodles fortified with 20% finger millet flour had the highest total sensory score of 9.46 on a 10-point scale. The 2, 2-diphenyl-1-picrylhydrazyl (DPPH) radical scavenging activity, 2, 2-azinobis (3-ethylbenzothiazoline-6-sulphonic acid) (ABTS+) radical cation scavenging activity, ferric-reducing antioxidant power (FRAP), and total phenolic content of the finger millet-fortified rice noodles increased to 223.53, 577.93, 442.06 µg/g Trolox equivalents (TE), and 8.72 µg/g GAE, respectively, from 100.50, 276.04, 194.5 g/g TE, and 0 g/g GAE in the control. Dahal et al. (2020) studied different proportions of cornflour (90, 100, and 110 g), proso millet flour (60, 85, and 110 g), and egg (25, 37.5, and 50 g) for the preparation of gluten-free dough for noodles. The process was optimized using RSM, Design-Expert Software, Version 10.0.0. The responses studied were the viscosity, pasting temperature, and colour (b* value only) of the dough. The highest viscosity of 347.33 rapid-visco analyzer unit (RVU) was observed for the dough prepared from a blend containing 100 g cornflour, 85 g proso millet flour, and 37.5 g egg. The highest pasting temperature of 95.4°C was obtained for a blend with 90 g cornflour, 110 g proso millet flour, and 37.5 g egg. The highest b* (+yellowness/−blueness) value of 25.05 was obtained for noodles prepared with 110 g cornflour, 60 g proso millet flour, and 37.5 g egg. Based on studied responses, the optimized concentrations of cornflour, proso millet flour, and egg for the preparation of dough were 98.77 g, 60 g, and 50 g, respectively. The optimized dough had a viscosity, pasting temperature, and b* value of 176.21 RVU, 68.65°C, and 21.73, respectively. Further, the effect of different concentrations of xanthan gum (0, 0.38, 0.75, 1.13, and 1.5 g) and guar gum (0, 0.38, 0.75, 1.13, and 1.5 g) on the functional properties of noodles was also studied. The responses studied were the water holding capacity, adhesiveness, and tensile strength of the noodles. Based on responses, the optimized concentrations of xanthan and guar gum were 1.5 g and 1.47 g, respectively. The noodles prepared using these concentrations of gums had a water holding capacity, adhesiveness, and tensile strength of 61.2%, 57.95 g.s., and 19.34 g, respectively. Shukla and Srivastva (2014) prepared noodles by the addition of different concentrations of finger millet flour (30, 40, and 50%) to refined wheat flour. The noodles prepared with the incorporation of 30% finger millet flour had the highest overall acceptability (8.6/9) which was equivalent to noodles prepared from 100% refined wheat flour. The noodles prepared using this blend also had a low GI of 45.13 compared to the control (62.59). Noodles were prepared using different ratios of wheat and malted finger millet flour (MFMF) (90:10, 80:20, 70:30, 60:40, and 50:50), and their sensory and nutritional characteristics were studied. The noodles prepared with a 70:30 ratio of wheat flour and MFMF along with vegetable oil (16%), wheat gluten (10%), guar gum (2%), glycerol monostearate (1%), and baking powder (0.5%) had the highest overall acceptability (4.8 on a 5-point scale) among the millet-based noodles. The noodles prepared using this blend also had high protein (15.66%), fibre (1.35%), calcium (160 mg/100 g), phosphorus (186 mg/100 g), and iron (6.61 mg/100 g) content (Kulkarni et al. 2012).

14.2.1.10 Weaning Foods

Weaning is the process of decreasing the dependence of infants on the mother's milk by the gradual introduction of solid and semi-solid foods into the diet to ensure their healthy growth (Dong et al. 2021). Dong et al. (2021) developed two weaning foods, i.e. roasted quinoa-millet complementary food (RQMCF) and extruded quinoa-millet complementary food (EQMCF) using roasted and extruded flours of quinoa and millet (7:3). To give prebiotic potential to the mix a mixture of fructo-oligosaccharides and oligosaccharides (1:1) was also added. The formula for the preparation of RQMCF and EQMCF was 70% quinoa-millet flour, 15% skim milk powder, 8% egg yolk powder, 3% fructo-oligosaccharides and galactooligosaccharides, 2% fruits and vegetable powder, and 2% minerals. The millet-based weaning foods, i.e. RQMCF and EQMCF, had moisture, carbohydrates, protein, fat, ash, total dietary fibre, and energy content of 7.76–8.01%, 55.59–55.98%, 18.94–19.11%, 11.64–11.70%, 2.87–3.03%, 2.56–2.61% and 404 kcal/100 g, respectively. In comparison with a commercially available millet complementary food (CMCF), it was found that RQMCF and EQMCF had less moisture and carbohydrates and more protein, fat, ash, total dietary fibre, and energy content than CMCF (9.06% moisture, 56.91% carbohydrates, 16.23% proteins, 10.21% fat, 2.43% ash, 2.16% dietary fibre, and 403 kcal/100 g). The content of minerals like calcium, phosphorus, potassium, zinc, iron, and manganese was also high in comparison to CMCF. The nutritional composition of the product was also close to the specification of weaning foods given by FAO/WHO, 1991. These products also had higher prebiotic potential and supported better growth of *Lactobacillus plantarum* and *Lactobacillus delbruekii* in comparison to the CMCF. A process for the development of weaning mix based on malted and extruded pearl millet and barley flour was optimized by Balasubramanian et al. (2014) using the central composite rotatable design (CCRD). The independent variables of the study were pearl millet extrudates (PME) (20–25%), pearl millet malt extrudates (PMME) (6–9%), barley extrudates (BE) (20–25%), and barley malt extrudates (6–9%). The constant variables were skim milk powder (SMP) (25%), WPC-70 (5%), sugar (6%), and refined vegetable oil (4 ml/100 g). The responses studied were lightness, peak viscosity (Pv), water solubility index, water absorption index, and overall acceptability. The software generated the optimized conditions for the development of the weaning mix as 20.77% PME, 7.39% PMME, 20.99% BE, and 6.53% BME. The optimized weaning mix had moisture, carbohydrates, protein, fat, and minerals content of 4.59%, 67.95%, 14.73%, 9.88%, and 2.85%, respectively. The calcium, phosphorus, and iron contents of the optimized weaning mix were 354 mg/100 g, 251.2 mg/100 g, and 5.92 mg/100 g, respectively. The product specifications were as per the guidelines published under the Prevention of Food Adulteration Act, 2004 for milk cereal-based weaning foods. Thathola and Srivastva (2002) prepared millet-based weaning food by mixing the malted flours of foxtail millet (30%), barnyard millet (30%), roasted soybean flour (25%), and skim milk powder (15%). The mix had a moisture content of 7.33%, carbohydrate content of 60.89%, protein content of 18.37%, crude fat content of 9.0%, and ash content of 4.0%, and produced 398 kcal/100 g. The calcium, iron, ascorbic acid, and niacin contents of the mix were 253.33 mg/100 g, 4.90 mg/100 g, 74.66 mg/100 g, and 2490 µg/100 g, respectively. The protein, calcium, zinc, ascorbic acid, and niacin contents of the mix were more than the minimum requirements of weaning foods published in 1991 under the Prevention of Food Adulteration Act, 1954. The levels of other nutrients like riboflavin, vitamin B12, folic acid, vitamin A, and vitamin D were increased by fortification. The protein efficiency ratio of mixed protein was 2.25. The score for the overall acceptability of millet-based gruel was 7.7/9 on a hedonic scale and this was comparable to the market available weaning gruel (8.4/9).

14.2.2 Fermented Products

14.2.2.1 Synbiotic Beverages

Synbiotics refer to a mixture of prebiotics and probiotics (Kumar et al. 2015). Prebiotics are the non-digestible or low-digestible components of food that act as a dietary supplement for the good microorganisms of the gut and support their growth. Probiotics are live microorganisms which when administered in an adequate amount confer health benefits to the host (Gibson and Roberfoid 1995). Prebiotics are selective in their action and hence a selective prebiotic can support only the growth of a specific type of microorganisms in the colon (Manning and Gibbson 2004). For example, fructo-oligosaccharides favour the growth of *Lactobacillus* and *Streptococci*, the fractionated galactooligosaccharides support the growth of *Bacillus lactis* and *L. rhamnosus*, inulin supports larger total bacterial counts, and xylo-oligosaccharides and lactulose support the growth of *Bifidobacterium* (Kumar et al. 2015). The presence of galactooligosaccharides is also associated with a decrease in the count of clostridia. Prebiotics also protect the probiotics against bile acid stress. *L. acidophilus* NCFM and *L. reuteri* NCIMB 11951 are found resistant against 2 mM cholic and taurocholic acid (Adebola et al. 2014) in the presence of lactulose (1%) or lactobionic acid (1%). Millets are a good source of prebiotics like arabinoxylans, arabinogalactan, beta-glucan, inulin, fructo-oligosaccharides, and hemicelluloses (Kumar et al. 2022), and their application in the development of probiotics has been studied by many researchers. Kumar et al. (2020) optimized the process for the preparation of a synbiotic beverage based on the lactic acid fermentation of nutricereals (finger millet + oats) and milk-based beverages. The finger millet grains were malted (germinated at 25°C for 84 hours and dried at 50 ± 2°C until a constant weight was obtained), coarsely ground, mixed with coarsely ground oat flour (60:40 finger millet malt flour:oat flour), water was added (300 ml in 100 g of flour mix), the mixture was mashed (as per the process of wort making) and filtered, and a malt drink was prepared. The malt drink was mixed with double toned milk (52:48) and a composite drink was prepared. Further, four beverages, i.e. sweetened plain milk (SPM), functional drink (FD), rose-flavoured drink (RFD), and marigold-flavoured drink (MFD), were prepared and inoculated with three activated probiotic cultures, i.e. *Lactobacillus acidophilus* NCDC14 (1%), *L. rhamnosus* NCDC14 (5%), and *L. casei* NCDC297 (5%). The cultured samples were incubated at

37°C for 3, 6, 9, and 12 hours, respectively. On the sensory evaluation of the drinks it was found that the most desirable sensorial properties (sensory score of 87.71/100 on a composite sensory scale) were obtained for RFD fermented with 1% activated culture of *L. acidophilus* for 6 hours. The optimized drink had an acidity of 0.65%, pH of 4.37, and probiotic count of 6.38 log CFU/ml. The prepared beverage could be stored for 15 days at refrigerated temperature without any noticeable loss of quality. A process for the preparation of a probiotic beverage from the flour of roasted finger millet was optimized by Fasreen et al. (2017). To prepare the beverage, finger millet grains were roasted for 3 minutes and ground to flour, and 25 grams of roasted flour was cooked with water (500 ml), cooled to 40°C, cultured with the commercial probiotic culture of *Lactobacillus casei* 431® at the rate of 0.031 gL^{-1}, and fermented at 37°C for 2, 4, and 6 hours. The fermentation time of 4 hours was found best for the preparation of the probiotic beverage. To the fermented finger millet mix, pasteurized milk, sugar, and cocoa powder were added at the rate of 150 ml, 46 g/litre, and 7.9 g/litre, respectively. The optimized product had a storage life of 5 weeks at refrigerated temperature (5 ± 1°C). Gupta (2017) prepared probiotic beverages using blends of pearl millet starch-free extracts and different concentrations (10–50%) of pineapple and apple juices. To obtain the pearl millet extract, the millet grains were steeped in water at 25°C for 24 hours, germinated at 30°C for 48 hours, and kilned at 55°C, the roots and shoots were removed, the grains were ground to flour, water was added (1:1), the mixture was boiled for 30 minutes at 80°C and cooled, and the millet extract was allowed to settle for 30 minutes. The millet extract was sterilized at 121°C for 15 minutes, blended with sterilized pineapple and apple juice (90:10, 80:20, 70:30, 60:40, 50:50), inoculated with an activated culture of *Lactobacillus plantarum* (2% v/v), and incubated at 37°C for 48 hours. On the sensory evaluation of various treatments, it was found that the probiotic beverage prepared by blending 50:50 of pearl millet extract and pineapple juice had the maximum scores for colour (15.55/20), flavour (16.45/20), consistency (15.55/20), taste (16.75/20), and overall acceptability (15.55/20). The viable bacterial count at this concentration was 2.8×10^8 CFU/ml. Stefano et al. (2017) prepared four pearl millet-based novel probiotic formulations, i.e. water-millet based, milk-millet-based, dried millet-based, and millet flour-based. To prepare the water-based formulation, 1 litre of water was added to different concentrations (4, 6, 7, 8, and 10%) of hulled pearl millet, the mixture was heated to 90–95°C for 60 minutes, 5% sugar or honey was added (in the last five minutes of heating), and it was cooled to 40°C and inoculated with a Fiti sachet containing 1 gram each of *L. rhamnosus* GR-1 and *S. thermophilus*. In the milk-millet fermented beverage, the concentrations of millet and the preparation steps were similar to the previous treatment except for the use of homogenized milk (3.25%), or a mix of water and milk (50:50), in the preparation of the formulation. Dried millet was prepared by the addition of 1 litre of water to 400 g of hulled pearl millet. The mixture was then boiled and covered, the heat was reduced, and the mixture was left to simmer until the water was completely absorbed. The mix was cooled (40°C) and the bacterial culture was added. The flour-based millet formulation was prepared by the addition of either 1 litre of milk or water or a mix of water and milk (50:50). To prepare the beverage 800 ml of milk was added to 152 g of millet flour. The mixture was brought to a boil and stirred constantly at a boil for 15 minutes, cooled, and cultured. After incubation for 9 hours and 12 hours, a 2 log cycle increase in the population of *L. rhamnosus* GR-1 and a 3 log cycle increase in *S. thermophiles* were found. A higher cell population was recorded in milk-based beverages in comparison to water-based formulations. The probiotic beverage containing 4% millet and 5% sugar in milk, fermented at 40°C for 12 hours, had the highest sensory scores. Non-dairy probiotic drinks were prepared by adding flours of sprouted wheat (7.86 g), barley, pearl millet, and green gram (0, 2, 4, 6, and 8%) separately to the soya milk and water base (soymilk:water 0:100, 30:70, 40:60, 50:50)

(Mridula and Sharma 2015). For sprouting, the grains were cleaned, washed, and soaked in water in the ratio of 1:2 of seed to water for 8 hours at 30°C. The grains like wheat, barley, and pearl millet were germinated at $35 \pm 2°C$ and 95% RH for 36 hours, while the green gram was germinated for 24 hours under similar conditions. Oats (6 g/100 ml liquid), guar gum (0.6 g/100 ml liquid), sugar (7 g/100 ml liquid), and cardamom (0.01 g/100 ml liquid) were also added to the drink, and fermentation was carried out using *Lactobacillus acidophilus*- NCDC14 (1 ml/100 ml). An increase in the probiotic count was observed in the drinks with an increased level of grain flour. Among the studied samples, the highest probiotic count (11.51 log CFU/ml) was found in the pearl millet-based probiotic drink. The probiotic drink containing 50:50 ml of soy-milk and distilled water with 4 g of pearl millet flour had the highest sensorial acceptability. The effect of cereals on the production of probiotic drinks was also studied by Hassan et al. (2012). In this study, rice and millet were soaked, milled, and filtered, and cereal milk was prepared. This was further heat-treated at 90°C for 20 minutes, cooled to 37°C, 5% honey was added, and the mixture was inoculated with 5% of commercial ABT-2 starter culture containing *L. acidophilus*, *Bifidobacterium* BB-12, and *S. thermophiles* and incubated at 37°C for 16 hours. The fermented milk was further categorized into two groups, i.e. control and fortified (with 10% seeds of pumpkin and sesame). The viable probiotic count on the completion of fermentation reached about 4.3×10^9 CFU/ml in the rice and millet-based fermented beverages. The sensory characteristics of both the rice and millet-based beverages were improved on fermentation with ABT-2 starter culture. However, the fortification of the beverages with pumpkin and sesame seeds decreased the overall acceptability of these drinks. The addition of pumpkin (9.73 log CFU/ml) and sesame (9.80 log CFU/ml) seeds significantly ($P < 0.05$) increased the *L. acidophilus* count in comparison to the rice- (9.66 log CFU/ml) based beverage. The count of *Bifidobacterium* BB-12 was significantly ($P < 0.05$) reduced, i.e. 9.66 log CFU/ml in pumpkin seed-fortified drink and 9.74 log CFU/lm in the sesame seed-fortified drink as compared to 9.87 log CFU/ml in the rice-based beverage. The developed probiotic drinks had a storage life of 15 days. Mohana (2007) optimized the process for the development of pearl millet-based ready-to-reconstitute *Rabadi* (a buttermilk and millet-based fermented beverage of the north-western states of India). The preparatory steps were the germination (24 hours) of pearl millet grains, grinding to flour, mixing of flour with hot (60°C) concentrated skim milk, pasteurization at 80°C for 20 minutes, cooling to 37°C, the addition of *dahi* culture NCDC 167 at the rate of 3%, and incubation for 10–12 hours. On the completion of fermentation, the mix was salted, homogenized, and spray dried. The dried powder had a total viable cell count of 36.5%. The count of viable lactic acid bacteria was 32.5% of the actual count in the beverage. To the dried powder, cumin, black pepper, and pectin were added. The prepared powder had total solids, proteins, fat, and ash content of 91.5%, 21.6%, 19.84%, and 10.6%, respectively.

14.2.2.2 Alcoholic Beverages

Koval Breweries Ltd., Chicago, produce a millet-based whiskey known as Koval single barrel millet whiskey (Mazlien 2015). In addition, several research studies have been also carried for the development of millet-based alcoholic beverages. Bano et al. (2015) optimized the process for the development of beer using a blend of the flours of malted finger millet, barnyard millet, and paddy. The effect of three independent variables, i.e. blend ratio (80:20:0, 80:10:10, and 80:0:20), concentration of α-amylase (0, 0.4, and 0.8%) and slurry ratio (1:5, 1:7, and 1:9), on the alcohol content, pH, and colour of beer was also studied. Among the three variables, blend ratio had the maximum effect on alcohol content, pH, and colour, enzyme concentration had the maximum effect on alcohol content, and slurry ratio had the maximum effect on the colour of beer. On statistical analysis, it was revealed that a blend ratio of 80:9.6:10.4, i.e. 56 g finger millet, 6.72 g barnyard millet, and 7.28 g paddy, 0.45% enzyme concentration, and a slurry ratio of 1:6.82 produced the beer with most desirable quality attributes. The beer produced using this blend had an alcohol content of 10.5%. Kumar et al. (2015b) optimized the process for the development of beer from malted (germinated for 30–36 h) grains of finger millet, barley, and a combination of both using Box Behenken Design, Design Expert Software, 8.0.6.1. The independent variables of the study were the blend ratios of grains (100:0, 50:50, 0:100), kilning temperature (50, 70, 90°C), and millet grain to water ratio (1:3, 1:5, 1:7), and their effect on the quality attributes like pH, colour, bitterness, and alcohol content was studied. All three variables had a significant effect on the studied quality attributes of beer. The optimized conditions for the development of beer were a finger millet to barley ratio of 68:32, a kilning temperature of 50°C, and a slurry ratio of 1:7. The refinement of the traditional technology of *sur* (a finger millet-based traditional alcoholic beverage of Himachal Pradesh, India) production was carried out by Kumar (2013). In the refined technology, the general beer-making steps like soaking, germination, kilning, mashing, addition of hops, pitching, siphoning, and bottling were used for the preparation of *sur* and their effect on the quality of *sur* was also studied. The finger millet grains were germinated for 96 hours, dried (50 ± 2°C), pulverized coarsely, and maize grits (30%) were added as adjuncts. To the grain mix, three times the volume of potable grade water was added, mixed properly, mashed, filtered, wort was collected and fermented using two inoculums, i.e. the traditional inoculum "*dhaeli*" used for *sur* production and a pure culture of *Saccharomyces cerevisiae* var. *ellipsoideus* used for beer production. The use of the pure culture of *S. cerevisiae* was found to produce beer with better sensory attributes in comparison to the traditional *sur*. A process for the development of finger millet- and pearl millet-based alcoholic beverages was optimized by Khandelwal et al. (2012). The millet grains were soaked in water for three time intervals, i.e. 24, 48, 72 hours, germinated (24 hours), dried, deculmed, and ground to flour. The flour was blended in the ratios of 70:30 and 50:50 with the juices of three fruits, namely green grape juice, black grape juice, and apple juice. The blends were fermented using 5% activated culture of *S. cerevisiae*. The alcoholic beverages produced using finger millet were found to have more sensorial acceptability in comparison to the pearl millet-based beverages. On further investigation it was revealed that the alcoholic beverages with best sensorial acceptability were produced from a blend of finger millet, green grape juice, and black grape.

14.3 MILLET-BASED PRODUCTS COMMERCIALLY AVAILABLE IN THE INDIAN MARKET

Millet-based products are commercially promoted as health foods with various health claims. Millet-based ready-to-eat products such as cookies, extruded snacks, bread, cakes, muffins, breakfast cereals, cereal bars, and noodles are readily available in the market. Other products such as health drink mix, weaning mix, breakfast porridge mix, and composite flour which requires minimal processing are also available in the market. Table 14.1 also includes detailed information on commercially available millet-based products. Figure 14.1 depicts images of some commercial millet-based products. The most common health claims of these products are high fibre, high protein, iron-rich, gluten-free, helps in digestion and weight loss, diabetic-friendly, and reduces the risk of heart diseases. Some uncommon health claims include that they are a good source of calcium, they provide antioxidants, they improve immunity, and they can help in the management of cholesterol levels.

TABLE 14.1   MILLET-BASED COMMERCIAL PRODUCTS AND THEIR HEALTH CLAIMS

Product name	Brand name	Claims and health benefits	Type of food
Diet millet cookies	Weleet	Provides adequate quantities of fibre and antioxidants, helps to maintain healthy glycaemic index, best snack for all who have type-2 diabetics and gluten free.	Snack
Iron shakthi bajra barnyard millet jaggery cookie	Early foods	Nutrient-dense snack.	Pregnancy snack
Foxtail millet cookies	FittR Bites	Protein and iron rich. Source of micronutrients – vitamins and minerals. Antioxidant activity. Lowers risk of heart disease and improves digestive health.	Snack
Millet health mix	Jeeni	Increases immunity power, reduces heart-related problems, reduces blood pressure, increases red blood cells, reduces asthma and thyroid problems.	Breakfast porridge mix
Health mix	Manna	Rich in protein, dietary fibre, and iron.	Health drink for children
Millet health mix	Harasu	Immunity booster, improves brain function, stamina booster, weight loss, strengthens heart functions, blood purification, controls menstrual cramps, controls hormonal imbalance, and relieves joint pain.	Health drink
Multi millet mix	Aashirvaad	Naturally gluten free, rich in dietary fibre, source of protein. Source of zinc which aids in normal cognitive function. Source of magnesium that helps reduce fatigue.	Composite flour
Foxtail millet noodles	Thanjai natural	Easy to digest. Rich in fibre and micronutrients.	Fortified noodles
Bajra noodles	Foodstrukk	Rich in protein and micronutrients like iron and folic acid. Good source of fibre.	Fortified instant noodles
Ragi/bajra/quinoa/jowar bar	Mindful	Gluten free, high in iron and magnesium, high in protein, rich in fibre, improves digestion and improves metabolic health. Low GI and controls blood sugar.	Energy bar
Millet crunchies cream and onion	Kiru	Healthy snack for fitness enthusiasts, weight watchers, health-conscious individuals. Abundant in fibre. Gluten free and zero trans-fat.	Extruded snack

(*Continued*)

TABLE 14.1 (CONTINUED) MILLET-BASED COMMERCIAL PRODUCTS AND THEIR HEALTH CLAIMS

Product name	Brand name	Claims and health benefits	Type of food
Mighty Munch	Slurrp Farm	Perfect anytime snack for all ages. Zero trans-fat. Low in sugar.	Extruded snack
Multi-millet bread	Pristine OvenOrg	High fibre for easy digestion and lower blood sugar. Reduces the risk of cardiovascular problems, stroke, and diabetes and also helps with weight control. Helps manage cholesterol levels. Is associated with reduced cancer risk.	Breakfast or snack
Sorghum Sliced Bread	Bon & Bread	High in fibre.	Breakfast or snack
Multi Millet Bread	Nroots	Good alternative to white bread. High in fibre content.	Breakfast
Multi millets choco muffin	Nroots	Wholesome snack. Healthy choice for kids and adults.	Snack
Multi millet muffins	Pristine OvenOrg	High fibre for easy digestion and lowering blood sugar. Reduces the risk of cardiovascular problems, stroke, and diabetes and also helps with weight control. Helps manage cholesterol levels. Is associated with reduced cancer risk.	Breakfast or snack
Multi millet nuts cake	Nroots	Guilt-free cake for health watchers. Keeps you full for a longer time. Good source of fibre and easy to digest.	Snack
Multi millet atta	Manna	High in protein and dietary fibre. Helps digestion. Aids in weight loss. Rich in iron, calcium, magnesium, and zinc.	Composite flour
Millet Muesli – fruit, nuts and seeds	Soulfull	High protein, high fibre, gluten free.	Breakfast cereal
Ragi flakes	Murginns	Source of calcium and fibre.	Breakfast cereal
Bajra flakes	Health Sutra	Protein rich, calcium rich.	Breakfast cereal

14.4 CONCLUSIONS AND FUTURE OUTLOOK

Millet-based food products are an emerging trend and an increased demand for millet-based products has been observed throughout the world. The fortification of staple foods with millets, the optimization of new millet-based recipes, and the utilization of millets in the development of novel foods are the most popular trends. But still, the range of millet-based foods is very limited and there is not a wide range of choices for customers. This limits the potential for the growth of the millet-based food market. Hence, more research is required to produce new millet-based products with better sensory properties and customer acceptability. There is also a crucial need for the dissemination of the developed technologies to food business operators so that the

Figure 14.1 Some of the commercially available millet-based food products.

maximum number of studies can be brought from the laboratory to reality. This will increase the availability of millet-based products in the market and will help to popularize millet-based foods.

REFERENCES

Adebanjo, L. A., G. O. Olatunde, M. O. Adegunwa, O. C. Dada, and E. O. Alamu. 2020. Extruded flakes from pearl millet (*Pennisetum glaucum*)-carrot (*Daucus carota*) blended flours-Production, nutritional and sensory attributes. *Cogent Food and Agriculture* 6(1):1733332. doi:10.1080/23311932.2020.1733332.

Adebola, O. O., O. Corcoran, and W. A. Morgan. 2014. Synbiotics: The impact of potential prebiotics inulin, lactulose and lactobionic acid on the survival and growth of lactobacilli probiotics. *Journal of Functional Foods* 10:75–84. doi:10.1016/j.jff.2014.05.010.

Agrawal, A., A. Verma, and S. Shiekh. 2016. Evaluation of sensory accessibility and nutritive values of multigrain flour mixture products. *International Journal of Health Sciences and Research* 6(1):459–65.

Anju, T., and S. Sarita. 2010. Suitability of foxtail millet (*Setaria italica*) and barnyard millet (*Echinochloa frumentacea*) for development of low glycemic index biscuits. *Malaysian Journal of Nutrition* 16(3):361–8.

Balasubramanian, S., J. Kaur, and D. Singh. 2014. Optimization of weaning mix based on malted and extruded pearl millet and barley. *Journal of Food Science and Technology* 51(4):682–90. doi:10.1007/s13197-011-0579-6.

Bano, I., K. Gupta, A. Singh, and N. C. Shahi. 2015. Finger millet: A potential source for production of gluten free beer. *International Journal of Engineering Research and Applications* 5(7):74–7.

Blaga, G., I. Aprodu, and I. Banu. 2018. Comparative analysis of multigrain and composite flours based on wheat, rye and hulled oat. *Scientific Papers: Series D, Animal Science-The International Session of Scientific Communications of the Faculty of Animal Science*, 61(1):261–65.

Business Standard. 2023. Why Sitharaman called millets 'Shree Anna' - The mother of all grains? https://www.business-standard.com/article/economy-policy/why-sitharaman-called-millets-shree-anna-the-mother-of-all-grains-123020300457_1.html (accessed Feburary 20, 2023).

Chen, J., L. Wang, P. Xiao, C. Li, H. Zhou, and D. Liu. 2021. Informative title: Incorporation of finger millet affects in vitro starch digestion, nutritional, antioxidative and sensory properties of rice noodles. *LWT-Food Science and Technology* 151:112145. doi:10.1016/j.lwt.2021.112145.

Choton, S., N. Gupta, J. D. Bandral, N. Anjum, and A. Choudary. 2020. Extrusion technology and its application in food processing: A review. *The Pharma Innovation Journal* 9(2):162–8.

Dahal, A., M. B. Sadiq, and A. K. Anal. 2020. Improvement of quality of corn and proso millet-based gluten-free noodles with the application of hydrocolloids. *Journal of Food Processing and Preservation* 45(2):e15165. doi:10.1111/jfpp.15165.

Dayakar Rao, B., K. Kalpana, K. Srinivas, and J. V. Patil. 2015. Development and standardization of sorghum-rich multigrain flour and assessment of its storage stability with addition of TBHQ. *Journal of Food Processing and Preservation* 39(5):451–7. doi:10.1111/jfpp.12250.

Desai, A. D., S. S. Kulkarni, A. K. Sahoo, R. C. Ranveer, and P. B. Dandge. 2010. Effect of supplementation of malted *ragi* flour on the nutritional and sensorial quality characteristics of cake. *Advance Journal of Food Science and Technology* 2(1):67–71.

Devani, B. M., B. L. Jani, M. B. Kapopara, D. M. Vyas, and M. D. Ningthoujam. 2016. Study on quality of white bread enriched with finger millet flour. *International Journal of Agriculture, Environment and Biotechnology* 9(5):903–7.

Diao, X., and G. Jia. 2017. Origin and domestication of foxtail millet. In *Genetics and Genomics of Setaria* (pp. 61–72). Cham: Springer. doi:10.1007/978-3-319-45105-3_4.

Dong, G., J. Dong, Y. Zhu, R. Shen, and L. Qu. 2021. Development of weaning food with prebiotic effects based on roasted or extruded quinoa and millet flour. *Journal of Food Science* 86(3):1089–96. doi:10.1111/1750-3841.15616.

Drábková, M., M. Hrušková, and I. Švec. 2017. The baking potential of wheat-fonio flour composites evaluated by Mixolab. *Maso International-Journal of Food Science and Technology* 1:43–8.

Fasreen, M. M. F., O. D. A. N. Perera, and H. L. D. Weerahewa. 2017. Development of finger millet based probiotic beverage using lactobacillus casei431®. *OUSL Journal* 12(1):128–38. doi:10.4038/ouslj.v12i1.7384.

Fathi, B., M. Aalami, M. Kashaninejad, and A. Sadeghi Mahoonak. 2016. Utilization of heat-moisture treated proso millet flour in production of gluten-free pound cake. *Journal of Food Quality* 39(6):611–19. doi:10.1111/jfq.12249.

Financial Express. 2018. Government renames millets as nutri cereals. https://www.financialexpress.com/market/commodities/government-renames-millets-as-nutri-cereals/1140338/. (accessed on November 17, 2021).

Florence Suma, P., A. Urooj, M. R. Asha, and J. Rajiv. 2014. Sensory, physical and nutritional qualities of cookies prepared from pearl millet (*Pennisetum typhoideum*). *Journal of Food Processing and Technology* 5(10):377. doi:10.4172/2157-7110.1000377.

Ganesan, V., B. Subbaiah, N. M. Stephen, M. Dhanushkodi, R. Kolandaivadivel, and R. Veeramani. 2021. Optimization of extrusion process using response surface methodology for producing squid and millet based extrudate. *Turkish Journal of Fisheries and Aquatic Sciences* 22(2):TRJFAS20041. doi:10.4194/TRJFAS20041.

Gibson, G. R., and M. B. Roberfroid. 1995. Dietary modulation of the human colonic microbiota: Introducing the concept of prebiotics. *Journal of Nutrition* 125(6):1401–12. doi:10.1093/jn/125.6.1401.

Goswami, D., R. K. Gupta, D. Mridula, M. Sharma, and S. K. Tyagi. 2015. Barnyard millet based muffins: Physical, textural and sensory properties. *LWT – Food Science and Technology* 64(1):374–80. doi:10.1016/j.lwt.2015.05.060.

Government of India. 2021. https://pib.gov.in/newsite/PrintRelease.aspx?relid=186206. (accessed November 18, 2021).

Gupta, M. 2017. Studies on development and evaluation of millet-based probiotic beverage. PhD Thesis, Faculty of Applied Sciences and Biotechnology. Shoolini University of Biotechnology and Management Sciences, Solan, H.P., India.

Hassan, A. A., M. M. A. Aly, and S. T. El-Haldidie. 2012. Production of cereal-based probiotic beverages. *World Applied Sciences Journal* 19(10):1367–80. doi:10.5829/idosi.wasj.2012.19.10.2797.

Hymavathi, T. V., V. Thejasri, and T. P. Roberts. 2019. Enhancing cooking, sensory and nutritional quality of finger millet noodles through incorporation of hydrocolloids. *International Journal of Chemical Studies* 7(2):877–81.

ICRISAT. 2021. Millets, sorghum, and grain legumes: The smart foods of the future. https://www.icrisat.org/millets-sorghum-and-grain-legumes-the-smart-foods-of-the-future/. (accessed November 18, 2021).

Inyang, U. E., E. A. Daniel, and F. A. Bello. 2018. Production and quality evaluation of functional biscuits from whole wheat flour supplemented with *acha* (Fonio) and kidney bean flours. *Asian Journal of Agriculture and Food Sciences* 6(6):193–201.

Jyotsana, R., C. Soumya, S. Sarabhai, and P. Prabhasankar. 2016. Rheology, texture, quality characteristics and immunochemical validation of millet based gluten free muffins. *Journal of Food Measurement and Characterization* 10(4):762–72. doi:10.1007/s11694-016-9361-9.

Kane-Potaka, J., S. Anitha, S., T. Tsusaka, R. Botha, M. Budumuru, S. Upadhyay, P. Kumar, K. Mallesh, R. Hunasgi, A. K. Jalagam, and S. Nedumaran. 2021. Assessing millets and sorghum consumption behavior in urban India: A large-scale survey. *Frontiers in Sustainable Food Systems* 5. doi:10.3389/fsufs.2021.680777.

Karuppasamy, P., and M. R. Latha. 2020. Assessment and comparison study of millet bar with farmer practice through on farm trial (2018–19). *Journal of Pharmacognosy and Phytochemistry* 9(4):381–4. doi:10.22271/phyto.2020.v9.i4Sg.12130.

Karuppasamy, P., D. Malathi, P. Banumathi, N. Varadharaju, and K. Seetharaman. 2013. Evaluation of quality characteristics of bread from kodo, little and foxtail millets. *International Journal of Food and Nutritional Sciences* 2(2):32–6.

Kavitha, B., R. Vijayalakshmi, C. R. Poorna, I. M. Yalagala, and D. Sugasini. 2018. Nutritional evaluation and cell viability of formulated probiotic millet fruit bar. *Journal of Food and Nutritional Disorders* 7(2):1000246. doi:10.4172/2324-9323.1000246.

Khandelwal, P., R. S. Upendra, U. Kavana, and S. Sahithya. 2012. Preparation of blended low alcoholic beverages from under-utilized millets with zero waste processing methods. *International Journal of Fermented Foods* 1(1):77–86.

Krishnan, R., U. Dharmaraj, R. S. Manohar, and N. G. Malleshi. 2011. Quality characteristics of biscuits prepared from finger millet seed coat based composite flour. *Food Chemistry* 129(2):499–506.

Kulkarni, D. B., B. K. Sakhale, and R. F. Chavan. 2021. Studies on development of low gluten cookies from pearl millet and wheat flour. *Food Research* 5(4):114–9. doi:10.26656/fr.2017.5(4).028.

Kulkarni, S. S., A. D. Desai, R. C. Ranveer, and A. K. Sahoo. 2012. Development of nutrient rich noodles by supplementation with malted ragi flour. *International Food Research Journal* 19(1):309–13.

Kumar, A., P. Kumari, and M. Kumar. 2022. Role of millets in disease prevention and health promotion. In eds. R. B. Singh, S. Watanabe, and A. A. Isaza,*Functional Foods and Nutraceuticals in Metabolic and Non-Communicable Diseases*, 341–57.London, United Kingdon: Academic Press.

Kumar, A., A. Kaur, K. Gupta, Y. Gat, and V. Kumar. 2021. Assessment of germination time of finger millet for the value addition in functional foods. *Current Science* 120(2):406–13.

Kumar, A., V. Tomer, A. Kaur, K. Sharma, and S. Dimple. 2020. Fermented foods based on millets: Recent trends and opportunities. In G. K. Sharma, A. D. Semwal, and J. R. Xavier (Eds.), *Advances in Fermented Foods and Beverages* (pp. 191–228). New Delhi: New India Publishing Agency.

Kumar, A., A. Kaur, V. Tomer, P. Rasane, and K. Gupta. 2020a. Development of nutricereals and milk-based beverage: Process optimization and validation of improved nutritional properties. *Journal of Food Process Engineering* 43(1):e13025.

Kumar, A., A. Kaur, and V. Tomer. 2020b. Process optimization for the development of a synbiotic beverage based on lactic acid fermentation of nutricereals and milk-based beverage. *LWT-Food Science and Technology* 131:109774.

Kumar, P., C. Kaur, S. Sethi, and H. K. Jamb. 2020c. Effect of extruded finger millet on dough rheology and functional quality of pearl millet-based unleavened flatbread. *Cereal Chemistry* 97(5):991–8. doi:10.1002/cche.10321.

Kumar, A., V. Tomer, A. Kaur, V. Kumar, and K. Gupta. 2018. Millets: A solution to agrarian and nutritional challenges. *Agriculture and Food Security* 7(1):31. doi:10.1186/s40066-018-0183-3.

Kumar, A., V. Tomer, A. Kaur, and V. K. Joshi. 2015a. Synbiotics: A culinary art to creative health foods. *International Journal of Food and Fermentation Technology* 5(1):1–14.

Kumar, S., A. Singh, N. C. Shahi, K. Chand, and K. Gupta. 2015b. Optimization of substrate ratio for beer production from finger millet and barley. *International Journal of Agricultural and Biological Engineering* 8(2):110–20.

Kumar, A. 2013. Refinement of the technology of the traditional sur production in Himachal Pradesh (Doctoral dissertation, M. Sc. Thesis UHF, Nauni).

Li, Y., J. Lv, L. Wang, Y. Zhu, and R. Shen. 2020. Effects of millet bran dietary fiber and millet flour on dough development, steamed bread quality, and digestion *in vitro*. *Applied Sciences* 10(3):912. doi:10.3390/app10030912.

Man, S., A. Paucean, S. Muste, A. Pop, and E. A. Muresan. 2016. Quality evaluation of bread supplemented with millet (*Panicum miliaceum L.*) flour. *Bulletin of University of Agricultural Sciences and Veterinary Medicine Cluj-Napoca: Food Science and Technology* 73(2): 161–2.

Manley, D. 1998. *Biscuit, Cookie and Cracker Manufacturing Manuals: Manual 3: Biscuit Dough Piece Forming*. Cambridge: Elsevier.

Manning, T. S., and G. R. Gibson. 2004. Prebiotics: Best Practice and Research. *Clinical Gastroenterology* 18(2):287–98. doi:10.1016/j.bpg.2003.10.008.

Marak, N. R., C. C. Malemnganbi, C. R. Marak, and L. K. Mishra. 2019. Functional and antioxidant properties of cookies incorporated with foxtail millet and ginger powder. *Journal of Food Science and Technology* 56(11):5087–96. doi:10.1007/s13197-019-03981-6.

Mazlien, T. 2015. Whiskey review: Koval single barrel millet whiskey. https://thewhiskeywash.com/whiskey-styles/american-whiskey/whiskey-review-koval-single-barrel-millet-whiskey/ (accessed 22 December, 2021).

McWatters, K. H., J. B. Ouedraogo, A. V. Resurreccion, Y. C. Hung, and R. D. Phillips. 2003. Physical and sensory characteristics of sugar cookies containing mixtures of wheat, fonio (*Digitaria exilis*) and cowpea (*Vigna unguiculata*) flours. *International Journal of Food Science and Technology* 38(4):403–10. doi:10.1046/j.1365-2621.2003.00716.x.

Mohammed, M. I., A. I. Mustafa, and G. A. Osman. 2009. Evaluation of wheat breads supplemented with teff ('*Eragrostis tef* (ZUCC.') Trotter) grain flour. *Australian Journal of Crop Science* 3(4):207–12.

Mohana, H. K. 2007. Technology of ready to reconstitute pearl millet based fermented milk beverage. M. Tech. Thesis. Division of Dairy Technology. National Dairy Research Institute (I.C.A.R) Karnal - 132001 (Haryana), India.

Naseer, B., V. Sharma, S. Z. Hussain, and J. Bora. 2022. Development of Functional snack food from almond press cake and pearl millet flour. *Letters in Applied Nanobioscience* 11(1):3191–207. doi:10.33263/LIANBS111.31913207.

Patil, S. S., S. G. Rudra, E. Varghese, and C. Kaur. 2016. Effect of extruded finger millet (*Eleusine coracana* L.) on textural properties and sensory acceptability of composite bread. *Food Bioscience* 14:62–9. doi:10.1016/j.fbio.2016.04.001.

Rahman, R., S. Hiregoudar, M. Veeranagouda, C. T. Ramachandra, M. Kammar, U. Nidoni, and R. S. Roopa. 2015. Physico-chemical, textural and sensory properties of muffins fortified with wheat grass powder. *Karnataka Ournal of Agricultural Sciences* 28(1):79–82.

Rai, S., A. Kaur, and B. Singh. 2014. Quality characteristics of gluten-free cookies prepared from different flour combinations. *Journal of Food Science and Technology* 51(4):785–9. doi:10.1007/s13197-011-0547-1.

Ranasalva, N., and R. Visvanathan. 2014. Development of cookies and bread from cooked and fermented pearl millet flour. *African Journal of Food Science* 5(5):327. doi:10.4172/2157-7110.1000327.

Różyło, R. 2014. New potential in using millet-based yeast fermented leaven for composite wheat bread preparation. *Journal of Food and Nutrition Research* 53(3):240–50.

Saha, S., A. Gupta, S. R. K. Singh, N. Bharti, K. P. Singh, V. Mahajan, and H. S. Gupta. 2011. Compositional and varietal influence of finger millet flour on rheological properties of dough and quality of biscuit. *LWT – Food Science and Technology* 44(3):616–21.

Samuel, K. S., and N. Peerkhan. 2020. Pearl millet protein bar: Nutritional, organoleptic, textural characterization, and in-vitro protein and starch digestibility. *Journal of Food Science and Technology* 57(9):3467–73. doi:10.1007/s13197-020-04381-x.

Sehgal, A., and A. Kwatra. 2007. Use of pearl millet and green gram *flours in biscuits and their sensory and nutritional quality. Journal of Food Science and Technology* 44(5):536–8.

Seth, D., and G. Rajamanickam. 2012. Development of extruded snacks using soy, sorghum, millet and rice blend–A response surface methodology approach. *International Journal of Food Science and Technology* 47(7):1526–31. doi:10.1111/j.1365-2621.2012.03001.x.

Shukla, K., and S. Srivastava. 2014. Evaluation of finger millet incorporated noodles for nutritive value and glycemic index. *Journal of Food Science and Technology* 51(3):527–34. doi:10.1007/s13197-011-0530-x.

Silva, E. C. D., V. D. Santos Sobrinho, and M. P. Cereda. 2013. Stability of cassava flour-based food bars. *Food Science and Technology, Campinas* 3(1):192–8. doi:10.1590/S0101-20612013005000025.

Singh, K. P., and H. N. Mishra. 2014. Millet-wheat composite flours suitable for bread. *Journal of Ready to Eat Foods* 1(2):49–58.

Singh, K. P., H. N. Mishra, and S. Saha. 2012. Changes during accelerated storage in millet-wheat composite flours for bread. *Food and Bioprocess Technology* 5(5):2003–11.

Singh, R., and M. S. Nain. 2020. Nutrient analysis and acceptability of different ratio pearl millet (*Pennisetum glaccum* (L.) R. Br.) based biscuits. *Indian Journal of Agricultural Sciences* 90(2):428–30.

Sinha, R., and B. Sharma. 2017. Use of finger millet in cookies and their sensory and nutritional evaluation. *Asian Journal of Dairy and Food Research* 36(3):264–6.

Sobana, R. M. 2017. Quality evaluation of millet based composite sports bar. *International Journal of Food Science and Nutrition* 2(4):65–8.

Stefano, E., J. White, S. Seney, S. Hekmat, T. McDowell, M. Sumarah, and G. Reid. 2017. A novel millet-based probiotic fermented food for the developing world. *Nutrients* 9(5):529. doi:10.3390/nu9050529.

Sukumar, A., and K. A. Athmaselvi. 2019. Optimization of process parameters for the development of finger millet based multigrain extruded snack food fortified with banana powder using RSM. *Journal of Food Science and Technology* 56(2):705–12.

Suri, S., A. Dutta, N. C. Shahi, R. S. Raghuvanshi, A. Singh, and C. S. Chopra. 2020. Numerical optimization of process parameters of ready-to-eat (RTE) iron rich extruded snacks for anemic population. *LWT: Food Science and Technology* 134:110164. doi:10.1016/j.lwt.2020.110164.

Swami, S. B., N. J. Thakor, and U. N. Shinde. 2013. Effect of baking temperature and oil content on quality of finger millet flour cookies. *Journal of Agricultural Engineering* 50(4):32–8.

Teshome, E., Y. B. Tola, and A. Mohammed. 2017. Optimization of baking temperature, time and thickness for production of gluten free biscuits from Keyetena teff (*Eragrostis tef*) variety. *Journal of Food Process Technolnology* 8(5):1000675. doi:10.4172/2157-7110.1000675.

Thathola, A., and S. Srivastava. 2002. Physicochemical properties and nutritional traits of millet-based weaning food suitable for infants of the Kumaon hills, Northern India. *Asia Pacific Journal of Clinical Nutrition* 11(1):28–32.

United Nations Digital Library. 2021. International year of millets, 2023: Resolution/adopted by the general assembly. https://digitallibrary.un.org/record/3904090?ln=en. (accessed November 17, 2021).

Ushakumari, S. R., S. Latha, and N. G. Malleshi. 2004. The functional properties of popped, flaked, extruded and roller-dried foxtail millet (*Setaria italica*). *International Journal of Food Science and Technology* 39(9):907–15. doi:10.1111/j.1365-2621.2004.00850.x.

Xu, J., Y. Zhang, W. Wang, and Y. Li. 2020. Advanced properties of gluten-free cookies, cakes, and crackers: A review. *Trends in Food Science and Technology* 103:200–13. doi:10.1016/j.tifs.2020.07.017.

Yadav, G. P., C. G. Dalbhagat, and H. N. Mishra. 2021. Development of instant low glycemic rice using extrusion technology and its characterization. *Journal of Food Processing and Preservation* 45(12):e16077. doi:10.1111/jfpp.16077.

Zięć, G., H. Gambuś, M. Lukasiewicz, and F. Gambuś. 2021. Wheat bread fortification: The supplement of teff flour and chia seeds. *Applied Sciences* 11(11):5238. doi:10.3390/app11115238.

Chapter 15

Millet By-Products and Their Food and Non-Food Applications

15.1 INTRODUCTION

The harvesting, milling, and processing of the millets produce a variety of by-products like straw, husk, bran, germ, etc. The straw, husk, bran, and germ yield in various millet varieties ranges from 20 to 70 quintal/ha, 10 to 30 g/100 g, 6 to 35 g/100 g, and 6 to 17 g/100 g, respectively (Hassan et al. 2021; FAO 2021, Chapke et al. 2018). These by-products can be utilized in several ways. The straw and husk can be used as animal fodder; these can also be broken down into fermentable sugars by acid or alkali treatment and can be used for the production of bioethanol or biogas. The husk can also be used for the development of thermoplastic composites, bricks, and in the partial replacement of cement. The use of husk in plastics as fillers improves their tensile strength (Hammajam et al. 2014). The bran of millets is a rich source of nutrients (especially protein, fat, and minerals), phytochemicals (phenolics and flavonoids), and dietary fibre. The millet bran can be added to popular snack foods like biscuits, cookies, bread, etc. and can improve their fibre as well as phytochemical content. Dietary fibre exhibits cholesterol-lowering effects and helps to control blood sugar levels. Millet fibre is also rich in prebiotics which selectively stimulate the growth of probiotics in the gut. Millet phytochemicals have antioxidant, antiageing, anticholesterol, anticancer, and antidiabetic properties. Pearl and foxtail millet bran can be used for the extraction of oil. This oil is rich in unsaturated fatty acids and hence can be used for cooking. The pigments can also be extracted from the coloured brans and can be used as natural food colorants. This chapter discusses the various by-products generated from millets and the possible mechanisms of their utilization.

15.2 MAJOR WASTE GENERATED FROM MILLET

The major waste generated from the millets is straw/stover on harvesting, husk on threshing and cleaning of the grains, bran, and germ on milling. This section discusses the various by-products produced from millets and the ways they can be utilized.

15.2.1 Stover/Straw

The harvesting of millets produces large quantities of biomass. The highest stover yields of 6–7 t/ha are reported in pearl millet and finger millet. This is followed by proso millet (5–6 t/ha), kodo millet (3–4 t/ha), foxtail millet (2–4 t/ha), little millet, and barnyard millet (2–2.5 t/ha each), respectively (Chapke et al. 2018). The stover can be used in various ways. The major applications of millet straw along with novel methods for its utilization are discussed in the following sections.

15.2.1.1 Animal Fodder

The most common use of millet straw is its application as fodder. The straw yield in the pearl millet varieties varies with the purpose for which pearl millet is grown and the application of fertilizer. Choudhary and Prabhu (2014) reported a stover yield of 8.57 t/ha in pearl millet variety AVKB-19 when grown for dual purpose (fodder + grain) and 10.91 t/ha when grown only for grain purposes. This yield is equivalent to maize (7.41–9.88 t/ha) stover (Chaudhary et al. 2012) and higher than other common cereals like wheat and paddy which produce a biomass of 2.51 and 6.78–7.59 t/ha, respectively (Chahal and Chhabra 2014). The yield of straw was increased to 10.83 t/ha in the dual-purpose crop on the application of a recommended dose of fertilizer at an increased level of 125%. This study reported a protein content of 3.6–4.0% in pearl millet stover. Finger millet stover contains nearly 61% digestible nutrients which makes it good fodder for ruminants. In a study conducted in Kenya on the fodder yield of three finger millet varieties (U-15, P-224, and a local check variety) it was reported that the yield of finger millet dry stover is affected by variety, region, and season of cultivation (Wafula et al. 2017). The stover yield was most responsive to phosphorus (P_2O_5) application and the highest yield of dry stover (11,616 kg/ha) was obtained for P_2O_5 application of 25 kg/ha in most of the places. However, in some regions, a P_2O_5 application of 37.5 kg/ha was recommended. The highest protein content reported in dried finger millet stover was 11%. The fodder value of millet straw is also provided in Table 15.1.

Foxtail millet produces nearly 3.21–8.64 t/ha of biomass. The use of its straw as animal fodder is limited as plants grown under high-stress conditions accumulate excess nitrate. Due to the presence of glucoside setarian, foxtail millet hay acts as a laxative to horses and can damage their liver, kidney, and bones (Sheahan 2014). Barnyard millet produces nearly 5 t/ha of dry fodder. Its straw contains up to 61% total digestible nutrients and more proteins and calcium in comparison to rice, oat, and timothy straw and is considered good fodder. Among the Japanese and Indian barnyard millet varieties, Japanese barnyard millet has higher fodder production potential (Sood et al. 2015). Proso millet does not produce quality forage as it has a low leaf-to-stem ratio. However, the proso varieties grown for grain retain most of the leaves at the time of harvest which produces straw of high feed value. This straw has been used as fodder in the United States of America (USA) for wintering dry and pregnant beef cattle (Nebraska-Lincoln Extension Educational Program and USDA 2008). In India, proso millet straw is considered poor fodder and it is often used for the bedding of cattle (Feedipedia 2020). Kodo millet straw does not make good fodder and has been reported to cause higher respiratory distress in dairy cows on feeding for one week (Kalaiselvan et al. 2020). Little millet straw can also be used for fodder (Seetharam and Gowda 2007) but detailed studies on the fodder application of little millet straw are scarce. The stover left after the grain harvest of teff is also fit for fodder purposes. It has been extensively used as fodder in Ethiopia, especially during the dry season. The yield of teff forage has been reported in the range of 11.11–23.22 t/ha in Nevada, USA, and 4.052–6.103 t/ha in Cheonam, South Korea. This is a good source of crude protein, i.e. 8–15%. The content of acid detergent fibre and natural detergent fibre has been reported in the range of 33–42% and 58–65%, respectively (Habte et

TABLE 15.1 COMPOSITION AND NUTRITIONAL VALUE OF STRAW OBTAINED FROM VARIOUS MILLETS

Parameter	Pearl millet straw	Finger millet straw	Proso millet straw	Kodo millet straw	Teff straw	Fonio straw
Dry matter (DM) (%)	93.1	93.1	94.5	–	91.6	94.1
Crude protein (%)	5.2	5.0	3.7	3.5	4.1	5.6
Crude fibre (%)	41.8	36.5	42.6	34.3	35.8	32.7
Neutral detergent fibre (%)	80.0	73.7	80.6	–	70.9	67.2
Acid detergent fibre (%)	52.9	45.9	48.3	–	41.7	44.7
Lignin (%)	10.7	5.1	6.9	–	5.8	6.6
Ether extract (%)	0.7	0.8	1.2	1.5	1.9	-----
Ash (%)	8.6	9.4	10.4	12.3	7.9	6.6
Gross energy (MJ/kg DM)	17.7	17.7	17.7	17.1	18.1	18.2
Calcium (g/kg DM)	2.5	8.4	–	–	3.0	2.9
Phosphorus (g/kg DM)	1.5	0.9	0.5	–	1.4	0.7
Zinc (mg/kg DM)	26	25	–	–	25	23
Copper (mg/kg DM)	7	7	–	–	10	4
Iron (mg/kg DM)	667	243	–	–	176	2903
Organic matter digestibility (ruminants) (%)	47.7	57.7	54.8	–	57	51.9
Energy digestibility (ruminants)	43.7	53.9	50.6	–	53.6	48.1
Digestible energy for ruminants	7.7	9.5	9.0	–	9.7	____
Metabolizable energy, ruminants	6.3	7.7	7.2	–	7.9	7.2
Nitrogen digestibility, ruminants	7.0	16.0	–	–	15.6	60

Source: Heuzé et al. 2019a, Heuzé et al. 2019b, Heuzé et al. 2019c, Tran 2017, Heuzé et al. 2015a, Heuzé et al. 2015b.

al. 2019). The fonio chaff and straw are also used as animal fodder (World Intellectual Property Organization 2022).

The digestibility of the stover/straw is not good and hence it is not considered good fodder. The digestibility of millet straw can be enhanced by physical or chemical methods. The physical methods, like grinding, break down the lignin bonds, increase the surface area of straw, and make celluloses and hemicelluloses more exposed to microbial fermentation. In a study on pearl millet straw, Mohammed et al. (2016) found that chopping the millet straw into relatively long sections of 1–10 cm and grinding it to 2–5 mm increases its digestibility. In the same study, the effect of urea treatment on the digestibility of straw was also studied. One hundred kilograms of ground millet straw was treated with a solution containing 4 kg of urea and 1% salt dissolved in

50 litres of water. The mix was packed into airtight bags and was allowed to ferment for 3 weeks at room temperature. The urea-treated straw had more protein (14.56–14.61%) and digestibility (64–65%) compared to untreated straw, i.e. 7.8–11.9% and 52–56%, respectively. A similar treatment is recommended by the National Dairy Research Institute, Karnal, India, to improve the quality of poor roughages (Indian Council for Agricultural Research 2020). Alemu et al. (2020) studied the effect of the treatment of urea molasses (UM) and effective microbiome (EM) on the nutritive value of finger millet straw (FMS) and the growth performance of Washera sheep in North-Western Ethiopia. The sheep were given four different diets, i.e. untreated FMS + 150 g wheat bran (WB) (T1), untreated FMS + 150 g WB + 150 g Noug seed cake (NSC) (T2), UM-treated FMS + 150 g WB + 150 g NSC (T3), and EM-treated FMS + 150 g WB + 150 g NSC (T4). In the UM treatment, 100 kg of straw was treated with a solution containing 5 kg of urea dissolved in 80 litres of water and 10 litres of molasses. For EM-treated FMS, 50 kg of straw was wetted by spraying it with 18 litres of chlorine-free water, 1 litre of EM, and 1 litre of molasses. After the treatment, the mix was transferred into two separate pits of 2 metres in length, 1 metres in height, and 1 metres in width lined with polyethylene plastic sheets. The bags were packed well to exclude air and were covered for 30 days. It was found in the study that UM- and EM-treated FMS had better digestibility in comparison to native straw. The total dry matter intake of straw was increased in UM-treated FMS to 501.44 ± 11.3 g/day in comparison to the control (434.8 ± 11.3 g/day). The percent digestibility of crude protein, neutral detergent fibre, and acid detergent fibre was also increased in UM-treated FMS (77.4, 73.2, and 64.2%) and EM-treated FMS (83.20, 67.4, and 71.4%) in comparison to control (22.8, 46.4, and 50.4%). These treatments increased the overall nutritional value of the straw and helped the lambs to gain more body weight in comparison to those lambs who were given a controlled diet.

15.2.1.2 Fermentable Sugars and Bioethanol

Packiam and Karthikeyan (2018) studied the effect of the chemical pretreatment of pearl millet biomass with different concentrations of orthophosphoric acid (4, 8, 12, and 16%), alkaline hydrogen peroxide (1.2, 1.8, 2.4, and 3%), and lime (0.7, 0.9, and 1.1%) for the recovery of fermentable sugars. The millet biomass was dried to a moisture content of 6–9% and the size of the straw was reduced to 0.2–10 cm. The solid loading of the dried biomass was done at three concentrations, i.e. 7.5, 10, and 12.5% at different temperatures and time intervals i.e. 100 and 121°C for 60, 120, and 180 minutes for the acid treatment and 80, 100, and 121°C for 60, 90, and 120 minutes for lime and hydrogen peroxide. For lime and hydrogen peroxide treatment, the pH was adjusted to 11.5. The pre-optimized conditions of total solids and chemical concentrations were then optimized at temperatures 140, 150, and 160°C and time intervals of 10, 20, and 30 minutes. The sugar was measured using the dinitrosalicylic acid assay (DNSA) method and, based on the results, modelling was developed for the optimization of various treatments. The highest sugar yield (41.80 g sugar/100 g biomass) was obtained with the pretreatment of biomass with 16% orthophosphoric acid, 12.5% solid loading, and heat treatment of 121°C for 180 minutes. Tekaligne et al. (2015) optimized the process for the production of bioethanol from finger millet straw. The process was composed of three major steps, i.e. pretreatment, hydrolysis of pretreated samples, and fermentation. In the pretreatment 50 g of finger millet straw was treated with 100 ml of 0.5% sulphuric acid to remove lignin, reduce cellulose crystallinity, and increase the porosity of the materials. The pretreated samples were further heated at 125–130°C for 1 h under a pressure of 25 psi. The hydrolysis conditions were optimized for time (1–5 days), temperature (25, 30, 35, 40, and 45°C), acid concentration (0–4%), and biomass concentration (6.25, 7.14, 8.33, 10.00, 12.5, and

16.61% w/v). The highest yield of sugar (68.72–70.65%) was obtained after 4 days for 10% biomass, 2% acid concentration, and 35°C incubation temperature. The hydrolysate was further fermented with *Saccharomyces cerevisiae*. The fermentation conditions were optimized for pH (4, 4.5, 5, 5.5, 6.0, and 6.5), yeast concentration (2, 3, 4, 5, and 6 g/L), the incubation period (2, 3, 4, 5, 6, and 7 days), and temperature (25, 27.5, 30, 32.5, 37.5, and 40°C). The highest ethanol concentration (6.92–7.28%) was obtained at pH 6.0, yeast concentration of 6 g/L, and incubation temperature of 32.5°C after 4 days of fermentation. The researchers also reported that finger millet straw can produce ethanol content comparable to ground hulls (6.2%), rice husks (5.5%), and mango juice (7–8.5%) and hence can be a good substitute for bioethanol production for these crops and food crops.

15.2.2 Husk

Husk is one of the most abundant by-products of millet cultivation with no or limited applications. The postharvest handling of the husk can be a challenge as it requires labour and results in additional costs to farmers. The burning of the husk further causes environmental hazards. The millet husk can be used as a valuable raw material in the development of composites and bricks or as a replacement for cement. Some of the major studies on the standardization of the processes for the utilization of millet husk are discussed in the following sections.

15.2.2.1 Biogas

Abba et al. (2014) studied the scope of the utilization of millet husk in biogas production. The millet husk was collected, air-dried, and then the final drying was carried at 110°C in an oven. The dried husk was ground to a fine powder, sieved for uniform size, and stored in plastic bags for further study. For the biogas production, 400 g of dried millet husk powder was mixed with 2000 cm^3 of water, and the slurry was transferred into a digester. The study was further divided into two parts, A and B. In part A, the slurry prepared by mixing dried millet husk powder and water was directly fermented in a digester, while in study B, the millet husk powder was mixed with cow dung in the ratio 8:1 before the addition of water and fermentation. The millet husk was reported to be a good feedstock for the generation of biogas. Digester A was reported to generate 5400 cm^3 of biogas and 4200 cm^3 of purified methane, while digester B produced 7333.33 cm^3 of biogas and 5733.33 cm^3 of purified methane. Seeding of the husk with cow dung increased biogas production. In a comparative study on the quantitative and qualitative efficiency of various materials, i.e. cow dung, millet husk, rice husk, sawdust, and paper waste, for biogas production it was found that the cow dung produced the highest amount of biogas followed by millet husk, rice husk, sawdust, and paper waste, respectively (Bagudo et al. 2011). The highest percentage of methane was in the biogas produced using paper waste followed by sawdust, rice husk, and cow dung. The millet husk had the highest carbon dioxide content and the lowest hydrogen sulphide content (Table 15.2).

15.2.2.2 Filler in Plastics

The use of pearl millet husk as a filler in high-density polyethylene (HDPE) was studied by Hammajam et al. (2014). In this study, millet husk fibres (MHF) were treated with 4% w/v sodium hydroxide (NaOH) solution for 1 h at 25°C. After the treatment, the MHF was cleaned with distilled water to remove NaOH completely. The cleaned fibres were dried at 105°C for 48 h and were ground to a size of 250 μm. The powdered MHF was blended with HDPE in concentrations of 10, 20, 30, and 40 wt%. To this blend, 2% of the coupling agent, i.e. maleic anhydride polyethylene,

TABLE 15.2 THE EFFECT OF VARIOUS SUBSTRATES ON THE QUANTITY AND QUALITY OF BIOGAS

Substrate	Total biogas (cm³)	Methane (%)	Carbon dioxide (%)	Hydrogen sulphide (%)
Cow dung	8545	66.00	33.00	1.00
Millet husk	6525	58.08	40.72	1.00
Rice husk	1386	64.97	33.00	2.00
Saw dust	974	68.79	29.65	1.53
Paper waste	476	72.59	24.27	3.14

Source: Adapted from Bagudo et al. 2011.

was also added to enhance adhesion. The moulding was carried out at 25 kN compression force for 10 minutes at a temperature of 165–170°C. The loading of HDPE with MHF produced heterogeneous results. The loading of up to 10% of MHF increased the tensile strength of HPPE in comparison to 100% HDPE. A decrease in tensile strength was observed at concentrations of 20 and 30%. A loss of strength was also observed at a concentration of 30% due to the presence of voids. However, the blending with 40% MHF increased the tensile strength, rigidity, and flexible moduli. The researchers concluded that MHF can be successfully used as a filler in HDPE to improve its properties. Hammajam et al. (2014) also studied the effect of filling 40 wt% millet husk on HDPE thermoplastic composites. It was found in the study that the addition of millet husk improved the bending strength and modulus. However, the impact strength decreased considerably with an increase in the loading of millet husk.

15.2.2.3 Soil Blocks and Concrete

Abdulwahab et al. (2017) studied the effect of millet husk ash (MHA) and bitumen on the strength and durability properties of lateritic soil blocks. It was found in the study that the addition of 30% MHA to the lateritic soil produced blocks with an optimum compressive strength of 10.8 N/mm². The development of blocks using 50% MHA blended with 14% bitumen solution produced watertight bricks. The replacement of 5% of cement with MHA in the cement concrete is reported to increase the compressive strength of the cement and can also be used to improve the hardening properties of concrete (Bheel et al. 2018).

15.2.3 Bran

Millets are small-seeded grains and a major part of the grain is constituted by the bran. The bran yield of various millets is provided in Table 15.3. The bran is a rich source of nutrients, phytochemicals, fibre, minerals, and oil. The physicochemical and phytochemical composition of millet bran has been studied by various researchers. The following section discusses in detail the composition and applications of millet brans.

15.2.3.1 Nutritional and Phytochemical Composition

Jha et al. (2015) separated the bran of pearl millet variety CO(CU)$_9$ and found that the yield of bran was 6.5–7 g/100 g in raw grains and 7.5–8 g/100 g in germinated steamed grains. The researchers found that the bran had more fat (11.75 g/100 g) in comparison to the endosperm (5.40 g/100 g).

TABLE 15.3 BRAN CONTENT IN DIFFERENT MILLETS

Millet type	Bran content (%)	References
Pearl millet	5.5–8	Satyavathi et al. 2017; Malathi et al. 2014
Foxtail millet	1.5–6	Liang and Liang 2019; Malathi et al. 2014
Proso millet	14.4–20.8	Lorenz and Howamg 1986
Finger millet	11	Malathi et al. 2014
Barnyard millet	6	
Kodo millet	7.5	
Little millet	6	

The bran also had high ash (2.51 g/100 g) content in comparison to endosperm (1.36 g/100 g). The bran layer was rich in polyphenols (1.42 g/100 g), flavonoids (0.18 g/100 g), phytic acid (1.02 g/100 g), dietary fibre (soluble – 1.63 g/100 g, insoluble fibre – 63.52 g/100 g), iron (25.71 mg/100 g), and zinc (6.56 mg/100 g). The bran of finger millet is also known as seed coat matter (SCM) and its content ranges from 11 to 26%. It can be obtained by milling, malting, and decortication. It is rich in dietary fibre, minerals, phenolics, and vitamins. The phenolic content is highest in the outer aleurone layer, testa, and pericarp (Chandra et al. 2016). Krishnan et al. (2011) studied the yield and nutritional and phytochemical composition of SCM obtained from native, malted, and hydrothermal-treated finger millet grains. Among the three treatments, SCM yield ranged from 13 to 26 g/100 g. The highest yield was obtained for malted finger millet (26.0 ± 3.00 g/100 g) and the lowest was obtained in hydrothermal treatment (13.0 ± 1.0 g/100 g). It was also found in the study that finger millet SCM is a rich source of protein (9.5–12.2 g/100 g), total dietary fibre (39.6–48.8 g/100 g), and ash (4.3–5.1 g/100 g). The content of polyphenols (3.3–4.8 g/100 g), calcium (707–864 mg/100 g), iron (5.5–7.5 mg/100 g), and zinc (2.2–2.7 mg/100 g) was higher in bran in comparison to the whole finger millet grain (Kumar et al. 2018). Amadou et al. (2011) studied the total phenolic content and antioxidant activity of distilled water and ethanol (30, 50, and 70%) extracts of defatted foxtail millet bran. The total phenolic content in the various extracts ranged from 21.49 to 29.39 mg GAE/100 g extract. The highest inhibitory concentrations (IC_{50}) values for 2, 2-diphenyl-1-picrylhydrazyl (DPPH) were obtained for the aqueous extracts (3.118 mg/ml), followed by 70, 30, and 50% ethanol extracts, i.e. 2.12, 0.652, and 0.131 mg/ml, respectively. The highest IC_{50} values for 2,2'-azino-bis(3-ethylbenzothiazoline-6-sulphonic acid) (ABTS) were obtained for 70% ethanol extract (1.88 mg/ml), followed by 50% (1.77 mg/ml) and 30% (1.03 mg/ml) ethanol and aqueous extract (0.80 mg/ml). The highest IC_{50} for superoxide was obtained for aqueous extract (2.07 mg/ml). The IC_{50} values for superoxide for the 70, 50, and 30% ethanol extracts were 1.06, 0.83, and 0.16 mg/ml, respectively. The difference in the phytochemical composition of the whole and dehulled proso millet was studied by Kalam Azad et al. (2019). The total phenols and flavonoid content in the whole barnyard millet flour was 295 ± 2.34 mg/100 g ferulic acid equivalent (FAE) and 183 ± 3.57 mg/100 g rutin equivalent (RE), respectively. This content was 43.38 and 15.84% more than the total phenols (167 ± 1.23 mg GAE/100 g) and total flavonoid content (154 ± 4.53 mg RE/100 g) of dehulled proso millet. The levels of syringic acid, gallic acid, 4-hydroxy benzoic acid, ferulic acid, sinapic acid, and catechin in the whole millet flour were 11.27, 9.37, 9.11, 6.87, 68.63, and 9.32 µg/100 g, respectively. The level of all these phenolic acids except sinapic acid was higher in whole proso millet in comparison to dehulled millet which had 5.34, 11.1, 9.11, 6.87, 38.36, and 12.87 µg/100 g of syringic, gallic, 4-hydroxy benzoic, ferulic, sinapic acid, and catechin, respectively. Barnyard millet bran is also a good source of minerals and phytochemicals. The unpolished grains of barnyard millet are reported to have significantly higher total dietary fibre (40.14%) and minerals (80%) and phenolic acids (8.68%) compared to polished grains (Rajesawari and Priyadarshini).Sarma et al. (2017) reported carbohydrate, protein, fat, total dietary fibre, and ash content in kodo millet bran to be 79.84, 4.92, 2.83, 48.42, and 5.55, respectively, at a moisture content of 7.07%. The phytochemical and mineral content of the bran of ten little millet varieties grown in Dharwad, Karnataka, India, was studied by Kundgol et al. (2014). In this study, the kodo bran was reported to have 133.66–248 and 2.49–4.80 mg/100 g of total polyphenols and α-tocopherol, respectively. The percent DPPH radical scavenging activity of millet bran ranged from 27.85 to 33.89%. The content of zinc, copper, manganese, and iron in the bran samples ranged from 0.14 to 0.60, 0.38 to 0.57, 0.25 to 0.47, and 2.58 to 4.01 mg/100 g, respectively. The phytochemical composition of the defatted kodo and little millet bran was studied by Chandrasekara et al. (2012). The authors reported the total phenolic content in the kodo millet bran to be 112 ± 1.37 µmol ferulic acid equivalent (FAE)/g defatted meal. The DPPH radical scavenging activity, hydroxyl radical scavenging activity, and hydrogen peroxide scavenging activity of the bran were 127 ± 1.62, 133 ± 23, and 84.0 ± 5.02 µmol FAE/g defatted meal, respectively. The oxygen radical absorption capacity and superoxide radical scavenging activity were 226 ± 1.54 µmol Trolox equivalent/g defatted meal and 0.90 ± 0.10 mmol FAE/g defatted meal. The total phenolic content of little millet hull was reported to be 26.5 ± 1.51 µmol FAE/g defatted meal. The DPPH radical scavenging activity, hydroxyl radical scavenging activity, and hydrogen peroxide scavenging activity of the little millet hull were 17.9 ± 1.09, 85.9 ± 4.47, and 77.4 ± 1.68 µmol FAE/g defatted meal, respectively. The oxygen radical absorption capacity and superoxide radical scavenging activity were 106 ± 2.33 µmol Trolox equivalent/g defatted meal and 0.55 ± 0.08 mmol FAE/g defatted meal. Teff grain is very small and is generally made into whole-grain flour because of its small size (Gebremariam et al. 2014). Information on the nutritional and phytochemical composition of fonio is also scarce. The phytochemical composition of various millet brans is also given in Table 15.4.

The dietary fibre and polyphenols from the finger millet SCM also exhibit blood glucose- and cholesterol-lowering, antimicrobial, antioxidant, and albumin glycation properties. SCM

TABLE 15.4 PHYTOCHEMICAL CONTENT OF MILLET BRANS (PER G DEFATTED MEAL)

Parameter	Millet type					
	Pearl millet	Foxtail millet	Proso millet	Finger millet	Kodo millet	Little millet
Total phenolic content (µmol FAE)	34.3 ± 1.69	22.8 ± 0.88	15.9 ± 0.5	19.1 ± 0.29	112 ± 1.37	26.5 ± 1.51
DPPH radical scavenging activity (µmol FAE)	38.3 ± 1.07	11.9 ± 0.16	14.9 ± 0.62	17.9 ± 0.40	127 ± 1.62	17.9 ± 1.09
Hydroxyl scavenging activity (µmol FAE)	196 ± 8.26	39.1 ± 6.57	18.1 ± 0.66	183 ± 29.6	133 ± 23	85.9 ± 4.47
Hydrogen peroxide scavenging activity (µmol FAE)	130 ± 0.83	42.1 ± 1.05	50 ± 0.13	131 ± 5.86	84 ± 5.02	77.4 ± 1.68
Oxygen radical absorption capacity µmol Trolox equivalent)	219 ± 7.35	234 ± 0.74	104 ± 4.46	132 ± 8.31	226 ± 1.54	106 ± 2.33
Superoxide radical scavenging activity (µmol FAE)	3.20 ± 0.20	0.81 ± 0.02	0.80 ± 0.03	0.44 ± 0.02	0.90 ± 0.10	0.55 ± 0.08

Source: Chandrasekara et al. 2012.

polyphenols control postprandial hyperglycaemia by inhibiting intestinal α-glucosidase and pancreatic amylase. This extract is also effective against *Bacillius cereus* and *Aspergillus niger* and inhibits fructose-induced albumin glycation (Chandra et al. 2016). However, the mineral bioavailability of the SCM is limited because of the presence of phytates. The fermentation of SCM for 24 hours with *Lactobacillus pentosus* CFR3 has been reported to reduce the phytate content by 56.70, 66.65, and 87.75% in native, malted, and hydrothermally treated SCM, respectively. Fermented SCM from native, malted, and hydrothermally treated SCM had a bioavailability of 28.40, 34.57, and 12.10%, respectively (Amritha et al. 2018).

15.2.3.2 Bran Oil

Millets are small-seeded grains and bran constitutes a large fraction (6–26%) of the grain. Shi et al. (2015) studied the effect of solvent extraction (SE), supercritical carbon dioxide extraction (SCE), and subcritical propane extraction (SPE) on foxtail millet bran oil and analyzed the yield, physicochemical properties, fatty acid profile, oil oxidative stability, and tocopherol composition. The yield of the oil was 17.14, 19.65, and 21.79%, respectively, in SE, SCE, and SPE. The oil had a density of 0.89–0.92 g/m^3, a refractive index of 1.45–1.48, a saponification value of 170.95–188.10 mg KOH/g oil, and a peroxide value of 1.40–7.05 meq of active oxygen/kg oil. All these values were similar to those of other vegetable-based edible oils. The peroxide value is also below the maximum permissible limit of 10 meq of active oxygen/kg oil (Codex Alimentarius 2019). The peroxide value is largely affected by the method of extraction. The highest peroxide value was obtained for SCF (7.05 meq/kg) and the lowest was obtained for SPE (1.04 meq/kg). Foxtail millet bran oil is also rich in linoleic (67.96–68.05%) and linolenic acid (2.19–2.21%). The oil obtained from all of the methods was rich in α and β tocopherol, i.e. 8.07–12.98 and 71.88–89.47 mg/100 g, respectively. Amadou et al. (2011) reported the content of monounsaturated, polyunsaturated, and saturated fatty acids in foxtail millet bran oil to be 87, 4, and 9%, respectively. The content of monounsaturated fatty acids in foxtail millet bran oil is higher than in common cooking oils such as soybean (24%), olive (77%), and rice bran (47%) oil. The content of saturated fatty acids is also lower compared to these oils. Devittori et al. (2000) studied the process of oil extraction from proso millet bran using soxhlet and supercritical carbon dioxide extraction. In the soxhlet apparatus petroleum ether was used as a solvent and the extraction was carried out at a temperature between 40 and 60°C. In the supercritical carbon dioxide extraction, the solvent requirement was 20–30 kg CO2 per kg of bran. The complete de-oiling of millet bran pellets was achieved within 200 to 500 minutes under 300 bar at 40°C with a specific solvent flow of 2–10/h. On comparison of the oils extracted using the two methods it was found that the amount and profile of fatty acids, tocopherols, and peroxides were similar in both oils; however, there was a difference in phospholipids, metals, and waxes content. The supercritical extracted oil was free of phospholipids and waxes. The oil obtained from both methods was also a good source of unsaturated fatty acids like linolenic (64%) and oleic acid (23%), and the content of free fatty acids was approximately 12%. The contents of unsaponifiable matter, peroxides, and tocopherols were in the range of 5.1–5.9%, 3–6 meq O$_2$/kg, and 350–400 mg/kg, respectively.

15.2.3.3 Bran Fibre

Dietary fibre can be purified from millet bran and it can be further added to foods to lower their glycaemic index. Li et al. (2020) purified the fibre from millet (millet type not mentioned in the study) by removing the fat with hexyl hydride and the starch and protein by sequential enzyme digestion with α-amylase, glucoamylase, and neutral amylase. After the removal of fat and the hydrolysis of starch and protein, the enzymes were deactivated by keeping the bran in a boiling water bath for 10 minutes and the dietary fibre was purified by centrifugation at 3500 rounds/minute for 20 minutes. This was dried overnight in a vacuum oven at 60°C. Li et al. (2020) further added the purified millet bran dietary fibre (0, 2, 4, 6, 8, and 10%) and steamed millet flour (25%) to wheat flour (75–65%) and studied the effect of their addition on dough development quality, the textural and sensorial properties of steamed bread, and *in-vitro* digestibility. The addition of millet flour and dietary fibre to the wheat flour decreased the water absorption capacity, dough development time, dough stability time, and farinograph quality number, and increased the softening degree of the mixed dough. An increase in hardness and a decrease in the elasticity and sensory acceptability of steamed bread were also observed with an increase in millet dietary fibre content. The increase in dietary fibre content also increased the content of rapidly digested and slowly digested starch. It also significantly decreased the hydrolysis rate, hydrolysis index, and glycaemic index of the bread. A gradual decrease in protein digestibility was also observed. It was concluded from the study that the addition of 2% millet dietary fibre produced bread with good sensory properties and a medium glycaemic index. Krishnan et al. (2011) also utilized the seed coat-based composite flour for the development of biscuits. It was concluded from the study that 10% of SCM from native and hydrothermally processed millet and 20% from malted millet could be used to produce sensorially acceptable biscuits.

15.2.4 Germ

15.2.4.1 Germ Oil
Pearl millet germ is a rich source of lipids. On average, the lipid content of pearl millet varies from 4.5 to 5.5% (Kumar et al. 2018). This oil is rich in both mono- (27.57%) and polyunsaturated (49.65%) fatty acids and the total share of unsaturated fatty acids in the pearl millet germ oil is 77.22%. The ratio of unsaturated to saturated fatty acids in pearl millet germ oil is 3.39. This value is higher than the standard value of 2.5 for cooking oils which are considered to have good nutritional value (Salma et al. 2020).

15.2.5 Spent Grains
The malting behaviour of the finger millet is only second to barley and it has been used for the preparation of local beers like *sur*, *channg*, and *kodo ko jannr* in India and *pombe* in Tanzania (Joshi et al. 2015; Ray et al. 2016; Kubo 2016). Koval Breweries Ltd., Chicago, also produces a millet-based single barrel whiskey with the name Koval (Mazlien 2015). The production of cereal malt-based drinks also yields spent grain residue (Kumar et al. 2020). This residue is reported to be high in nutrients. Exclusive studies on the composition of the millet-based brewer's spent grains (BSG) are not available; however, a recent review has reported the content of water, proteins, lipids, cellulose, and ash in BSG in the range of 75–80, 19–30, 10, 12–25, and 2–5%, respectively. The spent grains are also reported to be high in calcium, choline, niacin, pantothenic acid, and riboflavin, i.e. 1049, 1800, 44, 8.5, and 1.5 ppm, respectively (Karlović et al. 2020). Due to its high nutritional content, the BSG is generally used as animal feed. In recent decades, the spent grains have also found use as a medium for the growth of microorganisms, the extraction of phenolic compounds, the adsorption of heavy metals, and the production of bioethanol and microbial gums (Aliyu and Bala 2011). The fortification of the fresh or dried spent grain flour in human food products like biscuits and cookies is also reported.

15.2.5.2 Biscuits/Cookies

Guo et al. (2014) reported that good quality biscuits can be prepared by substituting wheat flour with 10% dried BSG of particle size 0.16 mm and adding 12% salad oil, 20–25% syrup, and 0.2% leaving agent. Petrović et al. (2017) studied the effect of the addition of various concentrations, i.e. 15, 25, and 50%, of fresh non-dried and non-milled brewer's spent grains to the wheat flour on the colour, microbial stability, and sensory characteristics of cookies. In this study, it was reported that the addition of 25% of brewer's yeast spent grain produced the cookies with the best sensory characteristics in terms of appearance, flavour, and grittiness. The cookies were microbiologically stable as they were free from *Enterobacteriaceae*, *Escherichia coli*, *Clostridium* spp., yeasts, and moulds.

15.2.6 Non-Food Grade Grains

15.2.6.1 Bioethanol

Millet grains are rich in carbohydrates and these can also be fermented into bioethanol if not used for food purposes. Wu et al. (2006) studied the potential of four pearl millet varieties (Tifgrain 102, 04F-303, 04F-106, and 04F-2304) for the preparation of bioethanol. The grain samples were ground to a fine powder (2 mm) in a cyclone sample mill and the bioethanol was produced using three major steps, i.e. liquefaction, saccharification, and fermentation. Different concentrations of pearl millet (20, 25, 30, and 35 g db) were taken in 250 ml volumetric flasks and an aliquot of 100 ml fermentation solution (3.0 g peptone, 1.0 g potassium dihydrogen phosphate, and 1.0 g ammonium sulphate) was added to each of the flasks. To this, high-temperature α-amylase was added at the rate of 3 Kilo Novo Unit/g of starch. The flasks were then kept for 45 minutes in a water bath shaker moving at 100–120 rpm and maintained at 95 °C. The temperature of the content was further reduced to 80 °C and the gelatinized and partially liquefied grains were further liquefied by adding liquozyme (3 KNU/g of starch) to each flask and the temperature was further maintained at 80 °C for 30 minutes. The temperature was then reduced to 60 °C and glucoamylase was added into each flask at 150 AGU/g of starch for saccharification. The flasks were maintained at this temperature for 30 minutes in a water bath shaker running at a speed of 120 rpm. The flask was removed and cooled to 30 °C and the pH of the mashes was adjusted to between 4.4 and 4.3 with 2N HCl before incubation. The prepared mashes were inoculated with 5 ml of yeast culture (*Saccharomyces cerevisiae* strain ATCC 24860). The fermentation was carried out for 72 h at 30 °C and a shaking rate of 150 rpm. The ethanol produced in various flasks containing 20, 25, 30, and 35 g db of pearl millet flour was 9, 11, 13–14, and 16–17%, respectively. The ethanol content was reported to be proportional to the starch content of pearl millet flour. The varieties used in this study had similar starch content so there was no varietal effect on the bioethanol production. Proso millet varieties were also studied for bioethanol production (Rose and Santra 2013). In this study, it was found that waxy proso millet varieties exhibit higher fermentation efficiencies than non-waxy proso millet varieties. The study concluded that highly fermentable lines of proso millet can be used as a promising feedstock for bioethanol production.

15.2.7 Prebiotics

Millets are a good source of prebiotics. The major prebiotics in millets are arabinoxylans, β-glucans, fructooligosaccharides, inulin, and hemicelluloses (Table 15.5).

Banerjee et al. (2017) reported that the maximum inulin yield (0.47 g/g) can be obtained from pearl millet at the optimized conditions of 70 °C, 0.8 M of HCl, and heating of 60 minutes. The

TABLE 15.5 PREBIOTICS IN MILLETS AND THEIR HEALTH BENEFITS

Millet type	Prebiotic components	Interrelation with probiotics	Other health benefits
Pearl millet	Arabinoxylans, arabinogalactan, β-glucan, inulin, fructooligosaccharides	Long chain water extractable arabinoxylans stimulate the growth of *Bifidobacterium longum* (Van den Abbeele et al. 2011)	Long-term administration of arabinoxylans in diabetic patients can restore the glucose and insulin response
Finger millet	Arabinoxylans, beta-glucans, fructans, inulin, xylooligosaccharides		
Proso millet	Arabinoxylans, arabinogalactan, β-glucan	A mixture of fructooligosaccharides and arabinogalactans has been reported to induce a significant increase in concentrations of *Enterococcus faecium* and *Streptococcus salivarius* spp. *Thermophilus* in dogs (Pinna and Biagi 2016)	All prebiotics help in weight management, control the release of blood sugar, and decrease colon cancer (Van den Abbeele et al. 2011)
Foxtail millet	Arabinoxylans (pentosans), β-glucan		
Barnyard millet	Arabinoxylans, β-glucan, hemicelluloses	Inulin promotes the growth of *Bifidobacteria* and *Lactobacilli* in the colon	
Kodo millet	Arabinoxylans, β-glucan, inulin, hemicelluloses	β-glucan can selectively promote the growth of *Lactobacillus* and *Bifidobacterium* spp. (Singh and Bhardwaj 2023)	
Little millet	Hemicelluloses	Hemicellulose-derived oligosaccharides play a significant role in the modulation of gut microbiota (Jana et al. 2021)	

inulin extracted from pearl millet also had a higher prebiotic activity score (3.2) in comparison to commercial inulin (1.0) for the growth of *Lactobacillus casei*. Prashanth and Muralikrishna (2014) extracted the water unextractable portion of the finger millet with saturated barium hydroxide and 1 M potassium hydroxide. The extracts consisted of arabinose and xylose in various concentrations, i.e. 1.0:0.8 in barium hydroxide extract (BE) and 1.0:1.2 in the case of potassium hydroxide extract (KE). The yield was 2.9% for BE and 2.1% for KE. The purified arabinoxylans of finger millet can also be used as immunomodulators (Prashanth and Muralikrishna 2014). Sharma et al. (2018) reported 5.78 g of crude β-glucan per 100 g of raw foxtail millet with 76.98% of purity. The yield of crude β-glucan in raw kodo millet was 6.12 g/100 g flour with a purity of 82.51%. The germination of both millets (40 h at 25°C for foxtail millet and 38.35 h at 35°C for kodo millet) decreased the β-glucan content per 100 g of flour; however, the purity increased to 79.95% and 88.50% in foxtail and kodo millet flour, respectively, on germination. Srinivasan et al. (2019) reported the percentage of phenolic acid-bound arabinoxylans in malted little and kodo millets to be 42.69 and 27.62% w/w, respectively. The ratio of xylose to arabinose in the extracted arabinoxylans was 1.48:1.0 and 2.26:1.0 for little and kodo millet flour, respectively.

15.2.3 CONCLUSIONS

Millets have diverse applications and each part of millet, i.e. stover, husk, seed, bran, germ, and husk, can be used for one or more purposes. The stover of all the millets except foxtail millet and kodo millet is good fodder for animals. The fodder value or the digestibility of the stover can also be enhanced by urea and urea-molasses treatment. The straw or husk can also be digested into fermentable sugars and can be used for the production of bioethanol and biogas. The husk is also used as fodder and filler in plastics. Husk ash is used to make thermoplastic composites, bricks, and the partial replacement of cement. The bran of millets is rich in nutrients and phytochemicals and can be used for the fortification of common foods and snacks. Pearl millet germ and foxtail millet bran are good sources of edible oil which is at par in quality with other edible oils. These oils are also rich in essential fatty acids. The millet grains of non-food grade can be used for the production of bioethanol. In addition to these, millets can also be used for the extraction of prebiotics. This shows that each part of the millet can be successfully used in one or another forms. This makes millet a waste-free crop with huge potential for cultivation in the future.

REFERENCES

Abba, A., U. Z. Faruq, U. A. Birnin-Yauri, M. B. Yarima, and K. J. Umar. 2014. Study on production of biogas and bioethanol from millet husk. *Annual Research and Review in Biology* 4(5):817–27.

Abdulwahab, M. T., O. A. U. Uche, and G. Suleiman. 2017. Mechanical properties of millet husk ash bitumen stabilized soil block. *Nigerian Journal of Technological Development* 14(1):34–8.

Alemu, D., F. Tegegne, and Y. Mekuriaw. 2020. Comparative evaluation of effective microbe–and urea molasses–treated finger millet (*Eleusine coracana*) straw on nutritive values and growth performance of Washera sheep in northwestern Ethiopia. *Tropical Animal Health and Production* 52(1):123–9.

Aliyu, S., and M. Bala. 2011. Brewer's spent grain: A review of its potentials and applications. *African Journal of Biotechnology* 10(3):324–31.

Amadou, I., T. Amza, Y. H. Shi, and G. W. Le. 2011. Chemical analysis and antioxidant properties of foxtail millet bran extracts. *Songklanakarin Journal of Science and Technology* 33(5):509–15.

Amritha, G. K., U. Dharmaraj, P. M. Halami, and G. Venkateswaran. 2018. Dephytinization of seed coat matter of finger millet (*Eleusine coracana*) by *Lactobacillus pentosus* CFR3 to improve zinc bioavailability. *LWT-Food Science and Technology* 87:562–6.

Bagudo, B. U., B. Garba, S. M. Dangoggo, and L. G. 2011. The qualitative evaluation of biogas samples generated from selected organic wastes. *Archives of Aapplied Science Research* 3(5):549–55.

Banerjee, D., R. Chowdhury, and P. Bhattacharya. 2017. Optimization of extraction process of inulin from Indian millets (jowar, bajra and ragi)—Characterization and cost analysis. *Journal of Food Science and Technology* 54(13):4302–14.

Bheel, N. D., F. A. Memon, S. L. Meghwar, and I. A. Shar. 2018. Millet husk ash as environmental friendly material in cement concrete. In *5th International Conference on Energy, Environment and Sustainable Development*, 153–8, Mehran University of Engineering and Technology, Jamshoro, Pakistan.

Chahal, S. S., and A. S. Chhabra. 2014. Crop biomass production and its utilization in Punjab: Some energy considerations. *Economic Affairs* 59(4):529–37.

Chandra, D., S. Chandra, and A. K. Sharma. 2016. Review of Finger millet (*Eleusine coracana* (L.) Gaertn): A power house of health benefiting nutrients. *Food Science and Human Wellness* 5(3):149–55.

Chandrasekara, A., M. Naczk, and F. Shahidi. 2012. Effect of processing on the antioxidant activity of millet grains. *Food Chemistry* 133(1):1–9.

Chapke, R. R., G. Prabhakar, Shyamprasad, I. K. Das, and V. A. Tonapi. 2018. Improved millets production technologies and their impact. Technology Bulletin. ICAR-Indian Institute of Millets Research, Hyderabad.

Chaudhary, D. P., A. Kumar, S. S. Mandhania, P. Srivastava, and R. S. Kumar. 2012. Maize as fodder? An alternative approach. Technical Bulletin. Directorate of Maize Research, Pusa Campus, New Delhi.

Choudhary, M., and G. Prabhu. 2014. Quality fodder production and economics of dual-purpose pearlmillet (*Pennisetum glaucum*) under different fertility levels and nitrogen scheduling. *Indian Journal of Agronomy* 59(3):410–4.

Codex Alimentarius. 2019. http://www.fao.org/fao-who-codexalimentarius/sh-proxy/en/?lnk=1&url=https%253A%252F%252Fworkspace.fao.org%252Fsites%252Fcodex%252FStandards%252FCXS%2B210-1999%252FCXS_210e.pdf (accessed May 9, 2021).

Devittori, C., D. Gumy, A. Kusy, L. Colarow, C. Bertoli, and P. Lambelet. 2000. Supercritical fluid extraction of oil from millet bran. *Journal of the American Oil Chemists' Society* 77(6):573–9.

FAO. 2021. Grains and their structure. http://www.fao.org/3/t0818e/t0818e02.htm (accessed May 9, 2021).

Feedipedia. 2020. Proso millet (*Panicum miliaceum*), forage. https://www.feedipedia.org/node/409 (accessed April 11, 2021).

Gebremariam, M. M., M. Zarnkow, and T. Becker. 2014. Teff (*Eragrostis tef*) as a raw material for malting, brewing and manufacturing of Gluten-Free Foods and beverages: A review. *Journal of Food Science and Technology* 51(11):2881–95.

Guo, M., J. Du, Z. A. Zhang, K. Zhang, and Y. Jin. 2014. Optimization of brewer's spent grain-enriched biscuits processing formula. *Journal of Food Process Engineering* 37(2):122–30.

Habte, E., M. S. Muktar, A. T. Negawo, S. H. Lee, K. W. Lee, and C. S. Jones. 2019. An overview of Teff (*Eragrostis teff Zuccagni*) trotter) as a potential summer forage crop in temperate systems. *Journal of The Korean Society of Grassland and Forage Science* 39(3):185–8.

Hammajam, A. A., Z. N. Ismarrubie, and S. M. Sapuan. 2014. Millet husk fiber filled high density polyethylene composites and its potential properties. *International Journal of Engineering and Technical Research* 2(11):248–50.

Hassan, Z. M., N. A. Sebola, and M. Mabelebele. 2021. The nutritional use of millet grain for food and feed: A review. *Agriculture and Food Security* 10(1):1–14.

Heuzé, V., G. Tran, P. Hassoun, and D. Sauvant. 2015a. *Pearl millet (Pennisetum glaucum), forage*. Feedipedia, a programme by INRAE, CIRAD, AFZ, and FAO. https://www.feedipedia.org/node/399 (accessed September 21, 2021).

Heuzé, V., G. Tran, and S. Giger-Reverdin. 2015b. *Scrobic (Paspalum scrobiculatum) forage and grain*. Feedipedia, a programme by INRAE, CIRAD, AFZ, and FAO. https://www.feedipedia.org/node/401 (accessed September 23, 2019).

Heuzé, V., G. Tran, P. Hassoun, and F. Lebas. 2019a. Finger millet (*Eleusine coracana*), forage. Feedipedia, a programme by INRAE, CIRAD, AFZ, and FAO. https://www.feedipedia.org/node/447 (accessed September 22, 2019).

Heuzé, V., G. Tran, P. Hassoun, and F. Lebas. 2019c. Fonio (*Digitaria exilis*). Feedipedia, a programme by INRAE, CIRAD, AFZ, and FAO. https://www.feedipedia.org/node/460 (accessed September 22, 2019).

Heuzé, V., H. Thiollet, G. Tran, and F. Lebas. 2019b. Tef (Eragrostis tef) straw. Feedipedia, a programme by INRAE, CIRAD, AFZ, and FAO. https://www.feedipedia.org/node/22033 (accessed September 23, 2021).

Indian Council for Agricultural Research. 2020. Urea treatment for poor quality roughages. https://krishi.icar.gov.in/PDF/Selected_Tech/animal/30-AS-Urea%20treatment%20of%20poort%20quality%20roughages.pdf (accessed April 16, 2021).

Jana, U. K., N. Kango, and B. Pletschke. 2021. Hemicellulose-derived oligosaccharides: Emerging prebiotics in disease alleviation. *Frontiers in Nutrition* 8:670817. https://doi.org/10.3389/fnut.2021.670817.

Jha, N., R. Krishnan, and M. S. Meera. 2015. Effect of different soaking conditions on inhibitory factors and bioaccessibility of iron and zinc in pearl millet. *Journal of Cereal Science* 66:46–52.

Joshi, V. K., A. Kumar, and N. S. Thakur. 2015. Technology of preparation and consumption pattern of traditional alcoholic beverage 'Sur' of Himachal Pradesh. *International Journal of Food and Fermentation Technology* 5(1):75–82.

Kalaiselvan, E., D. Desinguraja, R. Manikandan, and M. Dinesh. 2020. Use of Kodo millet straw ruins the cross bred cows amid COVID-19. *Journal of Entomology and Zoology Studies* 8(3):405–6.

Kalam Azad, M. O., D. I. Jeong, M. Adnan, T. Salitxay, J. W. Heo, M. T. Naznin, J. D. Lim, D. H. Cho, B. J. Park, and C. H. Park. 2019. Effect of different processing methods on the accumulation of the phenolic compounds and antioxidant profile of broomcorn millet (*Panicum miliaceum* L.) flour. *Foods* 8(7):230. doi:10.3390/foods8070230.

Karlović, A., A. Jurić, N. Ćorić, K. Habschied, V. Krstanović, and K. Mastanjević. 2020. By-Products in the malting and brewing industries—Re-usage possibilities. *Fermentation* 6(3):82.

Krishnan, R., U. Dharmaraj, R. S. Manohar, and N. G. Malleshi. 2011. Quality characteristics of biscuits prepared from finger millet seed coat based composite flour. *Food Chemistry* 129(2):499–506.

Kubo, R. 2016. The reason for the preferential use of finger millet (*Eleusine coracana*) in eastern African brewing. *Journal of the Institute of Brewing* 122(1):175–80.

Kundgol, N. G., B. Kasturiba, K. K. Math, and M. Y. Kamatar. 2014. Effect of preliminary processing on antiradical properties of little millet landraces. *International Journal of Farm Sciences* 4(2):148–54.

Kumar, A., V. Tomer, A. Kaur, V. Kumar, and K. Gupta. 2018. Millets: A solution to agrarian and nutritional challenges. *Agriculture and Food Security* 7(1):31. doi:10.1186/s40066-018-0183-3.

Kumar, A., A. Kaur, V. Tomer, P. Rasane, and K. Gupta. 2020. Development of nutricereals and milk-based beverage: Process optimization and validation of improved nutritional properties. *Journal of Food Process Engineering* 43(1):e13025.

Li, Y., J. Lv, L. Wang, Y. Zhu, and R. Shen. 2020. Effects of millet bran dietary fiber and millet flour on dough development, steamed bread quality, and digestion in vitro. *Applied Sciences* 10(3):912.

Liang, S., and K. Liang. 2019. Millet grain as a candidate antioxidant food resource: A review. *International Journal of Food Properties* 22(1):1652–61.

Lorenz, K., and Y. S. Hwang. 1986. Lipids in proso millet (*Panicum miliaceum*) flours and brans. *Cereal Chemistry* 63(5):387–90.

Mohammed, M., F. Babo, and M. Abdelkreim. 2016. Improvement of millet (*Pennisetum glaucum*) straw as animal feed using physical and chemical treatment. *International Journal of Innovative Science, Engineering and Technology* 3(5):87–93.

Malathi, D., N. V. Varadharaju, and G. Gurumeenakshi. 2014. Study on nutrient composition of millets and ways to minimize loss during processing and value addition. Post-harvest Technology Centre Agricultural Engineering College and Research Institute. Tamil Nadu Agricultural University Coimbatore, Tamil Nadu.

Mazlien, T. 2015. Whiskey review: Koval single barrel millet whiskey. https://thewhiskeywash.com/whiskey-styles/american-whiskey/whiskey-review-koval-single-barrel-millet-whiskey/.

Nebraska-Lincoln Extension Educational Program and USDA. 2008. Producing and marketing proso millet in the great plains. http://extensionpublications.unl.edu/assets/pdf/ec137.pdf. (accessed April 11, 2021).

Packiam, M., and S. Karthikeyan. 2018. Effective chemical pretreatment for recovery of fermentable sugars from pearl millet biomass. *Madras Agricultural Journal* 107:1–5.

Petrović, J. S., B. S. Pajin, S. D. Kocić-Tanackov, J. D. Pejin, A. Z. Fišteš, N. D. Bojanić, and I. S. Lončarević. 2017. Quality properties of cookies supplemented with fresh brewer's spent grain. *Food and Feed Research* 44(1):57–63.

Pinna, C., and G. Biagi. 2016. The utilisation of prebiotics and Synbiotics in dogs. *Italian Journal of Animal Science* 13(1):3107.

Prashanth, M. R. S., and G. Muralikrishna. 2014. Arabinoxylan from finger millet (*Eleusine coracana*, v. indaf 15) bran: Purification and characterization. *Carbohydrate Polymers* 99:800–7.

Ray, S., D. J. Bagyaraj, G. Thilagar, and J. P. Tamang. 2016. Preparation of Chyang, an ethnic fermented beverage of the Himalayas, using different raw cereals. *Journal of Ethnic Foods* 3(4):297–9.

Rose, D. J., and D. K. Santra. 2013. Proso millet (*Panicum miliaceum* L.) fermentation for fuel ethanol production. *Industrial Crops and Products* 43:602–5.

Slama, A., A. Cherif, F. Sakouhi, S. Boukhchina, and L. Radhouane. 2020. Fatty acids, phytochemical composition and antioxidant potential of pearl millet oil. *Journal of Consumer Protection and Food Safety* 15(2):145–51.

Sarma, S. M., P. Khare, S. Jagtap, D. P. Singh, R. K. Baboota, K. Podili, R. K. Boparai, J. Kaur, K. K. Bhutani, M. Bishnoi, and K. K. Kondepudi. 2017. Kodo millet whole grain and bran supplementation prevents high-fat diet induced derangements in a lipid profile, inflammatory status and gut bacteria in mice. *Food and Function* 8(3):1174–83.

Satyavathi, T. C., S. Praveen, S. Mazumdar, L. K. Chugh, and A. Kawatra. 2017. Enhancing demand of pearl millet as super grain - Current status and way forward. ICAR- All India Coordinated Research Project on pearl millet, Jodhpur. https://www.iari.res.in/files/Bulletins/PearlMillet%20_31012020.pdf (accessed April 6, 2021).

Seetharam, A., and K. T. Krishne Gowda. 2007. Production and utilization of small millets in India - Food uses of small millets and avenues for further processing and value addition, 1–8, Project Cordination Unit, All India Coordinated Small Millets Improvement Project, University of Agricultural Sciences, Bengaluru.

Sharma, S., D. C. Saxena, and C. S. Riar. 2018. Characteristics of β-glucan extracted from raw and germinated foxtail (*Setaria italica*) and kodo (*Paspalum scrobiculatum*) millets. International Journal of Biological Macromolecules 118(A):141–8.

Sheahan, C. M. 2014. *Plant Guide for Foxtail Millet (Setaria italica). USDA-Natural Resources Conservation Service*, Cape May Plant Materials Center, Cape May, NJ.

Shi, Y., Y. Ma, R. Zhang, H. Ma, and B. Liu. 2015. Preparation and characterization of foxtail millet bran oil using subcritical propane and supercritical carbon dioxide extraction. *Journal of Food Science and Technology* 52(5):3099–104.

Singh, R. P., and A. Bhardwaj. 2023. B-glucans a potential source for maintaining gut microbiota and the immune system. *Frontiers in Nutrition* 10:1143682. https://doi.org/10.3389/fnut2023.1143682.

Sood, S., R. K. Khulbe, A. K. Gupta, P. K. Agrawal, H. D. Upadhyaya, and J. C. Bhatt. 2015. Barnyard millet–a potential food and feed crop of future. *Plant Breeding* 134(2):135–47.

Srinivasan, A., S. P. Ekambaram, S. S. Perumal, J. Aruldhas, and T. Erusappan. 2019. Chemical characterization and immunostimulatory activity of phenolic acid bound arabinoxylans derived from foxtail and barnyard millets. *Journal of Food Biochemistry* 6:e13116.

Tekaligne, T. M., A. R. Woldu, and Y. A. Tsigie. 2015. Bioethanol production from finger millet (*Eleusine coracana*) straw. *Ethiopian Journal of Science and Technology* 8(1):1–13.

Tran, G. 2017. *Proso millet (Panicum miliaceum), forage*. Feedipedia, a programme by INRAE, CIRAD, AFZ, and FAO. https://www.feedipedia.org/node/409 (accessed September 22, 2019).

Van den Abbeele, P., T. Van de Wiele, and S. Possemiers. 2011. Prebiotic effect and potential health benefit of arabinoxylans. *Agro Food Industry Hi-Tech* 22(2):9–12.

Wafula, W. N., M. Siambi, H. F. Ojulong, N. Korir, and J. Gweyi-Onyango. 2017. Finger millet (*Eleusine coracana*) fodder yield potential and nutritive value under different levels of phosphorus in rainfed conditions. *Journal of Agriculture and Ecology Research International* 10(4):1–10.

World Intellectual Property Organization. 2022. Finding Africa's future cropp in the past. https://www.wipo.int/ipadvantage/en/details.jsp?id=5556 (accessed January 20, 2022).

Wu, X., D. Wang, S. R. Bean, and J. P. Wilson. 2006. Ethanol production from pearl millet using *Saccharomyces cerevisiae*. *Cereal Chemistry* 83(2):127–31.

Chapter 16

Prospects in the Production, Processing, and Utilization of Millets

16.1 INTRODUCTION

Millets have been grown for a very long time as staple crops in many parts of the world, feeding millions of people. Due to their adaptability and resilience, these small-seeded grasses have proven to be extremely useful in the fight against environmental problems and food insecurity. The importance of examining the prospects for millet production, processing, and use is growing as the world struggles with the effects of climate change, population growth, and malnutrition (Das et al. 2019). The cultivation and consumption of millets, which include species like sorghum, finger millet, foxtail millet, and pearl millet, have a long history. Since they thrive in harsh environmental conditions where other crops struggle to survive, they have been grown for thousands of years, especially in arid and semi-arid areas. They are the perfect crops for farmers with limited resources because of their tolerance for drought, quick growth cycles, and low input needs (Saxena et al. 2018).

Due to their nutritional value, millets have recently attracted new interest and attention. They are valuable dietary components for addressing malnutrition and dietary-related diseases because they are naturally abundant in essential nutrients, dietary fibre, and bioactive compounds (Foley et al. 2021). Millets are suitable for people with celiac disease or gluten sensitivity because they are gluten-free. They also have a low glycaemic index, which aids in controlling blood sugar levels and might help with better diabetes management (Asrani et al. 2022). The future of millets depends on several important factors, including genetic advancement, environment-friendly farming methods, value-added processing, nutritional fortification, market expansion, policy support, and research funding. These factors cover every step of the millet's value chain, from crop enhancement to consumer acceptance. Exploring advancements and innovations in each area can unlock the full potential of millets and ensure their sustainable cultivation, processing, and utilization (Khairuddin and Lasekan 2021).

A key factor in increasing millet production and quality is genetic advancement. Researchers have been able to pinpoint and modify the genes responsible for desirable traits like drought tolerance, disease resistance, and increased nutrient content thanks to developments in biotechnology and genomics (McSweeney et al. 2017). Farmers can increase yields, improve their ability to adapt to changing climatic conditions, and improve nutritional profiles by creating improved millet varieties (Gupta et al. 2017). For millet production to continue in the future, sustainable agricultural practices are essential. Utilizing remote sensing, geographic information systems (GIS), and data analytics, precision farming techniques can optimize irrigation, fertilizer application, and pest control, resulting in millet cultivation that is both resource-efficient and environmentally sustainable (Spielman et al. 2021). In areas where millet is grown, conservation agriculture techniques like minimum tillage, crop rotation, and cover crops can help maintain healthy soil, conserve water, and protect biodiversity. Additionally, combining millet farming with agroforestry systems can improve the environment and diversify farmers' sources of income (Talabi et al. 2022).

Utilizing millet for value-added processing is another important aspect. Although millet has traditionally been eaten as a whole grain, there is an increasing market demand for millet products that have been processed (Adebiyi et al. 2016). Foods made from millet can improve their sensory qualities, shelf-life, and nutritional value by processing techniques like milling, dehulling, extrusion, and fermentation. The production of pasta, bakery goods, breakfast cereals, and snacks made from millet can meet consumers' changing dietary needs and convenience requirements. Enhancing the nutritional value of millet requires both nutritional fortification and biofortification techniques (Birania et al. 2020). Although millets are already abundant in essential nutrients, fortification entails adding extra micronutrients to millet products, such as iron, zinc, and vitamin A, to address particular nutrient deficiencies common in some areas. Contrarily, biofortification concentrates on developing millet varieties with increased levels of target nutrients using conventional breeding techniques or genetic engineering. By addressing micronutrient deficiencies and associated health problems, these methods can help to improve public health outcomes (Mahajan et al. 2023).

Increased demand and consumption of millet-based products depend on market development and promotion. Consumer perceptions and preferences can be altered by increasing awareness of the nutritional advantages, culinary adaptability, and environmental sustainability of millets (Kane-Potaka et al. 2021). Millet can become a commonplace food choice by increasing consumer demand with the help of marketing campaigns, consumer education initiatives, and strategic market connections. Consistent product quality can be ensured by establishing quality standards, certifications, and labelling laws, which will also increase consumer confidence (Bhatt et al. 2023). Investments in research and changes to the policy are required to support millets' prospects. By providing financial incentives, subsidies, and encouraging policies, governments can significantly contribute to the promotion of millet cultivation (Thakur and Tiwari 2019). Farmers can be empowered to adopt better agricultural practices, invest in cutting-edge technology, and overcome obstacles associated with millet production and marketing by having access to credit, insurance, and training programmes (Singh and Burman 2019). Additionally, the development of the millet industry can be aided by research funding and cooperation between scientists, agronomists, nutritionists, and food technologists.

In conclusion, there is enormous potential for addressing global issues like food security, malnutrition, and environmental sustainability through the production, processing, and use of millet. Millets have the potential to become a sustainable and nutrient-dense solution for a world that is changing quickly through genetic improvement, sustainable agricultural practices, value-added processing, nutritional fortification, market development, policy support, and research investment. Unlocking millets' full potential and building a future in which these remarkable crops play a crucial role in sustaining communities and protecting the environment depend on stakeholders supporting innovation, collaboration, and knowledge-sharing.

16.2 FUTURE PROSPECTS IN THE PRODUCTION OF MILLETS

Millet production holds a lot of promise and potential for addressing global issues like climate change, population growth, and food security. As resilient and adaptable crops, millets have numerous advantages over traditional cereals, making them an appealing option for sustainable agriculture (Salej et al. 2019). In this section, we will look at the future prospects for millet production, with a focus on genetic advances, sustainable agricultural practices, and millet cultivation promotion.

16.2.1 Advancements in Genetics

Genetic advancements have enormous potential for improving millet production. Scientists have gained a better understanding of millets' genetic makeup through genetic research, paving the way for targeted breeding programmes and the development of improved varieties (Singh and Sood 2020). Traditional breeding techniques such as hybridization and selection have been used to improve desirable traits such as yield, disease resistance, and nutritional quality. Researchers now have access to advanced tools and techniques that can accelerate the genetic improvement of millets thanks to advances in biotechnology and genomics (Sabreena et al. 2021). Breeders can use marker-assisted selection (MAS) to identify specific genes or markers associated with desirable traits, allowing them to select and breed for those traits more efficiently. This method reduces the time and resources needed for traditional breeding methods, allowing for the faster development of improved millet varieties (Singh and Nara 2023). Another aspect of biotechnology, genetic engineering, holds promise for introducing novel traits into millets. Scientists can improve crop resistance to pests, diseases, and environmental stresses like drought and heat by incorporating genes from other organisms (Dhaka et al. 2021). Genetic engineering also provides opportunities for improving nutritional content and increasing millet productivity (Mbinda and Masaki 2020). However, it is important to consider the potential environmental and socioeconomic impacts of genetically modified crops and ensure thorough safety assessments and regulatory frameworks are in place.

16.2.2 Sustainable Agricultural Practices

Sustainable agricultural practices are critical for millet production in the future. As climate change increases agricultural challenges, millets' resilience and adaptability become even more valuable (Zhang et al. 2018). Sustainable practices can increase millet cultivation, productivity, and efficiency while reducing environmental impacts. Among the most important sustainable agricultural practices are crop diversity, combining millet cultivation with agroforestry systems, the use of precision farming technologies, climate-smart agriculture, and community-based approaches. The inclusion of crop diversity in millet cultivation systems can help with sustainability. Intercropping millets with legumes or other complementary crops can improve soil health by improving nutrient cycling and reducing pest and disease pressure. Crop rotation can also interrupt pest and disease cycles, reduce weed pressure, and maintain soil fertility (Abrouk et al. 2020). Combining millet cultivation with agroforestry systems has the potential to provide additional benefits. Agroforestry improves soil fertility, biodiversity, carbon sequestration, and income generation through the cultivation of trees and non-timber forest products. This method contributes to long-term millet production while also helping farmers' livelihoods (Bado et al. 2021). Precision farming technologies and tools can optimize resource use and improve input efficiency in millet production. Remote sensing, drones, and GIS can all provide useful data on crop health, soil moisture levels, and nutrient status (Tariyal et al. 2022). This information enables farmers to make more informed decisions about irrigation, fertilization, and pest control, resulting in more precise and targeted interventions. Adopting climate-smart agricultural practices in millet production contributes to climate change resilience. This includes techniques like agroecological zoning, which selects suitable millet varieties based on specific agroclimatic conditions. Climate information and early warning systems can assist farmers in making timely decisions and adapting their farming practices (Babele et al. 2022). Promoting community-based millet production can promote knowledge-sharing, collaboration, and collective action. Farmer field schools, participatory research, and farmer-led innovation networks allow farmers to share their experiences, learn from each other's successes and challenges, and work together to address common millet production issues (Scarpa et al. 2021). Governments and policymakers are critical in promoting sustainable agricultural practices for millet production. Farmers can be encouraged to adopt sustainable practices by developing supportive policies and providing incentives for their adoption (Barbon et al. 2017). Subsidies for environmentally friendly inputs, extension services, training programmes, and financial assistance for implementing sustainable practices are examples of policy measures. Furthermore, incorporating sustainable agriculture into national agricultural strategies and climate change adaptation plans can provide a comprehensive framework for promoting long-term millet production (Delele et al. 2021).

Overall, millets have enormous potential as a sustainable and resilient crop for agriculture's future. Millets are well-suited to adapt to climate change challenges. Millets can thrive in a variety of soil and climatic conditions due to their low water and input requirements, making them a dependable choice for farmers dealing with erratic floods, droughts, water scarcity, high temperatures, harsh winters, and altered rainfall patterns (Das et al. 2019; Kumar et al. 2018). Millet crops have a significant advantage in unpredictable growing conditions due to their short duration. Because of their ability to complete their life cycle quickly, farmers can adjust their cultivation plans in response to changing weather patterns and mitigate risks associated with climatic uncertainties (Ceasar et al. 2018). This inherent adaptability makes millet a valuable option for ensuring food security in the face of declining performance and yields of major crops projected in the coming years.

In addition to their resilience, millets are high in nutrients such as dietary fibre, protein, vitamins, and minerals. As the prevalence of lifestyle diseases such as obesity and type 2 diabetes rises, millets can play an important role in addressing these health issues (Shweta et al. 2018). Consumers, particularly in urban areas, are shifting towards healthier eating habits and seeking organically grown produce. Millets, with their low input requirements and low pest and disease incidence, can be an excellent choice for organic farming, meeting the demand for environmentally friendly and healthy alternatives (Mbinda and Masaki 2020). Furthermore, millets have a rich cultural and historical significance due to their use in traditional subsistence agriculture and tribal food systems. Reviving millet cultivation can not only provide nutritional security but can also help to preserve traditional knowledge and cultural heritage. Communities can feed their families a healthy diet while also earning extra money by reintroducing millet into their farming practices (Suparna et al. 2021).

Government support and assistance are critical in promoting millet cultivation in general. Policymakers can encourage farmers to grow millet by providing incentives, R&D funding, and policies that prioritize millet in government programmes (Bisht et al. 2018). This assistance should include educating farmers about the benefits of millets, facilitating access to high-quality seeds, and developing market connections. Finally, millets have a bright future in agriculture and food security. They are an appealing choice for sustainable and climate-smart farming practices due to their resilience, adaptability, nutritional value, and cultural significance (Malhotra et al. 2021). Farmers can navigate the challenges posed by climate change by embracing millet cultivation, consumers can access healthier food options, and communities can revitalize traditional agricultural practices. Millets, with adequate support and promotion, can significantly contribute to the development of a sustainable and resilient food system for the well-being of both people and the planet (Kala and Nautiyal 2023).

16.3 FUTURE PROSPECTS IN THE PROCESSING AND UTILIZATION OF MILLETS

The processing and utilization of millets have a bright future, thanks to increased awareness of their nutritional value, sustainability, and versatility. Millets have numerous applications, including processed food products, beverages, livestock feed, and industrial applications (Dias-Martins et al. 2018; Kumar et al. 2020). This section will look at potential advancements and opportunities in millet processing and utilization.

16.3.1 Diversification of Millet-Based Food Products

The diversification of millet-based food products has great potential for expanding the market and increasing consumer acceptance of millet. Millets can be incorporated into a variety of food products after being processed into various forms (Arora et al. 2023). One of the most commonly processed forms of millet is flour, which can be used as a direct substitute for wheat flour in baking or as a blend in composite flours for bread, cookies, and other bakery products. Millets can be ground into flakes, which are popular in breakfast cereals, muesli, and granola bars (Anitha et al. 2022). Flakes are a convenient and nutritious breakfast option for consumers looking for quick and healthy breakfast options. Grits are a type of millet that can be used in savoury dishes, porridges, and side dishes, making them a versatile alternative to traditional grains (Srilekha et al. 2019). Millet flour pasta is becoming popular as a gluten-free and nutritious alternative to wheat pasta. The development of high-quality millet pasta with desirable texture and cooking properties represents a significant opportunity for the pasta market to diversify (Hema et al. 2022). Snack foods like millet-based chips, crackers, and puffed snacks are also popular with health-conscious consumers. Millets are an appealing option for innovative snack formulations due to their distinct texture, crunch, and nutritional profile (Tomar et al. 2022). Furthermore, experimenting with new ingredient combinations and formulations can improve the taste, texture, and nutritional value of millet-based foods (Arora et al. 2023). Millets can be combined with other grains, legumes, and vegetables to create novel food products that cater to a wide range of culinary preferences and dietary requirements. Innovative processing techniques and technologies are critical in improving millet-based product quality. Advanced milling processes can be used to achieve finer particle size and optimize the functional properties of the flour (Amadou 2022). Extrusion technologies can produce expanded millet products with better texture and mouthfeel. Novel cooking and processing methods can be investigated to maximize the nutritional value of millets while improving their sensory properties. In conclusion, the diversification of millet-based food products creates significant millet processing and utilization opportunities. The market appeal and consumer acceptance of millets can be greatly increased by developing innovative processing techniques, refining ingredient combinations, and improving the sensory attributes of millet-based foods.

16.3.2 Value-Addition and Fortification

Millets must be fortified and value-added to improve their nutritional value and address specific nutrient deficiencies in populations. To improve their nutritional content and contribute to better health outcomes, millets can be fortified with essential micronutrients such as iron, zinc, and vitamins (Ramashia et al. 2021). Millets are fortified by adding specific nutrients during processing to address common deficiencies in the population. Iron and zinc supplementation can aid in the treatment of anaemia and improve overall health (Badejo et al. 2020). Fortification with vitamins such as vitamin A, vitamin B-complex, and vitamin D can help with immune function, eye health, and overall well-being. Fortification techniques vary depending on the type of millet being processed, such as flour, flakes, or grits, and can be accomplished through a variety of methods such as coating, blending, and extrusion (Sukumar and Athmaselvi 2019). In addition to fortification, the value of millets can be increased by incorporating them into composite flours or blends to improve the nutrient profile of staple foods. Millets can contribute to the overall nutritional quality of commonly consumed foods using this method (Sahu and Patel 2021). By combining millet flour with other grains, legumes, or vegetables, millet-based composite flours can be created, resulting in a nutritionally balanced and diverse food product. This can help people who rely on staple foods overcome nutrient deficiencies and improve their overall diet (Ramashia 2018). The development of cost-effective and scalable techniques for the value addition and fortification of millets holds great promise for the future. It is critical to ensure that fortified millet products are available and affordable to all demographics (Birania et al. 2020). This necessitates research and development of fortification methods that are appropriate for millet processing while not jeopardizing the sensory attributes or stability of the final products. Establishing quality control measures is also critical to ensuring the accurate and consistent fortification of millet products (Malla et al. 2021). The widespread availability of fortified millet products has the potential to significantly improve nutritional outcomes, particularly in areas where millets are a staple crop. Governments, policymakers, and food industry stakeholders all play an important role in promoting millet value addition and fortification through the development of policies and regulations, the provision of incentives, and funding for research and development. The nutritional value of staple foods can be increased by fortifying millets with essential nutrients and incorporating them into composite flours, addressing specific nutrient deficiencies and promoting better health outcomes. Continuous research, innovation, and collaboration are required to develop cost-effective and scalable techniques that ensure the widespread availability of fortified millet products. This contributes to improved nutrition and better overall well-being, particularly in populations where millets are a significant dietary component.

16.3.3 Functional Food and Nutraceutical Applications

Millets have enormous potential for the development of functional foods and nutraceuticals with specific health benefits due to their inherent functional properties. High fibre content, antioxidants, and bioactive compounds are among the properties that contribute to their nutritional value and health-promoting effects (Singh et al. 2019). Millets' high fibre content, which includes both soluble and insoluble fibres, can play an important role in promoting digestive health. By incorporating millet fibres into products such as bread, cereals, and snacks, millet-based functional foods can be developed to target specific gastrointestinal conditions such as constipation or irritable bowel syndrome (Amadou 2022). These products can help with bowel regularity, gut health, and overall digestive well-being. Millets contain antioxidants such as phenolic compounds and flavonoids, which may have health benefits such as lowering the risk of chronic diseases. Millet-based functional foods can be formulated to maximize antioxidant retention and bioavailability, resulting in products with improved antioxidant properties (Palaniappan et al. 2017). These functional foods can help manage and prevent conditions like cardiovascular disease, cancer, and oxidative stress-related disorders. Millet-based products can also be developed to control specific health conditions, such as diabetes. Millets have a low glycaemic index, which means they raise blood sugar levels more slowly than refined grains (Thakur and Tiwari 2019). This qualifies millet for inclusion in functional foods designed for diabetics or those seeking blood sugar control. Millet-based snacks, bread, and ready-to-eat meals can be formulated to provide sustained energy release and help manage blood sugar levels.

Furthermore, the unique combination of nutrients and bioactive compounds found in millet has the potential to promote cardiovascular health. Millet-enriched functional foods can be created to promote heart health by incorporating ingredients that help with cholesterol management, blood pressure regulation, and overall cardiovascular function (Dey et al. 2022). Millets' potential in functional food and nutraceutical applications will necessitate extensive research to identify and quantify the specific bioactive compounds responsible for their health benefits. This includes researching their bioavailability, bioactivity, and the mechanisms through which they affect the human body (Majid and Priyadarshini 2020). Furthermore, it is critical to develop innovative processing techniques that preserve the functional properties of millets while improving the sensory attributes and shelf-life of the final products. The development and commercialization of millet-based functional foods and nutraceuticals will require collaboration between researchers, food scientists, and the food industry (Anjali and Vijayaraj 2020). Conducting clinical trials and studies to validate the health benefits of these products, establishing quality control measures, and ensuring regulatory compliance are all part of this process. In conclusion, the functional properties and bioactive compounds found in millets present exciting opportunities for the development of functional foods and nutraceuticals that target specific health conditions. Prospects in this field include further research into the health benefits of millets, the development of novel processing techniques, and stakeholder collaboration to bring millet-based functional foods and nutraceuticals to market. This can help to promote overall health and well-being by offering consumers nutritious and beneficial food options.

16.3.4 Beverage Industry

The use of millet in the beverage industry opens up a world of possibilities for product innovation and diversification. Millets can be incorporated into a variety of beverages, catering to a wide range of consumer preferences and nutritional requirements. One promising area is the creation of millet-based milk substitutes (Amadou 2019; Kumar et al. 2018). With the growing demand for plant-based milk alternatives, millet can be a great substitute for traditional options like almond, soy, and oat milk. Millet milk is made by combining millet grains, water, and other ingredients to create a creamy, nutritious beverage (Taylor and Duodu 2019). These millet-based milk substitutes can be fortified with vitamins, minerals, and other functional ingredients to improve their nutritional profile and provide a viable plant-based option for people who are lactose intolerant or follow a vegan diet. Fermented beverages are another exciting application for millet (Rao et al. 2021). Millets have been used in traditional fermented beverages in many cultures for centuries. Fermented millet porridges and alcoholic beverages such as millet beer and traditional spirits are examples of these beverages (Rao et al. 2021). The fermentation process not only imparts distinct flavours and aromas to beverages, but also improves their nutritional value by producing beneficial compounds such as probiotics, enzymes, and organic acids. Innovative fermentation techniques and starter cultures tailored to millets can help to improve the quality and consistency of these beverages (Das et al. 2019).

Ready-to-drink millet-based beverages have market potential as well. Millet extracts or millet flours can be combined with other ingredients such as fruits, vegetables, or herbs to create refreshing and nutritious drinks (Byresh et al. 2022). These ready-to-drink millet-based beverages, which offer natural flavours, functional benefits, and clean label attributes, can be positioned as healthy alternatives to sugary sodas or artificial beverages. Exploring the potential of millets in the brewing and distilling industries also offers exciting opportunities (Taylor and Duodu 2019). Millet beers, for example, can be made by substituting millet grains for traditional malted grains during the brewing process. This not only gives the beer unique flavours and aromas but also meets the growing demand for gluten-free beer options. Millets can be used in the same way. Similarly, millet can be utilized in the production of distilled spirits, offering a distinctive and innovative alternative to traditional grain-based spirits (Cichońska and Ziarno 2021).

In summary, millet's future prospects in the beverage industry include the development of formulations that balance taste, texture, and nutritional attributes. Millets can be used to make milk substitutes, fermented beverages, ready-to-drink beverages, and even brewing and distilling. These innovative millet applications provide opportunities to create a diverse range of beverages that cater to different consumer preferences and dietary needs, contributing to the beverage market's overall growth and diversification.

16.3.5 Livestock Feed

The use of millet as livestock feed has the potential to improve animal nutrition and promote sustainable livestock production. When compared to conventional feed ingredients, millets have several advantages as feed ingredients, including high nutrient content, digestibility, and cost-effectiveness (Hassan et al. 2021). The optimization of feed formulations is one of the most promising future prospects for the use of millets as livestock feed. This entails creating balanced diets that include millet to meet the nutritional needs of various livestock species. Millets can be added to the recipe to provide energy, protein, fibre, minerals, and vitamins. The nutritional value of millets can be maximized by carefully formulating feed rations to support optimal growth, reproduction, and overall health of the animals (Das et al. 2019).

Millets are especially valuable in areas where conventional feed ingredients such as maize or soybean are scarce. Because of their adaptability to a variety of agro-climatic conditions, they are an appealing option for small-scale and subsistence livestock production systems (de Assis et al. 2018). Millets can be grown as a feed crop alongside other food crops, allowing farmers to make better use of their land resources and reduce the need for costly feed ingredients. Furthermore, incorporating millet into livestock diets can help to reduce the environmental impact of livestock production (Renganathan et al. 2020). Millets require less water and chemical inputs than conventional feed crops, making them more environmentally friendly and sustainable. Farmers can reduce their reliance on resource-intensive feed production systems and contribute to the environment by using millet as a feed ingredient (Sun et al. 2021).

Millets provide advantages in terms of feed storage and processing, in addition to nutritional benefits. Millets have a long shelf-life and can be easily processed into whole grains, pellets, or meals (Gebreyohannes et al. 2021). This allows for greater flexibility in feed preparation and allows farmers to tailor feed formulations to the specific needs of their livestock. Overall, the future of millets as livestock feed involves optimizing feed formulations to improve animal health, growth, and productivity (Packiam et al. 2018). Farmers can improve the nutritional value of feed, reduce reliance on conventional feed ingredients, and contribute to sustainable and cost-effective livestock production by incorporating millet into animal diets. The adoption of millet-based feed strategies has the potential to play a significant role in meeting the rising demand for animal products while minimizing the environmental impact of livestock production systems.

16.3.6 Market Development and Consumer Awareness

Market development and consumer awareness are critical for successful millet processing and utilization. The prospects in this area revolve around strengthening market ties and increasing consumer awareness and demand for millet-based products. Promoting millets as nutritious, sustainable, and versatile food options is critical for increasing consumer interest and market growth (Prathyusha et al. 2021). Millets' nutritional benefits can be highlighted through marketing campaigns that emphasize their high fibre content, rich micronutrient profile, and gluten-free status. These campaigns can also highlight millet cultivation's environmental sustainability, such as its low water and chemical inputs compared to other crops (Adekunle et al. 2018). Consumer perception and awareness of millets can be improved by effectively communicating these benefits.

Nutritional education is essential for raising consumer awareness and acceptance of millet. To educate consumers about millet-based products and their health benefits, workshops, cooking demonstrations, and educational programmes can be organized. Consumers can be inspired to include millets in their daily diets by showcasing innovative recipes, cooking techniques, and meal plans that incorporate millets (Chitra and Sulaiman 2017). Collaborations with nutritionists, dieticians, and culinary experts can help to promote millets' use in a variety of culinary applications. Collaboration with the food industry is critical for market growth. Engaging food manufacturers, processors, and retailers can result in the creation of a diverse range of millet-based products that cater to a variety of consumer preferences and dietary needs (Balakrishnan and Schneider 2022). These collaborative efforts can increase consumer trust, ensure product quality and safety, and foster a sustainable millet market ecosystem.

Finally, the future of millet processing and utilization is dependent on market development and consumer awareness. Through marketing campaigns, nutritional education, and collaboration with the food industry, millets can be promoted as healthy, sustainable, and versatile food options, which can drive consumer demand and market growth. Governments, non-governmental organizations, and industry stakeholders all play critical roles in creating favourable market environments and policy frameworks that support millet processing and utilization. Working together, we can realize the full potential of millet, benefiting both consumers and the agricultural sector.

16.4 QUALITY STANDARDS FOR MILLETS

In the context of grains, the concept of quality can be specified as food grain safety, compliance with trade specifications, and compliance with end-use requirements (Taylor and Duodu

2017). Quality standards are also essential to regulate the storage, distribution, and sale of the product. Knowledge of the quality attributes of the grains like size, shape, density, presence of foreign seeds, percent of damaged kernels, insect and microbial loads, and nutritional composition of the grains helps to determine the end use of the grains (Duodu and Dowell 2019). When grain needs to be stored, precise measurements of characteristics like moisture content and insect presence are needed to effectively manage the crop and prevent quality loss. The nutritional composition and microbiological safety are important to produce healthy and safe food and feed products. The establishment of national and international quality standards also improves trade. The quality requirements of millets are very similar to other common cereals; for example, the grains should be dried mature, free from added colouring matter, moulds, weevils, obnoxious substances, discoloration, poisonous seeds, and all other impurities, and should also be free from rodent hair and excreta. The quality standards for millets available worldwide are provided in Table 16.1. However, the quality standards are lacking for many minor millets and their is a crucial need to develop standards for such millets to promote their trade and utilization.

16.5 CONCLUSIONS

In conclusion, the prospects in the production, processing, and utilization of millets hold tremendous promise for addressing global challenges such as climate change, population growth, and food security. As resilient and adaptable crops, millets offer numerous advantages over conventional cereals, making them an attractive option for sustainable agriculture. Advancements in genetics, sustainable agricultural practices, and the promotion of millet cultivation are driving future prospects in millet production. Through genetic improvement, researchers are developing high-yielding and climate-resilient millet varieties that can thrive in diverse growing conditions. Sustainable agricultural practices such as conservation agriculture, water management, integrated pest management, and soil health management are enhancing the productivity and efficiency of millet cultivation while minimizing environmental impacts. The promotion of millet cultivation through farmer training, access to quality seeds, market development, and policy support is crucial for unlocking the full potential of millets. In the processing and utilization of millets, there are exciting prospects in diversifying millet-based food products, fortifying millets with essential micronutrients, developing functional foods and nutraceuticals, exploring opportunities in the beverage industry, and utilizing millets as nutritious livestock feed. These prospects are driven by increasing awareness of millets' nutritional value, sustainability, and versatility. Market development and consumer awareness play a critical role in realizing the future prospects of millet. By promoting millets as healthy, sustainable, and versatile food options through marketing campaigns, nutritional education, and collaboration with the food industry, consumer demand and market growth can be stimulated. Governments, NGOs, and industry stakeholders also have a crucial role to play in creating favourable market environments and policy frameworks that support millet processing and utilization. In conclusion, the prospects in the production, processing, and utilization of millets are promising. With ongoing advancements in research, technology, and market development, millets have the potential to make a significant contribution to sustainable agriculture, food security, and nutrition. By embracing the opportunities presented by millets, we can pave the way for a more resilient, inclusive, and sustainable future.

TABLE 16.1 WORLDWIDE QUALITY STANDARDS FOR VARIOUS MILLETS

Crop name/ Quality characteristic	Pearl millet (whole grains)	Pearl millet (Decorticated grains)	Finger millet	Proso millet	Teff
Moisture (% by mass)	Max. 13%	Max. 13%	Max. 12%	Max. 13%	Max. 12.5%
Extraneous matter (% by mass)	Not more than 1.0% of which • Mineral matter: Not more than 0.25% • Impurities of animal origin: Not more than 0.10%	Not more than 1.0% of which • Mineral matter: Not more than 0.25% • Impurities of animal origin: Not more than 0.10%	Not more than 1.0% of which • Mineral matter: Not more than 0.25% • Impurities of animal origin: Not more than 0.10%	———	
Other edible grains (% by mass)	Not more than 2.0%		Max. 2.0%		———
Damaged seed (% by mass)	Not more than 6% of which, Ergot affected grain: Not more than 0.05%	Not more than 6%	Max. 2.0%		———
Weevilled grains (% by count)	Not more than 6%	———	Max. 2.0%		———
Uric acid	100 mg/kg	———	100 mg/kg		———
Ash (on dry matter basis)	———	0.8 – 1%		4.2%	3.0–4.0%
Protein (N × 5.7) (% by mass on dry matter basis)	Min. 8%	Min. 8%			Min. 8%
Decortication	———	20–22%			
Crude Fibre (% by mass on dry matter basis)	3.0–4.5%	2%		3.0%	
Fat (% by mass on dry matter basis)	– 6%	2–4%		Max. 2.0%	

Sources: Codex 2020; FSSAI 2018; East African Standards 2011; Ethiopian Standards 2015

REFERENCES

Abrouk, M., H. I. Ahmed, P. Cubry, D. Šimoníkov, S. Cauet, Y. Pailles, J. Bettgenhaeuser, L. Gapa, N. Scarcelli, M. Couderc, L. Zekraoui, N. Kathiresan, J. Čížková, E. Hřibová, J. Doležel, S. Arribat, H. Bergès, J. J. Wieringa, M. Gueye, and S. G. Krattinger. 2020. Fonio millet genome unlocks African orphan crop diversity for agriculture in a changing climate. *Nature Communications* 11(1): 4488.

Adebiyi, J. A., A. O. Obadina, O. A. Adebo, and E. Kayitesi. 2016. Fermented and malted millet products in Africa: Expedition from traditional/ethnic foods to industrial value-added products. *Critical Reviews in Food Science and Nutrition* 58(3): 463–474.

Adekunle, A., D. Lyew, V. Orsat, and V. Raghavan. 2018. Helping agribusinesses—Small millets value chain—To grow in India. *Agriculture* 8(3): 44.

Amadou, I. 2019. Millet based fermented beverages processing. In *Fermented Beverages* (Grumezescu, A. M., and A. M. Holban, Eds.), Vol. 5, Woodhead Publishing, California, pp. 433–472.

Amadou, I. 2022. Millet based functional foods: Bio-chemical and bio-functional properties. In *Functional Foods* (Chhikara, N., A. Panghal, and G. Chaudhary, Eds.), Scrivener Publishing, Massachusetts, pp. 303–329.

Anitha, S., D. I. Givens, K. Subramaniam, S. Upadhyay, J. Kane-Potaka, Y. D. Vogtschmidt, R. Botha, T. W. Tsusaka, S. Nedumaran, H. Rajkumar, A. Rajendran, D. J. Parasannanavar, M. Vetriventhan, and R. K. Bhandari. 2022. Can feeding a millet-based diet improve the growth of children?-A systematic review and meta-analysis. *Nutrients* 14(1): 225.

Anjali, P., and P. Vijayaraj. 2020. Functional food ingredients from old age cereal grains. In *Functional and Preservative Properties of Phytochemicals* (Prakash, B., Ed.), Academic Press, Cambridge, pp. 47–92.

Arora, L., R. Aggarwal, I. Dhaliwal, O. P. Gupta, and P. Kaushik. 2023. Assessment of sensory and nutritional attributes of foxtail millet-based food products. *Frontiers in Nutrition* 10: 1146545.

Asrani, P., A. Ali, and K. Tiwari. 2022. Millets as an alternative diet for gluten-sensitive individuals: A critical review on nutritional components, sensitivities and popularity of wheat and millets among consumers. *Food Reviews International*. doi: 10.1080/87559129.2021.2012790.

Babele, P. K., H. Kudapa, Y. Singh, R. K. Varshney, and A. Kumar. 2022. Mainstreaming orphan millets for advancing climate smart agriculture to secure nutrition and health. *Frontiers in Plant Science* 13: 902536.

Badejo, A. A., U. Nwachukwu, H. N. Ayo-Omogie, and O. S. Fasuhanmi. 2020. Enhancing the antioxidative capacity and acceptability of Kunnu beverage from gluten-free pearl millet (*Pennisetum glaucum*) through fortification with tigernut sedge (*Cyperus esculentus*) and coconut (*Cocos nucifera*) extracts. *Journal of Food Measurement and Characterization* 14(1): 438–445.

Bado, B. V., A. Whitbread, and M. L. Sanoussi Manzo. 2021. Improving agricultural productivity using agroforestry systems: Performance of millet, cowpea, and Ziziphus-based cropping systems in West Africa Sahel. *Agriculture, Ecosystems and Environment* 305: 107175.

Balakrishnan, G., and R. G. Schneider. 2022. The role of amaranth, quinoa, and millets for the development of healthy, sustainable food products-A concise review. *Foods (Basel, Switzerland)* 11(16): 2442.

Barbon, W., R. R. John, and J. F. Vidallo. 2017. The promotion of climate-smart villages to support community-based adaptation programming in Myanmar. *CCAFS Working Paper*. https://reliefweb.int/report/myanmar/promotion-climate-smart-villages-support-community-based-adaptation-programming (accessed 22 March 2023).

Bhatt, D., P. Rasane, J. Singh, S. Kaur, M. Fairos, J. Kaur, and M. Gunjal. 2023. Nutritional advantages of barnyard millet and opportunities for its processing as value-added foods. *Deakin University Journal Contribution*. doi: 10779/DRO/DU:22044563.v1.

Birania, S., P. Rohilla, R. Kumar, and N. Kumar. 2020. Post harvest processing of millets: A review on value added products. *International Journal of Chemical Studies* 8(1): 1824–1829.

Bisht, I. S., P. S. Mehta, K. S. Negi, S. K. Verma, R. K. Tyagi, and S. C. Garkoti. 2018. Farmers' rights, local food systems, and sustainable household dietary diversification: A case of Uttarakhand Himalaya in north-western India. *Agroecology and Sustainable Food Systems* 42(1): 77–113.

Byresh, T. S., B. Malini, L. Meena, C. K. Sunil, D. V. Chidanand, R. Vidyalakshmi, and N. Venkatachalapathy. 2022. Effect of addition of pineapple peel powder on white finger millet vegan probiotic beverage. *Journal of Food Processing and Preservation* 46(10): e16905. doi: 10.1111/jfpp.16905.

Ceasar, A., T. Stanislaus, T. P. Maharajan, M. Krishna, G. V. Ramakrishnan, L. Roch, and S. Satish. 2018. Improvement: Current status and future interventions of whole genome sequence. *Frontiers in Plant Science* 9: 1054.

Chitra, D. I., and D. Sulaiman. 2017. A study on customer awareness and consumption of minor millets as a diabetic food product-with reference to madurai city. *International Journal of Advanced Scientific Research and Development* 4(1): 38–44.

Cichońska, P., and M. Ziarno. 2021. Production and consumer acceptance of millet beverages. In *Milk Substitutes - Selected Aspects* (Ziarno, M., Ed.), IntechOpen, London. doi: 10.5772/intechopen.94304.

Codex Alimentarius. 2020. CXS 170–1989. https://www.fao.org/fao-who-codexalimentarius/codex-texts/list-standards/en/ (accessed 26 August 2021).

Das, S., R. Khound, M. Santra, and D. Santra. 2019. Beyond bird feed: Proso millet for human health and environment. *Agriculture* 9(3): 64.

de Assis, R. L., R. S. de Freitas, and S. C. Mason. 2018. Pearl millet production practices in Brazil: A review. *Experimental Agriculture* 54(5): 699–718.

Delele, T., D. Atnafu, and T. A. Hailu Gebre. 2021. Roles of community-based seed system to mitigate seed shortage problem under Metekel zone, Western Ethiopia. *American Journal of Agriculture and Forestry* 9(5): 276–282.

Dey, S., A. Saxena, Y. Kumar, T. Maity, and A. Tarafdar. 2022. Understanding the antinutritional factors and bioactive compounds of Kodo millet (*Paspalum scrobiculatum*) and little millet (*Panicum sumatrense*). *Journal of Food Quality*: 1578448. doi: 10.1155/2022/1578448.

Dhaka, A., R. K. Singh, M. Muthamilarasan, and M. Prasad. 2021. Genetics and genomics interventions for promoting millets as functional foods. *Current Genomics* 22(3): 154–163.

Dias-Martins, A. M., K. L. F. Pessanha, S. Pacheco, J. A. S. Rodrigues, and C. W. P. Carvalho. 2018. Potential use of pearl millet (Pennisetum glaucum (L.) R. Br.) in Brazil: Food security, processing, health benefits and nutritional products. *Food Research International (Ottawa, Ont.)* 109: 175–186.

Duodu, K. G., and F. E. Dowell. 2019. Sorghum and millets: Quality management systems. In *Sorghum and Millets: Chemistry, Technology, and Nutritional Attributes* (Taylor, J. R. N., and K. G. Duodu, Eds.), Woodhead Publishing and AACC International Press, Cambridge, pp. 421–442.

East African Standards. 2011. https://law.resource.org/pub/eac/ibr/eas.89.2011.pdf (accessed 26 August 2021).

Ethiopian Standard. 2015. Teff - Flour specification. https://www.nrc.no/globalassets/pdf/tenders/ethiopia/dec-2020-procurement-of-different-emergency-food-and-nonfood-items/annex-1_es-3880-2015-ethiopian-standard.pdf (accessed 26 August 2021).

Foley, J. K., K. D. Michaux, B. Mudyahoto, L. Kyazike, B. Cherian, O. Kalejaiye, O. Ifeoma, P. Ilona, C. Reinberg, D. Mavindidze, and E. Boy. 2021. Scaling up delivery of biofortified staple food crops globally: Paths to nourishing millions. *Food and Nutrition Bulletin* 42(1): 116–132.

FSSAI. 2018. https://foodsafetyhelpline.com/fssai-notifies-amendments-to-standards-for-some-cereals-and-cereal-products/ (accessed 26 August 2021).

Gebreyohannes, A., H. Shimelis, M. Laing, I. Mathew, D. A. Odeny, and H. Ojulong. 2021. Finger millet production in Ethiopia: Opportunities, problem diagnosis, key challenges and recommendations for breeding. *Sustainability* 13(23): 3463.

Gupta, S. M., S. Arora, N. Mirza, A. Pande, C. Lata, S. Puranik, J. Kumar, and A. Kumar. 2017. Finger millet: A "certain" crop for an "uncertain" future and a solution to food insecurity and hidden hunger under stressful environments. *Frontiers in Plant Science* 8: 643.

Hassan, Z. M., N. A. Sebola, and M. Mabelebele. 2021. The nutritional use of millet grain for food and feed: A review. *Agriculture and Food Security* 10(1): 16.

Hema, V., M. Ramaprabha, R. Saraswathi, P. N. Chakkaravarthy, and V. R. Sinija. 2022. Millet food products. In *Handbook of Millets - Processing, Quality, and Nutrition Status* (Anandharamakrishnan, C., A. Rawson, and C. K. Sunil, Eds.), Springer, Singapore, pp. 265–299.

Kala, C. P., and S. Nautiyal. 2023. Traditional food knowledge of local people and its sustainability in mountains of Uttarakhand State of India. *Journal of Social and Economic Development* 25(1): 32–51.

Kane-Potaka, J., S. Anitha, T. W. Tsusaka, R. Botha, M. Budumuru, S. Upadhyay, P. Kumar, K. Mallesh, R. Hunasgi, A. K. Jalagam, and S. Nedumaran. 2021. Assessing millets and sorghum consumption behavior in urban India: A large-scale survey. *Frontiers in Sustainable Food Systems* 5: 680777. doi: 10.3389/fsufs.2021.680777.

Khairuddin, M. A. N., and O. Lasekan. 2021. Gluten-free cereal products and beverages: A review of their health benefits in the last five years. *Foods (Basel, Switzerland)* 10(11): 2523.

Kumar, A., V. Tomer, A. Kaur, V. Kumar, and K. Gupta. 2018. Millets: A solution to agrarian and nutritional challenges. *Agriculture and Food Security* 7(1): 1–15.

Kumar, A., A. Kaur, V. Tomer, P. Rasane, and K. Gupta. 2020. Development of nutricereals and milk-based beverage: Process optimization and validation of improved nutritional properties. *Journal of Food Process Engineering* 43(1): e13025.

Mahajan, M., P. Singla, and S. Sharma. 2023. Sustainable postharvest processing methods for millets: A review on its value-added products. *Journal of Food Process Engineering*: e14313. doi: 10.1111/jfpe.14313.

Majid, A., and C. G. P. and Priyadarshini. 2020. Millet derived bioactive peptides: A review on their functional properties and health benefits. *Critical Reviews in Food Science and Nutrition* 60(19) : 3342–3351.

Malhotra, A., S. Nandigama, and K. S. Bhattacharya. 2021. Food, fields and forage: A socio-ecological account of cultural transitions among the Gaddis of Himachal Pradesh in India. *Heliyon* 7(7): 07569.

Malla, J. K., S. Ochola, I. Ogada, and A. Munyaka. 2021. Nutritional value and sensory acceptability of M. oleifera fortified finger millet porridge for children with cerebral palsy in Nairobi County, Kenya. *Journal of Food Research* 10(5): 36.

Mbinda, W., and H. Masaki. 2020. Breeding strategies and challenges in the improvement of blast disease resistance in finger millet. A current review. *Frontiers in Plant Science* 11: 602882.

McSweeney, M. B., K. Seetharaman, D. D. Ramdath, and L. M. Duizer. 2017. Chemical and physical characteristics of proso millet (*Panicum miliaceum*)-based products. *Cereal Chemistry Journal* 94(2): 357–362.

Packiam, M., K. Subburamu, R. Desikan, S. Uthandi, M. Subramanian, and K. Soundarapandian. 2018. Suitability of pearl millet as an alternate lignocellulosic feedstock for biofuel production in India. *Journal of Applied and Environmental Microbiology* 6(2): 51–58.

Palaniappan, A., S. S. Yuvaraj, S. Sonaimuthu, and U. Antony. 2017. Characterization of xylan from rice bran and finger millet seed coat for functional food applications. *Journal of Cereal Science* 75: 296–305.

Prathyusha, N., V. Vijaya, and T. Lakshmi. 2021. Review on consumer awareness and health benefits about millets. *The Pharma Innovation Journal* 10(6): 777–785.

Ramashia, S. 2018. Physical, functional and nutritional properties of flours from finger millet (Eleusine coracana) varieties fortified with vitamin B2 and zinc oxide. Ph.D. Thesis, University of Venda, Limpopo Province, South Africa.

Ramashia, S. E., E. T. Gwata, S. Meddows-Taylor, T. A. Anyasi, and A. I. O. Jideani. 2021. Nutritional composition of fortified finger millet (Eleusine coracana) flours fortified with vitamin B2 and zinc oxide. *Food Research* 5(2): 456–467.

Rao, M., K. G. Venkateswara, C. K. Akhil, N. Sunil, and R. Venkatachalapathy. 2021. Effect of microwave treatment on physical and functional properties of foxtail millet flour. *International Journal of Chemical Studies* 9(1): 2762–2767.

Renganathan, V. G., C. Vanniarajan, A. Karthikeyan, and J. Ramalingam. 2020. Barnyard millet for food and nutritional security: Current status and future research direction. *Frontiers in Genetics* 11: 497319. doi: 10.3389/fgene.2020.00500.

Sabreena, N. M., B. A. Ganai, R. A. Mir, and S. M. Zargar. 2021. Genetics and genomics resources of millets: Availability, advancements, and applications. In *Neglected and Underutilized Crops - Towards Nutritional Security and Sustainability* (Zargar, S. M., A. Masi, and R. K. Salgotra, Eds.), Springer, Singapore, pp. 153–166.

Sahu, C., and S. Patel. 2021. Optimization of maize-millet based soy fortified composite flour for preparation of RTE extruded products using D-optimal mixture design. *Journal of Food Science and Technology* 58(7): 2651–2660.

Salej, D. C., A. K. Joshi, and A. Chandra. 2019. Phenomics and genomics of finger millet: Current status and future prospects. *Planta* 250(3): 731–751.

Spielman, D., E. Lecoutere, S. Makhija, and B. Van Campenhout. 2021. Information and communications technology (ICT) and agricultural extension in developing countries. *Annual Review of Resource Economics* 13(1): 177–201.

Saxena, R., S. Vanga, J. Wang, V. Orsat, and V. Raghavan. 2018. Millets for food security in the context of climate change: A review. *Sustainability* 10(7): 2228.

Scarpa, G., L. Berrang-Ford, S. Twesigomwe, P. Kakwangire, R. Peters, C. Zavaleta-Cortijo, K. Patterson, D. B. Namanya, S. Lwasa, E. Nowembabazi, C. Kesande, H. Harris-Fry, and J. E. Cade. 2021. A community-based approach to integrating Socio, cultural and environmental contexts in the development of a food database for Indigenous and rural populations: The case of the Batwa and Bakiga in southwestern Uganda. *Nutrients* 13(10): 3503.

Shweta, S., S. Sindhu, K. Sushma, and S. Sandhya. 2018. Little millets: Properties, functions and future prospects. *International Journal of Agricultural Engineering* 11: 179–181.

Singh, A. K., and R. R. Burman. 2019. Agricultural extension reforms and institutional innovations for inclusive outreach in India. In *Agricultural Extension Reforms in South Asia* (Babu, S. C., and P. K. Joshi, Eds.), Academic Press, Cambridge, pp. 289–315.

Singh, M., and U. Nara. 2023. Genetic insights in pearl millet breeding in the genomic era: Challenges and prospects. *Plant Biotechnology Reports* 17(1): 15–37.

Singh, R. B., S. Khan, A. K. Chauhan, M. Singh, P. Jaglan, P. Yadav, T. Takahashi, and L. R. Juneja. 2019. Millets as functional food, a gift from Asia to western world. In *The Role of Functional Food Security in Global Health* (Singh, R. B., R. R. Watson, and I. Takahashi, Eds.), Elsevier Science, Amsterdam, Netherlands, pp. 457–468.

Singh, M., and S. Sood. 2020. *Millets and Pseudo Cereals: Genetic Resources and Breeding Advancements: Genetic Resources and Breeding Advancements*, Woodhead Publishing, Cambridge, p. 208.

Srilekha, K., T. Kamalaja, K. Uma Maheswari, and R. Neela Rani. 2019. Nutritional composition of little millet flour. *International Research Journal of Pure and Applied Chemistry* 20(4): 1–4.

Sukumar, A., and K. A. Athmaselvi, K. A. 2019. Optimization of process parameters for the development of finger millet based multigrain extruded snack food fortified with banana powder using RSM. *Journal of Food Science and Technology* 56(2): 705–712.

Sun, M., C. Lin, A. Zhang, X. Wang, H. Yan, I. Khan, B. Wu, G. Feng, G. Nie, X. Zhang, and L. Huang. 2021. Transcriptome sequencing revealed the molecular mechanism of response of pearl millet root to heat stress. *Journal of Agronomy and Crop Science* 207(4): 768–773.

Suparna, R., U. Kapoor, A. Ghosh, S. Singh, and J. Downs. 2021. Pathways of climate change impact on agroforestry, food consumption pattern, and dietary diversity among indigenous subsistence farmers of Sauria Paharia tribal community of India: A mixed methods study. *Frontiers in Sustainable Food Systems* 5: 667297.

Talabi, A. O., P. Vikram, S. Thushar, H. Rahman, H. Ahmadzai, N. Nhamo, M. Shahid, and R. K. Singh. 2022. Orphan crops: A best fit for dietary enrichment and diversification in highly deteriorated marginal environments. *Frontiers in Plant Science* 13: 839704.

Tariyal, N., A. Bijalwan, S. Chaudhary, B. Singh, C. S. Dhanai, S. Tewari, M. Kumar, S. Kumar, M. M. S. Cabral Pinto, and T. K. Thakur. 2022. Crop production and carbon sequestration potential of Grewia oppositifolia-based traditional agroforestry systems in Indian Himalayan region. *Land* 11(6): 839.

Taylor, J. R. N., and K. G. Duodu. 2019. Traditional sorghum and millet food and beverage products and their technologies. In *Sorghum and Millets: Chemistry, Technology, and Nutritional Attributes* (Taylor, J. R. N., and K. G. Duodu, Eds.), Woodhead Publishing and AACC International Press, Cambridge, pp. 259–292.

Taylor, J. R., and K. G. Duodu. 2017. Sorghum and millets: Grain-quality characteristics and management of quality requirements. In *Cereal Grains*, Second Edition (Wringley, C., I. Batey, and D. Miskelly, Eds.), Woodhead Publishing, Cambridge, pp. 317–351.

Thakur, M., and P. Tiwari. 2019. Millets: The untapped and underutilized nutritious functional foods. *Plant Archives* 19(1): 875–883.

Tomar, M., R. Bhardwaj, R. Verma, S. P. Singh, A. Dahuja, V. Krishnan, R. Kansal, V. K. Yadav, S. Praveen, and A. Sachdev. 2022. Interactome of millet-based food matrices: A review. *Food Chemistry* 385: 132636.

Zhang, H., Y. Li, and J. K. Zhu. 2018. Developing naturally stress-resistant crops for a sustainable agriculture. *Nature Plants* 4(12): 989–996.

Index

A

Abrasion, 26
ABTS, 161
Adi, 31
Aglycones, 159
Agrarian requirements
    barnyard millet, 12
    finger millet, 12
    fonio, 14
    foxtail millet, 11
    kodo millet, 12–13
    little millet, 13
    pearl millet, 11
    proso millet, 11–12
    teff, 13
Ajeen, 233
Aleurone, 77–79, 81
Alkaloids, 140
Amino acids, 46, 105–108, 128, 157, 161
Amylases, 43, 169, 176, 258, 275
Amylopectin, 82, 100, 142, 175, 178
Amylose, 82, 100, 102, 142, 175, 177, 178
Angiotensin-1, 158
Angle of repose, 93
Animal fodder, 284
Antagonistic factors, 203, 204
Anthers, 3, 5–8
Anti-adipogenic, 162
Anti-inflammatory, 163
Antinutrients, 203, 209, 210
Antioxidant, 40, 120, 161, 191, 196
    endogenous, 159
    potential, 119, 161
Anti-proliferative, 130
Arabinoxylan, 155, 161–163
Aspergillus sojae, 147
Aspirator, 27
Attrition, 26

B

Baking, 44, 192, 235, 249
Basundi, 235
Battu, 54, 55
Belt dryer, 26
Ben-Saagla, 217
Bharola, 28, 29
Bioaccessibility, 209
Bioactive peptides, 128, 132
Bioethanol, 270, 277
Biogas, 271
Biopolymers, 194
Biscuit, 244, 277
Botanical Description, 3–8
Boza, 232
Bran, 77, 272
Bran fibre, 275
Bran oil, 275
Bread, 245
Bukhari, 30, 31
Burukutu, 234
Bushera, 217
By-products, 267, 271
β-glucan, 155

C

Cake, 247, 261, 270
Campesterol, 104
Cancer, 168
Carbohydrates, 99
Cardiovascular diseases, 164
Caryopsis, 75, 80, 83
Centrifugal dehuller, 27
Cephalin, 105
Cereal Bars, 248
Chakki, 54, 55
Chazz, 54, 72

Chelation, 202
Chymotrypsin, 140, 148
Cleaners, 61
Coating, 206
Coefficient of friction, 91
    external, 92
    internal, 91
Coleorhiza, 78
Colour sorting machines, 64
Combines, 56
Conditioning, 37
Cookies, 249
Cooking, 39, 42, 122, 125, 143, 147, 192, 224, 251
Crude fibre, 41, 43, 101, 223, 293
Cultivation, 11–15, 17, 24–28, 216, 243, 268, 271, 279, 283, 284, 286, 287, 291, 292
Curcumin, 196
Curing, 26
Cyclodextrins, 208

D

Dambi, 139, 146
Dambu, 235
Debranning, 37
Decortication, 26
Dehulling, 26, 53, 55, 67, 68, 112, 144, 146
Dehusker, 57, 66, 68
Density
    bulk, 88
    true, 89
Deoba, 233
Dephytinization, 140
Destoner, 65
Dextrinases, 176
Dhaeli, 217, 229, 234
Dhokla, 216, 222, 223
Dhoosi, 31
Diabetes, 99, 159, 167, 243, 261, 286
Diameter, 86
Dietary fibre, 103
Dimension, 84
Disulphide solvation, 190
Diversification, 287
Dosa, 71, 216, 219, 223, 229
DPPH, 83, 161

E

Embryonic tissue, 188
Encapsulated, 196
Endosperm, 81
Endotoxemia, 162

Enzyme hydrolysis, 177
Enzyme inhibitors, 143
Enzyme susceptibility, 178
Extruded products, 150
Extrusion, 45, 191, 192

F

Fat, 103–105
Fermentation, 46, 147, 216
Ferulic acid, 158, 162
Filler, 271
Flaking, 42
Flavones, 126
Flavonoids, 122, 124, 126, 127, 129, 131, 158
Flavonols, 127
Flavonones, 127
Foam capacity, 194
Fodder, 268
Fortification, 202, 207, 208
Fortified, 223–225
Fractionation, 189, 191
Free radical damage, 159
Frying, 42
Fumigation, 33
Functional food, 289
Fura/Fura da Nono, 223

G

Gadoli, 29
Garo, 31
Gelatinization, 101, 175
Genetics, 285
Genetic variation, 204
Geometric mean diameter, 6, 87
Germ, 83, 276
Germination, 132, 146
Germ oil, 276
Germplasm, 140
Germplasm accessions
    barnyard millet, 2
    finger millet, 2
    fonio, 3
    foxtail millet, 2
    kodo millet, 2
    little millet, 2
    pearl millet, 2
    proso millet, 2
    teff, 3
Gharat, 54, 55
Glucosidases, 176
Glume, 78

Glutathione, 161
Glutelin, 106
Gluten, 175
Glycemic index, 160, 181
Glycosylvitexin, 142
Grading machines, 62
Grain length, 86, 94
Grinding, 38
Growth cycle, 3

H

Halwa, 71, 235
Hapur Kothi, 30
Hapur Tekka, 32
Harvesting, 23–25, 32
Hectolitre, 28
High pressure processing, 191, 193
Hull, 112, 227, 274
Husk, 170, 267, 271, 272
Husk separator, 27
Hyperlipidemia, 164, 165
Hypertension, 164
Hypoglycemic, 162

I

Ibyer, 224
Idli, 71, 216, 219, 224, 229
Impaired glucose tolerance, 154
Infections, 169
Inflammation, 163
Inflorescence, 3, 5–8
Injera, 71, 225
Intercropping, 285
Interstice, 90

J

Jaandh, 233

K

Kafirins, 106
Kanaja, 29, 30
Kharif, 11–12, 24, 32, 140, 142
Kheer, 216, 235
Khori, 54, 55
Kindrimo, 223
Kodo ko Jaanr, 233
Koko, 225
Koozh, 226
Kothar, 30

Kothi, 30
Kurrakan Kanda, 237
Kuthala, 31

L

Laddu, 216, 236, 237
Lecithin, 105
Lipids, 103–105
Lipogenesis, 164
Livestock feed, 290

M

Macronutrients, 99
Malting, 43
Mandro, 233
Mangisi, 226
Masavusu, 237
Mazoria, 225
Mechanized Mills, 27
Mechanization, 53
Merissa, 233
Micronutrients, 110
Microstructural, 196
Milling, 23, 26, 36, 102, 146, 176, 206
Mineral metabolism, 205
Minerals, 112, 203
Modified starches, 182
Moosal, 54, 55
Morai, 29, 30
Muda, 3
Muffins, 252
Multigrain flour, 254
Mushuk, 234

N

Nahu, 31
Nixtamization, 142
Non-haem, 204
Non-starch polysaccharides, 143
Noodles, 254
Nutraceutical, 154

O

Obesity, 162
Ogi, 226
    baba, 226
    gero, 226
Okhal, 54, 55
Omalodu, 232

Osborne approach, 189
Oshikundu, 227
Oxalates, 142
Oxygenases, 163

P

Paddu, 227
Panicle, 25
    barnyard millet, 6
    finger millet, 6
    fonio, 8
    foxtail millet, 5
    kodo millet, 8
    little millet, 8
    pearl millet, 3
    proso millet, 6
    teff, 8
Panicle development stage, 3
Panicle initiation, 3
Parboiling, 191
Payasam, 235
Pearling, 38
Pectin, 204
Pericarp, 75–78, 80
Peru, 29, 30
Peti, 30
Phenolic compounds, 120
Phytates, 138, 139, 143, 146, 148, 149
Phytic acid, 138–140, 148
Phytochemicals, 120
Phytosterols, 127, 158
Pito, 234
Pneumatic drier, 26
Polishing, 38
Polypeptide, 193
Pongal, 216, 236
Popping, 40
Porosity, 90
Pounding, 27
Prebiotics, 277
Pro anthocyanidins, 158, 159
Production
    barnyard millet, 17
    finger millet, 17
    fonio, 17
    foxtail millet, 14
    kodo millet, 17
    little millet, 17
    pearl millet, 14
    proso millet, 14, 17
    teff, 17

Productivity, 14, 17, 24, 56, 138, 285, 291–292
Prolamines, 188–192
Prostaglandins, 195
Proteases, 201, 207
Protein, 105, 106, 189
    characterization, 190
    denaturation, 193
    digestibility, 109
    extraction, 189
Protocatechuic acid, 158
Pullulanases, 176
Pulverisation, 53, 55, 69
Pulveriser, 55

Q

Quality, 292

R

Rabadi, 228
Rabi, 11–13, 24, 100, 103
Radiation, 149
Resistance Starch, 155
Retrogradation, 101, 177, 179, 182
Rheology, 180
Roasting, 39
Roti, 71

S

Sandaka, 30
Sandook, 30
Scutellum, 78, 83
Segregation, 189
Septu, 233
Sheller rubber roll, 27
Silo, 84
Snacks, 181
Soaking, 143
Soil blocks and concrete, 272
Soil salinity, 11–12, 15–16
Sorting, 62
Spent grains, 276
Sphericity, 87
Spikelets, 3, 5–8
Stabilizer, 181
Standards, 292, 293
Starch, 178
    rheological properties, 180
    solubility, 179
    swelling power, 179

Steeping, 176, 182
Sterols, 104
Stover, 268
Supa, 54, 72
Superoxide dismutase, 161
Sur, 234, 259
Surface area, 90
Sustainable, 285
Synbiotic foods, 257
Syneresis, 177

T

Tannin, 140
    condensed, 140
    hydrolysable, 140
Taxonomy, 3
Tegmen, 78
Tempering, 23
Terminal velocity, 93
Testa, 78
Thiamine, 110, 111
Thickener, 181
Thousand kernel weight, 88
Thresher, 58
    horizontal, 60
    vertical, 59
Threshing, 23, 25, 26, 32, 53, 58, 60, 61, 65
Togwa, 228
Tramping, 23

Triglyceride, 176
Tuni, 30

U

Uji, 228
Ultrasonication, 196
Uthapam, 71, 216, 219, 229

V

Value addition, 288
Viscosity setback, 178

W

Weaning foods, 255
Windrowers, 55
Winnowing, 25, 27

X

Xiao Mi Jiao, 235
X-ray diffraction, 179

Z

Zeta potential, 196
Zoom-Koom, 229